The Routledge Handbook of Teaching Landscape

Written in collaboration with the European Council of Landscape Architecture Schools (ECLAS) and LE:NOTRE, *The Routledge Handbook of Teaching Landscape* provides a wide-ranging overview of teaching landscape subjects, from geology to landscape design, reflecting different perspectives and practices at university-level landscape curricula. Focusing on the didactics of landscape education, this fully illustrated handbook presents and discusses pedagogy, teaching traditions, experimental teaching methods and new teaching principles.

The book is structured in three parts: reading the landscape, representing the landscape and transforming the landscape. Contributions from leading experts in the field, such as Simon Bell, Marc Treib, Jörg Rekittke and Susan Herrington, explore landscape analysis, history and theory, design visualisation, creativity and art, planning studio teaching, field trips and site engineering. Aimed at engaging academic researchers and instructors across disciplines such as landscape architecture, geography, ecology, planning and archaeology, this book is a must-have guide to landscape pedagogy as it stands today.

Karsten Jørgensen is Professor of Landscape Architecture at the Department of Landscape Architecture and Spatial Planning at the Norwegian University of Life Sciences, Norway, and holds a Dr.-Scient. degree from NMBU, 1989, in landscape architecture. He was Founding Editor of *JoLA* – the *Journal of Landscape Architecture* – 2006–2015. Karsten Jørgensen has published regularly in national and international journals and books. He edited the volume *Mainstreaming Landscape through the European Landscape Convention* (Routledge 2016) together with Tim Richardson, Kine Thoren and Morten Clemetsen.

Nilgül Karadeniz is Professor of Landscape Architecture at Ankara University, Turkey. Her teaching and research interest focusses on participatory landscape planning and recently on landscape biography. She has been an editorial board member of SCI-expanded journals. She was Secretary General (2006–2009) and Vice President (2009–2012) of ECLAS. She is founding member of LE:NOTRE Institute and, since January 2016, she has been the chair of the Institute.

Elke Mertens is Professor of Garden Architecture and Landscape Maintenance at the Hochschule Neubrandenburg - University of Applied Sciences, Germany. She holds a Dr.-Ing. degree from the Technical University in Berlin (1997). She is Co-Chair of the German Hochschulkonferenz Landschaft (HKL), member of the board of LE:NOTRE Institute and has been active in the LE:NOTRE Thematic Network as well as in ECLAS as member of the executive boards.

Richard Stiles is Professor of Landscape Architecture in the Faculty of Architecture and Planning at Vienna University of Technology, Austria, having studied biology and landscape design at the Universities of Oxford and Newcastle upon Tyne and having previously taught at Manchester University in the UK. His teaching and research interests focus on strategic landscape planning and design in urban areas. He is a past President of the European Council of Landscape Architecture Schools and was Coordinator of the European Union co-funded LE:NOTRE Thematic Network in Landscape Architecture for 11 years, during which time he was closely involved in preparing recommendations for landscape architecture education.

Routledge International Handbooks

Routledge Handbook of Global Populism *ed. Carlos de la Torre*

Handbook of Indigenous Environmental Knowledge *ed. Thomas Thornton, Shonil Bhagwat*

The Routledge Handbook of Museums, Media and Communication *ed. Kirsten Drotner, Vincent Dziekan, Ross Parry, Kim Christian Schrøder*

Routledge International Handbook of Sex Industry Research *ed. Susan Dewey, Isabel Crowhurst, Chimaraoke Izugbara*

Routledge Handbook of Cultural Sociology 2e *ed. Laura Grindstaff, Ming-Cheng Lo, John R. Hall*

The Routledge Handbook of Latin American Development *ed. Julie Cupples, Marcela Palomina-Schalscha, Manuel Prieto*

The Routledge International Handbook of Embodied Perspectives in Psychotherapy: Approaches from Dance Movement and Body Psychotherapies *ed. Helen Payne, Sabine Koch, Jennifer Tantia, Thomas Fuchs*

The Routledge Handbook of Teaching Landscape *ed. Karsten Jørgensen, Nilgül Karadeniz, Elke Mertens, Richard Stiles*

The Routledge International Handbook of Spirituality in Society and the Professions *ed. Laszlo Flanagan, Bernadette Flanagan*

The Routledge International Handbook of Language Education Policy in Asia *ed. Andy Kirkpatrick, Anthony J. Liddicoat*

For more information about this series, please visit: www.routledge.com/Routledge-International-Handbooks/book-series/RIHAND

The Routledge Handbook of Teaching Landscape

Edited by Karsten Jørgensen, Nilgül Karadeniz,
Elke Mertens and Richard Stiles

LONDON AND NEW YORK

First published 2019
by Routledge
2 Park Square, Milton Park, Abingdon, Oxon OX14 4RN

and by Routledge
605 Third Avenue, New York, NY 10017

First issued in paperback 2020

Routledge is an imprint of the Taylor & Francis Group, an informa business

British Library Cataloguing-in-Publication Data
A catalogue record for this book is available from the British Library

Library of Congress Cataloging-in-Publication Data
A catalog record for this book has been requested

ISBN 13: 978−0−367−73160−1 (pbk)
ISBN 13: 978−0−8153−8052−8 (hbk)

Typeset in Bembo
by Swales & Willis Ltd, Exeter, Devon, UK

Contents

Contents

Biographies

Editors

Karsten Jørgensen is Professor of Landscape Architecture at the Department of Landscape Architecture and Spatial Planning at the Norwegian University of Life Sciences, Norway, and holds a Dr.-Scient. degree from NMBU, 1989, in landscape architecture. He was Founding Editor of JoLA - the Journal of Landscape Architecture - 2006–2015. Karsten Jørgensen has published regularly in national and international journals and books. He edited the volume Mainstreaming Landscape through the European Landscape Convention (Routledge 2016) together with Tim Richardson, Kine Thoren and Morten Clemetsen.

Nilgül Karadeniz is Professor of Landscape Architecture at Ankara University, Turkey. Her teaching and research interest focusses on participatory landscape planning and recently on landscape biography. She has been an editorial board member of SCI-expanded journals. She was Secretary General (2006–2009) and Vice President (2009–2012) of ECLAS. She is founding member of LE:NOTRE Institute and, since January 2016, she has been the chair of the Institute.

Elke Mertens is Professor of Garden Architecture and Landscape Maintenance at the Hochschule Neubrandenburg - University of Applied Sciences, Germany. She holds a Dr.-Ing. degree from the Technical University in Berlin (1997). She is Co-Chair of the German Hochschulkonferenz Landschaft (HKL), member of the board of LE:NOTRE Institute and has been active in the LE:NOTRE Thematic Network as well as in ECLAS as member of the executive boards.

Richard Stiles is Professor of Landscape Architecture in the Faculty of Architecture and Planning at Vienna University of Technology, Austria, having studied biology and landscape design at the Universities of Oxford and Newcastle upon Tyne and having previously taught at Manchester University in the UK. His teaching and research interests focus on strategic landscape planning and design in urban areas. He is a past President of the European Council of Landscape Architecture Schools and was Coordinator of the European Union co-funded LE:NOTRE Thematic Network in Landscape Architecture for 11 years, during which time he was closely involved in preparing recommendations for landscape architecture education.

Authors

Mick Abbott is Associate Professor at the Faculty of Environment, Society and Design, School of Landscape Architecture, Lincoln University, New Zealand. His research investigates how social, cultural and economic value can be built out of strategies that increase biodiversity. He has edited books on different landscape themes. He is also Director of Lincoln University's DesignLab (www.designlab.ac.nz), which focusses on the development and application of design-directed research methods in industry, government and community settings, including recent and ongoing projects with the Department of Conservation, Ngai Tahu Property, New Zealand Walking Access Commission, Air New Zealand, Eden Project UK, Rio Tinto, Conservation Volunteers NZ and Sport Taranaki.

Marc Antrop is Emeritus Professor and a geographer specialised in landscape sciences, remote sensing, GIS and planning at the University of Ghent, Belgium. He has published widely in international journals and books. His latest book is *Landscape Perspectives: The Holistic Nature of Landscape* with Veerle Van Eetvelde, Springer, 2017.

Jacky Bowring is Professor of Landscape Architecture at Lincoln University, New Zealand. Her area of expertise is in landscape architecture and her research and teaching interests circle cultural landscape, history, memory and emotion. She works through the processes of designing, critiquing and scholarly research. She is editor of the international peer-reviewed journal *Landscape Review* and author of *A Field Guide to Melancholy*, Oldcastle, 2008, and *Melancholy and the Landscape: Locating Sadness, Memory and Reflection in the Landscape*, Routledge, 2016.

Andrew Butler, PhD, is a Researcher at the Swedish University of Agricultural Sciences, Sweden. His interests lie in discourses on landscape; who creates them and by what means, issues of landscape identity, participation in landscape issues and landscape assessment as a means for forwarding these discourses. He is engaged with lecturing and supervision in the Landscape Architecture programme. The main focus of his teaching is landscape assessment as a tool for understanding a landscape, arguing for the values of a landscape and communicating the landscape values.

Peter M. Butler is Director, Associate Professor and Extension Specialist in Landscape Architecture, School of Design and Community Development, West Virginia University (WVU), USA. Peter's research includes community design process and design studio pedagogy. He integrates service-learning courses through the Community Engagement Lab and the Community Resources and Economic Development program (WVU).

Simon Colwill graduated in Landscape Architecture at the University of Greenwich, UK, in 1994. He then worked for ten years as a freelance landscape architect in several offices in Berlin. From 2004 he has worked as Assistant Professor at the Chair for Landscape Construction at the TU Berlin, Germany. He is currently working on a DFG-funded research project investigating the development of built landscapes through time.

M. Elen Deming, D.Des, FASLA, FCELA, is Professor of Landscape Architecture at North Carolina State University, USA, where she directs the Doctor of Design program – an interdisciplinary distance-education model for established design practitioners. She has taught design studios, design research, and topics in design history and theory for the past 25 years. Former editor of *Landscape Journal* (2002–2009), Deming co-authored *Landscape Architecture Research* (with Simon Swaffield, 2011), and edited *Values in Landscape Architecture and Environmental Design*, LSU Press, 2015 and *Landscape Observatory: The Work of Terence Harkness*, ORO/AR+D, 2017. She serves as Vice President for Research for the Landscape Architecture Foundation.

Noël van Dooren has a Master's degree in Landscape Architecture from Wageningen University, the Netherlands, 1991. Since graduating from Wageningen, his career has spanned the whole of landscape architecture. After working for five years at the office of H+N+S Landschapsarchitecten, he switched to his own practice. He was associated with the professional magazine *Blauwe Kamer* for many years. From 2004 to 2009 he led the landscape architecture programme at the Academy of Architecture Amsterdam and co-founded the international EMiLA programme. He continued as a Research Fellow at the Amsterdam School of Arts and defended his thesis "Drawing Time: The Representation of Change and Dynamics in Dutch Landscape Architectural Practice after 1985" early in 2017. He coordinated a professional practice programme, and was a member of the board of the *Journal of Landscape Architecture* for the critique section "Under the Sky". Currently, he holds the professorship in Sustainable Foodscapes in Urban Regions at VHL University of Applied Sciences in Velp, the Netherlands.

Lake Douglas is Associate Dean and Professor at Louisiana State University's College of Art + Design, USA, and he has published in America, Europe and China. Recent books include *Public Spaces, Private Gardens*, Louisiana State University (LSU) Press, 2011; *Steward of the Land*, LSU Press, 2014; and he co-authored *Buildings of New Orleans*, University of Virginia Press, 2018. He edits the Reading the American Landscape series for LSU Press.

Wenche Elisabet Dramstad has an MSc in Management of Natural Resources and a PhD in Landscape Ecology. She has been teaching landscape ecology at the Norwegian University for Life Sciences for nearly a decade, and works as a Senior Research Scientist at the Norwegian Institute of Bioeconomy Research, Norway. Her research has focussed on landscape change and she has a particular interest in landscape ecology as a bridge between landscape architects and planners and ecologists and biologists.

Shelley Egoz is a landscape architect and Professor of Place Making at the Norwegian University of Life Sciences (NMBU), Norway. Shelley's teaching experience spans landscape architecture programmes in the USA, New Zealand, Singapore and Norway for over 20 years. Her research interests are in landscape justice; she is the founder and head of the Centre for Landscape Democracy (CLaD), NMBU, and co-editor of the books *The Right to Landscape*, Ashgate, 2011, and *Defining Landscape Democracy*, Edward Elgar, 2018.

Luca Maria Francesco Fabris is a journalist and architect, has a PhD in Architectural and Environmental Technology, a Master's in Urban Planning and Environment, and is Associate Professor in Architectural Technology and Environmental Design at the Department of Architecture and Urban Studies (DAStU) of the Politecnico di Milano, Italy, where he has taught since 1997 at the Architecture Urban Planning Construction Engineering School at Master and Bachelor level. He is also Guest Professor at the North China University of Technology in Beijing. He focusses on research related to contemporary built environment, sustainability and landscape.

Ellen Fetzer has a Diploma and PhD in Landscape Planning from Kassel University, Germany. Since 2001 she has been working at the School for Landscape Architecture, Environmental and Urban Planning in Nürtingen (Stuttgart area), Germany. She is primarily coordinating an international Master's degree in Landscape Architecture (IMLA). The second focus of her work is on computer-supported collaborative learning.

Kasia Gallo teaches at the Department of Counseling, Educational Psychology and Foundations at Mississippi State University, USA. In addition to her Doctorate in Educational Psychology, she holds two design degrees (Bachelor of Architecture and Master of Landscape Architecture). Prior to joining her current department, she worked as an architectural designer, then taught at Mississippi State's Landscape Architecture Department. Her research reflects both of her backgrounds and focusses on fostering creativity and problem-solving, as well as on writing instruction and assessment. She presently teaches scientific writing, and a writing-heavy capstone course. Her doctoral dissertation centered on Prominent Feature Analysis of novice scientific writing.

Davorin Gazvoda is a Professor who has been teaching landscape design studios since 1989 at the Biotechnical Faculty, University of Ljubljana, Slovenia, and he is currently serving as a Vice-Dean of Academic Affairs. His international experience includes landscape design workshops at universities in Turkey, China, Serbia, Croatia, Russia and the USA. His research work includes published articles, papers at international conferences and research reports. In the field of landscape design he has accomplished dozens of landscape and urban design projects and competitions, most of them awarded.

Guido Granello is an architect specialising in landscape and participatory processes. After collaborations in different workshops with architects, garden planners and engineers, he is now a free researcher in land policies to support a different balance between the urban and peri-urban milieu and collaborates in on-field researches at the Department of Architecture and Urban Studies (DAStU) of the Politecnico di Milano, Italy.

Doris Gstach is Professor at the University of Applied Sciences Erfurt, Germany. She teaches in the field of spatial and landscape planning at the Department of Urban and Spatial Planning in the Faculty of Architecture and Urban Planning. Gstach studied Landscape Planning in Vienna and Manchester. She has worked as a Lecturer at the Vienna University of Technology, Austria, and at the Leibniz Institute for Structural Planning and Regional Development (IRS), Germany, and as an Assistant Professor at Clemson University in the United States.

Allan Gunnarsson is a landscape architect and university lecturer at the Swedish University of Agricultural Sciences, Sweden, with a focus on vegetation and park management. Most of his time is devoted to teaching in first cycle education. In addition, he conducts research in the field of vegetation, landscaping and gardening and cultural landscape history.

Roland Gustavsson is Professor Emeritus in Planting Design and Landscape Management at the Swedish University of Agricultural Sciences, Sweden. Gustavsson has dedicated his teaching, practice and research and their integration with studies in vegetation as an ecological, spatial and social structure. His work has informed and investigated vegetation as a tool for landscape architecture. His pedagogical work has developed and emphasised time studies, outdoor learning, action-oriented and place-contextual projects, the use of 'Shared Reference Landscapes' and the use of 'slow learning' in academic contexts. His articles have been published in *Urban Forestry & Urban Greening*, *Forest Policy and Economics* and *Landscape Ecology*, among others.

Marc Kirschbaum, Prof. Dr.-Ing. M.Arch./USA, is an architect, architectural theorist and publicist, and Professor for Architectural Theory and Design at SRH University Heidelberg, Germany. He is in charge of the transdisciplinary research project Reallabor

STADT-RAUM-BILDUNG. With Professor Kai Schuster, he is founder and partner of the research studio pragmatopia – architecture.city.life in Kassel.

Pinar Köylü is an Assistant Professor at the Faculty of Forestry, Department of Landscape Architecture, Düzce University, Turkey. She studied landscape architecture at Ankara University, Turkey, received a Master of Fine Arts degree in Interior Architecture and Environmental Design from Bilkent University, Turkey, and holds a PhD in Landscape Architecture from Ankara University. Her research interests include design theories, history of landscape architecture, and environmental psychology.

Karl Kullmann is a landscape architect, urban designer and Associate Professor at the University of California, USA, where he teaches design studios in landscape architecture and urban design, and courses in landscape theory and digital modeling and visualisation. Kullmann's scholarship and creative work bridges landscape architecture and urban design through diverse lenses, including urban topography, green infrastructure, urban wastelands, public gardens, urban decline, spatial orientation and disorientation, design modeling and visualisation, mapping and datascaping.

Janike Kampevold Larsen is Associate Professor at the Institute of Urbanism and Landscape at the Oslo School of Architecture and Design, Norway. She is project leader for Future North, a research project observing and mapping changes in settlements and territories in the Arctic and Subarctic region. She has worked as programme developer and coordinator for The Tromsø Academy of Landscape and Territorial Studies since 2011, and has been on the *JoLA* editorial team since 2015. Her recent publications include: *Routes, Roads and Landscapes*, Ashbury Publishing Ltd, 2012; "Global Tourism Practices as Lived Heritage: Viewing the Norwegian Tourist Routes", in *Future Anterior*, Vol IX, No. 1, Summer 2012; "Imagining the Geologic", in *Making a Geologic Now*, Punctum Book, 2013; and "Geologic Presence in the Twenty-First Century Wilderness Garden", in *Studies in the History of Gardens and Designed Landscapes*, John D Hunt (editor), 2014.

Ralf Löwner is Professor at the Hochschule Neubrandenburg, University of Applied Sciences, Germany, where he is teaching photogrammetry, remote sensing and GIS. His research focusses on web-based land management systems and GIS-based mapping methods. He has more than 20 years' experience in thematic field mapping, especially in Northwest and Northeast Africa. He also set up a geoportal with data on geology, hydrology, ecology and epidemiology while working for the German Research Center for Geosciences GFZ in Potsdam, Germany.

Christian Montarou is a painter and Assistant Professor at the Department of Landscape Architecture and Spatial Planning, The Norwegian University of Life Sciences (UMB), Norway. As a painter, Montarou has created numerous public artworks and is represented in several of the main art institutions in Norway. As a lecturer, he is responsible for a drawing teaching programme that makes up the basis of the three first compulsory years of landscape architecture education.

Yazid Ninsalam is a lecturer in Landscape Architecture at RMIT University, Melbourne, Australia. He worked as a landscape architect at National Parks Board Singapore, and received a doctorate from the National University of Singapore (NUS). He was a Researcher at Future Cities Laboratory (ETH) and holds a Master's degree in Landscape Architecture (NUS).

Thomas Oles is Professor of Landscape Architecture and Chair in Design Theory at the Swedish University of Agricultural Sciences, Sweden. He has taught landscape architecture and landscape history at the University of Oregon, USA, the Amsterdam University of the Arts (where he initiated and led the programme 'Living Landscape'), the Netherlands, Cornell University, USA, and the University of Edinburgh, UK. He is the author of two books, *Walls: Enclosure and Ethics in the Modern Landscape*, University of Chicago Press, 2014, and *Go with Me: 50 Steps to Landscape Thinking*, Architectura + Natura, 2013. He is currently writing a book on fieldwork in landscape design to be published by Routledge (forthcoming).

Ana Opriş has been a teaching assistant within the Urbanism & Landscape Design Department of UAUIM Bucharest since 2011, and holds a PhD in Urban Planning (2016), an MSc in Landscape and Territory (2010) and a BA in Urban Planning (2008), all three from UAUIM Bucharest, Romania.

Irina Paţa has a PhD degree in Urban Planning, in which she researched the relation between landscape and video/film representation and how this can be used in the teaching process. She has a Bachelor's degree in Landscape Architecture and a Master's degree in Landscape Planning.

Jörg Rekittke is a tree nursery gardener and landscape architect. He studied at TU Berlin and ENSP Versailles, and received a doctorate from RWTH Aachen University, Germany. He held academic positions at RWTH Aachen, University of Wageningen, the Netherlands, National University of Singapore and RMIT University, Australia. He founded the Berlin Megacity Laboratory.

Deni Ruggeri is Associate Professor of Landscape Architecture and Deputy Director of the Center for Landscape Democracy (CLaD) at the Norwegian University of Life Sciences (NMBU), Norway. He is the Chair-Elect of the board of EDRA, the Environmental Design Research Association. Dr Ruggeri's research and publications focus on sustainable urban development, place identity, landscape democracy and participation in design and planning.

Ingrid Schegk is Professor of Landscape Construction and Design in the Department of Landscape Architecture at Weihenstephan-Triesdorf University (HSWT), Germany. She lectures and carries out research projects in landscape design and construction, material science, technical planning and detailing, and runs design studios. She is Programme Director of the International Master of Landscape Architecture (IMLA) programme at HSWT. She also guest lectures and runs design studios in Austria, Spain, Ireland, Rome, Italy and Belgium.

Marc Treib is Professor of Architecture Emeritus, University of California, USA, and a historian and critic of landscape and architecture who has published widely on modern and historical subjects in the United States, Japan and Scandinavia. Recent publications include *Meaning in Landscape Architecture and Gardens* (editor), Routledge, 2011; *Austere Gardens: Thoughts on Landscapes, Restraint, and Attending*, ORO, 2016; *Pietro Porcinai and the Landscape of Modern Italy* (co-editor), Ashgate, 2016; and *Landscapes of Modern Architecture: Wright, Mies, Neutra, Aalto, Barragán*, Yale, 2017.

Mari Sundli Tveit is Rector at NMBU, The Norwegian University of Life Sciences, Norway. She is also Chair of Universities Norway and she is a member of the board of The European University Association. She has previously been Pro-Rector at the same university, and has served as a board member in a range of institutions within academia. Mari Sundli Tveit has a background in ecology and nature management, and is Professor of Landscape Architecture. Her main research interests are landscape ecology and landscape perception and preferences, and development of frameworks for landscape assessment.

Veerle Van Eetvelde is Associate Professor in the Department of Geography research unit Landscape Research at Ghent University, Belgium. She is a geographer and spatial planner and her research ranges from landscape ecology, landscape preference and historical geography to landscape and heritage management and planning. She is active as President in the European chapter of the Association of Landscape Ecology (IALE-Europe) and as Vice President of UNISCAPE, as well as in national and European projects related to landscape and cultural heritage.

Ed Wall is Academic Leader Landscape at the University of Greenwich, UK. His design and research work focusses on cities, landscape and public space. He has published articles for the journals *Urban* (in 2013), *Landscape* (in 2011 and 2012) and *Topos* (in 2011) and essays for books *Questo Metropolitan Architecture*, Maggioli Editore, 2015; *Revising Green Infrastructure: Concepts Between Design and Nature*, CRC Press, 2014; *Educating Architects*, 2014; and *Infrastructural Urbanism: Addressing the In-Between*, DOM, 2011. In 2010 Ed co-authored, with Tim Waterman, *Basics Landscape Architecture 01: Urban Design*, AVA Publishing, 2010. He is also co-editor, with Tim Waterman, of *Landscape and Agency*, Routledge, 2017.

Tim Waterman is Senior Lecturer and Landscape Architecture Theory Coordinator at the University of Greenwich, UK and a thesis tutor at the Bartlett School of Architecture, University College London, UK. He is also Non-Executive Director at the arts organisation Furtherfield and, until recently, Honorary Editor of *Landscape*, the journal of the Landscape Institute. Tim Waterman is author of *Fundamentals of Landscape Architecture*, which is now in its second edition from Bloomsbury, and, with Ed Wall, author of *Basics Landscape Architecture 01: Urban Design*, AVA Publishing, 2010. These books have been translated into seven languages. Also recently published are two edited collections for Routledge: *Landscape and Agency*, with Ed Wall (2017) and the *Routledge Handbook of Landscape and Food* with Josh Zeunert (2018).

Björn Wiström is a landscape architect, lecturer and researcher in vegetation and landscape care at the Swedish University of Agricultural Sciences, Sweden. The focus of his research is the care and design of urban forests and nature.

Foreword

Simon Bell, ECLAS President 2012–2018

This book is the latest in a series produced under the European Council of Landscape Architecture Schools (ECLAS) aegis and published by Routledge. It is part of the ECLAS strategy of helping to support and improve the quality of landscape architecture education across Europe and beyond. We are also pleased that we have developed such a good relationship with Routledge who publish, in their own right, an increasing number of landscape-orientated books, as well as being the publisher of the ECLAS-owned *Journal of Landscape Architecture* and, most recently, establishing a partnership agreement with many benefits for ECLAS member schools such as discounts on prices.

The European Council of Landscape Architecture Schools' first foray into book publishing comprised an edited collection of essays under the title *Exploring the Boundaries of Landscape Architecture*, which came out in 2011. Each chapter was written by an expert in a field other than that of "pure" landscape architecture – one of many (but by no means all) of the so-called "neighbouring disciplines" that frequently interact with landscape architecture and whose methods of analysis or research we often apply within landscape architecture (hence the idea of exploring the boundaries). Therefore, to a large extent, it was about theory and practice and trying to capture something of the extent of the disciplinary field. It included materials from well-known international practitioners from the natural sciences, social sciences, humanities and planning and design fields. As such, it offers an interesting way of looking at landscape architecture from the points of view of many different non-landscape architects. There is scope for a second volume covering subjects that were not included in the first one.

The second book, published in 2016, shifted the focus onto research in landscape architecture. This is a growing area for a disciplinary field dominated by practice for many decades but where research and research evidence in a wide range of areas is building momentum. One idea for that book was to see if it was possible to identify specific research methods of landscape architecture. The answer was no, it was not possible, and that many methods used are taken from other fields – similar to or the same as the neighbouring disciplines. We found that, in many respects, it is the tendency for landscape architecture to apply case study research using several methods and that the uniqueness lies in this application of methods, often from very different fields, all applied to solving a single (but usually) complex problem.

This now brings me to the subject of this particular volume – teaching landscape architecture. One of the specific corporate goals of ECLAS is to foster and develop the highest standards of landscape architecture education in Europe by, amongst other things, providing advice and acting as a forum for sharing experience on course and curriculum development, and supporting collaborative developments in teaching. For several years ECLAS members have been vocal in wanting something to support teaching. There is so little out there at the moment and the subject of pedagogy has been a popular theme at most recent ECLAS conferences, where papers are presented offering all sorts of teaching approaches and whose presenters look for feedback and encouragement from their peers. The European Council of Landscape Architecture Schools has worked for many years to provide guidance on landscape architectural education – in terms of the various "core", "subject specific" and "generic" competences that schools should teach. The Council has also been working recently with the European section of the International Federation of Landscape Architects (IFLA Europe) in order to harmonise the educational requirements of the profession with the guidance of academia. This recently (2018) culminated in a project called EULand21, which has produced a lot of practical materials such as "learning lines" and course modules suitable for a bachelor programme – work based on the ECLAS guidance and the IFLA requirements but converted into modes of teaching. As the disciplinary field expands and schools have to be more selective in what they can fit into a curriculum with many optional courses, some additional guidance and help in the form of examples of best practice is sorely needed.

Therefore, this book originated in a deeply felt need by all ECLAS members for up-to-date materials to help them to teach. It must be said at the outset that we do not want all schools to be alike and to teach exactly the same things in the same ways – we want to maintain diversity. This in part derives from regional and cultural traditions but also, perhaps most importantly, from the origins of the programmes in different universities at different times. Schools founded by and located within architecture departments, for example, have a different perspective from those located in, say, universities of life sciences. Therefore, none of the guidance and certainly none of the approaches presented here are set out to present or recommend a "best way" of teaching. We do hope, however, that the examples and models described in the book will be inspirational and set a standard to which every teacher should aspire.

Readers will note that the book is a "Handbook" – it sets out to be eminently practical and not too theoretical – although it is necessary for some pedagogic theory to be present. The book emerged in an interesting way: it was not conceived by the editors from a grand plan and structure whose chapters were then identified and authors sought to supply the materials. It was, conversely, a democratic, bottom-up approach. Abstracts that focussed on practical teaching experiences were invited as a kind of open competition by the editors. This attracted a large number of offers, some of which were quite similar in theme, so that the editors had to find a way of thinning these out to a number that could be included and that represented a good range of approaches. The authors were then asked to submit a chapter following a general guide and word length together with some illustrations. Once all these chapters came in and were reviewed it was also obvious that there were too many for the total allowance and some had, sadly, to be removed at a late stage (although I am sure these works will find their way to be published in some way).

The introduction covers, in more detail than I have space for, the essential aspects of the book, its structure, philosophy and intended readership. What I find especially strong here is the selection of the three sectional themes – and the dominance of the first one, "Reading the landscape". This is a subject dear to my heart.

This book is the beginning of a series of books to support teaching, expanding on the themes and areas not covered by this volume, such as, for example, the more theoretical aspects of landscape architecture pedagogy and didactics. However, the main methods of teaching landscape architecture's core competences of landscape analysis, planning, design and management through studio projects – which, according to the ECLAS Guidance on Landscape Architecture Education, should occupy some 50% of the curriculum – are covered here and in a second volume: *Teaching Landscape: The Studio Experience*, which is in the pipeline.

Introduction

Karsten Jørgensen, Nilgül Karadeniz, Elke Mertens
and Richard Stiles

Landscape studies are inherently multidisciplinary and multidimensional. There is an infinity of ways to describe and read landscapes, and landscapes may be sources of many different kinds of knowledge; from biology and ecology to archaeology, geography, social history and art, to mention but a few. The approaches to teaching in the disciplines related to landscape are thus as diverse and dynamic as the landscapes themselves.

This book is an attempt to present some of the diversity inherent in teaching landscape. The book is initiated by ECLAS – the European Council of Landscape Architecture Schools – and it responds to one of the main goals of the organisation to:

> Foster and develop the highest standards of landscape architecture education in Europe by, amongst other things, providing advice and acting as a forum for sharing experience on course and curriculum development, and supporting collaborative developments in teaching.[1]
>
> *ECLAS (n.d.)*

As a consequence, the book is likely to find much of its readership within the field of landscape architecture – in academia and in professional positions – as well as among students of the discipline, but it should also be of relevance to interested readers from other landscape-oriented fields such as landscape ecology, archaeology, planning and urban design. In line with the broad nature of the field, the themes covered can by no means be exhaustive; rather it is an attempt to open up a discourse on landscape teaching. In the call for contributions issued in January 2016, we opened up for a wide range of submissions:

> Landscape architecture curricula and teaching traditions and principles vary in many ways across the member schools of ECLAS and beyond, and they vary and develop over time. The idea of the book *Teaching Landscape* is to disseminate papers on landscape and design education, reflections on pedagogy, teaching traditions, experimental teaching methods, and new teaching principles, to mention just a few themes. [. . .] Central questions will be why and how we teach the different subjects.

The response to the call was overwhelming, and it demonstrated an enthusiasm among scholars longing for an opportunity to communicate their ideas about and experiences from teaching landscape classes of different kinds. Selecting and organising the many contributions proved to be a challenging task, and it ended with them being divided into four categories. The first, 'Reading the landscape', is concerned with how to understand the landscape; the deciphering of the fascinating and multifaceted phenomenon that is at the core of the landscape disciplines. The second, 'Representing the landscape', is about how to epitomise the landscape for different purposes and needs, and in different situations, whether it is to show the beauty of a scenery or the ecological state of affairs in a region. It involves methods ranging from technical drawings to scholarly or poetic descriptions. These types of actions are the purpose of a major part of landscape architects' activity; to prepare a common ground for discussion of landscape issues. The third category focusses on 'Transforming the landscape'. This is an activity that is to some degree specific to landscape architecture, and is what distinguishes landscape architecture from mere landscape studies. It is the first three of the four categories that make up the three parts in the book. This will be complemented by a second book based on the fourth category with the title *Teaching Landscape: The Studio Experience*, dealing with a range of approaches to the characteristic studio-based didactics found in landscape architecture schools all over the world. The studio experience book will appear (also published by Routledge) shortly after the publication of the present volume.

Although the subject of the book is teaching, the theme – landscape – permeates all the chapters. Didactic reflections are more often than not entrenched in the many stories about landscapes presented in the 30 chapters. As a result, many years of creative approaches to landscape teaching are to be found encapsulated in the chapters of this book.

Landscape is much more than mere static scenery, it is also about systems and processes and the people who perceive it. As a result, it is something dynamic, always changing. Therefore, future-oriented thinking is embedded in landscape architectural thinking. For landscape teachers, this means that they can never rely on yesterday's methods and philosophies for teaching, but are always on the lookout for new approaches, and cultivate a critical attitude to traditional methods. The approaches to the central theme of sustainability may help to explain the way in which critical awareness is key to landscape teaching. Landscape architecture is generally understood as being an environmentally responsible profession and discipline, one that is deeply rooted in and dependent on knowledge of soil, plants and ecological thinking. The main building materials used in landscape architecture are living material, and failure to master and comprehend these fields undermines the quality of the built landscapes. The discipline and the profession is thus based on the concept of sustainability. This view has been both confirmed and refined in the many documents from the UN and other institutions on the importance of sustainability in planning. The latest of these – the UN 'Sustainable development goals' – came into force in 2016 after having been adopted by world leaders in 2015. Of the 17 goals formulated, the majority are relevant to the work of landscape architects (O'Donnell 2015). Landscape architecture projects must clearly be weighed on scales of sustainability. However, landscape architects must also be wary of the political dilution of the sustainability concept and the many alternative agendas related to the moves towards a green economy we can see today. This emphasises the need for clear analyses of the different components of sustainability in landscape architecture in order to avoid supporting unsustainable agendas by merely green-washing suboptimal projects.

This state of critical awareness is necessary in all aspects of landscape teaching, for example in relation to questions of values. Landscape is a potent ideological concept and neither the academic discipline nor professional landscape architecture can avoid taking a political standpoint in public space, such that design can become an ideological instrument.

Our hope is that the present volume will stimulate creativity as well as critical thinking among teachers and students alike in the many schools in which landscape subjects are taught. Not for the sake of harmonising or tuning the approaches to teaching landscape but, on the contrary, to stimulate diversity and local initiatives to achieve the highest standards of scholarship and practical skills needed to tackle the challenges that will be faced by the landscapes of the future.

Note

1 ECLAS web page: www.eclas.org/index.php/about/goals-and-origins-of-eclas.

References

ECLAS (n.d.) Preamble at www.ECLAS.org (visited 25 May 2018).

O'Donnell, Patricia M. (2015) "Landscape Architects Can Help the World Achieve New Sustainable Development Goals" in *The Dirt*, a blog by ASLA, 23 September (visited 25 May 2018).

Introducing hope

Landscape architecture and utopian pedagogy

Tim Waterman

Teaching, landscape, and utopia

Teaching is a practice with its eye on both beginnings and futures: it is a work of constant introduction in an atmosphere of hope. The work of the teacher is always to prepare students to encounter new ideas, knowledge, or techniques, then to arrange the meeting. This meeting will then, hopefully, result in a dialogue between students, teachers, and the subject, which, in the case of landscape architecture, is often a site and its inhabitants, human and other species. The nature of these introductions is always complex and multi-layered. To complicate matters further, the idea of landscape within landscape architecture has changed radically in recent decades, meaning that the subject students are introduced to might be substantially different from the one they expect to meet.

The idea of landscape has recently had a new lease of life in a range of related disciplines, including geography, anthropology, archaeology, and many others, owing in part to the European Landscape Convention (Council of Europe, 2000). Its simple yet powerful language has become embedded and now serves as an organizing frame for many inter- and transdisciplinary discussions and collaborations: "landscape is part of the land, as perceived by local people or visitors, which evolves through time as a result of being acted upon by natural forces and human beings." Of course, what is omitted from this statement is that landscapes also shape their inhabitants, their languages, their everyday lives, and their structures of knowledge and practices of worlding—landscapes are worlds that make particular futures particular to themselves, and it is important that we acknowledge and examine those relationships and grow with them. Donna Haraway writes, "It matters what thoughts think thoughts. It matters what knowledges know knowledges. It matters what relations relate relations. It matters what worlds world worlds. It matters what stories tell stories" (2016: 35). Imagined futures are called forth from the dialogues between people and place, whether this is the creation of *near* futures such as those constructed in the design process with the end of being built, or the creation of *speculative* futures that might create whole new possible worlds. Indeed, utopianism, the creation of *impossible* worlds, used as a method for moving towards better worlds in combination with the creation of near futures, is the subject of this introduction. A dialogue such as this must be seen to be in a constant state of evolution and flux, mirroring the ecological conversations between people and their environments. Thus the idea of utopia as employed here is in direct opposition to the concept articulated, famously, by Isaiah Berlin (1991: 20):

> The main characteristic of most (perhaps all) utopias is that they are static. Nothing in them alters, for they have reached perfection: there is no need for novelty or change; no one can wish to alter a condition in which all natural human wishes are fulfilled.

As Judith Suissa so perceptively notes in countering this, "there is no such thing as the one finite, fixed form of social organization," and that the underlying principle of society, and by extension education, is "constant striving, improvement and experimentation" (2006: 141). The goals of utopian methodology in education are, or should be, permanently transgressive and transformative, as we shall see below.

Transgressive utopianism and its ends

It is commonly and erroneously considered, especially within the architectures,[1] which are steeped in high state modernist imagery and ideas, including Isaiah Berlin's, that utopia (the *eu*-topia—'good place'—that is *ou*-topia—'no place') exists as a totalizing blueprint written onto a clean slate, as a perfectly formed vision for an impossibly perfect world. This has pulled the architectures in two directions, the first towards the desire to create just such glitteringly perfect and static blueprints (and towards the disillusionment of the inevitably imperfect outcome), and, second, away from an architectural imperative to make the world better. The tension inherent between the false goal and the real goal, and the perception that any betterment must be discarded along with dreams of unchanging perfection, has led the architectures into a dreary impasse. On the one hand, a postmodern aversion to modernist optimism and its perceived failures leads to such outcomes as a concentration on empty form and concept, as exemplified, perhaps, by Charles Jencks's 'iconic building' (2005) now filling skylines with outlandish and enigmatic shapes. On the other hand, the neoliberal consensus, itself a bizarre, marketized utopia (or dystopia) pushes the form of building projects to be determined solely by 'the bottom line,' resulting in a dogsbody managerialism amongst designers that is focused merely upon satisfying a brief defined primarily by an extractive and deracinated profit motive.

This chapter argues, instead, that utopianism may usefully be seen and used as a process of becoming and a way of imaginatively prototyping that is rooted in the lived world and experience. In this, a utopian orientation has much in common with the work of education. Stuart Hall makes this parallel explicit, describing utopianism as "future-becoming," a work of both ontology and practice that is "not just an empty projection but is grounded in experiences that we can already have, and is seen as an incubation of prototypical relationships—trying to embed them as alternatives within an existing structure" (2007: 126). This mirrors Miguel Abensour's description of the evolution of such practical utopias as the "metamorphosis of utopia into a dialectical image" (2017 [2000]: 107). In this sense, landscapes (as places of situated experience) may be seen as places with utopian potential, but also as already containers of a plethora of utopian fragments or partially realised utopias, relics of past striving that still serve as essential (and loved) elements of lived space, and which are foundational to the collective work of 'future-becoming'. In this, landscapes are conceived also as collective works over time.

Utopian theorist Lucy Sargisson succinctly describes how (non-totalizing, practical, and critical) utopian thought works.

- It issues from political dissatisfaction and offers political critique.
- It articulates estrangement and offers an alternative perspective, from an alien (or new) space.

- It is creative and imaginative and often fictional.
- It has subversive and transformative potential.

(2000: 3)

She further defines utopian thought as, importantly, transgressive, seeking to break out of existing boundaries, to perpetuate a radical openness, and to create new space for thought. Thus she adds three further dimensions of transgressive, critical utopian thought.[2]

- It breaks rules and confronts boundaries.
- It challenges paradigms.
- It creates new conceptual and political space.

(2000: 4)

Each of these characteristics maps rather well to the motivations, processes, and ends of an emancipatory education, and has particular resonance with and bearing on the imaginative and creative studio education undertaken by architects and landscape architects. The goals of transgressive utopianism are directed towards the political agency of both individuals and collectives, and agency, too, is a critical outcome of architectural education; that designers are able to work with each other and with other disciplines and professions to effect positive change in landscapes, and, by extension, in the social, cultural, political, and ecological world in which those landscapes are situated and upon which they act. Both education and utopianism share that they are rooted in hope for the future and a dissatisfaction with the status quo. Tom Moylan writes that utopianism (and by extension, for my purposes here, education) is "rooted in the unfulfilled needs and wants of specific classes, groups, and individuals in their unique historical contexts" (1986: 1).

For the student of landscape architecture, those "unique historical contexts" inhere in the landscapes of everyday life, are at once medium and message, and are ecological in construction: an interdependent dialogue between people, place, climate, culture, other species, and much more. Elizabeth Meyer writes that landscape architecture's direction in the coming years should be to give "significant form and meaning to ecological processes through the making of landscape experiences" and that this should give rise to "design practices that engender more mature understandings of humanity's interdependence with nature, that stir ethical as well as aesthetic debates, and that do not sacrifice significant landscape form in the name of environmentalism" (2000: 244). The urge to do good in the world, and to mold its parts into more than mere functionality, but into art, is signally utopian and often transgressive.[3]

Finally, and importantly, the domain of both education and utopianism should be emancipation and autonomy. A utopian pedagogy should help students:

- to a position where they have agency in the world around them (and the agency to choose to what degree they wish to use that agency);
- to understand that democratic agency is *both* individual and collective, and simultaneously, inseparably so;
- to gain the confidence to make decisions in the world around them (designers, especially, are expected to make difficult aesthetic decisions, for example) on behalf of and at the behest of themselves, other people, other species, and even whole landscapes;
- to begin to see and understand the processes and forces through which the world is—or worlds are—made (and thus further to be prepared to identify and counter the obfuscations, misinformation, and oversimplifications on which bad decisions in the world are founded).

Gert Biesta helps to show how emancipation nests into education's additional domains, which might also be seen as utopian. The first is *qualification*, "which has to do with the acquisition of knowledge, skills, values, and dispositions." Then there is *socialization*, "which has to do with the ways in which, through education, we become part of existing traditions and ways of doing and being." And finally, *subjectification*, "which has to do with the interest of education in the subjectivity or 'subject-ness' of those we educate. It has to do with emancipation and freedom and with the responsibility that comes with such freedom" (2014: 4). What makes this approach not merely utopian, but transgressive, is the creation of the 'subject' and is not one geared towards statist models of citizenship requiring compliant subjects, but one that is equipped to make a critical way through the methods and human associations of a specialism and to act, individually and in concert with others, to effect positive change on its methods, associations, and outcomes to build a better world from within.

Learning hope

The domains of qualification, socialization, subjectification (and emancipation) all contain the purposes of education, and all of these are underwritten by hope and striving. Hope, write Michael Hardt and Antonio Negri, is "conceived as a temporal vector that points from the present into the future from a specific location, with a determinate direction and force" (2002: 201). Hope, thus, is not abstract and intangible, but specific to time, space, body and mind, and as such it inheres in places, and is acted out through the sorts of creative and imaginal supposition, simulation, and prototyping that are the stuff of design (for more on "the propositional imagination" see Waterman, 2017a; Nichols, 2006; Nichols and Stich, 2000). Ernst Bloch is explicit here, in the process of what he calls "learning hope" (1986: 3). He speaks of how hope is presented and performed in theatre and film, travel, dance, and fairytale. "Such things either present a better life, as in the entertainment industry, or sketch out in real terms a life shown to be essential." The goals of creating a better life and knowing how to get on in everyday life are interdependent and immanent in each of the three domains of education. Bloch goes on to describe how such imaginative sketchings-out of blueprints of the possible in such vehicles as film and theatre can emerge to exert force for change in the world, at which point, "we find ourselves for the first time among the actual, that is *planned or outlined utopias*." The blueprint, the prototype, the simulation makes its way from ideal to physical construction. Following this, it may be that a central problem with architectural utopias in the present day lies in much of the practice of architecture and landscape architecture, which is often divorced from physical acts of construction. Thus the problem for the architectures is that they are too abstractly utopian in that building exists only and first in the imagination, and then it is constructed by others as a sort of reification—the making-real of abstractions, which can often lead to the construction of cities as diagrams or buildings as 'icons'. If the contemporary architectures were more like, for example, Palladio's architecture, designed and built in a fluid, reiterative, dialogic way, it could be utopian but in a concrete way. The architectures consistently operate in ways that are neither *mise-en-scene* or *mise-en-sens* (particularly with digital representation privileging the visual). A few landscape architects, such as Martí Franch, who creates landscapes through design processes realized through landscape management practices, and architects such as Santiago Cirugeda, who creates guerrilla buildings with the people who need them, are challenging this, practicing forms of social and ecological dreaming and making that grow from sites and communities in truly dialogic and reflective ways (see Waterman, 2017b and de Sousa, 2014). Their hopes are all the more concretely utopian given that they are finding positive ways to act and imagine in Spain's bleak post-austerity reality. Their hopes are knowable and feel-able in time and space. Education, to be emancipatory and utopian, involves doing things and making things in and with the world so that the proof of the agency of the student may be felt, not just in the mind, but the body. While much of this making and doing is necessarily confined to the studio, there is still a proof of the efficacy of action in the world, and of the student's processes of worlding. Studio courses also often ground themselves in real sites with actual clients, and processes of site analysis and evaluation and community consultation have great potential to provide many opportunities for situated and embodied engagement.

Keeping hope tied to people, places and their positive transformation is important, because advocacy for hope is often derided, particularly amongst those who are in power and profiting from the status quo. It is also in the interest of hegemonic power in late capitalism—namely the super-rich and the politicians, nation-states, and media empires that serve them—to ensure education is directed towards shutting down, rather than opening up avenues for growth and emancipation. The phrase, from Alexander Pope's *Essay on Man* (1734), "hope springs eternal" is more often sarcastically used as a retort against quixotic quests or lost causes than it is to refer to a universal characteristic of human nature: an optimistic driver and an essential bulwark against the disappointments, setbacks, and losses that are inevitably presented by everyday life. It is natural to hope and strive for something better. However, the starry-eyed, the bleeding-hearts, the rose-tinted are inevitably given a kicking by those who see themselves as pragmatists or realists. Usually those 'realists' are in a position of privilege in which hope hardens into expectation, and they are dismissing those who would upset the order of things that supports their advantage. To deride hope as a delusional form of utopianism is a common tactic of power. Perhaps it's important here to be specific about the nature of hope. The hope to which I refer is not the same as desire for consumer goods or new experiences that might, for example, be fuelled by marketing, but rather the optimistic hopes that people have for the world that surrounds them: better lives and conditions for their friends and families, their communities, their societies, and, in a show of the remarkable extent of human empathy, for the fate of the planet at large and the species besides our own that are its inhabitants. Hope is rooted in landscape and, despite what anti-utopian 'realists' might say, in the real. Hope is not general and abstract, but is always *about* or *for* something, someone, or somewhere.

The real stuff of landscape is the stuff of our shared future: the grounds for human and planetary flourishing, the space in which the common good is to be found and grown. The contemporary student of landscape architecture encounters a profession that is becoming less concerned with scenography and the iconography of power than it is with a meaningful engagement with the dialogue between culture and ecology in the creation of sustainable landscapes. Thus a mastery of geometry in the service of display, written onto the landscape, has now been supplemented, and sometimes supplanted, by an immersive approach in which the designer comes into communication with all the processes and forces that shape a place. In many cases, what is required is a shift in a student's whole conception of how the built environment is constructed, and to what ends. Alberto Pérez-Gómez puts this well, arguing that "teaching the future architect the elements of praxis—how to articulate an appropriate and ethical position that becomes incorporated in the project—is paramount in view of this intertwining of embodied consciousness and world" (2015: 230). It is not just a practice of teaching the future landscape architect, but of preparing the future landscape architect to prepare for and build, develop, and grow an appropriate and ethical world.

Development

Both education and landscape architecture are directly concerned with growth and development. One of the great tragedies of our era, however, is an obsession with growth—I should add, an obsession with a particular idea of growth. Due in part to a modern and now neoliberal and dirigiste fascination with metrics and data, growth has come to be thought of as merely a measure of quantity. Anyone who has ever put a plant in the ground will know, however, that the quality of growth is as important, if not more important, than quantity. A beanstalk that yields a copious harvest is of no use at all if it crowds out other valuable plants and if its beans are hard, bitter, and inedible. Quantity, in this case, only matters to those who merely count beans. Development is the process of guiding and managing growth toward the best possible outcome. Terms such as 'development', 'growth', and 'investment' have been so co-opted and corrupted by the narrow definitions provided by finance and government that it has come to make sense to opponents of neoliberal hegemony on both ends of the political spectrum to arrange themselves against these very concepts rather than seeking to redefine them or broaden the definition of them. Doreen Massey, in her essay "The Vocabularies of the Economy" (which is rooted in her understanding of economy as whole 'household' management, as articulated against a purely 'market' defined, financialized model of the economy) asks "'What is an economy for?' 'What do we want it to provide?'" (2013: 10). Of course, we may also ask this question of education and of landscape, and her further questions are no less pressing in this regard:

> Investment implies an action, even a sacrifice, undertaken for a better future, while speculation (here in the financial rather than intellectual sense) immediately arouses a sense of mistrust. And while investment evokes a future positive outcome, expenditure seems merely an outgoing, a cost, a burden.
>
> *(2013: 10)*

Education's marketization reduces it to a quantitative cost or burden in precisely this way, and this is a serious ill to be vigorously resisted. Resistance might thus be included in the category of 'investment' as an action or sacrifice for a better future. Finally, Massey discusses growth:

> insofar as the dominant model of growth leads to increased inequalities, as it does, we now know also that it is a prime generator of ill health, crime, and social suffering, compared with what might be the case in a more equal and fair society.
>
> *(2013: 9–10)*

Education and society, of course, do not begin with higher education, but rather higher education is part of the collective work of places, societies, and mutual aspiration that spans lifetimes and geographic scales. Raising and educating a human to adulthood and responsibility is a delicate, careful, and multifaceted process. It involves physical, quantitative growth, of course, but mostly only in childhood. Development continues after the growth of the body to mature size. Of primary importance to human development is recognizing a person's capabilities, talents, and limitations and working with them, encouraging experimentation and learning, building upon past work and knowledge. Many people are involved in this process: families, teachers, communities, friends, writers of books, creators of films, games, TV, and so on. All this care can be thwarted, though, by a difficult environment, whether this is the home and its immediate locale, the community, or even whether the nation in which the child lives is prosperous or not. The good life is not automatic; it is constructed. The good life is mutually, collectively constructed in physical landscapes, which are themselves developed.

Development in education and development of cities and landscapes come together in the practice and theory of landscape architecture. What if we applied the same ethic of care and striving that we apply to child-rearing or education, at their best, to the processes of development by which we, as landscape professionals, are led? What if a property developer's ambitions by definition[4] involved getting to know a place and its people, recognizing their capabilities and helping them to be the best they can be in the moment and setting forth processes of perpetual betterment; taking painstaking stock of ecological functions and biodiversity; ensuring that local, regional, and global contexts were considered in all their historical, political, and cultural dimensions? That work would go on continuously over time—in fact indefinitely—along with the slow process of making a place better with and for its people. Given that such processes, where they exist, presently take place *despite* contemporary development processes rather than arising from them, these aspirations may seem to be quite impossibly utopian. Again, though, utopianism, employed as a method, sets in train a process driven by a vision of *what ought to be*, not a so-called 'realist' assessment of *what is*, especially in light of the fact that *what is* is often unacceptable.

Insurgent democracy, insurgent architectures, insurgent education

If landscapes, societies, cultures, professions, communities, and individuals (such as students) are all works over time involving considerable individual and collective effort and hope, then revolutionary upheaval of all sorts must be seen to be irredeemably negative. Such upheavals at all scales, from trendy 'disruption' to corporate 'restructuring' to the national coup d'état, attempt the rewriting of a totalizing utopia upon a fresh erasure, rather than taking a practical utopian form of insurgency—a 'surging up' of change that seeks to transform only those elements of any system that need transforming. Insurgency, unlike the insurrectionary urge to overthrow, is ongoing and open-ended (see Abensour, 2011). This insurgency is akin to the essential and ongoing function of dissent in democratic government. Dissent and insurgency both require a firm knowledge and understanding of processes and forces as they are, with an eye to what they ought to be. The tragedy, perhaps, is that revolution and erasure can be effected with simple messages that require little sophistication and which prey upon people's fears and prejudices, whereas the understanding necessary for insurgency is difficult, complex, and nuanced, and often place-specific. This, however, is where practical, transgressive utopias can be particularly useful, as they work through narrative and allegory, and they allow an engaging encounter with complexity, nuance, and otherness. Getting people hooked on a good, transformative story is far more satisfying and productive for all concerned than inciting them to repeat a slogan.

Narrative is central to the propositional imagination and the democratic imagination when applied both to politics, society, and governance, and to the built landscape. Narrative, as a key part of scenario-making, is how democratic ideas are imaginatively rehearsed. Scenario-making is also foundational to practice in the architectures. As Joshua Zeunert notes, writing of landscape architecture as a practice focused on sustainability (a mode of practical utopianism), "[l]andscape architects forecast and anticipate future scenarios: political, economic, social, and environmental" (2017: 299). Miguel Abensour (2011: xxiii–xxiv) writes that democracy

> is not a political regime but primarily an action, a modality of political agency, characterized by the irruption of the *demos*, or the people, onto the political stage in their struggle against those whom Machiavelli calls the *grandees* and for the establishment in the city of a state of non-domination.

Those Machiavellian 'grandees' are those now known as 'the 1%' (see Graeber, 2013), and they occupy a position of hegemonic domination in terms of both wealth and power, and they should be the primary target of democratic resistance. The form of an appropriate resistance is that of 'insurgent democracy', which Abensour proposes,

> involves the birth of a complex process, where the social is instituted and the institution directed at non-domination, one permanently inventing itself to better perpetuate its existence and to defeat the counter-movements that threaten to annihilate it and to effect a return to a state of domination.
>
> *(2011: xxiii–xxiv)*

There are fascinating resonances, if not a direct scaffolding, between the model of practice espoused by Abensour and David Harvey's elaboration of the idea of 'insurgent architectures', in particular as the built environment is one of the key areas where the social is both instituted and where it takes specific spatial form. "[T]he architect," who here we will take to describe a figure from any of the architectures,

> can (indeed must) desire, think and dream of difference. In addition to the speculative imagination which he or she necessarily deploys, she or he has available some special resources for critique, resources from which to generate alternative visions as to what might be possible. One such resource lies in the tradition of utopian thinking.
>
> *(Harvey, 2004: 237–8)*

The capacity of 'the architect' to create critical, practical utopian visions—and narratives—has immediate application not as an insurrectionary or revolutionary force, but rather insurgently within the processes, forces, and structures of practice.

Harvey speaks of 'doing-architecture' (2004: 204):

> It takes a huge exercise of the imagination to design an office tower, a residence, a factory, a leisure park, a city, or whatever. The architect has to imagine spaces, orderings, materials, aesthetic effects, relations to environments, and deal at the same time with the more mundane issues of plumbing, heating, electric cables, lighting, and the like. The architect is not a totally free agent in this. Not only do the quantities and qualities of available materials and the nature of sites constrain choices but educational traditions and learned practices channel thought. Regulations, costs, rates of return, clients' preferences, all have to be considered to the point where it often seems that the developers, the financiers, the accountants, the builders, and the state apparatus have more to say about the final shape of things than the architect. The process of 'doing architecture' entails all these complications. 'Doing architecture' is an embedded, spatiotemporal practice. But there is, nevertheless, always a moment when the free play of the imagination—the will to create—must enter.

'Doing-architecture'—which I would pluralize as 'doing-architectures'—as an engaged and relational process, in which theory and (spatial) practice cannot be separated, but are mutually constitutive, is directly analogous to the epistemologies developing in the field of practice theory. Practice theory holds immense potential for theorising landscape as ecological and sociopolitical process (see Waterman, 2018). Practice theory is embedded in the performative and enactive idea of 'doing-architectures', and it has been described in resonant ways by others: anthropologist Stephen Gudeman (2001) speaks of 'reason-in-action'; practice theorist Theodore Schatzki (2010) elaborates a theory of 'activity timespace'; and in describing the complex relations in the sociology of food, sociologist Jean-Pierre Poulain (2017) refers to 'food social space'. This last model of food and cooking also appears in Michel de Certeau, Pierre Mayol, and Luce Giard's concept of 'doing-cooking' (1998 [1994]), which perhaps clarifies that practice theory's genealogy is rooted in theories of everyday life. The complexity of growing, procuring, cooking, and serving food is analogous to the work of the architectures (and of Harvey's doing-architecture), and instructive when assessing how the model of the totalizing utopia versus the critical, practical utopia are applied as an overarching paradigm. Jean-Robert Pitte describes how the modernist "cook-inventor" is driven by a totalizing utopia: "They make themselves happy, they believe, in hatching some novel thing, not grown under natural sunlight but in the neon of their hodgepodge ideas." The true chef-gastronome applies the same measure of hope, but grounded in past, present, and in performance. "Rather than privilege originality at any cost, the cook must attempt to capture what remains of a heritage, to continually question himself about culinary idiosyncrasies, to try to know the deep motivations behind them" (2002: 171). The first is dancing to the thudding one-two beat of a simple slogan, whereas the second luxuriates in a rich orchestration pitched towards the best possible outcome and in scenario-making's narrative engagement with everyday life. To exploit the architect/cook analogy just slightly more, it's worth returning to architecture's disengagement from building: the chef in the restaurant is intimately engaged with his or her product from conception to realization to consumption, and where estrangement and transgression (through insurgent cookery?) most successfully drive innovation is when the customer eating the meal can delight in the familiar made strange—and made better.

Bringing all of these ideals, and using practice theory towards transformatively positive and transgressive ends through insurgent means, finally, must also be applied to education as a whole. So far, I have spoken of education and its role in teaching and its larger sociopolitical role, but education is also concerned with its own institutions. Insurgent education must be highly critical of, and act upon, the domination and hegemony present in its own institution (see Terranova and Bousquet, 2004). The academics, students, and staff of the University of Aberdeen have launched a manifesto and a drive to transform (not overthrow) their institution, called "Reclaiming the University". This has been in response to increasing marketization and financialization of education in Britain following neoliberal models similar to those employed in business (such as property development). These processes have at best compromised and at worst silenced the voices of those very people who comprise the institution—teachers and students—as they seek to uphold and defend the goals of emancipatory education. The defence of these processes, at Aberdeen as at so many other British institutions, has been stymied by a schooled practice of ignorance amongst its senior managers which hews to a neoliberal and dirigiste ideology and which privileges the quantitative over the qualitative when so very many of education's ends are explicitly and necessarily qualitative. The manifesto's invigoratingly utopian and insurgent goals are

> for fundamental reform of the principles, ethos and organisation of our university, in order (1) that it should be restored to the community to which it belongs and (2) that it can fulfil its civic purpose in a manner appropriate to our times, in the defence of democracy, peaceful coexistence and human flourishing.

And further

> But the manifesto's message is one of hope. It reaffirms the University as a democratic community of scholars, students, and support staff, united in the pursuit of truth for the public good. The motivations for academic work are moral, not instrumental: universities are charities, not businesses.
>
> *(Academics and Students of the University of Aberdeen, 2016)*

This shows that the propositional and democratic imagination of the whole institution can be differently constituted, and can be founded upon hope. Hope, through the action of utopia employed as method using the propositional imagination for scenario-making, is essential to teach for, design, and build a future that is substantively better and permanently geared toward further betterment.

Notes

1 'The architectures' describes all the various interlinked and interdependent professional platforms for the design and creation of built spaces.
2 Sargisson is explicit about her debt to Tom Moylan's (1986) definition of critical utopianism: "[t]he critical utopia, says Moylan, is critical in two senses: first in the Enlightenment sense of critique and second in the nuclear sense of critical mass" (2000: 4).
3 While Elizabeth Meyer does not employ the language of utopia or utopianism, her work is clearly grounded in a tradition that unites human being and becoming—including its ethical, aesthetic, and artistic dimensions—with the ecological world in which it is situated. Further, this epistemic culture of making and participating is fundamentally future-oriented. Meyer's work (see 2008), as so much current landscape architectural theory, including my own, finds itself firmly in debt to the writings of Catherine Howett and Anne Whiston Spirn (see particularly Howett 1987, and Spirn 1984, 1988).
4 When seeking the definition of the goals of property development and investment, here are some of the statements I found in two of the most popular British textbooks on the subject: *Real Estate Concepts: A Handbook* tells us, "developers are in essence entrepreneurs who identify opportunities and are prepared to take risks in order to deliver a completed property development scheme in anticipation of the requirements of the market in return for profit" and that, "no matter what the end product, the developer's principal objective is to generate profit" (Jowsey 2015: 119–120). The second edition of *Property Investment* reinforces this: "At a fundamental level property investment is a process of siphoning off a proportion of the value which has been created by other entities. This creates a faint suspicion that property investment is fundamentally a parasitical activity" (Isaac and O'Leary 2011: 290). The coy understatement of a bald fact in the last sentence underscores the fact that property development's current 'principal objective' is quite plainly unethical (i.e., it is *not okay* to be a parasite). There is also a larger discussion to be had about whether cities are to be thought of as 'landscape', which implies cities are shared resources and settings, or as 'property'. Another discussion to be had is how landscape architects engage with or drive existing or emerging ethical development models. Alas, here is not the place for this.

References

Abensour, Miguel (2017 [2000]) *Utopia from Thomas More to Walter Benjamin*. Translated by Raymond N. MacKenzie. Minneapolis, MN: Univocal Publishing.

Abensour, Miguel (2011) *Democracy Against the State: Marx and the Machiavellian Moment*. Cambridge: Polity Press.

Academics and Students of the University of Aberdeen (2016) *Reclaiming our University* website. Available online at: https://reclaimingouruniversity.wordpress.com (accessed 11 May 2017).

Berlin, Isaiah (1991) *The Crooked Timber of Humanity*. London: Fontana.

Biesta, Gert J.J. (2014) *The Beautiful Risk of Education*. London: Paradigm Publishers.

Bloch, Ernst (1986) *The Principle of Hope* (3 vols). Translated by Nevill Plaice, Stephen Plaice and Paul Knight. Oxford: Basil Blackwell.

Council of Europe (2000) "The European Landscape Convention (Florence, 2000)", *European Landscape Convention* website. Available online at: www.coe.int/en/web/landscape/the-european-landscape-convention (accessed 11 October 2017).

de Certeau, Michel, Luce Giard and Pierre Mayol (1998 [1994]) *The Practice of Everyday Life Vol. 2: Living and Cooking*. Translated by Timothy J. Tomasik. Minneapolis, MN: University of Minnesota Press.

de Sousa, Ana Naomi (2014) "How Spain's 'guerrilla architect' is building new hope out of financial crisis", The *Guardian*, 18 August. Available online at: www.theguardian.com/cities/2014/aug/18/santiago-cirugeda-guerrilla-architect-spain-seville-financial-crisis (accessed 11 October 2017).

Graeber, David (2013) *The Democracy Project: A History, A Crisis, A Movement*. London: Penguin.

Gudeman, Stephen (2001) *The Anthropology of Economy: Community, Market, and Culture*. Oxford: Blackwell Publishers.

Hall, Stuart (2007) "Universities, Intellectuals, and Multitudes: An Interview with Stuart Hall", interviewed by Greig de Peuter. In Mark Coté, Richard J.F. Day and Greig de Peuter (Eds) *Utopian Pedagogy: Radical Experiments Against Neoliberal Globalization*. Toronto: University of Toronto Press, 108–128.

Haraway, Donna J. (2016) *Staying with the Trouble: Making Kin in the Chthulucene*. Durham, NC: Duke University Press.

Hardt, Michael and Antonio Negri (2002) "Subterranean Passages of Thought: *Empire*'s Inserts". Compiled by Nicholas Brown and Imre Szeman. *Cultural Studies*, 16(2), 193–212.

Harvey, David (2004) *Spaces of Hope*. Edinburgh: Edinburgh University Press.

Howett, Catherine (1987) "Systems, Signs, Sensibilities: Sources for a New Landscape Aesthetic". *Landscape Journal*, 6(1) Spring, 1–12.

Isaac, David and John O'Leary (2011) *Property Investment*, 2nd edition. Basingstoke, UK: Palgrave Macmillan.

Jencks, Charles (2005) *The Iconic Building: The Power of Enigma*. London: Frances Lincoln.

Jowsey, Ernie (Ed.) (2015) *Real Estate Concepts: A Handbook*. Abingdon, UK: Routledge

Massey, Doreen (2013) "Vocabularies of the Economy". In Stuart Hall, Doreen Massey and Michael Rustin *After Neoliberalism? The Kilburn Manifesto*. London: Lawrence and Wishart, 3–17. May be read for free online at: www.lwbooks.co.uk/soundings/kilburn-manifesto.

Meyer, Elizabeth K. (2008) "Sustaining Beauty. The Performance of Appearance: A Manifesto in Three Parts". *Journal of Landscape Architecture*, 3(1), 6–23.

Meyer, Elizabeth K. (2000) "The Post-Earth Day Conundrum: Translating Environmental Values into Landscape Design". In Michel Conan (Ed.) *Environmentalism in Landscape Architecture – Dumbarton Oaks Colloquium on the History of Landscape Architecture XXII*. Washington, DC: Dumbarton Oaks.

Moylan, Tom (1986) *Demand the Impossible: Science Fiction and the Utopian Imagination*. London: Methuen.

Nichols, Shaun (Ed.) (2006) *The Architecture of the Imagination: New Essays on Pretence, Possibility, and Fiction*. Oxford: Oxford University Press.

Nichols, Shaun and Stich, Stephen (2000) "A Cognitive Theory of Pretense". *Cognition*, 74, 115–147.

Pérez-Gómez, Alberto (2015) "Mood and Meaning in Architecture". In Sarah Robinson and Juhani Pallasmaa (Eds) *Mind in Architecture: Neuroscience, Embodiment, and the Future of Design*. Cambridge, MA: The MIT Press, 219–235.

Pitte, Jean-Robert (2002) *French Gastronomy: The History and Geography of a Passion*. Translated by Jody Gladding. New York, NY: Columbia University Press.

Pope, Alexander (1734) "An Essay on Man: Epistle I". *Poetry Foundation* website. Available online at: www.poetryfoundation.org/poems-and-poets/poems/detail/44899 (accessed 24 May 2017).

Poulain, Jean-Pierre (2017) *The Sociology of Food: Eating and the Place of Food in Society*. Translated by Augusta Dörr. London: Bloomsbury Academic.

Sargisson, Lucy (2000) *Utopian Bodies and the Politics of Transgression*. Abingdon, UK: Routledge.

Schatzki, Theodore R. (2010) *The Timespace of Human Activity: On Performance, Society, and History as Indeterminate Teleological Events*. Lanham, MD: Lexington Books.

Spirn, Anne Whiston (1988) "The Poetics of City and Nature: Towards a New Aesthetic for Urban Design". *Landscape Journal*, 7(2) Fall, 108–126.

Spirn, Anne Whiston (1984) *The Granite Garden: Urban Nature and Human Design*. New York, NY: Basic Books.

Suissa, Judith (2006) *Anarchism and Education*. Abingdon, UK: Routledge.

Terranova, Tiziana and Marc Bousquet (2004) "Recomposing the University". *Mute*, 1(28), 72–81.

Waterman, Tim (2018) "Taste, Foodways, and Everyday Life". In Joshua Zeunert and Tim Waterman (Eds) *The Routledge Handbook of Landscape and Food*. Abingdon, UK: Routledge.

Waterman, Tim (2017a) "Making Meaning: Minds, Bodies, and Media in Architectural Education". In *Imaginaries of the Future 01: Bodies and Locations*. Cambridge: Open Library of the Humanities.

Waterman, Tim (2017b) "It's About Time: The *Genius Temporum* of Martí Franch's Girona Landscapes". *Landscape Architecture Magazine*, January 23. Available online at: https://landscapearchitecturemagazine.org/2017/01/23/its-about-time (accessed 11 October 2017).

Zeunert, Joshua (2017) *Landscape Architecture and Environmental Sustainability: Creating Positive Change Through Design*. London: Bloomsbury.

Part I
Reading the landscape

Landscape is increasingly an area of academic interest and concern for a wide range of disciplines, something that has been furthered by the growing awareness resulting from the widespread adoption of the European Landscape Convention. Indeed, following directly from its definition as 'an area, as perceived by people . . .', the European Landscape Convention more or less mandates us all to engage in reading the landscape. Reading may be one step beyond mere perception, but – without perception – no landscape.

The list of fields that explicitly espouse landscape in their names ranges from the humanities to the natural sciences, this time reflecting the other part of the Landscape Convention's landscape definition '. . . whose character is the result of the action and interaction of natural and/or human factors'. The academic fields concerned include landscape archaeology through landscape ecology to landscape urbanism, although for each of these the 'landscape' part of their titles represents a particular specialism within a wider discipline, in the form of an integrated holistic or spatial means of framing its particular field of enquiry, rather than a sole area of concern. Furthermore, each discipline tends to view landscape in its own particular manner – when considering the same scene, each will be actually perceiving something different and specific to its way of seeing.

This is the point made in the classic essay by D.W. Meinig, 'The Beholding Eye: Ten Versions of the Same Scene', quoted by Ed Wall in the opening chapter in this section, and it is perhaps even more true now, thanks to the broadened academic awareness stimulated by the European Landscape Convention, than it was when Meinig originally wrote it. Clearly, when teaching and learning about landscape, it is important to be aware of the multitude of disciplinary viewpoints from which the landscape can be read in order to understand it fully.

Within this first part of the book only a few of these disciplinary perspectives can be illuminated, with the stress being primarily on the 'natural' rather than the 'human factors'. These include what is perhaps the oldest landscape discipline of geography (Marc Antrop and Veele van Eetvelde), geology (Ralf Löwner) and ecology (Wenche Elisabet Dramstad and Mari Sundli Tveit). These are complemented by an outline of teaching landscape reading to architecture students (Luca Maria Francesco Fabris and Guido Granello), in which the 'human' factors dominate.

There is, though, one exception to this list of fields of enquiry fronted with the 'landscape' epithet, as pointed out at the beginning of Shelley Egoz's chapter, stressing the fact that landscape is more than the sum of its deconstructed parts as dissected by different individual disciplines. Here is it asserted that although 'everyone' seems now to be talking about landscape,

landscape architecture is the only discipline 'in which landscape is not a choice of scholarly approach to be adopted, but the essence of the discipline':

> If landscape architecture is to justifiably live up to this unique status, then it is vital that its students, at least, are aware of and embrace the many different approaches and are prepared to understand how the landscape is viewed by all of the disciplines which use the concept primarily as a means of framing and focussing their studies.

However, in teaching landscape it is not sufficient to take on board the multitude of different specialist ways of reading the landscape. Again, it is the Landscape Convention that makes it clear that the landscape is the concern of everyone and not merely a playground for the 'experts'. The growing concern with 'landscape democracy' as part of a wider movement for environmental rights is the focus of Deni Ruggeri's contribution, which highlights another facet of reading the landscape – namely the question of through whose eyes is it being read, and how can we integrate wider viewpoints into our landscape teaching and learning?

It is not just who is reading the landscape that must be considered, but also how and with the help of which media or techniques the reading is being done as the medium will clearly influence the message. Using film as a means of helping to see and interpret the landscape is not just about recording impressions of spaces, but also an aid to perceiving and understanding space, which is the concern of Irina Paţa and Ana Opriş' contribution.

Thomas Oles, by comparison, is wary of capturing landscape space on film, but rather argues for finding ways of teaching students to read the landscape through a process that might be termed subjective poetic immersion in which feelings are given at least equal value with facts.

Such approaches call for a direct interaction with real landscapes in the 'great outdoors', and the long-term development of approaches making use of Swedish islands as live landscape models to introduce students to real-world situations and to sharpen the perceptions of students in learning to read landscapes is the focus of the chapter by Roland Gustavsson, Allan Gunnarsson and Björn Wiström.

The use of a circumscribed landscape of a small island as a teaching ground contrasts with a second Scandinavian approach to reading landscapes on the ground at a different scale as defined by a whole climate zone. The understanding and interpretation of changing Arctic and Subarctic landscapes is the focus for a new teaching programme at universities in northern Norway, which is explained by Janike Kampevold Larsen.

Critique and critical approaches to both landscapes and to landscape teaching are common themes linking the final three chapters in this section, but dealing with very different aspects of reading the landscape. In the context of teaching landscape planning, Andrew Butler considers the possibilities for using landscape assessment as a vehicle for educating students to take a critical approach to the ideas associated with the landscape values that are embedded their work.

The role of design critique in reading landscapes, rather than in landscape planning, is the subject of Jacky Bowring's chapter, which considers how teaching analytical critique of contemporary projects can help students to understand the process of discourse about design and how this can develop and change over time. Elen Deming's chapter on values and transformative learning follows up on this point, showing the potential of a history class in elucidating the students' hidden or subconscious value sets.

Recognising the importance of the historic canon of landscape projects as a basis for teaching students to read and value both designed and vernacular landscapes lies behind Marc Treib's discussion of changing approaches to teaching landscape history.

This contribution rounds off the rich and diverse selection of ways of reading the many dimensions of landscape as reflected in the European Landscape Convention in this first part of the book.

2

'What . . . is landscape?'
Asking questions of landscapes through design drawings

Ed Wall

Several years ago, one of my students encountered an architect who dismissively asked, 'What the fuck is landscape?' He was taken aback by the tone of the question – but was reassured that his architectural colleague had found a place of uncertainty from within his general overconfidence. Although the student and I could easily disregard such blunt questioning, the architect had asked a question that, as designers and academics of landscapes, we frequently ask: what is landscape? The subjectivity and plurality of landscapes, as they are differentially experienced, open up more questions than they provide answers, creating rich environments for critical enquiry, exploration, speculation and practice. Such approaches to landscape contrast with many architectural traditions that focus on positions, not processes, and that can lead to attempts to provide solutions rather than facilitate investigations. In answering, rather than questioning, designers can accept misplaced assurances that they can solve landscapes rather than merely inform their future change.

More recently, a former colleague explained to me: 'The problem is that you ask your students what landscape is. You should just tell them what landscape is and get on with it'. While presented to me more politely than the question posed to my student, I was concerned by his limiting pedagogical advice. My colleague had been referring to an exercise that I had initiated with first year students where we[1] investigated meanings of landscape. Referencing D.W. Meinig's essay 'The Beholding Eye: Ten Versions of the Same Scene' (1979), we examined diverse definitions of landscape across a range of scales and of contrasting conditions. During the design studio we discussed the short landscape manifestos that students wrote, each accompanied by a single drawing, and throughout the year we developed and returned to these declarations of landscape as we explored site-specific, process-focussed design projects – asking 'what is *this* landscape?'

As part of these landscape investigations we would read contemporary landscape texts (such as Cosgrove 1984; Jackson 1984; Corner 1999b) alongside seminal and historic projects that explored relations of landscape (such as Geddes 1915; Howard 1902). We would also undertake detailed field studies of designed and unplanned landscapes as a means of grounding our research and design investigations: challenging our understandings through contrasting experiences of scale, unique qualities of place and processes of production and decay that revealed landscape conditions obscured by the abstracted frames of desk studies.

Students would also draw, to incrementally combine and synthesise site data, form conceptual statements and experiment throughout the development of projective designs. During the exercise, I would propose that landscape is a creative practice – defined by asking questions through dialogic relations of inventively engaging with our environments.

This chapter is focussed on the importance of asking questions. It considers landscape as a creative practice that investigates multi-scalar, site-specific relations between people and their environments, with the intention of developing proposals for future landscapes. I highlight what can be learned from investigating landscapes, experimenting with design drawings and critically questioning what is produced. As Perry Kulper explains in his essay 'A World From Below':

> rather than framing the possibilities of drawing as related to problem solving, or limiting the role of the drawing to a metrical description of a project, ideas are augmented through an emerging visual field of study that is discovered in the act of constructing a drawing.
>
> *(Kulper 2013:59)*

Based on over ten years of experience teaching the design of landscapes and cities, and a reflective review of the design projects produced by students, I discuss practices of design and how we can learn from emerging techniques.[2] I reject the premise of providing answers to complex landscape conditions and I defend the opportunity to challenge and reinvent landscapes through critical design practices. The chapter is not focussed on addressing the question of what landscape is, or even 'what the fuck' landscape is. Instead, I explore how the continued asking of questions can be facilitated through strategic composite drawings and how decisions can be made throughout site-specific design projects that enquire: 'what is *this* landscape?'

This chapter also reveals my frustration with representational conventions of architectural projections (such as plans, sections and elevations) that prioritise spatial forms and landscape renders that emphasise visual qualities. Often, such drawing approaches limit opportunities to investigate relational dimensions of landscapes, experiment with shifting ecologies or examine social concerns as central issues of landscape projects. Neil Spiller writes: 'Euclidean purity is a myth, the landscape of today reveals and secretes hidden archaeologies' (2000:87). I advocate in this chapter the creation of composite representations at key stages of the design process in order to ask questions, while collecting data, analysing findings, proposing new landscapes and communicating how these projects are produced. I highlight three common stages of design processes, hinges in design projects that can be effectively articulated through specific composite representations.[3] First, how we collate and analyse site information, synthesised into *base drawings*, to establish clear statements from research findings and a foundation from which to develop proposals. Second, how we communicate the active working landscape of design projects in *operational drawings*. And, third, how we can represent sequences of actions and events as *scenes* rather than aestheticised and objectified landscape images.

Landscape?

Investigating the etymology of the term 'landscape' frequently forms the point of departure for research into what landscape is. From J.B. Jackson's oft-cited description of the historical origins of landscape, in his essay 'The Word Itself' (1984), to John Stilgoe's more recent treatise *What is Landscape?* (2015), opportunities to define what appears to be a simple term can be tempting. But the perceived clarity of landscape is misleading. Highlighting the illusory capacity of

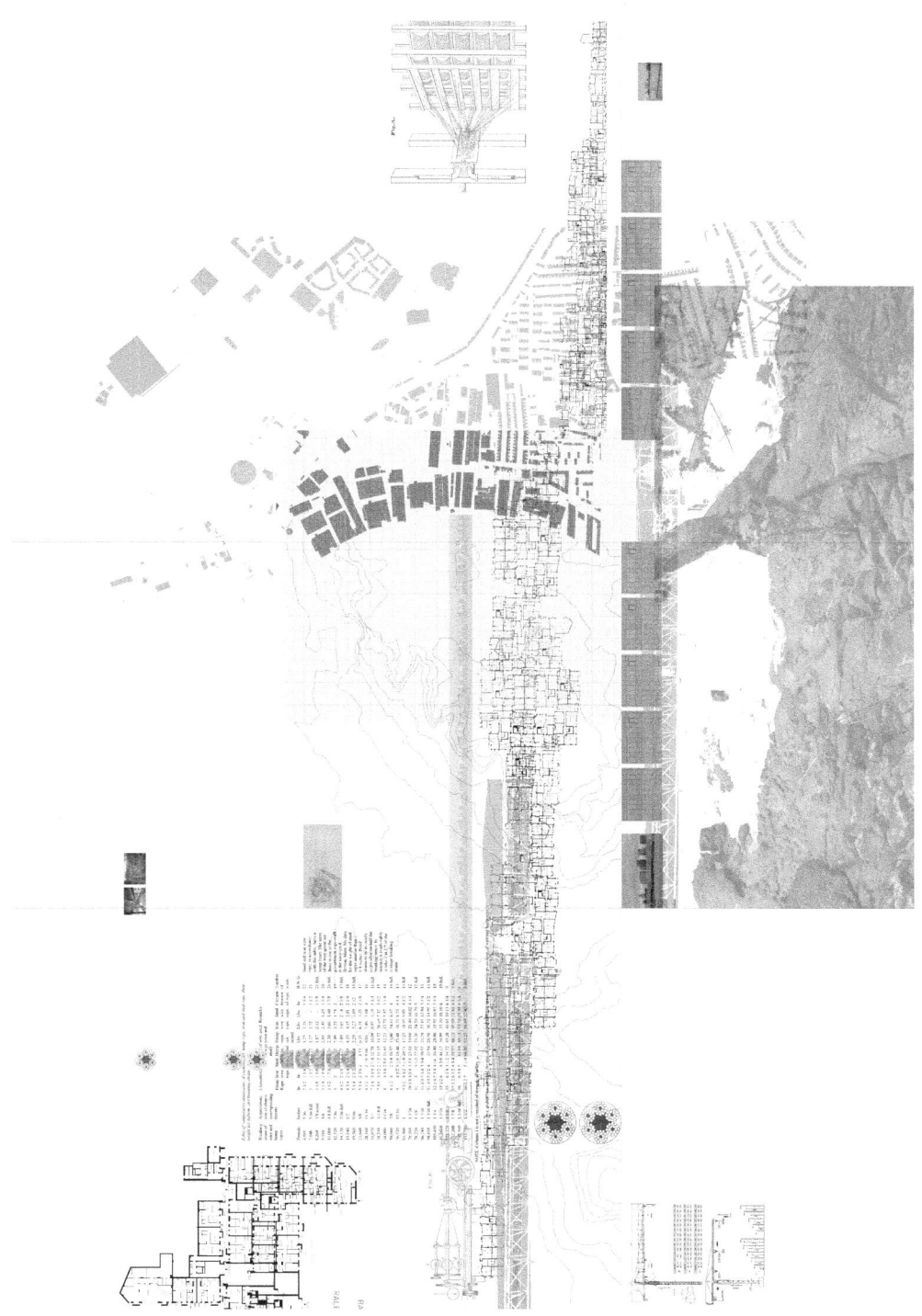

Figure 2.1 (Re)working waterfront by Iona Meldrum, 2015

landscape, Denis Cosgrove refers to John Berger's description of landscape as a 'way of seeing' (1984:55), to describe landscape as:

> a composition and structuring of the world so that it may be appropriated by a detached, individual spectator to whom an illusion of order and control is offered through the composition of space according to the certainties of geometry.
>
> *(1984:55)*

Such Euclidian approaches to space are employed in colonial appropriations, agricultural enclosures, urban extensions, residential subdivisions and privatisations of public space. These are places mapped, occupied and commodified, first as representations in landscape paintings and then as designed spaces. However, as Barbara Bender emphasises (1993:1), there are other relations with the land that are not based on centuries-old priorities for visual images, positions of power and control:

> In the contemporary western world we 'perceive' landscapes, we are the point from which the 'seeing' occurs. It is thus an ego-centred landscape, a perspectival landscape, a landscape of views and vistas. In other times and other places the visual may not be the most significant aspect, and the contemplation of the land may not be ego-centred.

Through exploring these 'other' landscapes we can expose the contradictions of western relations with the land and establish opportunities to challenge culturally fixed ideas of landscape and reinvent through theorising and design.

When working with students to investigate such alternative landscapes I have often begun with reading Meinig's essay (1979). I have employed Meinig's ten scenes of landscape (as nature, habitat, artefact, system, problem, wealth, ideology, history, place and aesthetic) as an invitation for students to explore other definitions that relate to specific sites, conditions and times. I have also discussed *Is Landscape . . .?* (Doherty and Waldheim 2016), a compilation of essays that resonates with Meinig's scenes and that take Gareth Eckbo's essay 'Is Landscape Architecture?' (1983) as a starting point to explore contrasting practices of landscape. Gareth Doherty and Charles Waldheim pose the question: *Is Landscape . . .?* (2016) to enquire of 13 different contributors landscape's association with literature, painting, photography, gardening, planning, urbanism, infrastructure, technology, history, theory, philosophy and life. While I recognise usefulness in what can become more abstract explorations, during practices of teaching the design of landscapes I have asked students to focus on site-specific questioning of existing and proposed landscapes. In such a way the unique conditions of landscapes become the basis from which places are understood and the medium from which new futures are imagined.

Such questioning of landscapes is not about answering problems, finding solutions or claiming certain truths. Dictionaries do not define 'questions' solely as devices from which to produce answers, rather they prioritise the role of questions in raising topics to be enquired about, discussed or debated (*OED* 'Question' 2017). When students investigate landscapes, attempting to identify their unique site conditions, they adopt practices of research and design that do not merely result in a designed artefact but rather they work with landscapes as entities continually being made and remade, planted and maintained, eroded and aggregated (see Bender 1993:3). Landscapes are not static spaces but are relations in tension across spatial scales and temporal rhythms, intertwined with geologic forms, climatic conditions, economic imperatives and unpredictable daily lives. The questioning of landscapes by students unfolds through incrementally layered mappings and composite forms of collages and maquettes. The representations

Figure 2.2 Sleep Estates by Anushka Athique, 2016

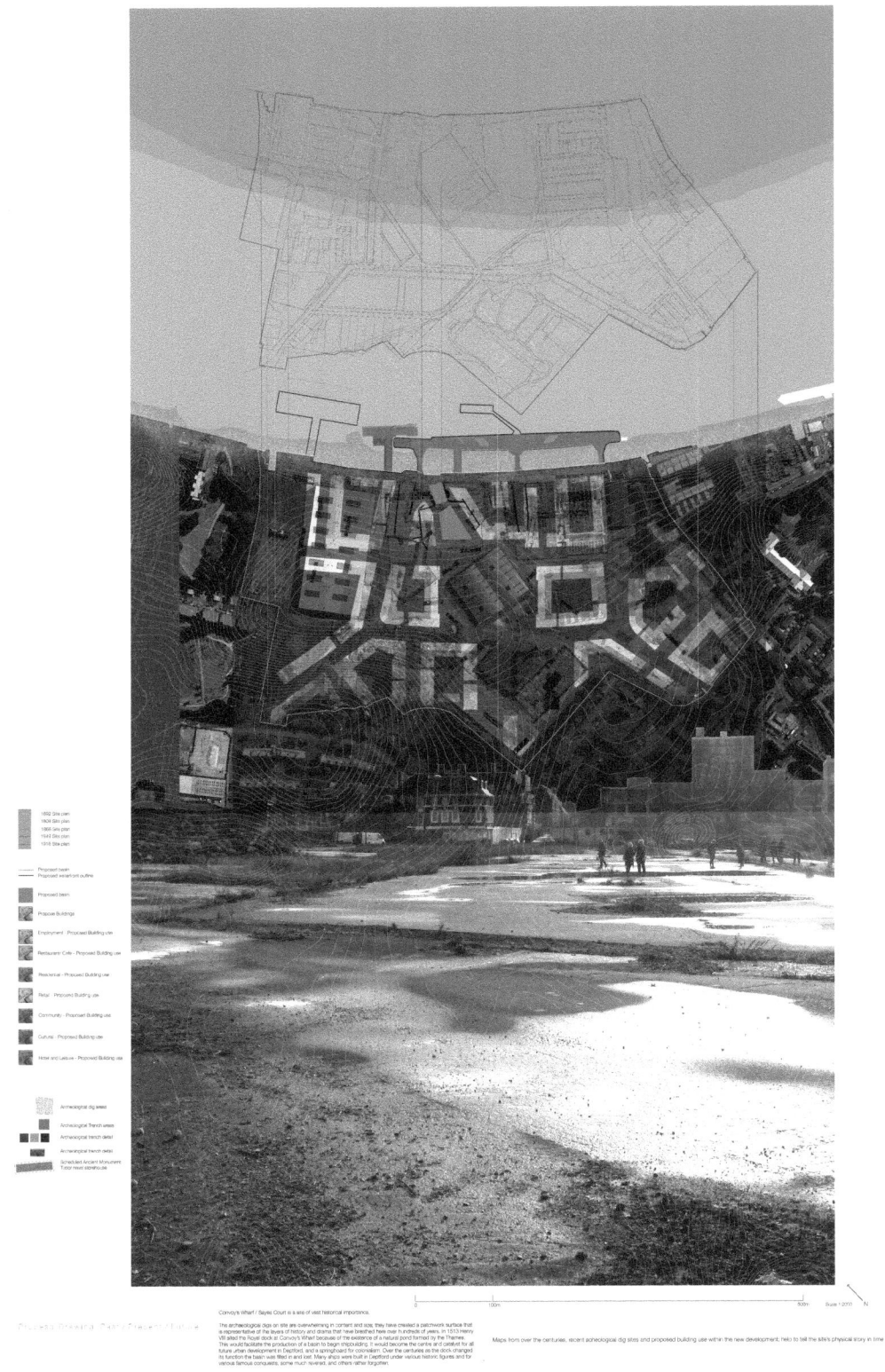

Figure 2.3 Drawing the Line/Cloud Chamber by Max Barnes, 2016

This figure is shown in colour in the plate section of this book.

create biographies of design processes, where sheets in portfolios present the stories of researching sites, precedent that has informed ideas, analysis of patterns and designs developed. Each composite drawing that the students create reflects backward to question earlier research stages of projects while projecting forward to ask questions of future designs. The creation of the *base drawing, operational drawing* and *scene* provide moments of intensely questioning what is found and proposed as well as making decisions and clarifying the future direction of the project.

Design?

The development of this approach to composite drawings is based on reviewing over 500 student design projects between 2007 and 2017.[4] Common amongst the projects have been investigations into site-specific urban geographies, primarily in London, Milan, Vienna and New York, and the development of landscape and urbanism proposals. Consistent approaches include the questioning of existing site conditions, analysis of what is found and advancement of creative proposals based on the issues identified. From the first stage of projects students employed a range of methods including mapping, modelling, photographic surveys, document surveys and interviews. Research frequently extended beyond the site boundaries as students traced the material and social relations that had come to form post-industrial waterfronts, disused public spaces, dilapidated buildings and complex infrastructures. Working between fieldwork, that in some cases involved hanging out in neighbourhood pubs to speak with residents or travelling through rural landscapes in canal boats, and desk studies of library, archive and internet research revealed arrays of narratives from which students attempted to make sense of their project sites.

As more site information was collected and synthesised through analytical drawings, students adopted a diversity of ways of generating new forms. We identified that while some students made more logical steps through their projects – through research, conceptualising and then proposing transformations to a specific site – in order to develop design proposals based on their research, other students made more speculative leaps before reflecting back on how the narratives of their projects fitted together. While encouraging students to employ a diversity of design and research methods, colleagues and I began to ask students to highlight five priorities during project development. We asked students to: first, emphasise landscape processes as well as spaces; second, model sites as the basis to develop situated proposals; third, to synthesise research into single, scaled drawings; fourth, to communicate operational qualities of proposals; and, finally, to demonstrate how their designs work as well as how they look.

The latter three priorities have come to be defined by three composite drawings that make explicit the decisions made and the core priorities in the projects, highlighting: conclusions of the research edited and synthesised into a single scaled *base drawing* from which designs are developed; how the design works, how it operates and how it is constructed in an *operational drawing*; how what we see and experience as part of proposed landscapes is composed and constructed, creating *scenes*. The three composite drawings do not define methods of research or approaches to design but rather they emphasise important moments in projects, what Spiller describes of his work *Landscape Matters* as 'new augmented ways of seeing and understanding landscape' (2000:87). To develop this language of landscape compositions we have explored and borrowed from experimental designers, in particular from individuals whose approaches we anticipated could be adapted to represent our priorities for landscape: techniques of layering and collaging provide for incremental development of drawings, they can represent the palimpsest and recombinatory conditions of landscape and they can be contextualised within historic and contemporary landscape architecture drawings; in contrast, vector-based diagrammatic drawings can represent the workings of landscapes, operations that are often relegated as background infrastructures or obscured by visually aesthetic priorities. Our references to Perry Kulper (Figure 2.4), Bryan Cantley (Figure 2.5) and Alexander Daxböck (Figure 2.6) reveal our interest in intersecting these two drawing approaches and to enquire beyond leading contemporary landscape architects (such as Corner, Mathur and Da Cunha). Such references embrace representational techniques that we feel can facilitate the development of landscape projects, the hinges in design processes, from reflective research to physical proposals and from spatial or aesthetic dimensions to operational qualities.

Base drawing

The base drawing is intended to support students to progress from reflective site-focussed research to design propositions. We ask students to gather information through a range of methods, from observations and conversations during fieldwork to archival and library research. The base drawing is the conclusion of an exploration of the elements and layers of the landscape, bringing together what James Corner describes, in 'Agency of Mapping' (1999a), as 'extracts' (1999a:229). Subsequent 'processes of gathering, working, reworking, assembling, relating, revealing, sifting and speculating' (1999a:218) create the base drawing to establish a framework for analysis and later proposition. We encouraged students to draw what they find, to spatialise information from interviews and to create maps that can be analysed alongside other visual representations. In order to analyse such a diversity of site information collected from contrasting methods students draw and edit what they find before ordering, layering and combining into a single base drawing composition. What Kulper terms his 'strategic plots' (2013:59) are important references for developing base drawings, as they 'represent conceptual frameworks, territories, actions and relations that are delineated, or plotted over and through time' (see Figure 2.4). The main composition of the base drawing is a scaled projection, usually in plan or sectional elevation, which combines important information gathered while researching the project sites to create a specific site profile. It is a complex research drawing that combines layers of maps, that includes spatialised elements of non-spatial data (such as information from photographic surveys and written documents), and often includes selected three-dimensional forms (see Figure 2.4). The base drawing can be understood as the unique biography of the existing site, written and edited through the students' knowledge, experiences and contrasting priorities.

The base drawings composed by students highlight frequencies of flooding in areas of low-income housing, fragmented networks of public spaces and communities overwhelmed by urban development. Undertaking the base drawing does not limit methods of research – rather it acts as a hinge within projects to assist students to progress from research to design proposal and for them to evidence the conclusions of their research. The base drawing, prepared once fieldwork and desk studies are almost complete, marks the culmination of project research from site-based fieldwork alongside desk-based inquiries. The drawing is composed and edited, representing key aspects of the research – but not all the information collected. It highlights spaces and processes. It is analytical, bringing together, editing out and redrawing information to reveal the unique conditions of landscapes. Corner explains that such 'tactics of appropriation, collage, abstraction, imaginative projection' are both 'strategies used to prompt free association' and provide 'liberatory mechanisms' (1992:160). The development of base drawings questions what has been found and communicates the conclusions that have been reached – identifying the narratives that are considered important of the landscape and providing a foundation from which design ideas can be drawn.

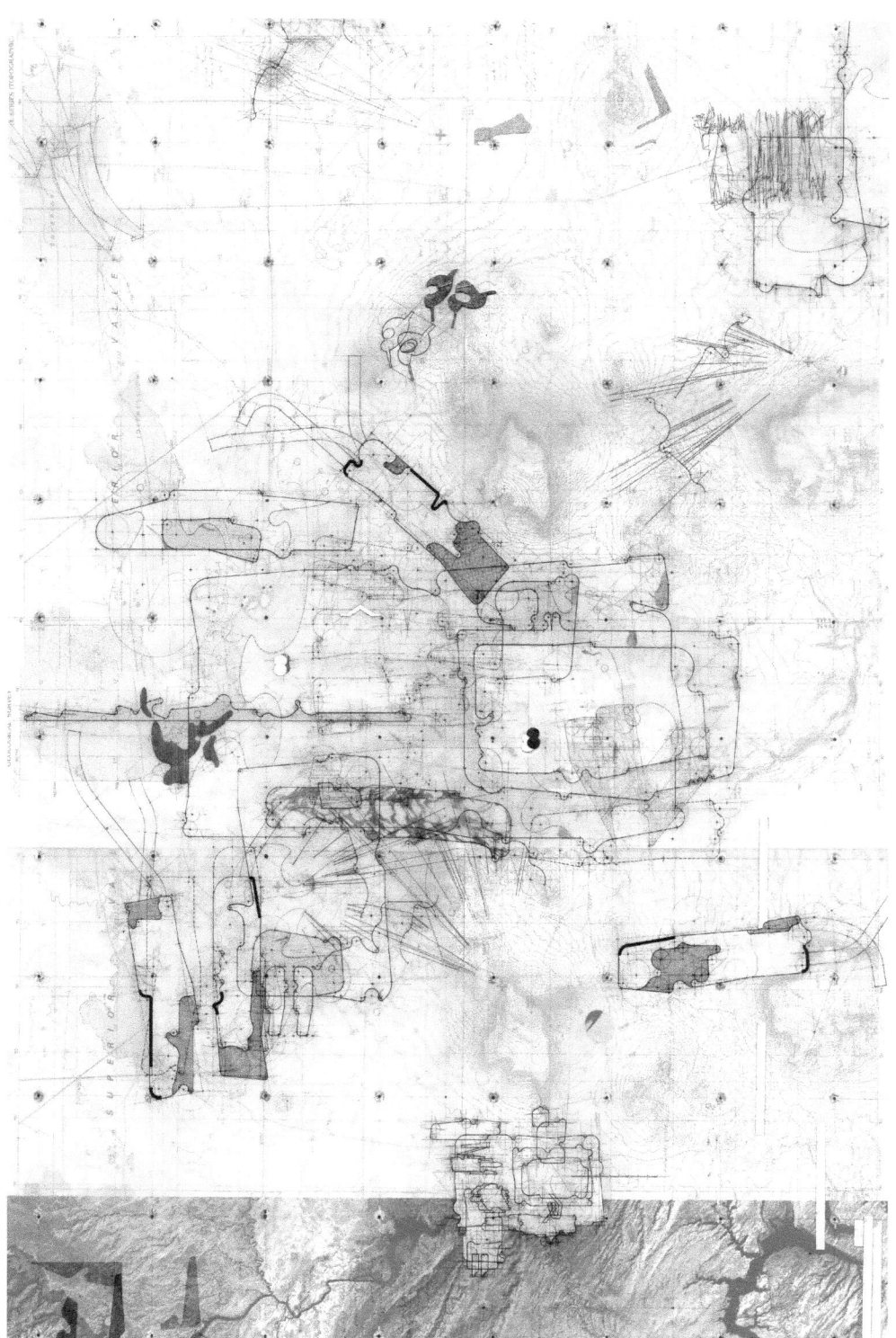

Figure 2.4 Fast twitch, site plan, Perry Kulper, 2004

The base drawing is part of a projective action. As students begin to develop design proposals the base drawing becomes a tool over which to trace responses to the conclusions of the site. As the base drawing includes elements of existing spaces and processes the proposal is able to directly relate to the past and present conditions of the landscape. As Kulper realises in his drawings, 'Interests can be derived through graphic exploration, and breeding latent and unpredictable opportunities, then visualised and capitalised upon towards design speculations' (2013:59). The iterative drawing approach is one of discovery as both unfamiliar compositions are formed and new ideas are generated. Kulper describes that working 'through lines and composited layers rather than through the logics of construction allows [his] work to incorporate both necessary and unexpected cultural and natural considerations' (2013:63). From the base drawing many design drawings are spawned. Thus the base drawing acts as a catalyst for an expanded repertoire of investigations, through questioning, testing and design development.

Operational drawing

The second composition of the operational drawing highlights how landscape projects work (see Figure 2.5). The operational drawing is designed to address Bender's claim of contemporary western landscapes that are 'visual' and 'ego-centred' (1993:1) by emphasising working landscapes. The composite operational drawing reveals how landscapes are produced, contested and lived. It shows opening and closing, growth and decline, planting and harvesting, flooding and drought. We identified that by focussing on asking questions through the design process students could highlight how proposed designs work (and how the project is made and how the landscape is maintained) rather than how they look and that such an approach could be employed to productive means. The operational drawing prioritises processes and highlights relations of landscapes as spatial and temporal entities.

Like the base drawing, the operational drawing is a combination of drawn elements (plans, views and diagrams) but recognisable as a single composition. The operational drawings reveal that the actions of making are not limited to architectural and construction practices but also include the rhythms of events scheduled in public spaces, the daily erection and dismantling of market stalls and the everyday occupation of landscapes. Operational drawing attempts to represent what Cantley terms 'The Thing Going Through Change' (2013:39). These landscapes advance and recede; they accelerate and slow; they illuminate and shade (see Figure 2.5). Landscapes can be intensely managed and they can be out of human control. The operational drawing reveals these workings of landscape – as they relate to core aspects of the design proposal.

Although we recognised that students continuously considered the operations of their projects throughout the design process, we found that the operational drawing was often most effectively completed as the material forms became more defined in order to finalise the working aspects of proposals. Through the progress of their projects all students developed drawings and visual representations, including traditional projects of plans, sections and axonometric projections that were complimented by diagrams and rendered perspective drawings. Reflecting the research stage of the projects, the students adopted a range of methods to developing proposals. As design propositions were formed through periods of intense design drawing and decision-making, the range of workings from which the designs were composed were expressed in the operational drawing: how events are planned; how creative maintenance regimes are expressed; how flooding is managed or confronted.

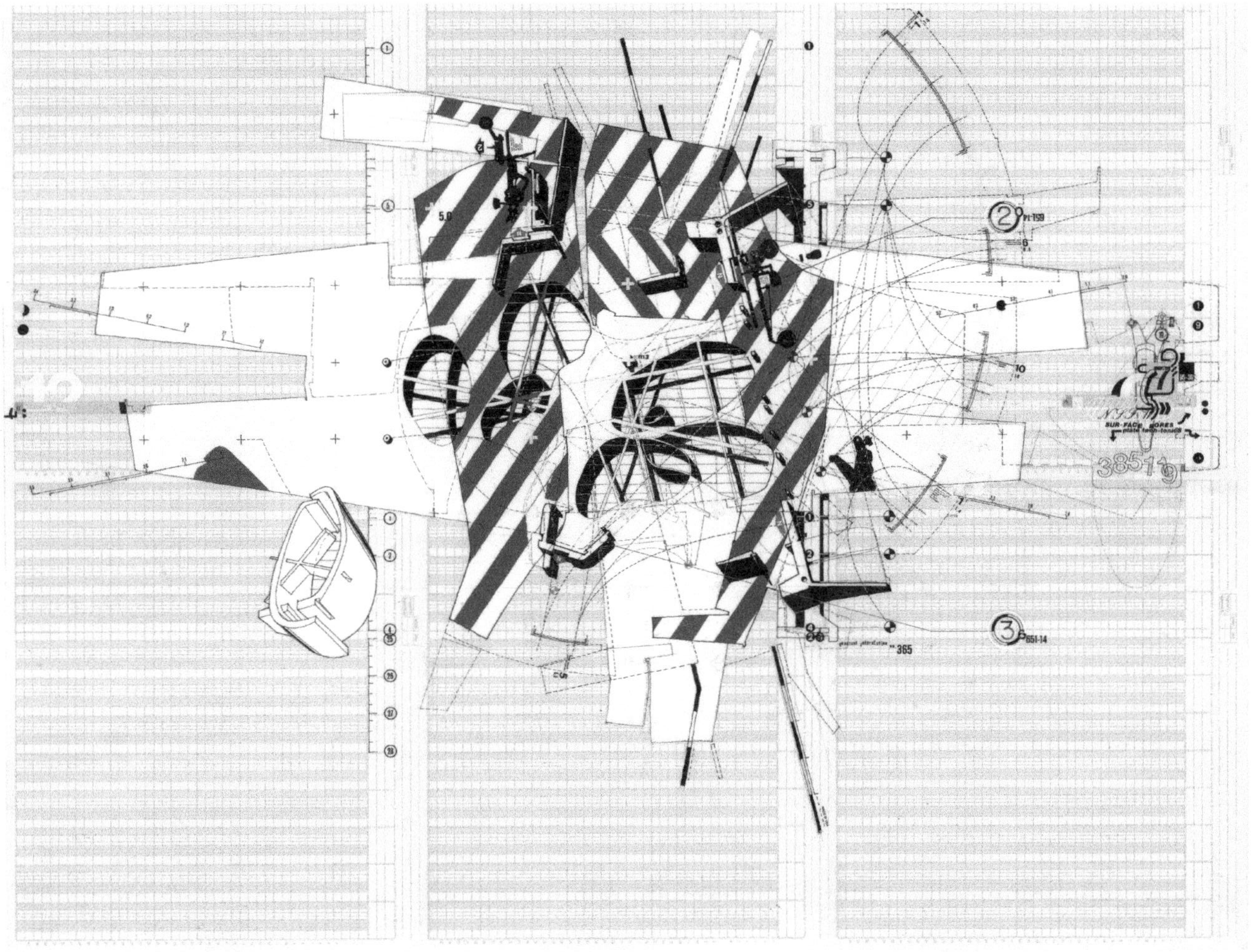

Figure 2.5 Sur-Face Bores by Bryan Canley/Form:uLA, 2012

Scene

The operational drawing, like the base drawing, represents the entire landscape being studied. The latter emphasises key aspects of the existing conditions while the former demonstrates how the proposed landscape will work. As neither composition reveals so clearly how the landscape is seen or experienced we proposed a third drawing, the *scene*, as a way of simultaneously recognising and challenging the significance of visual frames and ego-centred positions in prevailing landscape design. Scenes highlight an interest in how we experience places and events. Scenes do not merely represent a view of the project, they explain places or setting for real-life or fictional events, they highlight elements from which such settings are composed and they incorporate sequences of actions and events (*OED*, 'Scene' 2017). Scenes are not defined by being the most attractive representation of the project; the Kodak moment or the money-shot that defines a project's success. The scene shows us the construction of such events from a position of how it is experienced (see Figure 2.6). Through overlays of text and diagram the scene attempts to communicate complex qualities of experience, in the sequence of the arrival or the array of sights, smells and sounds. It provides an opportunity to represent how the view can be understood through its component parts and it can present the dynamic qualities of how and at what pace the viewer experiences the designed landscape.

The scene is also a layered composite drawing that explains the background to what we see and experience. It unpacks the construction of the image and the sequence of events that are often captured in a single image. We reference Alexander Daxböck's collage for the *Urban Satellite – from fragments to centrality* (see Figure 2.6), a drawing that presents multiple perspectives and positions in addition to quantitative dimensions of time and space. The scene represents, as Meinig explains:

> The land, the trees, roads, buildings, and man are regarded not as individual objects, ensembles of varied elements, or classes of phenomena, but as surficial clues of underlying processes.
>
> *(Meinig 1979:33–48)*

Through overlaying otherwise invisible information on the image the scene makes visible the image frame, the actions, the trajectory of processes and the relations between elements. Scenes may highlight the potential multiplicity of positions of the viewer, or viewers, and they may include within them the viewing and transformative landscape devices. The historic role of devices used to record (e.g., cameras), measure (e.g., theodolites) and view (e.g., black mirrors) landscapes can be read in the technological interventions of landscapes, from toposcopes to mobile phones. The scene provides an opportunity to reveal the usefulness of such devices and their relation to both experiences and wider infrastructures.

Scenes, most significantly, also open up questions. Unlike contemporary eye-level computer renders that smooth over gaps and juxtapositions of collaged drawings, scenes expose these situations. Scenes are compositions where the visual elements within a view are only as important as the gaps between and the information in the background. The scene is about relations: it suggests constructions, it reveals experiences, it presents routes through and it indicates velocities. As the base drawing and operational drawing provide information, through employing layers and fragments (often incorporating collaged elements), the scene presents information but also opens up the proposed landscape to further questioning.

Figure 2.6 Urban Satellite – from fragments to centrality by Alexander Daxböck and Georgia Papathanasiou, 2013

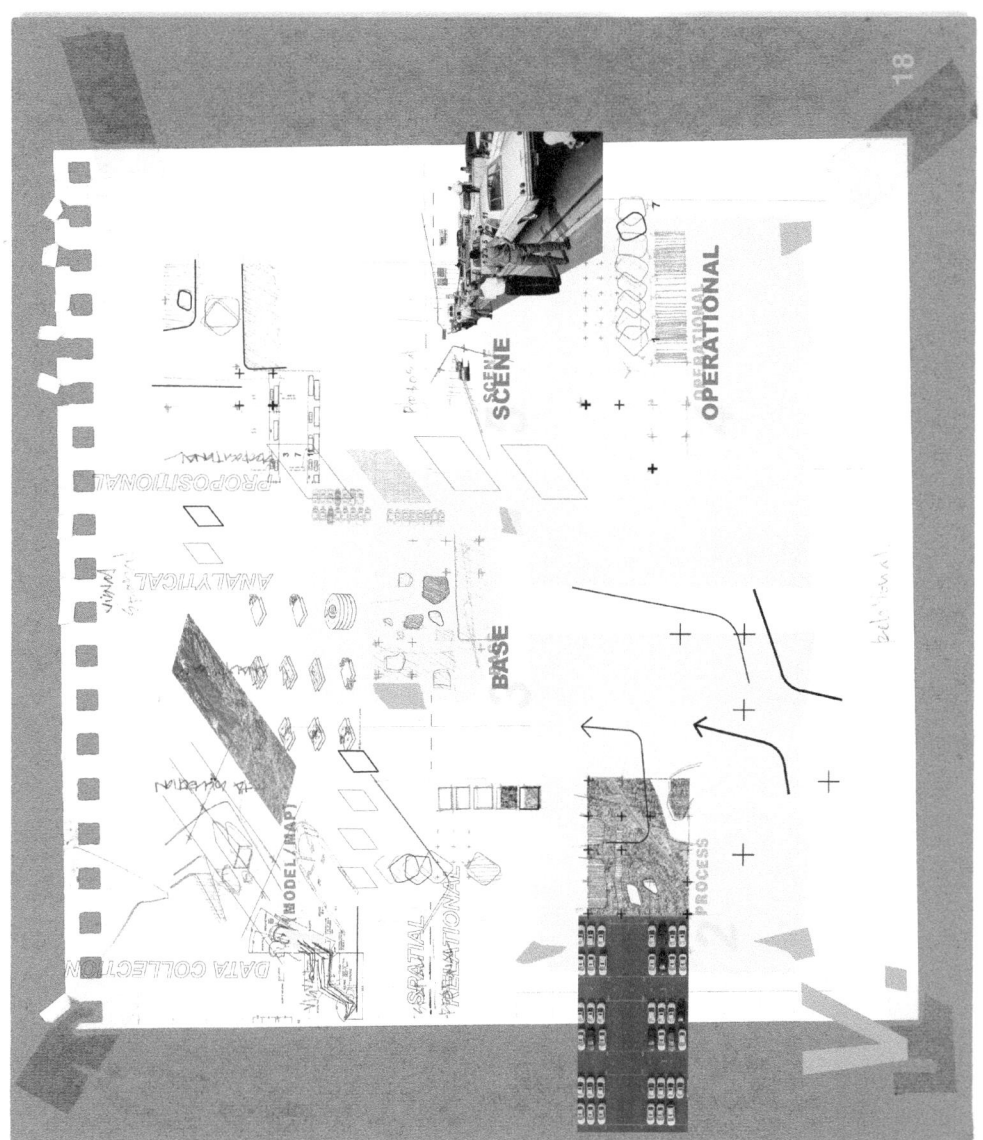

Figure 2.7 Five ways of working by Ed Wall, 2018

Conclusions

I began this discussion with a blunt question asked to one of my students, a proposition by a colleague to ask fewer questions of students and a description of a site-specific approach to questioning landscapes. These scenarios provide a context for over ten years of teaching where I have attempted to elevate the role of asking questions. Through my research and teaching practices I have defended the importance of asking what landscape is, of critically enquiring of project sites, of challenging design proposals and of rethinking how we represent them. It could seem ironic that the conclusions that I draw from these experiences lead me to ask for specific forms of representation from my students. In *Drawing Architecture* (2013) Spiller reminds us that to teach is 'not to pollute the young mind with the fetishes and guilty vices from which the tutors themselves suffer, but to lead the students to their own Elysian Fields' (2013:17). I argue that base drawings, operational drawings and scenes open up such possibilities of landscape to provide devices of questioning and decision making in the often intractable journey of design projects. As Kulper describes of his approach (2013:59):

> Design in this sense is fluid, weaving heterogeneous ideas, discussing one disciplinary set of questions in relation to another, and through the rehearsing of design skills in the drawings themselves, fusing visualisation and thinking as a relational and synthetic practice.

I have found that there is a need to draw and model inventively in order to critically question, develop and communicate new landscapes. Approaches to base drawings, operational drawings and scenes have three consistent qualities: they are developmental, in that they advance design ideas rather than merely represent them; they are composite representations that are formed incrementally through repeated collaging, layering, cutting and editing; and they provide an opportunity to represent alternative landscape relations that deny western tendencies for visually dominated landscapes, understood from ego-centred positions and prescribed by narrowly defined frames – challenging the 'primacy of the European "viewpoint"' (Bender 1993:1). They are also interrelated representations (see Figure 2.7) that can intersect more objective qualities of drawings with subjective experiences and they can make inseparable spatial and relational qualities of landscape.

I would argue that new forms of representation are therefore required, that open up the uniqueness of individual design projects rather than fixing alternative positions of viewing. Instead of further dissecting spatial forms of our existing or proposed landscapes (such as with plans and sections), adapting the projections (such as axonometric to perspective) or establishing new positions from which to view our landscapes (such as from eye-level to satellite), strategically constructed composite drawings of enquiry are required. Spiller proposes that we teach students to 'design with dexterity and an understanding of what might have gone before, but also with an imperative conditioned from an understanding of what might come after' (2013:17). The base drawing, the operational drawing and the scene mark analytical as well as representational moments in these process of designing landscapes, experimenting with site-specific conditions to test, expose and advance understandings of what landscape is.

Notes

1 For practical purposes, I use the plural pronoun 'we' to express the collective design work and pedagogical practices developed with students and colleagues, including when I have only informed the work of students and the teaching practices of colleagues.
2 This chapter is the result of many discussions with experienced colleagues and inspiring students, of which there are too many to list here.

3 The chapter is illustrated with drawings of graduate students (Iona Meldrum, JJ Watters, Anushka Athique, Max Barnes) and the works of designers who have inspired the approaches discussed (Perry Kulper, Bryan Cantley and Diller Scofidio + Renfro).

4 The conclusions in this chapter are based on reviewing design projects of over 500 students across ten years of teaching, since 2007. In London, at Kingston University (2007–2013) and at University of Greenwich (2013–2017), the landscape architecture and urbanism students included Bachelor and Master's students; projects ranged from one-week workshops to 12-month investigations. At Politecnico di Milano (2010–2017) projects included architecture and urbanism works from Bachelor, Master's and PhD students. At TU Wien (2017), interdisciplinary projects involved students from multiple institutions and programmes.

References

Bender, B. 1993. *Landscape: Politics and Perspectives*. Ann Arbor, MI: University of Michigan.

Berger, J. 1972. *Ways of Seeing*. London: Penguin.

Cantley, B. 2013. Two Sides of the Page: The Antifact and the Artefact. In: Spiller, N. (ed.) *Drawing Architecture*. Hoboken, NJ: Wiley.

Corner, J. 1999a. Agency of Mapping. In: Cosgrove, D. *Mappings*. London: Reaktion.

Corner, J. 1999b. *Recovering Landscape: Essays in Contemporary Landscape Architecture*. New York, NY: Princeton Architectural Press.

Corner, J. 1992. Representation and Landscape: Drawing and Making in the Landscape Medium. In: *Word and Image: A Journal of Verbal/Visual Enquiry* Vol. 8, No. 3, 144–165.

Cosgrove, D. 1984. Prospect, Perspective and the Evolution of the Landscape Idea. In: *Transactions of the Institute of British Geographers* Vol. 10, No. 1, 45–62.

Doherty, G. and Waldheim, C. 2016. *Is Landscape . . .? Essays on the Identity of Landscape*. Abingdon, UK: Routledge.

Eckbo, G. 1983. Is Landscape Architecture? In: Doherty, G. and Waldheim, C. *Is Landscape . . .? Essays on the Identity of Landscape*. Abingdon, UK: Routledge.

Geddes, P. 1915. *Cities in Evolution*. London: Williams and Norgate.

Howard, E. 1902. *Garden Cities of Tomorrow*. London: Swan Sonnenschein & Co., Ltd.

Jackson, J.B. 1984. *Discovering the Vernacular Landscape*. London: Yale University Press.

Kulper, P. 2013. A World From Below. In: Spiller, N. (ed.) *Drawing Architecture*. Hoboken, NJ: Wiley.

Meinig, D. 1979. *The Interpretation of Ordinary Landscapes*. Oxford: Oxford University Press.

Oxford English Dictionary (OED). Question (n.) [definition], www.oed.com/view/Entry/156343?rskey= xPN4sg&result=1#eid [accessed 6 December 2018].

Oxford English Dictionary (OED). Scene (n.) [definition], www.oed.com/view/Entry/172219?rskey=W gRePJ&result=1#eid [accessed 6 December 2018].

Spiller, N. (ed.) 2013. *Drawing Architecture*. Hoboken, NJ: Wiley.

Spiller, N. 2000. *Maverick Deviations*. Hoboken, NJ: Wiley.

Stilgoe, J. 2015. *What is Landscape?* Boston, MA: MIT Press.

3

From teaching geography to landscape education for all

Marc Antrop and Veerle Van Eetvelde

Introduction

This chapter discusses the education and teaching of landscape-related subjects in relation to the development of the study of landscape in the Western world. Education and teaching about the landscape reflect the changing concepts and focus in landscape research and the attitude toward the environment we live in. For the purpose of this chapter, we define education and teaching as follows. Education is the process of facilitating learning and the acquisition of knowledge, skills, values, beliefs, and habits in the context of a given culture. More specifically, landscape education is the continuous process of learning about the land, environment and society that are manifested in the landscape. From childhood on, we built a mental map of the geographical space we experience and learn to orient ourselves in it. According to the specific goals, different methods are used. Teaching is one of the methods and refers to any practice that helps others to develop knowledge or skills in a systematic and structured manner and for a specific purpose, e.g., a specific profession or training of experts. Thus, teaching about the landscape is the part of education giving a planned and organised activity to transfer knowledge, skills and attitudes with a specific purpose in mind. It involves different forms of learning and training. Many methods for studying the landscape have a pedagogic significance and are worthwhile to implement in landscape education.

First, we summarize the development of landscape study in the Western world and explain how this influenced education and teaching. Second, the specificity of teaching about the landscape will be discussed using four themes, all based upon the basic characteristics of the landscape, i.e., its dynamical, holistic nature and the relation to the perception and experience by the observer. This gives three fundamental issues concerning the teaching goals, the target groups and the necessary skills of the teacher, which are discussed in the third section. Next, two principles for teaching about the landscape are presented: the spatial and temporal scale dimensions and the landscape as a primary source of information. These are presented as general guidelines in the fourth section. In the fifth section, six methods with specific significance for teaching about landscape are discussed. Finally, a tentative synthesis of teaching and education about the landscape is given.

The landscape is accessible to and experienced by all; hence education about the landscape affects everyone (Cosgrove 2008). In general, the landscape is considered a common good, significant for the history and identity of communities, and a human right. This is often expressed in preferences, as an assessment of its beauty, or by assigning heritage values and designating areas for protection. The landscape is a holistic phenomenon and hence complex, multi-layered and having a very diverse composition. From the research perspective, many disciplines are involved in the study of the landscape. They range across such different scientific domains as natural sciences, social sciences and applied sciences. Each discipline has its proper theories, concepts, goals, language and methods. Most often they still do not work together in an inter-disciplinary, or better in a transdisciplinary way. Each discipline has its own way of teaching the students to become 'landscape experts' in their domain.

The European Landscape Convention (ELC) (Council of Europe 2000) considers landscape a basic component of the European natural and cultural heritage that consolidates the European identity. All landscapes should be considered, including urban areas and the country-side, degraded areas as well as in areas of high quality, areas recognized as being of outstanding beauty as well as everyday areas. The landscape contributes to human well-being and quality of life. Consequently, several measures for awareness-raising, training and education are proposed (Art. 6). However, the ELC does not refer explicitly to teaching about the landscape – in the meaning as described above – nor does it give suggestions as to how this should be done. It only makes a distinction between training for 'specialists', i.e., practitioners, and 'multidisciplinary training' for decision-makers, and asks for the devoting of attention to landscape issues and values in school and university courses 'in the relevant subject areas', mainly in relation to 'their protection, management and planning' (Council of Europe 2000; Council of Europe 2014).

The beginnings: naturalist's explorations and geography

The empirical study of the landscape started with the systematic descriptions during the naturalistic explorations of the 18th and 19th centuries. In these early days of natural sciences, the distinction between different disciplines was vague; the approach holistic and methods were mostly descriptive (Antrop 2013). The lectures and writings of Alexander von Humboldt and Charles Darwin, of Immanuel Kant and many others, were triggers to raise the public's interest in the landscape and can be considered as masters' discourses (Wulf 2015). Many young aristocrats were inspired to explore foreign landscapes and voyages as the 'grand tour' became popular. It fitted in with the romantic adoration of the sublime nature and wilderness (Schama 1995). With this élan, most of the geographical societies were established during the 19th century, e.g., the Royal Geographical Society in Britain in 1830 and the National Geographical Society in the USA in 1888 (Antrop 2013). The landscape became also popular in arts, in particular in painting and gardening (Cosgrove 2002; Olwig 2002). The colonization and the industrial revolution and many associated processes (urban sprawl, the enclosure of common land, the agricultural invasion by the import of new products, the up-scaling and mechanization) created new landscapes that erased existing ones. Scenic and symbolic meanings became more important and the idea of 'national landscapes' symbolizing the identity of the nation emerged (Schama 1995; Olwig 2002). Around the beginning of the 20th century, the degradation of nature and urban-industrial encroachment on the countryside initiated movements of protection of monuments, sites, nature and landscapes in most Western countries (Antrop 2005). The landscape became accepted as a common heritage and laws for protection were issued. The public became interested in the landscape for its values and engaged in the protection (Antrop and Van Eetvelde 2017). Exemplary is the foundation of the National Trust (NT) in 1895 in Britain.

In the second half of the 19th century, inspired by the ideas of the Age of Enlightenment, Rousseau, Pestalozzi and others introduced innovations in education and teaching. Geography, biology and history became core subjects and 'direct field observation' was an essential new method to explore the local environment. Wilhelm von Humboldt, the brother of Alexander, introduced the concept of holistic academic education, based on combining current research and teaching, on unbiased knowledge and analysis, and allowing students to choose their own course of study (Hohendorf 1993; Albritton 2009). Guided by these principles and the humanistic ideals and free thought, he founded the University of Berlin (now the Humboldt University).

Most theoretical thinking about geography and landscape in the academic world occurred in Germany from the mid-19th century to the Second World War (Wardenga 2006). In all European countries, the nationalist thinking influenced the development of science and arts and is also reflected in the geographical landscape studies (Cosgrove 2004). Most landscape studies were monographs aiming to make a synthesis of the natural and cultural character and identity of regions.

Sauer (1925) introduced the (German) concept of the landscape in the US. He made the cultural landscape the cornerstone of the cultural geography and emphasized the damaging impacts of the human way-of-life on the environment (Antrop and Van Eetvelde 2017). He argued that to understand a culture one must learn to read the landscape (Denevan and Mathewson 2009).

The collapse of the status of geography in Europe, in particular in Germany since the Second World War, led also to the decline of *Landschaftskunde*, which lost its societal significance as a the discipline (Paffen 1973).

Continuous specialization in science during the 1960s and 1970s resulted in the 'new orientation' in geography, focusing on quantitative modelling. In many countries geography split into historical, human and physical geography; hence regional geography and landscape studies became obsolete. This lost synthesis was regained with the re-introduction of landscape ecology in the 1980s (Antrop 2013).

Toward the end of the 20th century, the pace and magnitude of societal and environmental changes increased hugely, causing rapid and dramatic changes that became manifest in the landscape. Thus, the interest in landscape grew again in policy, planning, management and academic research, and in landscape ecology and heritage protection in particular. In 1992, the UNESCO World Heritage Convention added Cultural Landscapes as a new category to protect. In 2000, the European Landscape Convention of the Council of Europe opened a new perspective on the landscape in all domains of the society. The new focus on the landscape also appeared in the university curricula and in secondary education (Bogers et al. 2007; Antrop 2011).

The specificity of teaching about the landscape

The landscape as a subject of teaching is challenging since the landscape is omnipresent and everyone experiences it in their own way. The tangible landscape 'out-there' is reflected in a personal mindscape. Consequently, the landscape is not a well-defined subject to teach and its dynamical and holistic nature adds to the complexity. Some environments are even not recognized as a landscape by some. Transitions from one landscape to another are sometimes vague and fuzzy and distinct landscape types and units are not obvious. According to one's background and way of seeing, a different language will be used to describe and value the landscape. Hence, the first step to make the landscape a subject of study or to teach it is to explain its holistic nature, to accept its dynamic character and learning to deal with change, and finally, to define holism in the perspective one is using. Consequently, teaching goals and methods have to be adapted to the different target groups. All this will demand specific teaching skills.

Teaching holism – bridging nature and culture

The landscape is essentially a holistic phenomenon, offering a synthesis of natural and human features that compose the landscape (Antrop and Van Eetvelde 2017). The landscape is the manifestation of the interaction between nature and culture. A lot of the human attitude towards the environment they create is reflected in the landscape. The Gestalt expression 'the whole is more that its composing parts' implies that it is difficult or even impossible to grasp the whole using reductionist methods and highly specialized studies of the parts. To understand a landscape fully, an integrated study combining various disciplines is needed. Landscape ecology recognizes this and sees the landscape as a dynamical, hierarchical and multiscale system of embedded holons that function more or less autonomously. Antrop and Van Eetvelde (2017) discuss examples of such landscape holons, which are called 'ecodevices' by Van Leeuwen (1982) and 'Pandora boxes' by Zonneveld (2005). This way of seeing offers a solution to the dilemma of not being able to grasp the whole. All that is needed to obtain comprehensive and reliable knowledge about a landscape is to define a priori the scale and context of the study and to select the necessary data accordingly. By doing so, holism becomes a way of seeing (Cosgrove 2002), an attitude rather than a fact or method.

There will be always landscape – so what?

The landscape is dynamic and not a static painting. Changes happen . . . naturally. So what? Landscapes transform to become more adapted to new situations and needs, ecologically, socially and economically. Landscapes change continuously under a wide range of driving forces. Handling change and understanding processes will be essential topics when teaching about the landscape, in particular when it comes to assessing values. Landscape changes are often perceived as deterioration and even a loss of existing qualities relating to naturalness, diversity, identity and heritage, i.e., properties of landscape as a common good.

Beauty is in the eye of the beholder

Most people have an affectionate attitude toward the landscape, which is expressed as preferences and judgments. A typical example is the discussion about landscape aesthetics, such as beauty. The following fundamental philosophical question arises: is the aesthetic quality inherent (or intrinsic) to the landscape or is it in the 'eye of the beholder', thus a mental or social construct? (Lothian 1999).

This should be a basic topic in teaching about the landscape, as this discussion is essential when dealing with the assessment of landscape qualities and values. Often, 'experts' will assess landscape qualities, but the democratic principles demand also the participation of the general public. However, for most people, the landscape is not continuously an issue. The landscape is and will always be there, but it is for only certain iconic landscape views and during certain periods, i.e., holidays, that the interest awakens.

Teaching about the landscape – to whom?

People relate to the landscape in very different ways and sometimes have conflicting interests. For example, farmers see a different landscape than naturalists and tourists. The 'public' involved in a landscape can be extremely heterogeneous. Most people, i.e., 'lay-people', have different attitudes towards the landscape. Selman (2006) groups the actors into the insiders (locals) and the outsiders, who can be temporary visitors such as recreants and tourists, or distant and often 'invisible' policy and decision makers affecting the landscape of the locals. People have also different competencies towards the landscape. Hägerstrand (2001) makes the distinction between territorial and spatial competence. Those who have the power to make real changes in the landscape possess the 'territorial competence'. Essentially, these are the landowners, who have the usufruct of their land in a large degree of freedom. These can be locals, but also outsiders. Non-landowners may have special use rights on the land, such as the right of passing. Public authorities only have territorial competence for the public land they own and manage. Otherwise, they have 'spatial competencies', i.e., indirect power to regulate land use, set legal constraints on land, e.g., for conservation purposes, and formulate planning and design rules.

Different groups of actors ask very different questions and need very different knowledge and skills (Figure 3.1, Table 3.1). The public, the locals in particular, have expectations, preferences and rights. Their main question is whether changes will be for the better or the worse. Decision makers and managers set goals and need scenarios to evaluate their actions and options. They ask mainly 'What if . . .?' questions. Researchers and professionals are asked to build models, scenarios and to formulate predictions. Their main questions refer to scientific reliability and significance.

Consequently, each of the groups have different demands and goals regarding education, training and teaching. This implies that teachers or trainers need to possess, besides common knowledge about the landscape, also specialized knowledge and skills.

It should be noticed that we do not use the word 'expert' in this context. Academic researchers and practitioners are often regarded as 'experts', but so should be the local insiders who have the first-hand deep knowledge of the landscape that no one else might have. They are the source of most of the valuable and uncharted narratives and the oral history.

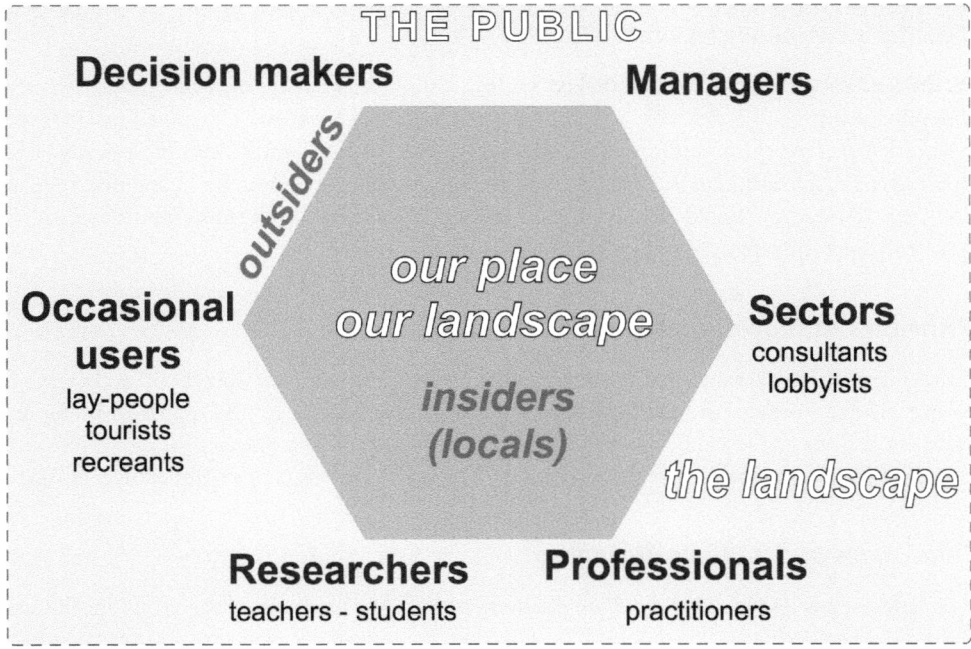

Figure 3.1 Learner groups for landscape teaching

35

Table 3.1 Actors and the landscape: demands and core questions

Group	Demands	Core questions	Common activities
The public (insiders and outsiders); occasional users; sectors	Demanding information and participation	Better or worse?	• Hearings • Discussion (in the field) • Awareness-raising • Participation
Decision makers and managers	Facts Objectives Alternatives Risk assessment	What if . . .?	• Organizing hearings, training, • Making decisions, follow-up • Monitoring, follow-up decisions
Researchers and professionals	Clear objectives Reliable data General background knowledge	What is significant? How to assess? How to predict?	• Field work, surveying • Data collection, mapping • Analysis, modelling • Assessment • Reporting

Landscape researchers and professional designers read landscapes with an analytical purpose. Their aim is to understand the landscape in very diverse aspects: its composition and configuration, its multi-layered and multi-scale structure, hierarchies of landscape units, its history and functioning. They seek knowledge to interact with the landscape in order to maintain or alter it and even to recreate it. Assessment of various values and judgment of actions expressed as negative, neutral and positive are necessary. Goals in landscape planning and management can be very different and can even be conflicting. Often, it is not possible to define the 'good' or 'optimal' landscape, and certainly not in a sustainable way.

Teacher's competencies and skills

As the landscape is a complex and diverse subject, a multidisciplinary, or, even better, an interdisciplinary approach is desirable in teaching as well. Probably several specialized teachers will be needed to cover the whole subject. However, each of them must think holistically. This demands a flexible attitude, the ability to change views and scales (zooming in and out, in space and time). Also, they should possess a basic common knowledge about landscape science and its basic concepts and language in particular.

Principles of teaching about the landscape

Teaching principles can be used as general guidelines to introduce the subject of landscape and to organize and present the teaching topics. Two are discussed here: the spatial and temporal scale dimensions of the landscape and the use of the landscape as a primary source of information. The knowledge of both of them should be inherent to the teacher's competences and skills

Dimensions and multiple disciplines

The landscape is a spatial and dynamic phenomenon and its two main dimensions are the spatial scale and the time scale (Figure 3.2). The time scale stretches from the distant past to the uncertain future. The spatial scale varies from the local 'micro'-landscapes to the global issues affecting all landscapes. They can be studied as separate time slices or as a delimited area at a

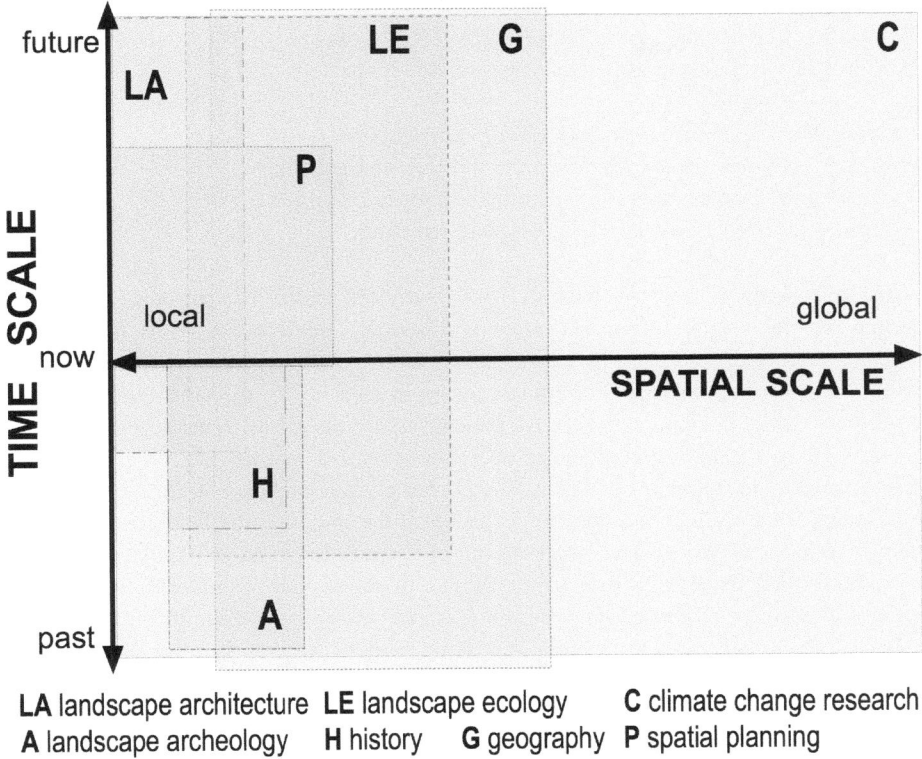

Figure 3.2 Time–space dimensions and disciplines involved in landscape teaching

fixed scale, but also as time series describing landscape paths or trajectories, and as a hierarchical spatial system that can be explored by zooming in and out. The landscape disciplines cover different domains and each has a special focus in this space–time. The ranges of the domains vary between the disciplines, but most overlap, indicating again the need for a common conceptual framework and language

The landscape as primary source – reading the landscape

The landscape is considered the primary source of knowledge offering basically two approaches: one of direct observation and experience and one of indirect information gathering through all kinds representations, photographs and maps and descriptions. In both cases, conscious 'reading the landscape' is necessary for full understanding.

One landscape view is insufficient to understand the landscape in its multi-scale complexity. Comprehension gradually emerges from mentally combining many different perspectives. Travelling through the landscape creates one's mental map. Prospects from high viewpoints help to see the complex spatial structure. Cognition of the landscape is gradually constructed from different perspectives. To widen the field of view, indirect observations are essential. The bird's eye perspective by satellite and aerial imagery allows us to vary the scale and to place individual details in a broader spatial context. Maps allow abstraction and deconstructing the landscape in thematic layers. Historical maps allow us to go back in time. Basic knowledge of different disciplines becomes necessary: geology, soil science, land use, ecology, agriculture, forestry, history, archeology, architecture, etc.

Four main layers in reading the landscape can be recognized:

- a scene offering a sensory (mainly visual) experience;
- a natural, physical system that forms the substrate of the land;
- a social system with places and territories and forms of land use; and
- a history revealed in successive, incomplete remnants.

The first layer is the scenery, which can 'read' as a work of art, as a painting, although it is more like experiencing a theatre play, a scene with action. The following layers demand a more systematic, 'scientific' reading. The natural, physical system formed the substrate that offered opportunities and gave restrictions to humans to live in the land and to shape it into a landscape. The substrate carries the cultural layers. All these layers are essentially dynamic and transform in different ways, speed and scale influenced by a series of interacting driving forces. Consecutive time layers make the palimpsest of the unique history of each landscape unit.

The metaphor 'reading the landscape' is based on the idea that the landscape is composed of very diverse elements forming larger meaningful structures, such as letters forming words and words forming sentences (Spirn 1998). Historical geographers consider the landscape as a palimpsest. Reading such a manuscript demands careful deciphering of the text fragments, often written in different languages and spelling and trying to fill in the gaps. Landscapes are built from successive time layers and each new one partially uses and changes the older ones.

Reading the landscape demands knowing the letters, words and grammar of the language. Letters and words of the landscape language are tangible elements and observable features: a tree, a house or settlement, roads, fields, rivers, crops, etc. Some words refer to abstract concepts such as diversity and beauty. Sentences express properties, relations and processes between the composing parts. Once a word or sentence is understood, it must be interpreted in a broader context, often demanding reading in between the lines and filling in the gaps. This is, for example, explicit in the reconstruction of past landscapes from relics and from historical maps.

To reduce the complexity, the landscape is deconstructed into thematic layers or models. Some themes are continuous phenomena (relief, soils), others consist of discrete objects or elements (fields, buildings), which have specific properties and functions.

Different models are used depending on the discipline and the goal of the study. Landscape ecologists often use the patch-corridor-matrix model. Elements of interest for the species studied are grouped in patches (i.e., habitats) and corridors (potential moving paths) and the rest of the landscape is considered a hostile matrix. Geographers studying settlements and regions will order these in types according to different criteria according to the goal of the analysis.

Teaching methods

Some selected teaching methods with a particular meaning to understanding the landscape are discussed below.

The power of going in the landscape – excursions and fieldwork

As the landscape is the primary source of knowledge, it is obvious that the perfect place to teach about the landscape is the landscape itself. This is achieved by excursions, fieldwork, project works and workshops. Depending on the size of the group and practical facilities, excursions can vary from a 'class on wheels' to field walks.

The power of language – narratives and the master discourse

A momentary view of the landscape or its representation does not reveal everything. Much of what is significant to know resides in narratives and stories. This can be the result of research lectured on and published by scholars, the stories told by explorers and travellers, and also the oral history remembered by locals that is hidden in place names and landmarks.

Narratives about landscapes have multiple facets. Here, we focus on the narratives that inspire readers and audience and not the factual descriptions as in many geographical descriptions in encyclopedia and atlases. The descriptions by the early explorers were not recordings of mere facts, they were also vivid descriptions of the landscape character, evoking the functioning and story of a landscape. The natural and human worlds were integrated. A special literary genre emerged with huge success: travel guides. In some cases, they consist of the recordings of the local oral history, offering valuable information for further research. *The Discovery of France* by Robb (2011) is illustrative of the immense amount of knowledge lost since dramatic changes from the end of the 19th century. Many other examples of such narratives exist (Schama 1995; Palang et al. 2004; Wulf 2015).

The master discourse is a powerful teaching method to introduce complex matter in a short time and in a comprehensive way. It allows us to integrate visual experience with existing knowledge and to demonstrate spatial and temporal relationships at various scales. The basic communication tool is language, often supported and enhanced with appropriate visualizations.

The power of imagination – representations and visualization

All descriptions of the landscape are represented in various ways, which can be ordered from concrete to abstract and according to the senses involved to interpret the representation (Figure 3.3). Thus, recordings and visualizations become valuable indirect information sources about the landscape as well. Special methods and techniques were developed to exploit them.

The power of observations – recording and analytical skills

Studying the landscape often involves recording observations using mapping, sketching and photography. These demand basic training in various technologies, such as cartography, image interpretation, geographical information systems (GIS), spatial analysis and visual representations. It is not essential to master all these techniques oneself but it is essential to be able to interpret and use them with a critical sense. After all, it is easy to lie with maps and statistics. . . .

The power of learning together – teaching to work in interdisciplinary ways

As a complex and holistic phenomenon, the landscape is studied by several disciplines, varying from the natural sciences to the humanities, from philosophy to practical designs. Full understanding of the landscape demands an interdisciplinary approach, and, when it comes to practical

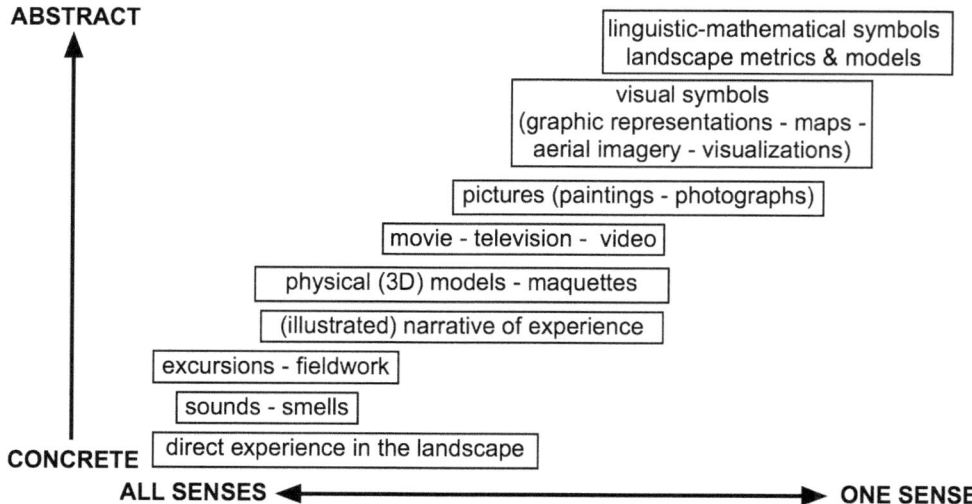

Figure 3.3 Landscape teaching methods classified according to degree of abstraction and use of senses

applications, even a transdisciplinary approach involving the democratic participation of the public. These approaches are advocated by the Aarhus Convention (UNECE 1998) and the European Landscape Convention (Council of Europe 2000). Tress et al. (2005) describe inter- and transdisciplinary research as 'integrative research'.

Interdisciplinarity implies an attitude to using a common language by all participating disciplines and cooperation from the start. At the academic level, this can be achieved by interdisciplinary research, where several disciplines focus on solving one common problem and contribute with their proper skills and to enriching the landscape teaching. Transdisciplinarity implies the participation of policy makers, administrators and laypeople, demanding that an adapted language be used. Here different methods of participation have to be employed.

Landscape experts involved in these integrative approaches need attitudes of co-operation, willingness to listen and to accept different perspectives, and to express their proper views clearly, in a way that is understandable to others. If they are in charge of organizing integrative projects, they need to master the necessary methods and techniques of communication and participation.

The power of iterative learning – from desk to field and back again

Reading and understanding the landscape is an iterative learning process. It consists of five phases, alternatively working on the desktop and visiting the landscape (Figure 3.4).

The first step consists of gathering information for the first visit. This includes studying maps and aerial imagery, pictorial information such as photographs and paintings, reading guides and descriptions and existing studies. Thus, the first field trip can be planned.

The second step is the first reconnaissance in the landscape itself. This is a phase for further orientation, careful observation and recording notes, making sketches, taking photographs, talking to locals and gathering new data.

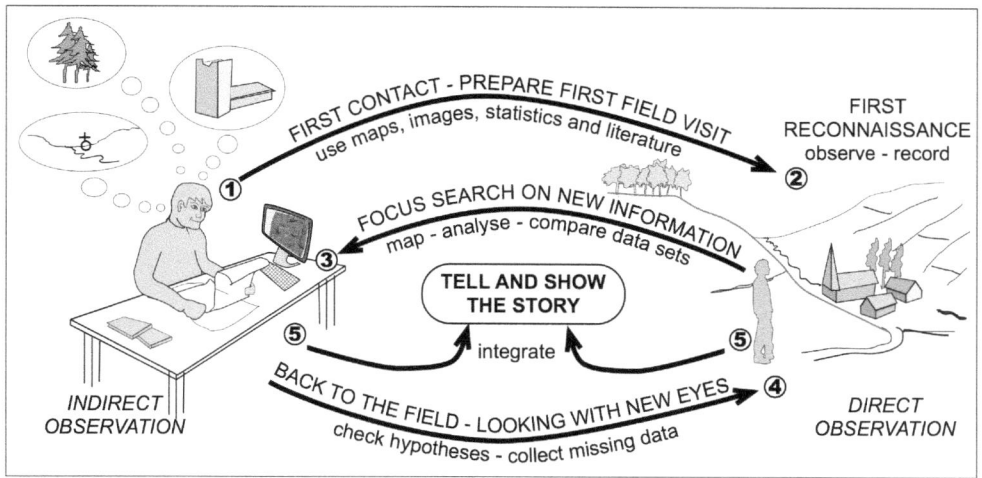

Figure 3.4 The process of iterative learning in studying the landscape

The third phase starts back at the desktop. The initial documents are studied again and compared with the direct observations made in the field. New data are mapped and integrated with the existing ones. Spatial and statistical data are analyzed if appropriate. Interviews are transcribed and compared with the literature. Hypotheses are formulated about the landscape structure, its development and dynamics, its character and identity, its values, etc.

The fourth phase is going back to the field looking for answers on remaining questions, to collect additional information, to check the maps and analyses made and to test the hypotheses.

Sometimes several iterations between the third and fourth phases will be needed to attain an acceptable knowledge in order to move to the fifth phase.

The fifth and final phase consists of integrating all desktop work and fieldwork and to tell the whole story about this landscape. A narrative is the best way to make a synthesis using landscape as means of communication. It allows explaining of the coming to be of the landscape one experienced, the why of its identity and character and the coherence between the composing parts.

Conclusion – the landscape of teaching landscape

As with the landscape itself, teaching about the landscape is also a complex endeavor. Figure 3.5 gives a tentative overview of the landscape of landscape teaching. All the aspects are organized around the following basic questions: What to teach? To whom and why? How? When and Where? The What? consists of the teaching contents and is elaborated in more detail. Special attention is given to basic concepts considered essential to understand the landscape in a comprehensive way. The grey-shaded area indicates the ideal common curriculum of teaching landscape to achieve a common language facilitating interdisciplinary and transdisciplinary practice, which is essential for good management of our landscapes. The selection of the indicated methods and teaching conditions (When? Where? and How?) is made with the same purpose.

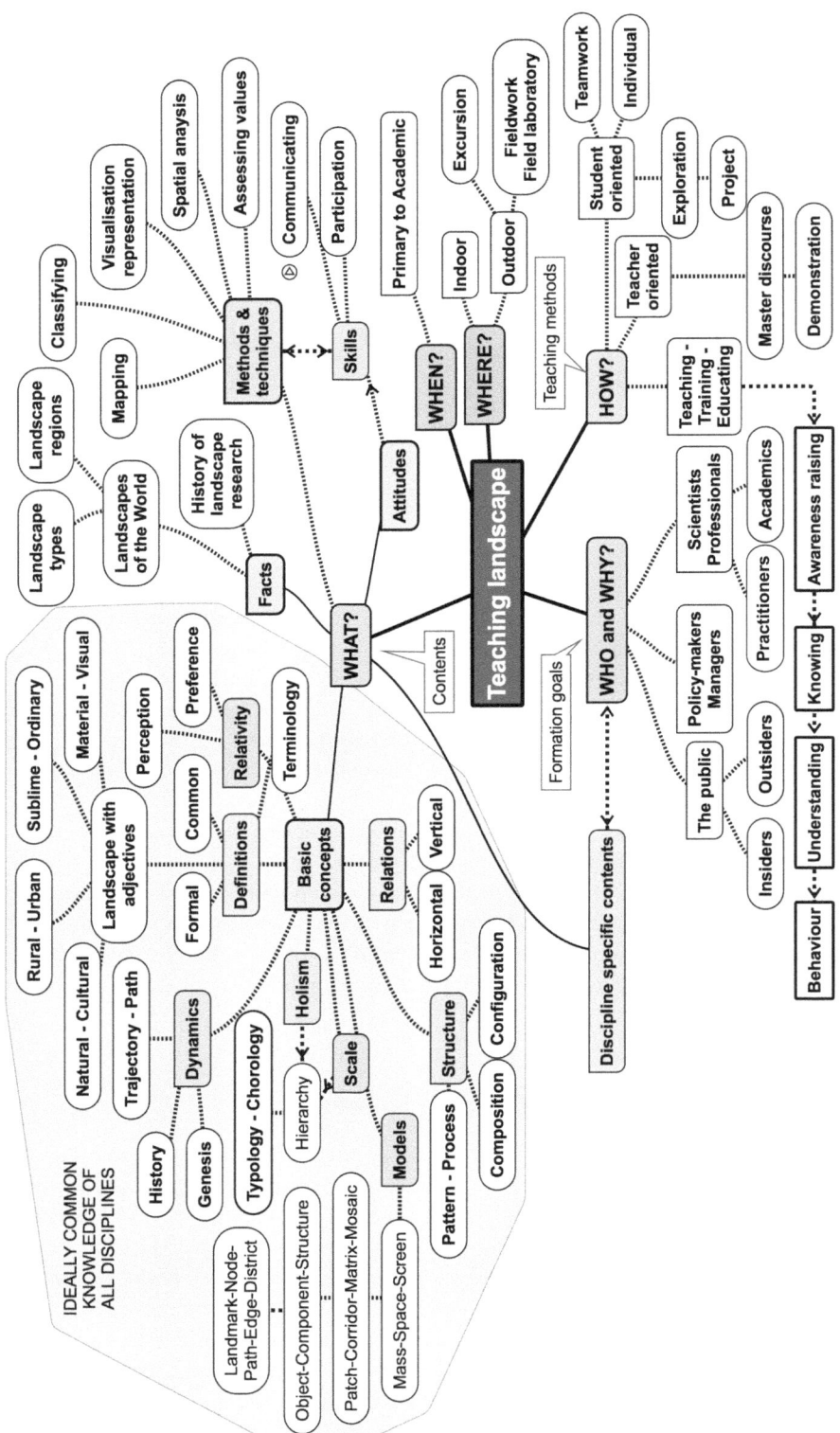

Figure 3.5 The landscape of teaching landscape

To this basic package, various disciplines can contribute with their proper concepts, methods and skills. They offer the possibility for fine-tuning the teaching according to the formation goals of different target groups (Who? and Why?).

References

Albritton, F.P. (2009), 'Humboldt's unity of research and teaching: influence on the philosophy and development of higher education in the U.S.'. *The FedUni Journal of Higher Education* 4/1: 7–19.

Antrop, M. (2005), 'Why landscapes of the past are important for the future'. *Landscape and Urban Planning* 70/1–2: 21–34.

Antrop, M. (2011), 'Moet er nog geografie zijn?'. *De Aardrijkskunde* 35: 85–95.

Antrop, M. (2013), 'A brief history of landscape research', in: P. Howard, I. Thompson, E. Waterton (eds), *The Routledge Companion to Landscape Studies* (Abingdon, UK: Routledge), 12–22.

Antrop, M. and Van Eetvelde, V. (2017), *Landscape Perspectives: The Holistic Nature of Landscape* (Heidelberg: Springer, Landscape Series, 23).

Bogers, M., Oonk, C., Palang, H., Pedroli, B. and Stastna, M. (2007), *Road Map for Education and Training on Land Use and Sustainability*. Action for Training in Land Use and Sustainability, Coordination Action FP6-018543 (Tallinn: Huma, 36).

Cosgrove, D. (2002), 'Landscape and the European sense of sight – eyeing nature', in: K. Anderson, M. Domosh, S. Pile and N. Thrift (eds), *Handbook of Cultural Geography* (London: SAGE Publications), 249–268.

Cosgrove, D. (2004), 'Landscape and landschaft'. *German Historical Institute Bulletin* 35: 57–71.

Cosgrove, D. (2008), 'Geography is everywhere: culture and symbolism in human landscapes', in: T.S. Oakes and P.L. Price (eds), *The Cultural Geography Reader* (Abingdon, UK: Routledge), 177–185.

Council of Europe (2000), *European Landscape Convention*. Firenze, 20.10.2000. http://conventions.coe.int/Treaty/en/Treaties/Html/176.htm.

Council of Europe (2014), 'Recommendation CM/Rec(2014)8 of the Committee of Ministers to member States on promoting landscape awareness through education'. https://search.coe.int/cm/Pages/result_details.aspx?ObjectID=09000016805c5138.

Denevan, W.M. and Mathewson, K. (2009), *Carl Sauer on Culture and Landscape: Readings and Commentaries* (Baton Rouge, LA: Louisiana State University Press).

Hägerstrand, T. (2001), 'A look at the political geography of environmental management', in: A. Buttimer (ed.), *Sustainable Landscapes and Lifeways: Scale and Appropriateness* (Cork: Cork University Press), 35–58.

Hohendorf, G. (1993), 'Wilhelm Von Humboldt'. *Prospects: The Quarterly Review of Comparative Education* XXIII/3–4: 613–623.

Lothian, A. (1999), 'Landscape and the philosophy of aesthetics: is landscape quality inherent in the landscape or in the eye of the beholder?'. *Landscape and Urban Planning* 44/4: 177–198.

Olwig, K. (2002), *Landscape, Nature, and the Body Politic. From Britain's Renaissance to America's New World* (Madison, WI: The University of Wisconsin Press).

Paffen, K. (ed.) (1973), *Das Wesen der Landschaft. Wege der Forschung, Band XXXIX* (Darmstadt: Wissenschaftliche Buchgesellschaft), 514.

Palang, H., Sooväli, H., Antrop, M. and Setten, S. (2004), *European Rural Landscapes: Persistence and Change in a Globalising Environment* (Dordrecht: Kluwer Academic Publishers).

Robb, G. (2011), *The Discovery of France* (London: Pan MacMillan).

Sauer, C.O. (1925), *The Morphology of Landscape*. University of California Publications in Geography, 2; reprinted in Leighly, J. (ed.), (1974) *Land and Life* (Berkeley, CA: University of California Press) 315–350.

Schama, S. (1995), *Landscape and Memory* (New York: A.Knopf).

Selman, P. (2006), *Planning at the Landscape Scale* (Abingdon, UK: Routledge).

Spirn, A.W. (1998), The Language of Landscape (New Haven, CT: Yale University Press).

Tress, B., Tress, G. and Fry, G. (2005), 'Defining concepts and the process of knowledge production in integrative research', in: B. Tress, G. Tress, G. Fry and P. Opdam (eds), *From Landscape Research to*

Landscape Planning: Aspects of Integration, Education and Application (Wageningen UR: Frontis Series), 13–26.

United Nations Economic Commission for Europe (UNECE) (1998), *Aarhus Convention. Convention on Access to Information, Public Participation in Decision-Making and Access to Justice in Environmental Matters*, Aarhus, 25.06.1998. http://ec.europa.eu/environment/aarhus.

Van Leeuwen, C.G. (1982), 'From ecosystem to ecodevice', in: S.P. Tsjallingii and A.A. de Veer (eds), *Perspectives in Landscape Ecology* (Wageningen: Centre for Agricultural Publishing and Documentation), 29–34.

Wardenga, U. (2006), 'German geographical thought and the development of Länderkunde' (Lisboa: Edições Colibri), *Inforgeo* 18/19, 127–147.

Wulf, A. (2015), *The Invention of Nature: Alexander von Humboldt's New World* (New York: A.A.Knopf).

Zonneveld, I.S. (2005), 'The land unit as a black box: A Pandora's box?', in: J.A. Wiens and M.R. Moss (eds), *Issues in Landscape Ecology, Studies in Landscape Ecology* (Cambridge: Cambridge University Press), 331–345.

4

The importance of geology in landscape architecture education

Ralf Löwner

Introduction

Landscapes are invariably shaped by the interactions between natural and cultural phenomena, and between biotic and abiotic systems, namely geology or geomorphology. In addition, virtually all landscapes on Earth have been shaped to some degree by human activity and entire ecosystems have become domesticated (Kareiva et al., 2007). Methods for quantifying the human footprint are widely discussed (e.g., Woolmer et al., 2008) and on a global scale it has to be assumed that only a few areas on earth are without direct or indirect human influence (Sanderson et al., 2002). This clearly indicates that science-based sustainable land management is required.

Cultural elements not only shape natural environments, but are also shaped by them and their prevailing conditions (Figure 4.1)—an easily observable bidirectional relationship. Geology plays a central role in determining the characteristics of both the local and regional anthropogenic development of a region, as well as regional developmental potential and constraints. Throughout the historical record, one can find a plethora of examples of geology's central role

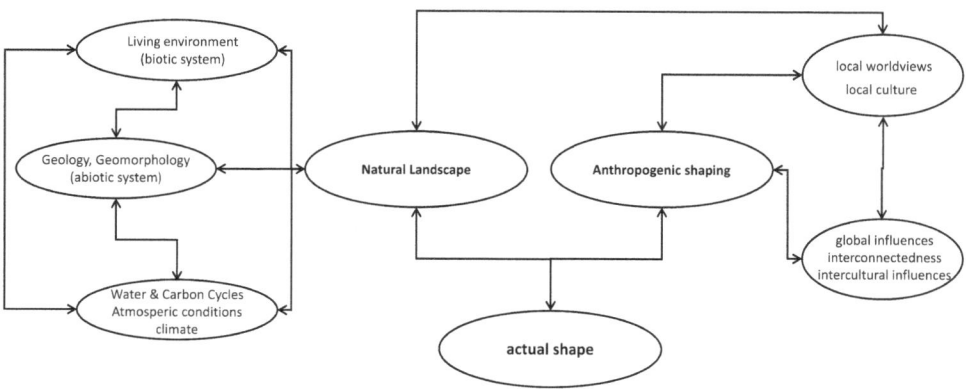

Figure 4.1 Mutual relations between the different aspects influencing humanized landscapes

Figure 4.2 Typical Persian garden architecture in Shiraz (Qavam House), Iran, with *qanat*-irrigated water basins

in the anthropogenic development of natural environments (e.g., Crouch, 2003: Greco-Roman patterns; Haworth 2003: the influence of geology on the shaping of Sydney).

Typically, geology's central role even applies to important cultural breaks, such as the Migration Period in Central Europe, and, in Asia, the Islamization of the Persian Empire. The famous Persian gardens in Iran were built in accordance with the available water supply system (Figure 4.2). Subterranean canals (*qanats* in Persian) are the foundational principle behind the gardens' architecture. Detailed knowledge of the region's geology was required by the gardens' designers in order to identify the water sources. Moreover, it was necessary to make use of the substratum and stratigraphic layers to obtain the subsurface's constant slope with extreme accuracy (Boucharlat, 2017). The gardens' existence has been documented since the middle of the second millennium BC and remain today virtually unchanged. Additionally, the gardens' *qanats* were built in consideration of the region's tectonic fault lines (Boucharlat, 2017).

Another example could be the anthropogenic-shaped landscape in the High Atlas Mountains of Morocco. Before and after the Early Muslim conquest, construction materials, settlements, agricultural areas, and irrigation methods had not changed substantially (Peyron, 1976) in certain areas up to the present.

Today, as in the past, it remains crucial to integrate geological knowledge and methods to arrive at a clear understanding of Earth-shaping dynamics. The effect of humans on the natural environment and landscape shaping can be analyzed using traditional geological methods. For understanding the effects of the earliest human cultures lake sediments are often a key to such investigation (Giguet-Covex et al., 2014). Landscape architecture must take into consideration the challenges faced by natural environments due to a constantly changing Earth. Geological knowledge can illuminate how anthropogenic landscape shaping affects natural environments and the role of climate in future human development.

The effect of geology on landscapes

As described above, the landscapes we see are the product of both past and present dynamic geological processes, ecological processes, weather, and human activity. Any detailed examination of the effect of geology on landscapes should consider these three primary factors:

- the present geological conditions;
- the dynamic Earth; and
- soil and other natural resources.

Factor 1 considers the continental crust as a dynamic three-dimensional system. This structural approach can be used to illustrate the effects of the continental crust on the Earth's surface and, therefore, on natural landscape formations. The geomorphology of any particular region can be considered as the interface between the continental crust and the region's topography. The continental crust is not only responsible for the topography of a region, but also for multiple other geological characteristics, including the stability of slopes, and the suitability of the ground for basement construction, aquifers for consumption and/or agriculture, and natural caves for settlements. Earth's dynamic geology directly impacts cultural development and landscape architecture, and knowledge of Earth's geological past is essential for understanding how landscapes evolve.

Factor 2 concerns the fourth dimension—time, and its effects on the natural environment and, consequently, on the living conditions of humans. In terms of geology, time comprises sedimentary and erosional phenomena, tectonic and volcanic events, and the life cycles of liquids and gases. Humans have always and continue to answer these geological challenges by integrating defense strategies into landscape shaping. As such, human activity releases potentials, e.g., land reclamation via isostasy.

Factor 3 includes soil, which, in the broadest sense, constitutes the substratum and surface of the landscape upon which all human activity occurs. In addition to soil, all natural resources are directly connected to cultural activity, as it is human requirements that define things as resources. Together, they have a direct effect on landscape management; for instance, implementation of a sustainable irrigation regime in a region with a finite groundwater supply. The natural resources in any given region determine how the landscape is shaped, and what types of architecture are designed and constructed. In all regions ever inhabited by humans, the landscape has been dominated by the use of soil and water, and locally available sedimentary, igneous, and metamorphic building materials.

The discovery of such resources as water, geothermal sources, fertile soil, and fossil fuel facilitates the establishment of new settlements, resulting in a geological resource-based landscape architecture. Examples of this phenomenon include the shaping of the Ore Mountains in Central Europe due to the exploitation of metal and mineral ore deposits (Thomasius, 1994), and of the Harz Mountains in Germany via the effect of mining activity (Bartels, 1996). In this specific case—landscape shaping via mining—contemporary landscape management must implement sustainable strategies that take into account all relevant geological and ecological conditions.

What to teach about geology

The teaching of geology should be adapted to the needs of students of landscape architecture and should concentrate on the natural environment as the target of sustainable, resource-oriented landscape management. Students should be introduced only to the basic principles of geoscience and analytical methods; therefore, curricular content can be based on the following three primary aspects of geological influences on the landscape.

1   The structural concepts of geology and the three-dimensional setting of landscapes.
2   The historical dimension of geology and the dynamic aspect of landscape evolution in time.
3   The global aspects of lithology: the nature and composition of geological units of hard rocks and soil.

Structural geology is important for students of landscape architecture, as knowledge about the subsurface and topography of an area or region helps one to understand how both contribute to a landscape's character. As such, course material on structural geology should provide students with knowledge about the geometrical build-up of the Earth's crust in three-dimensional space. This requires basic knowledge of stratigraphic layers and formations as fundamental units for understanding and modeling the subsurface and topographic surface of Earth. Additionally, the basics of general geology, regional geology, cartography, and scientific geological methods are crucial. The main educational objective should be the students' ability to understand three-dimensional surfaces and solid figures, and their geological modeling and interaction with the topographic surface.

The historical dimension of geology concerns the time dimension and the dynamics of landscape evolution, and instructors need to encourage students to examine in greater detail natural science, and the principles of geology and the fundamental natural laws. Teaching the historical dimension of geology can be divided into three parts. Instructors can begin with the evolution of Earth over time, providing an overview and a sense of geological scale in relation to the contemporary environmental situation. Next, the basic concepts of geology, such as the 1669 "Law of Superposition" by Nicolaus Steno and the 1830 concept of actualism by Sir Charles Lyell, most commonly expressed by the term "the present is a key to the past." The third part should cover the dynamic processes on Earth at all time scales, from long-term events such as tectonic movements and mountain building to such short-term events as volcanic eruptions. In this context, the circulation of fluids and gases should be introduced and include such topics as the water cycle and carbon cycle. Teaching the historical dimension of geology should conclude with the classical geological fields, including stratigraphy, paleontology, physical geology, and general/regional geology, based on available time. The end result should be students with a deeper understanding of the principles of time-related geological processes.

Careful examination of the material of the lithosphere in isolation facilitates an understanding of the nature and composition of the numerous geological units in detail. Independent of the structural concepts of geology and the historical dimension of geology, lithology completes the two preceding aspects and provides more practical knowledge of geology. In terms of structure, facies can be introduced as part of geological formations, which can lead to a detailed exploration of rock types—sediment, and igneous and metamorphic rocks. Furthermore, erosional products, especially soil building, should be discussed. If necessary, this part can include some aspects of general geology, petrology, regional geology, cartography, and scientific geological methods, as well as engineering geology. Students of landscape architecture should learn to differentiate different types of rocks and soils, and should develop a sense for the different materials and surfaces on Earth.

Teaching geology to landscape architecture students

The classical methodology of teaching geology is highly evolved and must be adapted to the needs of students of landscape architecture. In general terms, geology education for landscape architecture can be comprised of three parts: fieldwork, laboratory work, and classroom lectures and exercises. Advancements in education methodology show that active learning strategies are very important (Macdonald & Korinek, 1995), including student participation, discussion, and assessment, which are essential for applying newly acquired knowledge (Silberman, 1996).

Classroom work can begin with developing students' three-dimensional sense of the environment. This sense can be developed by showing how different geological bodies are situated in a three-dimensional space using the definition of the strike and dip of a geological body's surface. It can also be developed using analogue models, and landscape architecture students might even use their designing skills to build three-dimensional models as an exercise. Another option is the use of specialized 3D software. Geological bodies can be considered as geological layers and their position in space are possible beddings (horizontal, vertical, flat, steep, and inverted) that become visible and comprehensible using 3D software. Subsequently, from this three-dimensional visualization, an abstraction must be made to interpret the model from a bird's eye view, as it appears on a two-dimensional geological map.

In moving from three to two dimensions students can recognize the intersection of topography and geological bedding, and, therefore, perceive the landscape as the impression of the geological structure on the Earth's surface. Empirically, the relationship between structural geological models, topography, and maps is the most difficult thing for students to understand regarding mapping. This approach to teaching geology simply considers structural units as three-dimensional bodies without the necessity of delving into more nuanced geological knowledge. It supports a learning process that yields an understanding of the natural development of a region independent of other geological parameters and processes, and of previous knowledge of the natural sciences.

As geology is a natural science its teaching should be based on practical exercises and as much fieldwork as possible. The first extra-classroom activity should be an introductory fieldwork, in which the students play an active role in observation, drawing and analysis, and interpretation. The observation of nature—"real world" details at any scale—should be the first practical part of a foundation in geology. Observation is a skill that must be developed, especially of field colors, layering of bedrock, ledges and slopes of a rock wall, and of such characteristic natural landmarks as rivers, cliffs, as well as vegetation.

In addition, rather than photograph field observations, students should learn to observe critically by being made to visually record their observations on paper. To do so, all that is required is a pencil, an eraser, and a field book or a sheet of paper (Figures 4.3a and b). Ideally, students should situate themselves in front of an outcrop, a panoramic mountain view, or a characteristic landscape, and take as much time as necessary to draw and sketch what they see, without any advice from the instructor. Some students may require extra encouragement to record their observation because they think they cannot draw well, but the technique empowers them and pushes them to improve their capacity to observe, which is a life-long process. Landscape architecture students may be more motivated to perform this fieldwork exercise than students of geology, as the development of drawing techniques in central to landscape architecture.

Following observation and visual record making, students must learn to analyze and interpret their recordings. By comparing the students' sketches, the most commonly recorded geological characteristics can be discussed. During such discussions students can be encouraged to begin to consider why and how the natural setting they observed is as it is. For instance, students need to

Figure 4.3a and b Fieldwork with students in the High Atlas Mountains, Bin El Ouidane Lake, Morocco

consider the causes of a red-colored field, horizontal bedding, and the lack of vegetation in an observed area. This is also the time that rock types are identified, differentiated, and discussed, and the general geological situation is explained. One option is for the students to begin such a discussion in pairs, followed by group discussion with the instructor. This approach facilitates the active participation of all students and such experiential learning is invaluable. Lastly, the information and knowledge obtained by the student during the observation/sketching/analysis exercise can be generalized and applied in another area, simply by turning the group around 180 degrees or walking to a different location.

After this three-step-method, thoroughly selected outcrops could be investigated in order to underline and deepen the gained knowledge (Figure 4.4). Here, some details could be explained depending on students' interest. Also measurements of surface directions or fault structures could be undertaken by the use of a compass.

Additionally, samples could be taken for laboratory work, a good occasion for teaching sampling methodology.

Generally, the observation/sketching/analysis method of fieldwork described above can also be used to integrate geography, biology, and such cultural factors as agriculture, construction, and settlements. It is even possible that the "real world" can be observed via representative and abstract landscape paintings (Moll, 2017) in order to extract structures and natural patterns. Nonetheless, fieldwork is the best approach for teaching geology as it evokes a sense of place (Gruenewald, 2003) and environmental awareness. Such awareness can even include that of local environment-related cultural aspects and characteristics. For example, some rock outcrops in Montana are considered sacred by the indigenous population, whereas others may see only coal (Cohn et al., 2014). In extreme cases when fieldwork is not possible the described observation/sketching/analysis method can be used with carefully chosen photographs of landscapes with clearly observable geological characteristics (Reynolds & Peacock, 1998).

The laboratory is where students should be introduced to lithology, sedimentology, and engineering geology—the analysis of building materials, and the stability of foundation soil and slopes. In the laboratory setting, contemporary, easy geological investigation methods can be performed with or without sophisticated equipment. In this setting observation and team work also play a major role: samples, including stones and soils, can be analyzed with the naked eye to consider their lithological and paleontological aspects. In addition, the relative hardness of the samples should be determined according to the Mohs scale of mineral hardness. Next, the instructor should lead a discussion designed to help the students understand the paleo environment, formation history, and potential uses of the observed samples. If sieves are available, then

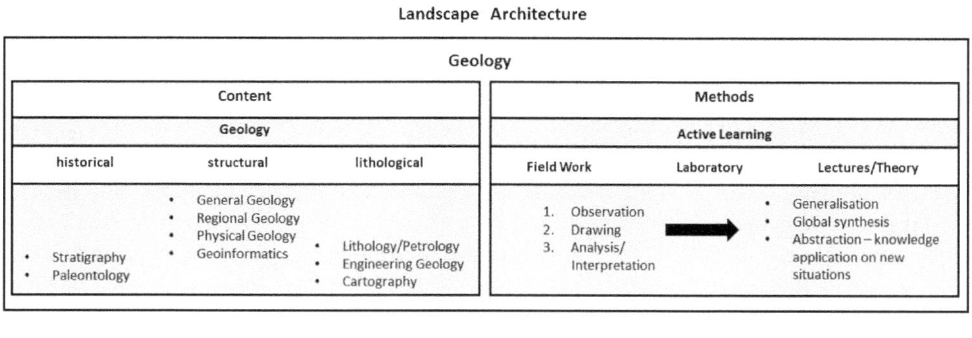

Landscape Architecture

Geology					
Content			**Methods**		
Geology			**Active Learning**		
historical	structural	lithological	Field Work	Laboratory	Lectures/Theory
• Stratigraphy • Paleontology	• General Geology • Regional Geology • Physical Geology • Geoinformatics	• Lithology/Petrology • Engineering Geology • Cartography	1. Observation 2. Drawing 3. Analysis/ Interpretation		• Generalisation • Global synthesis • Abstraction – knowledge application on new situations

Figure 4.4 Suggestion of a teaching concept for geology studies within landscape architecture education including the three-step model for field experiences

granulometric analysis can be conducted. Laboratory work should introduce students to the most important and prevalent sediments, and volcanic and metamorphic rock types.

Back in the classroom, the knowledge obtained via fieldwork and the laboratory can form the basis for further study: the students' field sketches, and additional observations and measurements, can be used to elaborate detailed tectonic cross sections. Furthermore, outcrop investigation and laboratory work could be used to establish synthetic sedimentological cross sections.

Teaching geology using geoinformatics

Whereas haptic experiences are crucial and indispensable for the study of the natural sciences, novel approaches made possible by technological advancements should not be overlooked. Computer technology, geoinformatics, and multimedia facilitate the teaching and visualization of geology in all four dimensions. Such tools, when available, should be used to augment students' fieldwork experiences. Nonetheless, some technologies are more appropriate than others for use in the classroom; what follows is a description of those that are most appropriate.

A geographic information system (GIS) is the most appropriate visualization and analysis tool for use in the spatial-based natural sciences. Use of a GIS does not require special training and most such systems are very affordable. Moreover, there are many high-quality free and open-source software (FOSS) programs for use with a GIS. For example, GIS can be used in the classroom for fieldwork preparation by collecting, storing, and analyzing available maps and data for a given region, and then for the subsequent visualization of findings obtained during the fieldwork. In such a scenario, students can learn how to search for and acquire available free and open data (data mining), and how to efficiently use online spatial services (e.g., www.onegeology.org, an international initiative of the geological surveys). Numerous web-based GISs for numerous spatial thematic data sets are available, such as the basic data provided by the OpenStreetMap (OSM) repository.

Satellite raster data sets are another important tool for exploring geological topics. The most obvious and well-known online destination for such data is GoogleMaps, which can be considered a good first step before moving on to more sophisticated image enhancement tools. Here again, FOSS is the recommended software for analyzing satellite data; for instance, the European Space Agency (ESA) provides a bundle of specialized FOSS tools for learning how to use remote sensing technologies.

Of particular importance for the study of geology is 3D modelling tools, which can be used to improve students' spatial perception and thinking. For example, 3D modelling can replace the classical analogue sand box experiment used to explore sedimentology and structural geological concepts. That said, even if simple 3D modelling is integrated into GIS products and FOSS is available, use of 3D modelling requires greater effort, expertise, and financial resources than use of a GIS and satellite data. With access to the necessary resources, it is even possible to engage in highly sophisticated 3D-model visualization of cave settings; however, free access to web-based 3D virtual geological fieldworks is something to hope for in the near future (Craven et al., 2017).

Free and open-source software is highly recommended for educational purposes, especially the OSGeo suite, which can be obtained from the Open Source Geospatial Foundation (www.osgeo.org). Research has highlighted the advantages of the open approach to software and data (among others, Brocco & Frapolli, 2011). This approach provides students with a rich learning experience that motivates them to develop technical and cognitive skills (Wurst et al., 2014). Additional advantages of the open approach are low start-up cost, platform independence, open standards and formats, and the support of a very active community.

In addition to its importance for teaching geology, geoinformatics is essential for the practice of landscape architecture. The most common geoinformatics tools are GIS and geoinformation services, both of which are necessary for sustainable and structured land management, and they are central to the modern concept of landscape shaping. Optimally, a broad-based education in landscape architecture should not only integrate geology, but also geoinformatics as an autonomous module. The combined exploration of landscape architecture and aspects of geology, such as mapping, can be integrated into a basic geoinformatics course. Above all, all other relevant topics can be taught in close relation to geoinformatics, due to the fact that landscape architecture focuses on spatial concepts.

Objectives of a foundation in geology

Geological knowledge and skills are necessary for landscape architecture students in order to integrate a well-grounded sense of the natural environment and for understanding in detail their processes. The application of geological expertise to landscape-shaping issues sharpens the professional's view of spatially related problems and feasibilities, and positively affects the analysis and decision-making in landscape management practice. The application of active learning concepts and, most importantly, fieldwork characterized by the observation/sketching/analysis method facilitates students' development of critical thinking skills. On the one hand, both foster spatial visualization ability and an understanding of a dynamic three- and four-dimensional Earth. With a sound understanding of the natural aspects of the landscape, abstracted information can be integrated into a multi-dimensional context. This achievement consists of a spatial understanding and, ultimately, a heightened awareness of the natural environment. On the other hand, active learning and observation/sketching/analysis fieldwork result in the ability to differentiate field observations and subsequent interpretation, and to differentiate objective facts and scientific analysis.

Lastly, group discussions and team work strengthen students' communicative competencies and ensures they are actively involved in the learning process. Although not previously mentioned, one factor cannot be underestimated: fun. In the end, students should come away with an active interest in geology and motivation to pursue additional study.

References

Bartels, C. (1996), Mittelalterlicher und frühneuzeitlicher Bergbau im Harz und seine Einflüsse auf die Umwelt. *Naturwissenschaften*, 83(11), 483–491.

Boucharlat, R. (2017), Wasserversorgung und Gärten im Iran, in B. Helwig (eds), *Iran – frühe Kulturen zwischen Wasser und Wüste* (Bonn: Hirmer Verlag), 228–236.

Brocco, A. & Frapolli, F. (2011), Open source in higher education: Case study computer science at the University of Fribourg. Diploma of Advanced Studies, Centre de Didactique Universitaire, University of Fribourg, Switzerland.

Cohn, T. C., Swanson, E., Whiteman Runs Him, G., Hugs, D., Stevens, L. & Flamm, D. (2014), Placing ourselves on a digital Earth: sense of place geoscience education in Crow Country. *Journal of Geoscience Education*, 62(2), 203–216.

Craven, B., Lloyd, G., Gordon, C., Houghton, J. & Morgan, D. (2017), *Fieldwork Skills in Virtual Worlds*. In EGU General Assembly Conference Abstracts (Vol. 19, p. 16432).

Crouch, D. P. (2003), *Geology and Settlement: Greco-Roman Patterns*. Oxford: Oxford University Press.

Giguet-Covex, C., Pansu, J., Arnaud, F., Rey, P. J., Griggo, C., Gielly, L., Domaizon, I., Coissac, E., David, F., Choler, P. & Poulenard, J. (2014), Long livestock farming history and human landscape shaping revealed by lake sediment DNA. *Nature Communications*, 5, 3211.

Gruenewald, D. A. (2003), Foundations of place: A multidisciplinary framework for place-conscious education. *American Educational Research Journal*, 40(3), 619–654.

Haworth, R. J. (2003), The shaping of Sydney by its urban geology. *Quaternary International*, 103(1), 41–55.

Kareiva, P., Watts, S., McDonald, R. & Boucher, T. (2007), Domesticated nature: shaping landscapes and ecosystems for human welfare. *Science*, 316(5833), 1866–1869.

Macdonald, R. H. & Korinek, L. (1995), Cooperative-learning activities in large entry-level geology courses. *Journal of Geological Education*, 43(4), 341–345.

Moll, H. F. (2017), Painting as data: a new way of analyzing the landscape. Landscape Architecture Undergraduate Honors Thesis, 8. University of Aransas, Fayetteville.

Peyron, M. (1976), Habitat rural et vie montagnarde dans le Haut Atlas de Midelt (Maroc). *Revue de géographie alpine*, 64(3), 327–363.

Reynolds, S. J. & Peacock, S. M. (1998), Slide observations: Promoting active learning, landscape appreciation, and critical thinking in introductory geology courses. *Journal of Geoscience Education*, 46(5), 421–426.

Sanderson, E. W., Jaiteh, M., Levy, M. A., Redford, K. H., Wannebo, A. V. & Woolmer, G. (2002), The human footprint and the last of the wild. *BioScience*, 52(10), 891–904.

Silberman, M. (1996), *Active Learning: 101 Strategies To Teach Any Subject*. Des Moines: Prentice-Hall.

Thomasius, H. (1994), The influence of mining on woods and forestry in the Saxon Erzgebirge up to the beginning of the 19th century. *GeoJournal*, 32(2), 103–125.

Woolmer, G., Trombulak, S. C., Ray, J. C., Doran, P. J., Anderson, M. G., Baldwin, R. F. & Sanderson, E. W. (2008), Rescaling the human footprint: a tool for conservation planning at an ecoregional scale. *Landscape and Urban Planning*, 87(1), 42–53.

Wurst, K., Postner, L. & Jackson, S. (2014), Teaching open source (software). In *Proceedings of the 45th ACM Technical Symposium on Computer Science Education* (pp. 734–734). New York: ACM.

5

Teaching (landscape) ecology

Wenche Elisabet Dramstad and Mari Sundli Tveit

Sustainable development as a backdrop for teaching ecology

The future that awaits our current students will be different from everything we have seen so far. It is well documented and widely acknowledged that our one and only planet is facing a large number of severe challenges. The effects of a changing climate are not yet fully understood, but it will affect species distributions, food production, droughts and flooding and the outbreaks of pests and parasites (FAO 2011; IPCC 2014). Pollution is damaging marine environments, and our oceans are expected to have more plastic than fish by 2050 (WEF 2016). We are not succeeding in meeting our goals regarding halting biodiversity loss (UN 2017). The Food and Agriculture Organisation (FAO) of the United Nations (UN) report (2011) highlights the need for increased food production to feed the growing population, while the state of our land resources is declining due to erosion, pollution, salinisation and urban sprawl (FAO 2011). The 4th industrial revolution, with exponential technological development, digitalisation and automation, is creating a pace of change affecting the labour market as well as the skills required to succeed in it. In general, the extent and diversity of the anticipated developments makes it difficult to see any future career not being affected one way or the other. This must influence our current teaching – and the teaching of ecology is no exception.

The challenges are proof of the fact that we have not yet succeeded in making our development and use of the planet's resources sustainable. The concept of sustainable development became world-famous following the work of the UN World Commission on Environment and Development and is described in the report entitled *Our Common Future*. This report, published already in 1987 (WCED 1987), defined sustainable development as: ' . . . development that meets the needs of the present without compromising the ability of future generations to meet their own needs'. Since it was first introduced, the idea and concept of sustainability has been repeatedly criticized for being difficult to define, and accordingly a wide range of definitions has been forwarded (see, e.g., Hansen 1996; White 2013). Recently, however, a set of 17 sustainable development goals (SDGs, see Figure 5.1) has been developed as part of a new sustainability agenda. These goals have been adopted by 193 countries (UN 2015). The most recent progress report (UN 2017) expresses concern that the rate of progress in many areas is far slower than needed to meet the targets by 2030.

Figure 5.1 The sustainable development goals (SDGs)

Ecology: 21st-century skills

To be able to meet the challenges that await us and to reach the SDGs, knowledge of ecology and other scientific fields is, of course, not all that is needed. The National Academy of Sciences (2010) outlined a number of so-called 21st-century skills, including adaptability, complex communication and social skills, non-routine problem-solving skills, self-management and self-development, and systems thinking. Looking at the description of these skills, we see that also, e.g., creativity and the ability and willingness to continuously learn new tasks, technologies and procedures are included (Bybee 2009). The Norwegian White Paper *Quality Culture in Higher Education* also highlights generic future skills that are key to the future labour market, including critical thinking, self-reflection, social and emotional skills, team work and communicative skills, and creative problem solving (Meld. St. 2016). We are convinced that the ability to work across disciplines and to collaborate in teams with people from a range of backgrounds and with a wide range of competences also belong on the list. In our experience, 'outside the box'-thinkers and risk-takers tend not to be easily rewarded in academia, however, and neither is failure commonly understood as a necessary part of the learning process towards success. If we accept that these are skills that will become more important in the future, this too should influence our teaching.

Looking at the teaching of ecology through the knowledge-centred lens, a key question forwarded by Donovan and Bransford (2005) is 'What it is important for the students to know and be able to do?' Similarly, Nordlund (2016) suggests beginning development of any course or lecture by defining the intended learning outcome. Our overall aim in teaching ecology is threefold: 1) make the students aware of the severity of the current situation; 2) provide them with a theoretical foundation contributing to their understanding of what is happening and why; and 3) equip them with a range of skills that we believe will be important when they are to meet these current and future challenges, including the Sustainability Development Goals (SDGs).

In this we are inspired by, e.g., Day et al. (2009), who outline the critical role to be played by ecology in this process, by suggesting that a primary role of ecologists will be to help define how to ensure a sustainable and efficient management and use of both natural and managed ecosystems as a key to maintaining human welfare. Similarly, Bybee already in 1991 (Bybee 1991) cautioned us about the state of the planet and underlined the importance this should have in education. In his paper entitled 'Planet Earth in crisis: how should science educators respond?' Bybee (1991, p. 146) stated:

> Science educators should pause and ask – what does this mean? What should we do? My answer is direct – we should educate students in ways of ameliorating the planetary crisis. The imperative to survive justifies teaching about our place in natural systems and it obligates us to act in ways that ensure sustainable growth.

Teaching for different ways of learning

Landscape ecology bridges disciplines in its very essence, and the bridge connects very different disciplines and approaches to learning and learning outcomes, such as those found in the natural sciences and in landscape architecture and design. This also brings together different kinds of learners, and the realization that there will be some visual learners, some auditory learners, some who are kinaesthetic learners and some who learn through reading and writing in any group of students (Kolb 2014; Tanner and Allen 2004), has also influenced the development of our teaching. This insight has been particularly relevant in our development of the various activities we include in the course (see Table 5.1).

To us, building on what students bring to the classroom involves paying attention to the starting points of the individual students, their backgrounds and cultural values, as well as their abilities. This is to apply what Donovan and Bransford (2005) describe as a learner-centred approach or the student-centred approach as described, e.g., by Nordlund (2016). One of our objectives is to make the students see that ecology is 'everywhere and in everything'. Our aim then is to make them think landscape ecology in their everyday interaction with landscapes and nature; that they think about barrier effects when they cross the motorway or that they think about patches and corridors when they walk to the train or sit in a plane looking down at the landscape. In this context, we also want the students to integrate knowledge from our teaching in their other courses, and to bring themes that are relevant to them and their studies back to the landscape ecology course. When we succeed in this endeavour, our next objective is to have the students share their observations, insights, experiences and points of view. Many of the students bring interesting and relevant experiences, e.g., based on previous work or personal experience. One such discussion, where there commonly are very different opinions, is whether nature in urban settings, or designed nature, is 'true nature' and can be compared with wilderness.

Further, in applying a learner-centred approach, we believe there is a need to better incorporate teaching that considers these different ways of learning (see, e.g., Kolb 2014). As forwarded by Donovan and Bransford (2005), it is important to help students assess the types of strategies they are using to learn and solve problems. Similarly, Nordlund (2016) forwards the importance of variation to be able to reach more students. We believe some of the available tools can help students understand more about their own learning, and that this knowledge can help them find their best learning strategy. In doing this, we have found tools such as the VARK assessment useful (Kolb 2014; Tanner and Allen 2004).

Our students are a heterogeneous group in ways other than merely regarding learning strategies, however. Some have completed their bachelor's degree and are on their way towards a master's degree, while others are second-year university students. Two study programmes dominate in the group, natural sciences with an emphasis on ecology, and landscape architecture and land-use planning. Adding to the heterogeneity of the group is the fact that a fairly large proportion of the students are international students. They typically have a range of different backgrounds, and while land-use planning, landscape architecture and ecology tend to be well represented, also, e.g., agro-ecology, hydrology and engineering have been included. Having these students share some of their experiences with the rest of the group is something we consider incredibly valuable, and it adds something we could not easily have done ourselves.

Table 5.1 Examples of elements used in our teaching aimed at the different types of learners

Type of learners	Elements of our teaching
Visual	Presentations with a visual emphasis and presentation style, using maps and aerial photos actively in the classroom, encouraging students to take photos and submit these to the class collection
Auditory	Oral presentations, discussions in classroom and on excursions
Kinaesthetic	Assignments in class and outdoors involving drawing, walking, finding cards that match, cake illustrating fragmentation, chocolate quiz awards
Reading and writing	A curriculum book, a library of scientific papers, powerpoints from lectures available to students

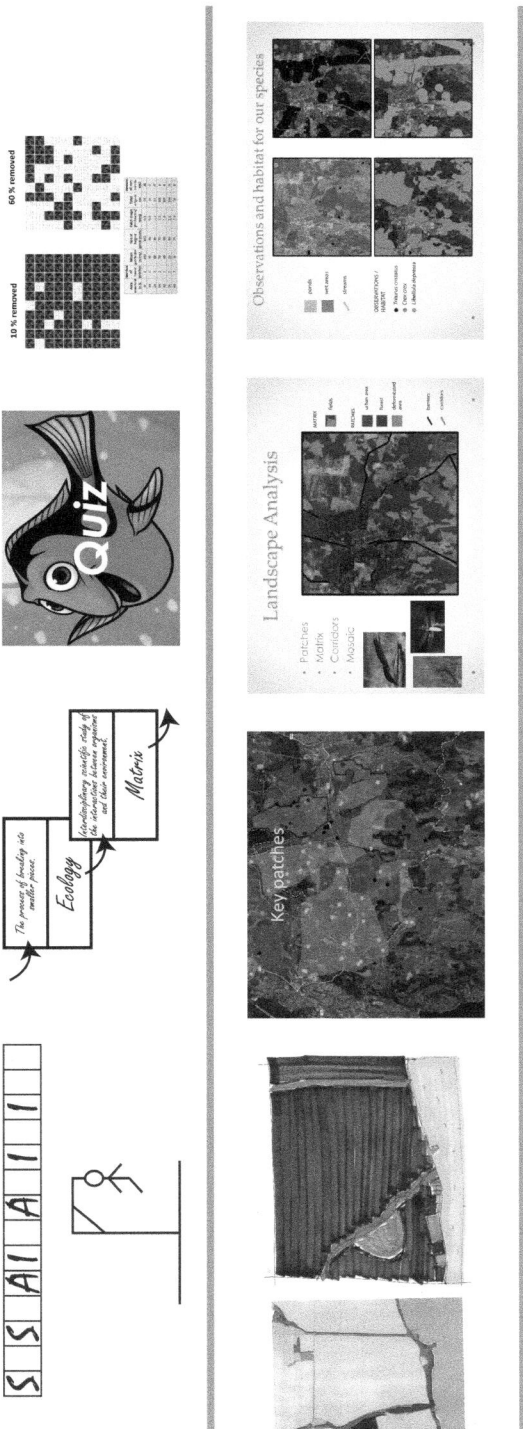

Figure 5.2 An illustration of how we see different assignments in the course representing a progression, from developing a conceptual common ground and learning tools (e.g., hangman game, definitions domino, quizzes), to doing desktop exercises on real landscapes (drawing on maps and aerial photos, adding and losing habitat assignment) to real-world exercises (plan for housing in real-world landscape, assessing urban habitat).

This figure is shown in colour in the plate section of this book.

Teaching a heterogeneous student group – creating a conceptual common ground

The non-uniformity of the student group presents some challenges and some benefits. We focus on establishing a classroom norm that encourages student thinking. Still, there are aspects of cross-disciplinary respect and also cultural aspects of different disciplines that need to be acknowledged and handled with humility. While it may sound self-evident at first, students must acknowledge that other disciplines possess different expertise – and that exploring and taking advantage of this can contribute towards a better end result. This, we find, is something the disciplinary orientation of the universities does not necessarily promote, as many students interact mainly with students and staff within the same discipline. As students progress through their study plans, they risk becoming more and more uniform in their thinking and problem-solving approaches, which stands in contrast to the 21st-century skills outlined above. The reason may be that their faculty and institute tend to be dominated by people with similar backgrounds and similar ways of thinking. This limits the degree to which students' perspectives and thinking is challenged. At the end of their studies then, students may consider their understanding of the world, or their field of expertise, as representing the 'correct one'. Here the heterogeneity of our student group is highly beneficial and something we do our best to explore, mainly through ensuring all groups doing assignments consist of students from differing backgrounds.

Otherwise our approach can be described as being based on the BSCS 5Es Instructional Model (see, e.g., Bybee 2014), which emphasizes Engagement – Exploration – Explanation – Elaboration and finally Evaluation. In our adaptation of this model (Table 5.2), we have also found it useful to make the students respond to a short (anonymous) questionnaire in the very first lecture. Here we ask students to state their expectations of the course and to say a few words about their backgrounds. Our objective is both to make them consider what their own expectations are, as well as to enable us to guide and correct possible misconceptions at an early stage. In addition, upon discussing this, we use the opportunity to clarify our expectations, e.g., that we expect them to get involved, to discuss and to contribute their experiences and point of view in the classroom.

Table 5.2 How we use the BSCS 5Es Instructional Model

Engaging, making it relevant	News stories analyzed in a landscape ecological context (see details below), cartoon presentations and illustrations on relevant themes, news from international organizations like FAO, UN, Convention on Biological Diversity (CBD), etc.
Explore	Assignments requiring exploration of maps and databases on groups of species and selected areas, as well as ideas, theoretical principles and scientific findings
Explanation	Theory introduced through lectures and presentations, scientific literature, reports, access to experts, bringing in guest lecturers with relevant expertise
Elaboration	Assignments working with real-life situations and working across disciplines to solve problems, present lines of thinking and argue for choices made
Evaluation	Term paper on a topic of their own choice and an oral exam

Doing this, we make it explicit that we aim to develop classroom norms that encourage the expression of ideas. We follow the recommendation by Donovan and Bransford (2005) in that mistakes are 'viewed not as revelations of inadequacy, but as helpful contributions in the search for understanding' (p. 20). In addition, students are continuously encouraged to question, we raise questions, and we focus on exploring incorrect or naïve answers to questions with interest. In general, our approach is to place students in a situation where their only 'escape' is to learn – continuously. To achieve this, we use a lot of practical assignments and tasks, where students analyze, discuss and use what they learn.

As can be seen in Table 5.1, there is a course-book, there are 'traditional' lectures focussed on key topics, and seminars arranged to discuss the different themes and topics. In developing these, we have learned from Sæbø (2016) who claims that learning is very much about attention, and attention is drawn towards surprise. Sæbø (2016) further states: 'A good presentation should therefore strive to be somewhat unpredictable, because it is the discrepancy between the students' beliefs and what I actually say, the surprise, which carries the potential of new learning'.

Coming up with these elements of surprise can be challenging, but also fun. As an example from our teaching, we now introduce our students to species interactions through the characters of Dory, Nemo, Lady and the Tramp and 'colleagues' from various cartoons. We do believe this has made some rather theoretical ecological concepts accessible to a larger group of students in a much more memorable way, compared with a mere reading about them or listening to us lecturing about them.

In fact, we have found that using cartoon drawings can be a good help to get a discussion going, in addition to being something that helps students remember the topic. For example, the spotted owl (*Strix occidentalis*) is a classic landscape ecology example of an interior type of species, sensitive to edge effects. These are important topics in teaching landscape ecology. Logging of timber in its main habitat pushed the spotted owl towards extinction. This caused a rather heated debate between the loggers and timber industry on the one hand and conservationists on the other, which was also communicated in a number of ways (see Figure 5.3a). In our experience, using such cartoon representations of a real-world problem is a good way to engage students in the discussions. Another example we find useful to initiate a classroom discussion can be seen in Figure 5.3b.

Figure 5.3a and b   (continued)

(continued)

Figure 5.3a and b   Examples of cartoons used to invite the students to contribute their thinking and points of view into the discussion. (Published with permission; ©Adrian Raeside and ©Justin Bilicki.)

Engaging through relevance

As outlined by, e.g., Bybee (2009), making the subject relevant is a good way of generating interest. It is beneficial when students see a need or a reason – and see relevant uses of the knowledge. This helps them make sense of what they are learning (Bybee 2009). Similarly, Chamany et al. (2008) argue for bringing social context into the teaching of biology and to 'teach students to make connections between what they learn in the classroom and what they see in everyday life' (p. 267).

Our aim is for our students to see the landscape, their future tasks and work with a 'landscape ecological filter'. We thus make an effort to continuously bring in relevant happenings and stories from the news and discuss them in class, with a landscape ecological approach to the issue, as is also recommended by Nordlund (2016). Some recent examples are given in Table 5.3.

To start the development of a 'common language', and a conceptual common ground, we begin through introducing the students to the concept of sustainability in the very first

Table 5.3 Examples of how current news articles can be brought into the teaching and seen through a 'landscape ecological lens'. (The location where the news article can be explored and the date it was published in brackets.)

In the news	Landscape ecological topic(s) involved and discussed in class
Lice problems in Norwegian salmon farms (*Barentswatch* July 2016), geese problems on airport (*New York Times* January 2009)	Salmon farms representing a resource patch (Forman and Godron 1986) to lice, and airports representing resource and habitat patches to geese. Similar discussion is relevant, e.g., to pests in agricultural crops.
Reindeer avoid crossing underneath powerlines (The *Guardian* March 2014)	Barrier effects and the difference between our perceptions of structural versus functional connectivity (Forman 1991). Isolation.

Moose and deer being killed in traffic accidents on roads and railways (*Yle News* September 2017, The *Guardian* November 2017)	Roads and railways as barriers and constructions fragmenting habitat (Forman et al. 2003; EEA 2011) while promoting connectivity for us.
Arctic foxes on Svalbard found to be related to arctic foxes on the mainland (Norwegian Polar Institute undated/ Forskning.no September 2012), fragmentation of panda habitat (*New York Times* September 2017)	Fragmentation (Collinge 2009). Temporal connectivity (e.g., due to ice) and proof of functional connectivity. Metapopulation theory (Hanski 1991).
Costly combat of invasive species (CBS News April 2017, *Burlington Free Press* August 2016)	Invasive species. Immigration and competition (Krebs 2009).
New male orangutan born in zoo (*Daily Mail* December 2017)	Zoos working to prevent inbreeding (AZA SAFE 2016). Fragmented populations. Metapopulation theory.

lecture. This is done in a way they appear to find both fun and engaging – as we play a game of 'hangman' on the whiteboard. The student being the first to solve the puzzle receives a piece of chocolate from our 'prize box', as inspired by Hodges (2016). We then continue by discussing the concept in class, focussing in particular on how sustainability relates to management of land and landscapes. Further, we bring up numbers, figures and illustrations (e.g., from the Consultative Group for International Agricultural Research (CGIAR), FAO and the UN) demonstrating that we are not meeting these ambitions yet. Our end point being that they, the students of today, do need to meet these goals.

Adding tools to the tool-box – desktop exercises

In practice, when teaching landscape ecology, we focus on demonstrating a variety of tools, and we especially encourage understanding of skills and tools that are typically mastered and dominating in other fields to promote a cross-disciplinary understanding. We introduce students to the scientific literature, if they are not already experienced users. We invest a lot of time in a wide range of small and medium projects and activities throughout the course period. In doing these, we promote and require cross-disciplinary teams working together. The output is to be presented as teamwork, using different techniques; poster and Powerpoint presentations, a term paper, a transect map, as well as written submissions. We also bring multiple practical tools into the learning process, e.g., Geographic Information Systems (GIS), Fragstats™, literature referencing software, as well as manual analyses of maps and aerial photos.

The different tasks and assignments represent a progression through the course, along with the theoretical progression in the lectures and literature presented. As illustrated in Figure 5.2, we begin with simple tasks in the classroom, such as the game of hangman, a definitions domino game, reading and drawing maps, analysing maps and aerial photos, and doing a grid-based fragmentation exercise.

At the next 'level' we focus on real landscapes, doing assignments where students work with a few species and a real-world area. The assignments are still desktop based and theoretical, however, in that no practical (e.g., related to land ownership or economy) aspects are brought into the exercise and 'everything is allowed'. The task involves using available online tools on species occurrences (e.g., Global Biodiversity Information Facility) and selecting three species to focus on. The species have to be different in terms of habitat requirements, and can be, e.g., one plant, one bird and one insect species. An additional requirement is that at least one should be on the list of threatened species. Having gathered the data and information necessary, they need to find a landscape where these three species co-exist. Their challenge, then, is twofold. In this landscape they have to a) remove 20% of the habitat and then b) add 20% of habitat. In doing this they are to think about landscape ecological concepts, such as connectivity, barriers, edge effects, etc. In addition they need also to consider other interests, e.g., recreation, visual aspects, cultural heritage, but they do not have to consider practical aspects related, e.g., to land ownership, costs or physical aspects. They present their result as a Powerpoint presentation to the group, where they argue for the choices they have made, and the class discuss this and give feedback (Figure 5.4a and b). There is no wrong or right answer, of course, but interesting discussions and alternative solutions result. In the context of the BSCS 5Es Instructional Model, the assignment brings in elements of exploration and elaboration.

Tackling real-world landscape ecological challenges – in the field exercises

At the final 'level' we go on excursions to be in the landscapes, emphasizing practical challenges related to ecology and planning. The first excursion focusses on the urban landscape and ecology. During this excursion the students again work in groups, and their challenge is to assess how the urban landscape, habitats and species change as we move through the city. In particular, we focus on the patchiness of this landscape, we discuss corridors and barriers, continuity and structure – horizontal as well as vertical.

Galium sterneri
(Bakkemaure)

Crex crex
(corncrake, åkerrikse)

Lissotriton vulgaris
(smooth newt, småsalamander)

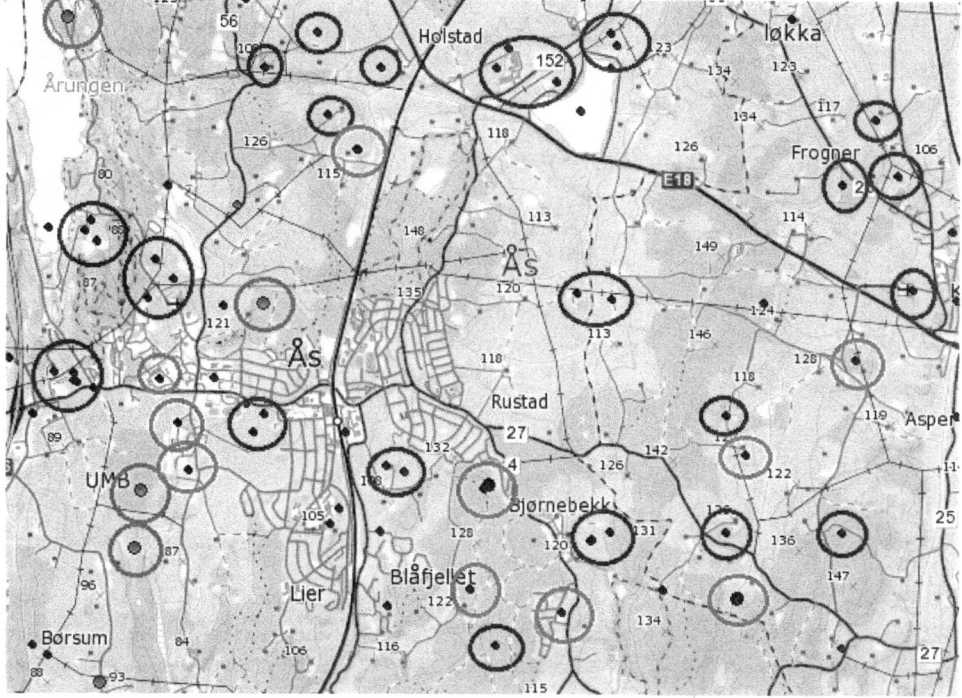

Figure 5.4a and b An example of the output from the assignment where students have selected three different species to focus on, identifying their potential habitat through records of occurrences available in databases, and then adding and removing habitat

The second assignment is analyzing and planning for real-life changes in a real-world landscape. In practice, the students are to plan for new housing. The location of the exercise is an island with a large number and variety of qualities (nature reserves, red-listed species, recreation interests, landscape protection area), on top of a naturally limited total area. A number of practical problems are relevant, e.g., how to manage a red-listed species protected by law, but that has begun to spread rapidly to the extent that it is causing problems in people's gardens. Another practical problem is related to the spread of invasive species and how to prioritize areas for management when the budget is limited and the workload is huge. In the assignment, the students need to bring these aspects and the existing situation and species and interests into consideration. Again, there is no wrong or right solution, but the students have to document the foundation and arguments underlying their choices. The end result is presented as a poster to the class, and groups are required to comment on solutions presented by other groups.

The final assignments can thus be described as problem-based learning (PBL) (e.g., Genareo and Lyons 2015). Our aim is that this will help prepare our students to think and act like professionals in their future careers. This is because we agree with Nordlund (2016), who states: 'If we want our students to possess skills that you need as a professional ecologist, for example collaboration, interdisciplinary thinking and strong communications skills, they need to practice those skills' (p. 180). We do not expect our students to become professional ecologists, however, rather we strongly promote collaboration across disciplinary boundaries. Still, we do think the skills Nordlund (2016) mentions are necessary in many careers in the 21st century. We further consider it important for professionals in many different fields to be able to communicate, ask the important questions and understand the answers. We believe this may bring us one step closer to a more sustainable development.

Conclusion

Our approach to teaching landscape ecology is based on a strong conviction that we need to revitalize teaching. Our current students have a large number of wicked problems and challenges ahead of them, and to meet these they will need a range of skills. Their ability to work in teams, to work across-disciplines, to be creative, think out of the box and look for new solutions will be important in achieving, e.g., the sustainable development goals.

We believe landscape ecology has a lot to offer in this endeavour. Being cross-disciplinary by origin, it is well equipped to bridge disciplines. The emphasis on landscapes and spatial patterns further helps. After all, the landscape is where many of the challenges are seen and have to be met. We thus strive to teach landscape ecology in a way that makes it accessible and available to students from a range of different backgrounds and disciplines, and to different types of learners. We focus on learning throughout the course rather than going for an exam period peak. We aim for this through using assignments and projects and in-class exercises, using differing tools and requiring different output formats. We require cross-disciplinary teamwork and aim to introduce students to the typical real-world prioritizations and complications they are likely to meet in their future careers.

Overall we aim to prepare our students to meet the challenges ahead. In our perspective that requires new ways of thinking and teaching – also about something as deep-rooted as ecology.

References

AZA SAFE (2016), [website], www.speakcdn.com/assets/2332/aza_arcshighlights_2016_final_web.pdf.

Barentswatch (2016), 'Salmon lice' [website], www.barentswatch.no/en/articles/Salmon-lice, accessed 4 June 2018.

Burlington Free Press (2016), 'Vermont's invasive species battle costs thousands' [website], www.burlingtonfreepress.com/story/news/2016/08/01/invasive-species-plants-vermont/86872240, accessed 4 June 2018.

Bybee, R.W. (1991), 'Planet Earth in crisis: how should science educators respond?' *The American Biology Teacher*, 53/3: 146–153.

Bybee, R.W. (2009), *The BSCS 5E instructional model and 21st century skills*. (Colorado Springs, CO: BSCS).

Bybee, R.W. (2014), 'The BSCS 5E instructional model: Personal reflections and contemporary implications'. *Science and Children*, 51 (April/May): 10–13.

CBS News (2017), '"Weird and wired" tech helps battle invasive species' [website], www.cbsnews.com/news/robots-tasers-technology-join-battle-against-invasive-species, accessed 4 June 2018.

Chamany, K., Allen, D. and Tanner, K. (2008), 'Making biology learning relevant to students: Integrating people, history, and context into college biology teaching'. *Cell Biology Education*, 7 (Fall): 267–278.

Collinge, S.K. (2009), *The ecology of fragmented landscapes*. (Baltimore, MD: The Johns Hopkins University Press).

Daily Mail (2017), 'Watch baby Sumatran orangutan's precious first moments as it clings to its mother at Chester Zoo and the birth is hailed as a boost for endangered species breeding programme' [website], www.dailymail.co.uk/news/article-5199121/Sumatran-orangutan-born-Chester-Zoo.html, accessed 4 June 2018.

Day Jr, J.W., Hall, C.A., Yáñez-Arancibia, A., Pimentel, D., Martí, C.I. and Mitsch, W.J. (2009), 'Ecology in times of scarcity'. *BioScience*, 59(4): 321–331.

Donovan, S. and J.D. Bransford (2005), *How students learn history, science, and mathematics in the classroom.* (Washington, DC: National Academy Press).

EEA (2011), *Landscape fragmentation in Europe.* (Copenhagen: European Environment Agency), 87.

FAO (2011), *The state of the world's land and water resources for food and agriculture (SOLAW) – Managing systems at risk.* (Rome: Food and Agriculture Organization of the United Nations and London: Earthscan).

Forman, R.T.T. (1991), 'Landscape corridors: from theoretical foundations to public policy'. *Nature Conservation*, 2: 71–84.

Forman, R.T.T. and Godron, M. (1986), *Landscape ecology.* (New York: John Wiley & Sons Inc.).

Forman, R.T.T., D. Sperling, J.A. Bissonette, A.P. Clevenger, C.D. Cutsall, V.H. Dale, L. Fahrig, R. France, C.R. Goldman, K. Heanue, J.A. Jones, F.J. Swanson, T. Turrentine and Winter T.C. (2003), *Road ecology. Science and solutions.* (Washington, DC: Island Press).

Forskning.no (2012), 'Fjellreven rasket over isen' [website], https://forskning.no/dna-rovdyr-polarforskning-zoologi/2012/09/fjellreven-rasket-over-isen, accessed 4 June 2018.

Genareo, V.R. and Lyons R. (2015), 'Problem-based learning: six steps to design, implement, and assess'. *Faculty Focus*, 30 November. Retrieved 18 February 2016.

Hansen, J.W. (1996), 'Is agricultural sustainability a useful concept?' *Agricultural Systems*, 50(2): 117–173.

Hanski, I. (1991), 'Single-species metapopulation dynamics: concepts, models and observations'. *Biological Journal of the Linnean Society*, 42: 17–38.

Hodges, K.E. (2016), 'Enhancing student engagement and learning via the optional Biodiversity Challenge'. *Global Ecology and Conservation*, 5: 100–107.

IPCC (2014), 'Summary for policymakers'. In: Field, C.B., V.R. Barros, D.J. Dokken, K.J. Mach, M.D. Mastrandrea, T.E. Bilir, M. Chatterjee, K.L. Ebi, Y.O. Estrada, R.C. Genova, B. Girma, E.S. Kissel, A.N. Levy, S. MacCracken, P.R. Mastrandrea and L.L. White (eds) *Climate Change 2014: Impacts, Adaptation, and Vulnerability. Part A: Global and Sectoral Aspects. Contribution of Working Group II to the Fifth Assessment Report of the Intergovernmental Panel on Climate Change*, pp. 1–32. (Cambridge: Cambridge University Press).

Kolb, D.A. (2014), *Experiential learning: Experience as the source of learning and development.* (Upper Saddle River, NJ: FT press).

Krebs, C.J. (2009), *Ecology.* Sixth edition. (San Francisco, CA: Pearson Education).

Meld. St. (2016), *Kultur for kvalitet i høyere utdanning.* [Quality Culture in Higher Education] (Meld. St. 16, 2016–2017) Government White Paper [website] www.regjeringen.no/en/dokumenter/meld.-st.-16-20162017/id2536007.

National Academy of Sciences (2010), *Exploring the intersection of science education and 21st century skills: a workshop summary.* National Research Council (US) Board on Science Education. (Washington, DC: National Academies Press).

New York Times (2009), 'Geese pose big risk at airports in region' [website], www.nytimes.com/2009/01/17/nyregion/17birds.html, accessed 4 June 2018.

New York Times (2017), 'Pandas are no longer endangered. but their habitat is in trouble' [website], www.nytimes.com/2017/09/25/science/pandas-habitat-china.html, accessed 4 June 2018.

Nordlund, L.M. (2016), 'Teaching ecology at university—Inspiration for change'. *Global Ecology and Conservation*, 7: 174–182.

Norwegian Polar Institute (n.d.), 'Arctic fox (*Vulpes lagopus*)' [website], www.npolar.no/en/species/arctic-fox.html, accessed 4 June 2018.

Sæbø, S. (2016), 'The statistics of effective learning' [website], http://blogg.nmbu.no/solvesabo/2016/08/the-statistics-of-effective-learning, accessed 4 June 2018.

Tanner, K. and Allen, D. (2004), 'Approaches to biology teaching and learning: learning styles and the problem of instructional selection—engaging all students in science courses'. *Cell Biology Education*, 3(4): 197–201.

The *Guardian* (2014), 'Animals see power lines as glowing, flashing bands, research reveals' [website], www.theguardian.com/environment/2014/mar/12/animals-powerlines-sky-wildlife, accessed 4 June 2018.

The *Guardian* (2017), 'More than 100 reindeer killed by freight trains in Norway "bloodbath"' [website], www.theguardian.com/world/2017/nov/27/reindeer-killed-by-freight-trains-norway, accessed 4 June 2018.

UN (2015), *Global sustainable development report.* Available from: https://sustainabledevelopment.un.org/globalsdreport/2015. (New York, NY: United Nations).

UN (2017), *The sustainable development goals report 2017*. (New York, NY: United Nations).

WAZA (2011), *Towards sustainable population management*. (WAZA Executive Office, IUCN Conservation Centre, Switzerland). Available from www.waza.org.

WCED (1987), *Our common future*. World Commission on Environment and Development. (New York, NY: United Nations).

WEF (2016), *The new plastics economy. Rethinking the future of plastics*. Geneva: World Economic Forum, REF 080116.

White, M.A. (2013), 'Sustainability: I know it when I see it'. *Ecological Economics*, 86: 213–221.

Yle News (2017), 'Moose on the loose: Peak crash season approaches' [website] https://yle.fi/uutiset/osasto/news/moose_on_the_loose_peak_crash_season_approaches/9841536, accessed 4 June 2018.

6

Learning-by-filming

A method to introduce non-LA students to landscape reading

Luca Maria Francesco Fabris and Guido Granello

Preamble

This chapter describes the birth and development of a perceptual–analytical method of representation and analysis. We structured first an elective theoretical course of environmental design followed by an ethnographical unit strictly based on the moving image (video clips). The intent of the second unit was to introduce students with no training in sketching and drawing techniques to an alternative, accessible method of representation and analysis of landscape.

The landscape-reading challenge

During past academic years we have been teaching an elective semester-long theoretical course on environmental design in the third year of the Bachelor in Architecture. After a few lectures, we found out that all the students attending the course were literally hungry for contemporary landscape issues, a subject they never explored during the first two years of the Bachelor course work. In addition to that, most of the students showed little or no familiarity with basic notions such as *sustainability*, *environment*, and *landscape*.

Listening to their voices and their passionate requests and considering their need to investigate a subject they felt so crucial to their understanding of contemporary life, we decided to reformulate the programme of the course by focusing on the territorial scale more than the built one and introducing a more collaborative and participative method to introduce these students to studies regarding landscape.

Taking account of the students' complete lack of theoretical background in landscape architecture, we instructed them to interpret landscape as the 'ultimate infrastructure', giving them an extremely elementary interpretation of the European Landscape Convention (2000) – we used the English version as the Italian one presents very contradictory differences in its translation – which states '"Landscape" means an area as perceived by people, whose character is the result of the action and interaction of natural and/or human factors' (Art. 1, a) and, most important, 'It concerns landscapes that might be considered outstanding as well as every-day or degraded landscapes' (Art. 2).

Authors are indeed convinced that

> When landscape will be recognized as the primary, basic structure, that we all need to preserve. Finding new ways to design good architecture and good planning to improve the environmental conditions not only for the present, but also for the future is the commitment. Thinking like this means localism has the same value of globalism, as the world is a whole and we should live (well and better) taking care of generations to come.

(Fabris, 2009: 36)

Actually, we noticed that students, born from the generation that thought, proposed, accepted, and developed all these meanings, seemed completely unaware of all that their parents have done. Taking this entire heritage for granted, they did not seem to feel the need and urgency to continue supporting and developing this legacy day by day. A lot of work needs to be done in this respect.

To provide students with some background about the topic, we gave them a multidisciplinary literature list and asked them to choose at least six essays for individual studying – under tutored supervision – according to a preferred sub-subject theme. This assignment, utilizing a long list of contemporary scientific texts in which landscape and landscape architecture are presented together with subjects known by the students as architecture, planning, construction technologies and so on, shows them that the multidisciplinary way to process a design solution has to be the preferred one both in the didactic and in the professional approach.

During the course, two thought-provoking essays containing many national and international case studies were introduced with the purpose of illustrating how, in the last 30 years, landscape architecture influenced and revolutionized the way we perceive the environment.

The first essay is *La Speranza Progettuale* (The Project is Hope) by Tomás Maldonado (1970), a tiny book that, even though written 47 years ago, is still full of ideas, questions and notions that help the reader in formulating his/her capacity of interpreting subject matter as timeless as environment and its relationship with society. Even though full of data and references related to the past, nowadays this fascinating book – the first one presenting environmental design as a subject in Italy where Maldonado founded the homonymous Chair at Politecnico di Milano in 1985 – is a guide to explore the relations between human behaviour and nature. And we know the outcome from this relationship is called *landscape*. How can we ignore the statement: 'in the context of a nature in crisis, of a nature with all the symptoms of praecox senility, the society deprives fatally itself of any tension towards the future' (Maldonado, 1970: 143).

The other essay we suggested the students read was *Le Manifeste du Tiers Paysage* by Gilles Clément (2003), a book that in Italy is considered fundamental to interpret the role of everyday landscape and its relation with contemporary architecture and planning. Students also had the chance to meet and learn from Clément, who had been invited to open the School of Architecture academic year with a lecture presenting his visions and his work. His lecture provided a great opportunity for the students to understand directly from the author *what* could be considered landscape and *how* landscape architecture can combine with architecture and planning both at the narrow or the wide scale.

The importance of maintaining attention to the possibilities of working with a multi-scaled approach is another factor we underline in our lectures. The reading of *The Third Landscape Manifest* actually opened the minds of our students. In addition, most of what is expressed in the European Landscape Convention's text became clear to them thanks to Clément's writing. The ability of the French writer to talk about environment, ecology, humanity and nature through different dimensions and approaches, not least the political one, was the key to help students read their 'ordinary' landscape as a 'unique' landscape – a landscape to protect, preserve, understand and improve.

Lectures supporting all these theoretical aspects presented case studies where the connections between landscape architecture, architecture and planning are very tight and linked to many other practical issues in terms of sustainability, resilience, and recycling. Throughout the course we considered the strategies, the projects, and the plans developed during the IBA Emscher Park (1989–1999), the IBA Fürst Pückler Land – See (2000–2010), the IBA Stadtumbau (2000–2010), and the IBA Hamburg (2007–2013) in Germany, as well as the Boscoincittà (Wood-into-Town, 1975–ongoing) reforestation park in Milan and the recovery and participated project for the new Parco delle Cave in Brescia (Park of Pits, 2014–ongoing).

To translate these notions into practice, we assigned students to shoot short video clips, asking them to keep in mind what they had learned from the given literature and lectures. In this process, authors saw the simplest translation from 'observing and analyzing' to 'describing and solving' for students, with an urgency to put into practice what they had learned in such a short time.

We asked students to decode their own living landscape (at home, at university, wherever . . .) through the lens of their smartphones, taking account of the new notions and interpretations learned. After having watched the first results, we asked the students to organize the videos into three-minute screenplays. Written work included a combination of phrases from assigned readings and/or keywords that, together with the moving images, would express a concept informed by the students' own ideas. As a result, we obtained brief documentary films showing various landscapes, their troubles, their beauty, and their neglected potentials. The intent of each film is readily comprehensible.

Creating a landscape-reading method

During the weeks of the Athens Programme, an exchange programme led by Paris Tech University participated in by Politecnico di Milano together with 14 other European universities, we had the opportunity to try a new multimedia experiment and to define a kind of ethnographic-based methodology for reading landscape. We think these experimental didactic experiences could well demonstrate the utility of a simplified – but not simplistic – way to get in contact with landscape architecture theories for students not involved at all in landscape architecture.

This weekly course titled *Milan, the unexpected green-growing city* (Director: prof. Luca Maria Francesco Fabris; 30 international students participating in each edition, two editions for academic year – fall and spring sessions) is based on a series of morning lectures held by various experts and teachers and, in the afternoons, dedicated to guided visits of Milanese locations where the students had the opportunity to see the practical results of the notions and the design projects explained during the morning classes. Methodologically, the morning lectures brought students the notions and the afternoon walks provided examples of what was presented in class, always united with presentations and comments by scholars or professionals directly involved in the project or in the visited location. First-day lectures touched on topics such as the history of gardens and parks in Italy and Milan, so in the afternoon the visit included the Brera Botanical Gardens, the Via Marina and the Public Gardens (the second public garden in Europe for foundation). On the following day students were introduced to the horticultural garden system in Milan, visiting parks such as Boscoincittà and Parco Nord (Northern Park) where public and social orchards are present. Then it was the turn of the Milanese contemporary urban and peripheral parks, most of them derived from former industrial areas, so students were guided to visit the Bicocca Quarter Gardens (designed by A. Kipar/LAND for V. Gregotti), the

Rubattino Park (by A. Kipar/LAND), and the newest Parco Tre Torri (Three Towers Park by Gustafson and Porter), Portello Park (by Topotek and Ch. Jenks with A. Kipar/LAND), and Franca Rame Gardens (by F. Giorgetta with Onsite). At the end, students were introduced to the green strategies applied in the new financial Milanese downtown core, with guided visits to the Porta Nuova – Garibaldi Area, which includes the famous Bosco Verticale (Vertical Forest) by Stefano Boeri Associati.

So, taking into account that there was not time enough to implement the students' knowledge about landscape design theory with any sort of literature, even if reduced 'in pills', we opted to underline frequently during the lectures the principal meaning and contents of this subject, actually using them as a mantra during the afternoon visits. Also on this occasion, from the first guided visit, we asked students to read the landscape reality through the lens of their camera, and then to keep their curious eyes open whenever they walked alone through the urban mineralized landscape of Milan. This simplified ethnographic method provided students with the opportunity to formulate an everyday-landscape critique evaluation, to engage with people, and even to plan the basis for a design strategy.

The students, divided into groups of five individuals possibly coming from different universities and careers (mostly from the Engineering disciplines), were invited to develop their report about Milanese landscape with the aid of a simple screenplay. A kind of open canvas helping to give a structure to their critique discourse created by the moving images. To help them in this, we called upon some of their former students, now professionals who, in two brief sessions during the week and one longer session on the last day, introduced students both to video-editing programmes and to layout concepts and notions to improve the final output. This work process allowed students to avoid sketches and drawings completely and pushed students to express themselves through sets of selected video frames from the original screenplay.

Results obtained

The proposed approach, now actually a method, is very practical, even if based on soft analyses, and can be considered as a new kind of learning-by-filming, which directly facilitates relating to the digital-born generation and permits students to merge their knowledge with visions and realism.

It is good to clarify that this process depends largely on the format chosen. Its characteristics are: time (one week); participants (students coming from different scientific backgrounds and not involved in landscape architecture); a tight rhythm (theory lectures in the mornings, long guided walks in the afternoons). This makes the final results comparable with Kevin Lynch's mental maps upgraded to contemporaneity. All sensations and perceptions are recorded directly in the video clips, demonstrating that this technique absorbs both settled methodologies such as the frontal lessons and the novel 'going wandering' experience (Careri, 2006) joining everything in a moving polaroid. The final result is not taken for granted and a posterior critical analysis allows us to verify an equivalence between the theoretical inputs provided and the understanding coming from the direct experiences provided during the guided tours.

This method, even if deliberately simplified, has the same basis as the most innovative landscape mapping projects where the integration of drone-taken images and the use of instant-device photographs is applied for collective participation and collective resolution of conflicts, as documented by the researches of the MIT Senseable City Lab coordinated by Carlo Ratti or the UCL-Bartlett's project 'cLIMA sin Riesgo'.

We have collected here a series of collages done with the frames taken from the video clips realized by the students attending the Athens Programme week in different sessions (Figures 6.1 to 6.9). Through the images and the brief text explaining the plot of each brief movie, we

Figure 6.1 *Milan, the green growing city*, by Ceren Karaman, Iwona Pawlak, Juraj Petrik, Elgar Slooten, Veronika Vanclovà (Athens, Fall 2016 session), video-clip frames collage by G. Granello

Figure 6.2 *Stories of a Nature* by Tamara Kuen, Carolyn Meyer, Thu Nguyen, Urszula Prokop, Lorraine Sijbrandij (Athens, Fall 2016 session), video-clip frames collage by G. Granello

Figure 6.3   Problems caused by water in cities by Renan Luis De Souza Silva, Gabriel Delage e Silva, Diane Desjardins, Aurore Mattio, Piotr Zalewski (Athens, Fall 2016 session), video-clip frames collage by G. Granello

Figure 6.4 Between green and grey, another viewpoint on Milan by Johanna Falkensteiner, Philipp Kollo, Malgorzata Lenart, Fabian Leonhardsberger, Aleksandra Machol (Athens, Fall 2015 session), video-clip frames collage by G. Granello

Figure 6.5 *Green Police* by Mathilde Pellizzari, Zakaria Coppeaux, Florian Cailleteau, Victor Guan (Athens, Spring 2016 session), video-clip frames collage by G. Granello

Figure 6.6 What if you were a duck? by Zhuqing Cui, Théo Nguyen, Wouter Parys, Barbara Plonczynska, Guanchu Wang (Athens, Spring 2016 session), video-clip frames collage by G. Granello

Figure 6.7 *Milano, the green city* by Polina Veleva, Harshal Patil, Laura Millan Mayoral (Athens, Spring 2016 session), video-clip frames collage by G. Granello

Figure 6.8 Intimacy of Green Spaces, learning from children by Alexis Geisler, Aydin Nazlican, Valerio Rigamonti, Nilufer Sueda Uludag, Xiaomin Li, Sheng Zhou (Athens, Spring 2017 session), video-clip frames collage by G. Granello

Figure 6.9 *What a greenful world* by André Dibé, Barbara Doroszuk, Camille Fuzier, Johannes Jell, Jakob Scheithe, Vincent Vinel (Athens, Spring 2017 session), video-clip frames collage by G. Granello

believe it's possible to evaluate how much and what students have learned about landscape as a text to be analyzed and interpreted.

Milan, the green growing city. Plot: students decided to show the journey in Milan by filming places and their details, focussing on the workers and machinery working in Milan. The meaning of their reflection is about who builds the city. Students: Ceren Karaman, Iwona Pawlak, Juraj Petrìk, Elgar Slooten, Veronika Vanclovà (Athens, Fall 2016).

Stories of a Nature. Plot: the leitmotiv of the video is a hand-made drawing showing an urban skyline becoming more and more complex. The movie is organized in four chapters. The Edge: the modern and old city of Milan. In Between: details of trees and grass showing the balance between leaves, branches and the filtered sunlight. Motion: images of daily life in parks and green areas. Texture: focussing on the materiality of open space and the differences between the diverse patterns. Students: Tamara Kuen, Carolyn Meyer, Thu Nguyen, Urszula Prokop, Lorraine Sijbrandij (Athens, Fall 2016 session).

Problems caused by water in cities. Plot: students decided to focus on the important relationship between water and urban fabric in the Milanese landscape. After some examples of flooding in cities like Bangkok, Venice, Paris and Calgary, they searched for examples and solutions to manage water in Milan. But water interacts with the daily life in Milan with beautiful fountains, canals, gardens, and reservoir. Without water, there is no fun and Milan has chosen to grow green thanks to the collaboration between architects and engineers. Students: Renan Luis De Souza Silva, Gabriel Delage e Silva, Diane Desjardins, Aurore Mattio, Piotr Zalewski (Athens, Fall 2016 session).

Between green and grey, another viewpoint on Milan. Plot: using the stop-motion method, students used a plastic glass as an actor, having the opportunity to film from the ground level, describing the Milan landscape from a different point of view. This kind of shooting emphasizes proportions, dimensions of the open spaces and puts the attention on the texture and materials that make the city. Students: Johanna Falkensteiner, Philipp Kollo, Malgorzata Lenart, Fabian Leonhardsberger, Aleksandra Machol (Athens, Fall 2015 session).

Green Police. Plot: in the video clip students perform as actors according to a screenplay about the 'Green Police' that serve to preserve and increase nature and green spaces in Milan. The different students/actors' backgrounds offer different ways to see the same reality, trying to answer questions about the green landscape in Milan. At the end of their journey, actors/students, with minds full of new green knowledge, join the Green Police to protect not only the Lombardy capital greenery, but also the greenery present in their home-place. Students: Mathilde Pellizzari, Zakaria Coppeaux, Florian Cailleteau, Victor Guan (Athens, Spring 2016 session).

What if you were a duck? Plot: the video clip begins by showing on a map the different places visited and the distances from Piazza Duomo (Milan centre) and the time to reach them by walking or by public transport. But what if you were a duck or . . . a stranger lost in a city? How could you find those green areas? Students ask people on the road where to find parks or green areas, testing the residents' knowledge about them. Students: Zhuqing Cui, Théo Nguyen, Wouter Parys, Barbara Plonczynska, Guanchu Wang (Athens, Spring 2016 session).

Milano, the green city. Plot: the video clip shows in a very fresh way the different feelings that a journey can provoke. Using images and music, students transmit a sense of frenzy and chaos about the built city against the relaxation related to parks and nature. At the end, the students chose some ambient music to underline a playing moment between a father and his sons in the new Porta Nuova district of Milan. Water in public open spaces gives life a sense of playground. Students: Polina Veleva, Harshal Patil, Laura Millan Mayoral (Athens, Spring 2016 session).

Intimacy of Green Spaces, learning from children. Plot: this video clip assembles short videos shot during the students' field trips to show how people can interact with the green spaces. Using the contrast between children's freedom and the over-connected life of the young people, the first part shows individuals addicted to their digital devices. Selfies, photos, social network messaging, and ear-buds isolate and break individuals away from reality. A girl leaves her smartphone on a bench and watches the reality with new eyes. Looking at the landscape as a child, she understands a new way to approach nature. Milan's typical fountains with their typical tap become a water game, as the seesaw to play with or a tree to climb, to touch. A hill becomes a natural slide, a meadow a kind of open-air gym. Students: Alexis Geisler, Aydin Nazlican, Valerio Rigamonti, Nilufer Sueda Uludag, Xiaomin Li, Sheng Zhou (Athens, Spring 2017 session).

What a greenful word. Plot: students wanted people to reflect freely on landscape and green spaces with this video clip, showing us a series of ideas and associations. The movie describes the values of parks, green spaces and nature through the faces of their users and their descriptions in a kind of spontaneous survey that shows how words can depict a spatial concept or indicate a meaning, describing actions taking place in the urban landscape. Students: André Dibé, Barbara Doroszuk, Camille Fuzier, Johannes Jell, Jakob Scheithe, Vincent Vinel (Athens, Spring 2017 session).

Conclusions

The brief movies realized by students not only present a peculiar way to see Milan and its urban landscape but, by using a fresh language that mixes pictures and moving pictures with short texts, they are able to document a scenery and also to comment on it with a proper critique.

Asking the Athens Programme students, after the final public presentation of their short movies, what they would take home from this experience, we were enthusiastic in listening to them reporting that they had comprehended that landscape – no matter what it is like – is part of their life and should be supervised, protected and improved by themselves without waiting for others to do it for them and that landscape architecture reflects very complex and multifaceted issues that could be easily interfaced with the subjects they are now studying.

Overall, the results obtained are a perfect overview of a real contemporary urban everyday landscape, as taken through the lens of youth that has yet to confront the habit of compromise. For sure this learning-by-filming method has proven that the less the 'film-makers' are involved in the landscape architecture theories, the more they can absorb its contents without prejudices, submitting very unexpected contributions to interpret the urban open spaces.

We are convinced that this simplified method may be used by urban and landscape professional designers as a tool for the participated-design approach, for example in charrettes, permitting citizens, having no education in landscape architecture subjects and design methods, to illustrate their perception of the everyday landscape and to interact with designers and stakeholders expressing their critiques and proposals to the design debate.

References

Careri, F. (2006), *Walkscapes. Camminare come pratica estetica* (Torino: Einaudi).

Clément, G. (2003), *Le Manifeste du Tiers Paysage* (Montreuil: Editions Sujet/Objet).

Council of Europe (2000), *European Landscape Convention.*

Fabris, L.M.F. (2009), *Tecnonatura* (Santarcangelo di Romagna: Maggioli Editore).

Fabris, L.M.F. and Li, X. (2016), 'Milan, a history in recreational spaces', *Środowisko Mieszkaniowe/Housing Environment*, vol. 17 "Recreation": 107/112.

Lynch, K. (1960), *The Image of the City* (Cambridge, MA: MIT Press).

Maldonado, T. (1970), *La Speranza Progettuale* (Torino: Einaudi).

Waldheim, C. ed. (2006), *The Landscape Urbanism Reader* (New York, NY: Princeton University Press).

7

Landscape is more than the sum of its parts

Teaching an understanding of landscape complexity

Shelley Egoz

Introduction

Studying landscape as a humanistic multidisciplinary concept has origins in the discipline of cultural geography. In the past decades, landscape studies have permeated into several disciplines, such as sociology, history, archaeology, anthropology, cultural studies, and others where the geographical term has expanded to become a framework for analysis and interpretation of scholarly work. As the Swiss scholar Michael Jakob (cited in Deriu and Kamvasinou 2014: 1) said,

> [Landscape is] no longer an exclusive concept defining the aesthetic worldview of a cultural elite, [but] this notion has permeated all aspects of social life as well as academic and art practices, leading to a "landscape babel" that extends from specialist discourses to everyday language.

Nonetheless, it is only the discipline of landscape architecture that explicitly bears the word landscape in its name, and in which landscape is not a choice of scholarly approach to be adopted but the essence of the discipline and profession. As such, understanding the term 'landscape' is imperative to educating landscape architects. The term itself is controversial—it has different meanings and interpretations, and therefore is a complex term that encapsulates both physical material entities and intangible processes.

Landscape architecture is often described as a merger of art and science (see, for example, ASLA 2016) and it requires skills that necessitate a wide spectrum of understandings. Landscape is a complex term: the theoretical grounding for landscape architecture relies on multidisciplinary knowledge that extends beyond the obvious hard sciences, such as geology, soil and plant sciences, and ecology, or the social sciences such as sociology and psychology. A traditional landscape analysis concerns the physicality of a site. Sometimes the socio–economic context is also examined. A particular pedagogic challenge is how to educate students about complexity, which sometimes can be interpreted as having 'fuzzy' edges in a profession that deals with very concrete situations.

The teaching and learning experience

The UK Design Council has recognised that design thinking is the key to a productive economy and good standard of living. The Council encourages across-the-board (not only in design institutions) education for creativity, innovation, flexibility and adaptation in a rapidly changing environment (UK Design Council 2010). The pedagogic discourse about the value of design thinking as a particularly relevant mode of thinking for addressing twenty-first century challenges has filtered into several arenas beyond the traditional design disciplines (Wrigley and Straker 2017; Retna 2016).

For landscape architecture,[1] design thinking extends the traditional focussed methodologies used by engineers and architects. As a future-oriented practice tackling the intricate relationship between natural resources, the built environment, and human beings, the breadth of issues includes 'wicked problems' (Buchanan 1992): the unpredictability of how ecological processes, social systems, and political dynamics will affect landscapes.

In this chapter I present and reflect on my experiences from four years of co-teaching an undergraduate (second year) joint landscape architecture and urban planning twenty-credit semester course[2] in place development (i.e., place making) at the Department of Landscape Architecture and Spatial Planning of the Norwegian University of Life Sciences. An analysis of an urban quarter in Oslo was introduced through the lens of landscape.

My involvement with landscape-architectural studio education in the past twenty years spans several programmes in different countries (in Israel, the USA, New Zealand, Singapore and, since 2013, in Norway). The curricula with which I am familiar bases studio teaching, including site analyses, on a model that matures from simple sites to more complex ones. The first projects the students deal with are small-scale sites with a straightforward brief, developing into more complex landscape sites and problems in each year. This model might begin with the task of designing a garden, or a neighbourhood park, perhaps followed at the next year level with a schoolyard design, or placement of buildings on steep topography, and progressing to projects such as residential and urban developments at a larger scale. There is usually an attempt to integrate the technical skills that have already been taught into each of the levels, i.e., planting design, construction, grading, ecology, visualisation (both hand- and digital-drawing skills) and others. The last year is when students are challenged to synthesise the acquired knowledge and skills, including theory, into a project.

This stepping-stones model assumes that there is a hierarchy that needs to be followed and that understanding and tackling complexity must be gradual. Nonetheless, from my own anecdotal experience, very few students actually grasp the significance of the complexity of landscape problems even upon their graduation. In the past two decades, the paradigm of sustainability has permeated landscape architecture. Sustainability is a rich holistic concept, but, in my experience, it is mainly the technical dimensions that students end up addressing. Thompson (2000) recorded a similar attitude in his study of professional landscape architects' values. He identified a response to sustainability through 'technocratic accommodation': i.e., technical engineering solutions for healthy environments through the design of rain-gardens, green roofs, surface storm-water design and management, wetland construction, native plantings, and more. Notwithstanding the contribution of such physical interventions towards supporting environmental sustainability, the understanding of landscape it presents is a simplification that reduces it to an engineering problem. If complexity is such a profound characteristic of landscape, perhaps we should try introducing lateral thinking from the beginning.

An alternative approach

The idea behind this pedagogic endeavour is to highlight complexity at the early stages of professional education and to offer a type of 'landscape tasting' exercise, where students explore the multifaceted aspects of an urban site at a municipal scale, and represent the analysis in a visual atlas-like document constructed according to various themes. Students spend the first half of the semester collating, analysing and evaluating relevant information as a basis for future visions, and create plans accordingly during the second half of the semester.

The requirement is to process the information and reflect on it, rather than presenting data that could easily be accessed through an internet search.

Exposing the students to the knotty nature of planning and designing human environments sheds light on the challenges of so-called 'informed decisions' and the uncertainties of both ecological and social processes. The pedagogic aim is to encourage and develop unconventional and innovative analytical and visionary thinking. The primary aim was to introduce all this at an early stage, before students formulate a set way of addressing landscape problems. In this way, we would like to instil an understanding that landscape solutions are not mere mechanical answers to a problem—that one cannot offer students one methodology or formula for addressing issues as if answers are always black or white. Such an ambiguous, confusing, and messy process is meant to elicit an awareness of nuances in any situation and to achieve depth of thinking. The presumption was that, by introducing abstract concepts within a planning and design situation at an early stage, one could avoid the need to de-learn a simplistic approach later. The goal was to stimulate students' thinking about some questions that may be trivial to a professional who has already internalised the knowledge, but that are rarely addressed in studio: e.g., what does it mean when a planner applies colours in various areas on a municipality map? Alternatively, when a landscape architect draws a path and some trees on their plan, what are the consequences? What is landscape? How do we interpret the landscape and evaluate an area in a meaningful and relevant way?

Course structure

The course was mandatory for planning and landscape architecture second-year students. Each year between sixty and seventy-five students attended from both programmes. The ratio between the students from these programmes, as well as the gender balance, was usually in favour of a slight majority of women landscape architecture students. The work was done in thirteen to sixteen groups of four or five students. Each group was mixed-gender and included students from both programmes. The ratio between students and teachers was twenty-five to one.

There were three main components to the course.

1   Phase A lasted six weeks and focussed on exploration and analysis. In thematic groups, students compiled an atlas and presented it to the whole class in short condensed presentations.[3] The completed atlas was then made available to the entire class as a collective resource.

2   An intermediate phase lasted two weeks and focussed on reflections and individual work: each student wrote a short essay on their understanding of landscape, reflecting on how they perceived landscape before the atlas exercise and their understanding of landscape after completing the task, watching the presentations, and gaining access to the compilation as a whole. A second task was to identify three possible sites within the municipality that would benefit from an intervention through planning and design, based on the information from the atlas and their developed familiarity with the study case. The teachers then reshuffled students into new groups based on individual choices.

3 Phase B lasted six weeks and the groups had to reach a collective vision for their site, and then create visionary plans developing scenarios according to principles of sustainable design. In this chapter, I do not elaborate on the planning and design stages and the scenario thinking in design processes—these would merit another pedagogic paper. Nonetheless, in the conclusions I will address how the results related to the first phase of exploration.

Group work

Group work was set as a goal for training and experience in teamwork—a real-life scenario that will develop essential skills for future professionals. It was also an opportunity to utilise a variety of skills, talents, and types of knowledge, so that students learn from one another in order to address complex problems. Teamwork is a challenge in itself. An external expert in team management dynamics delivered two half-day training workshops.

Phase A: exploration and atlas creation

In contrast to a traditional landscape site analysis, during phase A students explore different themes relating to the whole area of the municipality, and collectively create an A3-format atlas of various maps that describes and evaluates the qualities of the study site as stated in this brief:

> Each group will focus on a subject using a particular theme as their lens to explore that landscape. The expected result will be a rich detailed graphic description and analysis of the findings, keeping text descriptions to the bare minimum. The atlas is an exercise in creative interpretations and evaluations of information and its presentation. The idea is to gain experience in compiling, analysing, and evaluating useful information that will inspire grounded and visionary thinking at the same time. The themes are not, however, mutually exclusive: the expectation is that there will be overlaps and diverse interpretations of similar issues and processes.

One way to look at this task is to view it as a visual essay of an assigned topic/theme. The various themes are frameworks for interpretations, challenging students to perform a lateral analysis of landscape.

Students are reminded that:

> Change is inherent to landscape and all themes are to be addressed in terms of past (history), present and future(s) including on-going dynamic processes (climate where relevant) and social and economic variables. [. . .] All themes have a tangible landscape expression.

Themes

The array of themes included:

> Physical characteristics and landscape spatial morphology; landscape of movement; the political landscape; the landscape of blue-green infrastructures; the edible landscape; the landscape of industry; the landscape of dwelling; the landscape of retail; the landscape of heritage and culture; the landscape of recreation and leisure; the public and private landscapes; the people and social infrastructures in the landscape. Each group of students was to choose a theme and represent it in 6–8 A3 page sides. Figures 7.1 and 7.2 are a couple of examples of single pages.

Colours of the season - the Vestby farmland

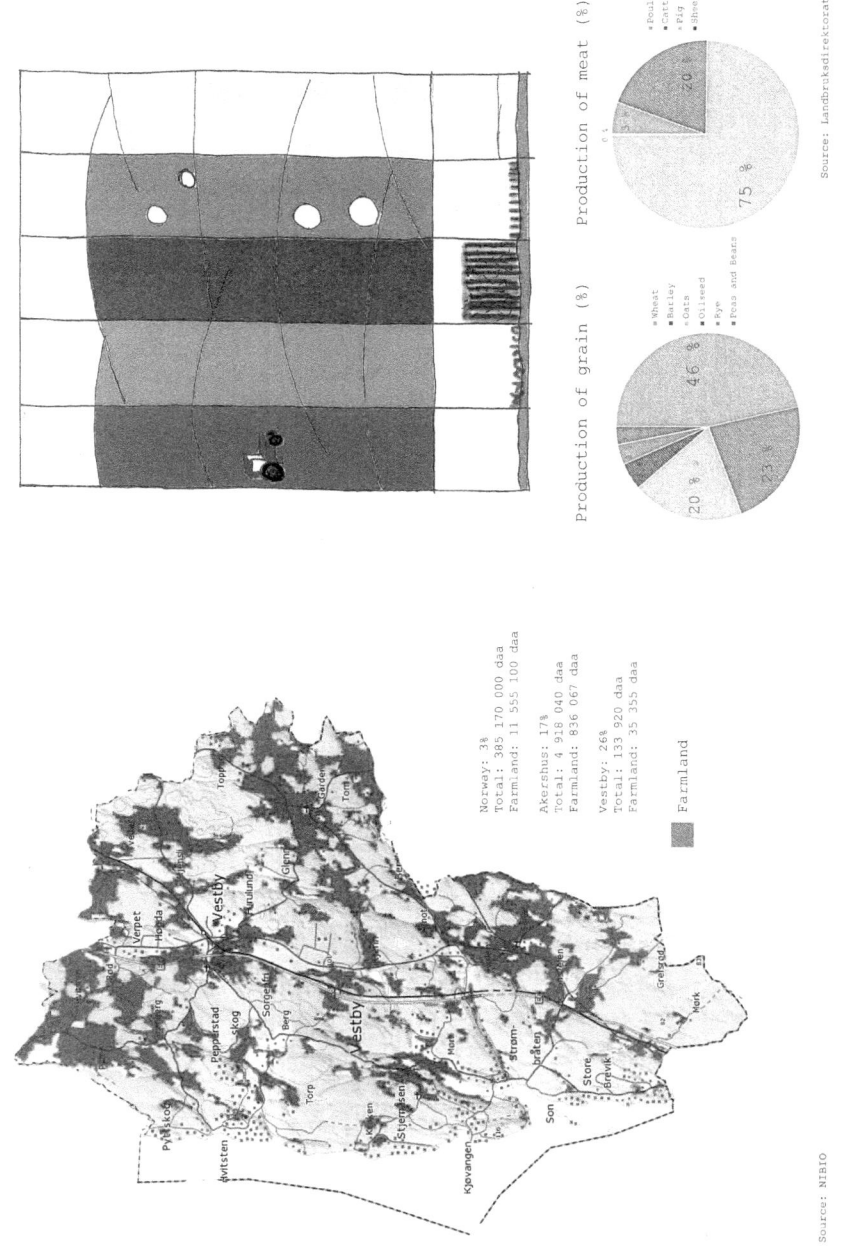

Figure 7.1 One page from the 'Edible Landscsape' theme (Atlas Vestby, Akershus, 2015)

This figure is shown in colour in the plate section of this book.

Historical development of housing

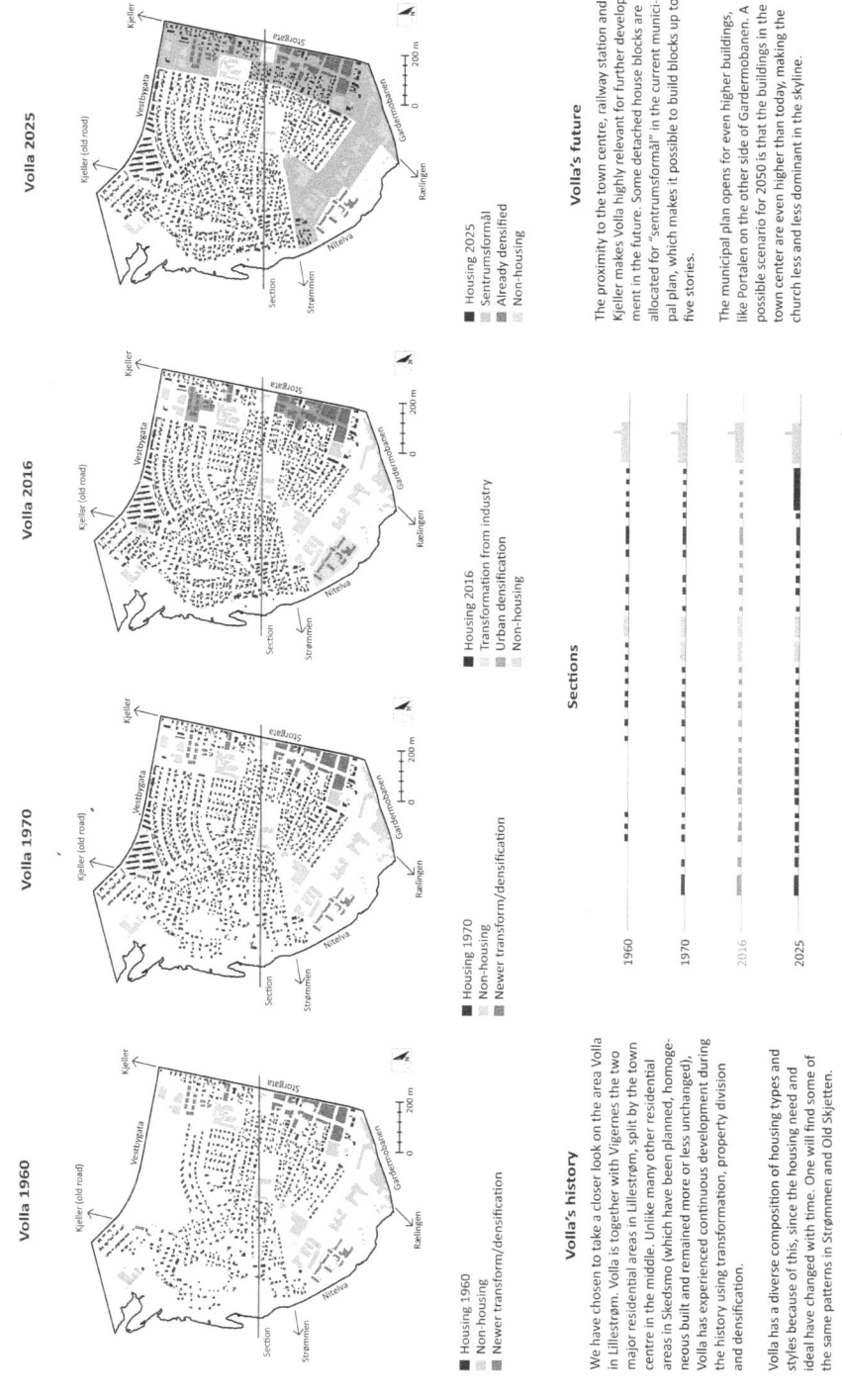

Figure 7.2 One page from the 'Landscape of Dwelling' theme (Atlas Skedsmo, Oslo, 2016)

Examples

The following are a few examples of the type of guidance offered to students during tutorials about how to begin exploring the landscape through the lens of a particular theme.

The physical landscape: *characteristics and spatial morphology*

What can you include?

- Geomorphology, geography, ecology (biology, vegetation) and their combination with the built environment (e.g., architecture, roads, bridges) presented at various scales.

What can you ask?

- What is the physical landscape made of? How would you characterise it? Is it the big geological structure and its land formations, e.g., mountains, valley, ocean, etc.? Or perhaps particular physical materials and vegetation types, e.g., forest, meadow, settlement, roads, asphalt, gravel, grass, trees, buildings, etc.?
- What are the various scales and especially the relationships between the scales and forms—both natural and human made—that create the mosaic we call the landscape of this study area?
- In what ways have the natural and cultural forms and materials defined the identity of the municipality (study area)?

Landscape of movement

What can you include?

- Transport means and their infrastructures, their effect on the landscape in terms of development and vice-versa. Circulation: local roads, railways, waterways and microcirculation: pathways and cycle ways.

What can you ask?

- How do people and goods move to and from the study area? How do they move within the study area?
- How does transport infrastructure (roads, railways, parking areas, cycling paths, walking paths and so on) shape, affect and influence the landscape?

The political landscape

What can you include?

- Legal frameworks, land ownership, institutions, hierarchies, etc.

What can you ask?

- Can we identify any particular 'political' tangible landscape, i.e., places that express governance? What is their presence (real or symbolic) in the landscape?

- Can we observe any other physical clues that tell us that this is a governed community?
- Regarding policies and regulation (plans, guidelines, pressure groups, decisions): what are their tangible expressions in the landscape?

The edible landscape

What can you include?

- Food production and marketing, e.g., farming and agricultural areas, food storage and sales.

What can you ask?

- How is the food landscape presented?
- What are the various scales and types of food production shaping the landscape?
- How does the landscape of food production/storage/sales contribute to the study area's identity?
- What other tangible elements related to food exist in the area? For example, storage, retail outlets, infrastructure logistics.

Synthesis

An editorial team is responsible for making sure the associations, connections, and influences on and between themes are addressed, and each group must actively reflect on those. For example:

- What are the relationships between transport and industry, food production and so on?
- What are the relationships between heritage and identity and the political landscape?
- How do demographic structures (age groups, ethnicities, gender) and lifestyles physically shape dwelling and habitation?
- Are there correlations between physical characteristics and planning?

The editorial group's responsibilities include collating the atlas and making decisions on graphic style and format. In addition, this group writes an introduction explaining the content and how it reflects an understanding of landscape, and prepares a diagram that portrays the relationships between the various themes. Figures 7.3 and 7.4 are examples of how the students translated the complex relationships into abstract diagrams.

Theory

At the beginning of the course I present a theoretical lecture that introduces the different ways in which the concept of landscape has been interpreted in the past four decades. Some key readings are also provided (e.g., J.B. Jackson, Yi Fu Tuan, D.W. Miening, K.R. Olwig, D. Cosgrove). Before listening to this lecture the students are asked to write down their own interpretation of the word 'landscape'. After completing the atlas, the students are asked to write a short essay in which they reflect on their understanding of landscape, comparing it to the reflections they wrote six weeks earlier.

Throughout the course, the students listen to lectures on an array of relevant topics and are advised to read articles to expand their knowledge of the multifaceted aspects of landscape.

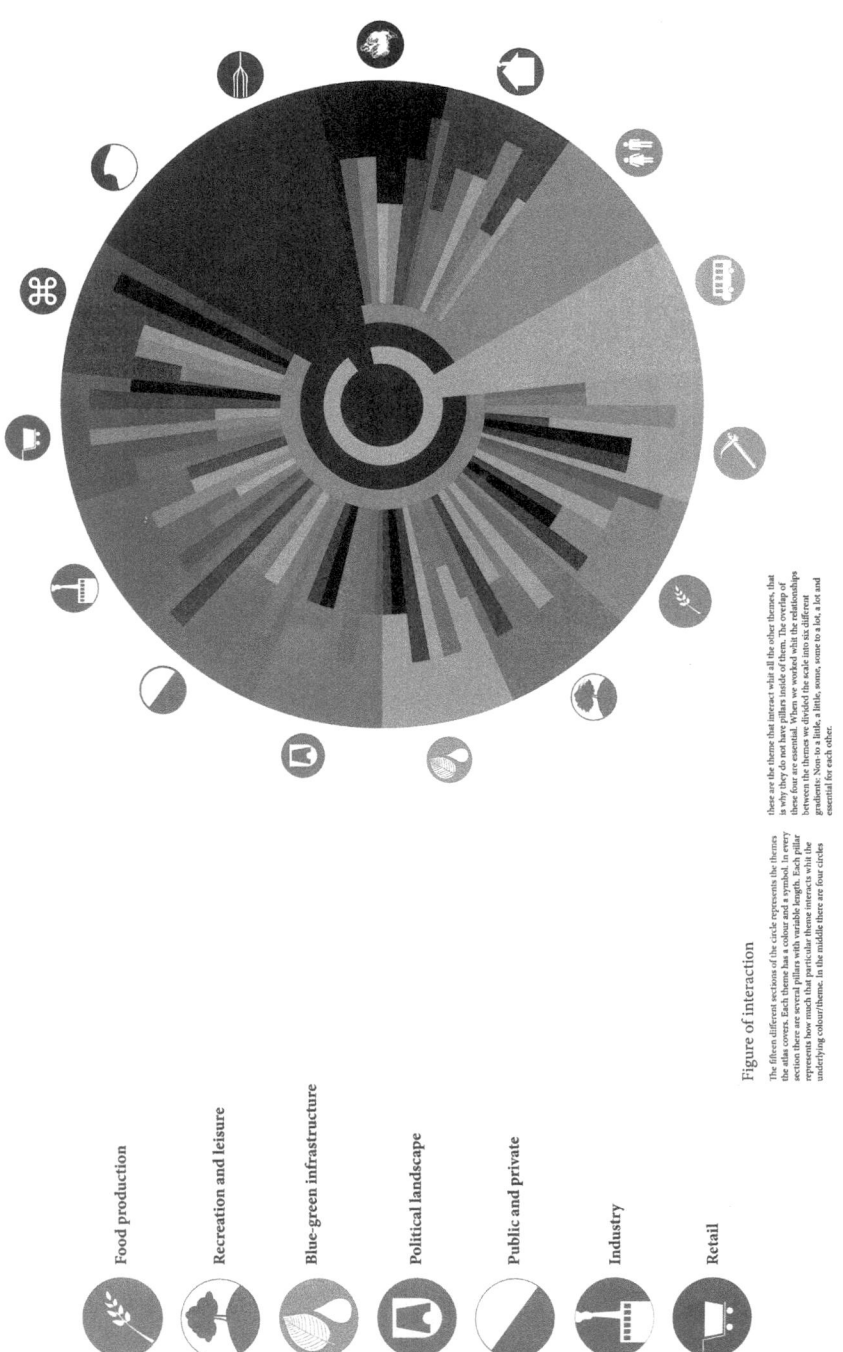

Food production

Recreation and leisure

Blue-green infrastructure

Political landscape

Public and private

Industry

Retail

Figure of interaction

The fifteen different sections of the circle represents the themes the atlas covers. Each theme has a colour and a symbol. In every section there are several pillars with variable length. Each pillar represents how much that particular theme interacts with the underlying colour/theme. In the middle there are four circles

there are the theme that interact what all the other themes, that is why they do not have pillars inside of them. The overlap of these four are essential. When we worked wiht the relationships between the themes we divided the scale into six different gradients: Non- to a little, a little, some, some to a lot, a lot and essential for each other.

Figure 7.3 Diagram representing the relationships between themes in Bærum, Oslo (Atlas 2014)

This image doesn't fully explain all intentions of the student's work, but shows how they translated the complex relationships into abstract diagrams.

This figure is shown in colour in the plate section of this book.

PULIC AND PRIVATE
HERITAGE AND CULTURE
REGINAL IDENTITES
MOBILITY
HABITATION AND DWELLING
THE POLITICAL LANDSCAPE
RECREATION AND LESIURE
NATURAL RESOURCES
PHYSICAL CHARACTERISTICS
PLANNING POLICIES AND REGULATION
RETAIL
PEOPLE AND SOCIAL INFRASTRUCTURE
BLUE AND GREEN INFRASTRUCTURE
THE EDIBLE LANDSCAPE

Figure 7.4   Diagram representing relationships between themes in Skedsmo, Oslo (Atlas 2016)

This figure is shown in colour in the plate section of this book.

Key textbooks were *Go with Me* (Oles 2014), and *Sustainable Landscape Planning* (Selman 2012). Both books are seminal to landscape architecture and give a valuable account of the intricate understandings of landscape. Oles' book became particularly popular as it managed to engage beginning students with the mysteries of landscape. Both books were recommended as resources to cherish and refer to throughout the course of education and professional practice.

Process and students' reflections

Writing an essay on the processes that they underwent enabled reflections and insight for both the students themselves and for teachers. The topic was 'landscape as an area as perceived by people: a personal account of the learning experience from the atlas assignment'.

A common thread in almost all the essays was that working on the atlas, and especially listening to all the themes presented at the end of phase A, significantly changed students' perception of what landscape means into a wider abstract notion. What follows are a few examples of excerpts from essays written during the 2016 course:

> At the beginning of the semester I wrote that to me, a landscape is everything that surrounds us outside . . . my perception was mainly a focus on what we can *see* [emphasis in original] when we are outside . . . [Now I realise] that the word holds many other features that we cannot necessarily see at first glance . . . This was clearly emphasised during the

PowerPoints presented by all the groups in the analysis phase. Landscape as an idea holds not only the vegetation, buildings, roads as described in my first assignment, but it also displays social relationships between people who live in the area, private and public spaces, infrastructure, retail, food production, cultural heritage, etc. There are, in other words, multiple actors that influence the landscape in different ways.

S.S.M., October 2016

First, I decided that it must be everything surrounding the buildings . . . and it hit me, maybe landscape is not just everything around buildings, but a landscape might as well be a city full of buildings . . . I realised that the term is much wider than I thought . . . I forgot to define landscape as an area . . . Also, it is important to notice that landscape without human perception is just land.

S.H., October 2016

My basic idea is still the same, but it has become more complex and now includes some perspectives I did not have before . . . it is important to understand the forces that have shaped an area or a society. To be aware of what you may be taking away and the fact that when something *is* [emphasis in original] changed, this change may not be possible to reverse . . . I think my work with the atlas, especially the presentations of other themes, has started to open my eyes to the inherent complex and interesting understanding of landscape.

H.M.O., October 2016

This course has not only given me a wider perspective on an essential topic in the continuation of my architectural path—but also perspective on perspective itself, and perhaps even heightened awareness for, but also frustration over, the many occurrences ahead of me where I will miss out on important information because I have my eyes set elsewhere and because my preconceptions blind me . . . The atlas experience left me with a larger understanding of something I thought was 'simple' . . . my eyes have opened up for the political landscape and how it influences the physical landscape. Personally, the term landscape has grown from being a word that makes me think about aesthetics and qualities of nature to a word that makes me think of conflicts between humans.

M.S.A., October 2016

Discussion and critical reflections—the bitter and the sweet

Although the essays articulate a very fruitful and enjoyable learning process, students' evaluations of the course were persistently poor in all years. Students' typical complaints were that teachers' communication was poor, the course was too vague and unclear about what they were supposed to do, and group work was frustrating.

From the teachers' point of view, this course had its hurdles and limitations. One of the obstacles was the large number of students: it was difficult to manage such a big class and several students prioritised personal activities (such as jobs or holidays abroad) over attending class. Making study material accessible on the intranet is a double-edged sword because it encourages students to skip the one-on-one interaction with the discussions, arguments and negotiations in class with fellow students and teachers that are essential for developing critical design thinking.

However, the overall results of students' work, both in phase A (the atlas) and phase B (visionary plans), were extremely rewarding for the teachers. External examiners who attended and provided feedback to the students at final presentations (different reviewers each year) were also very complimentary, commenting that the depth of thought and maturity exhibited through the work was extraordinary at second-year level.[4]

Another rewarding experience has been students approaching me in the following years to say 'now I understand what you meant'. While at the time they did not find the course useful, in hindsight, when taking more advanced studios, they realised how much they had gained from it.

Teaching landscape complexity at any level is complicated and challenging. It is probably also true that in education we cannot utilise a 'one-size fits all' pedagogic approach, and this course has been more useful for some students than for others. In an ideal world, I would have liked to work with up to thirty students (with three teachers) and assign themes to teams of two or three persons, emphasising ongoing interaction between the various themes.

Acknowledgements

I am very grateful to my colleague, Associate Professor in urban and regional planning and architect Marius Grønning, from whom I took over the course. Marius had first developed the basic ideas for the course structure, with a conceptual and strategic orientation that stressed the awareness of spatial complexity. The approach included assigning the compilation of the atlas as a basis for scenario thinking in design processes. During 2013 and 2014 we co-taught this course with Marius's full endorsement for framing the atlas work and theory within a landscape perspective.

Thank you too to the anonymous reviewers for their insightful comments.

Notes

1 Town planning is challenged by similar complexities.
2 A full load semester is thirty credits and courses can vary from five to twenty credits. A semester lasts fourteen weeks.
3 Each presentation is made up of twenty slides projected for only twenty seconds each.
4 External examiners commented that the standard of graphic visualisation was lower than in their programmes.

References

ASLA (2016), 'Landscape Architecture Education and Career Development' [website], www.asla.org/ContentDetail.aspx?id=12206&PageTitle=Education&RMenuId=54, accessed 5 April 2016.
Buchanan, R. (1992), 'Wicked Problems in Design Thinking', *Design Issues* 8/2: 5–21.
Deriu, D. and Kamvasinou, K. (2014), *Emerging Landscapes: Between Production and Representation* (Farnham, UK: Ashgate).
Oles, T. (2014), *Go With Me: 50 Steps to Landscape Thinking* (Amsterdam: Architectura & Natura).
Retna, K.S. (2016), 'Thinking about "design thinking": a study of teacher experiences', *Asia Pacific Journal of Education* 36/1: 5–19.
Selman, P. (2012), *Sustainable Landscape Planning: The Reconnection Agenda* (Abingdon, UK: Routledge).
Thompson, I. (2000), 'Aesthetics Social and Ecological Values in Landscape Architecture: A Discourse Analysis', *Ethics Place and Environment* 3/3: 269–287.
UK Design Council (2010), 'Multi-Disciplinary Design Education in the UK' [website], www.designcouncil.org.uk/sites/default/files/asset/document/multi-disciplinary-design-education.pdf, accessed 7 July 2017.
Wrigley, C. and Straker, K. (2017), 'Design Thinking Pedagogy: The Educational Design Ladder', *Innovations in Education and Teaching International* 54/4: 374–385.

8

The studio as an arena for democratic landscape change

Toward a transformative pedagogy for landscape architecture

Deni Ruggeri

Introduction

Future cities should be sustainable and designed with people's well-being in mind. Their landscapes should be accessible, well connected, and made to last and adapt to change. They should provide residents with landscapes that allow them to exercise their biophilic spirit. They should be co-designed to foster greater identity, civic pride, and stewardship. Over the past few decades, landscape architecture has claimed a new role in envisioning new urban development that can respond to these complex and highly interrelated goals. Landscape architects have a key role to play in the promotion of new processes of sustainable, democratic landscape transformation, yet their education offers limited opportunities to practice the partnerships and collaborations required to initiate such change. This chapter discusses a novel approach to landscape architecture education as the training ground for a transformative landscape architecture practice that may be able to address the wicked problems that contemporary society faces. The master's course "The Urban Landscape as a Social Arena" at the Norwegian University of Life Sciences (NMBU) illustrates the workings of an integrated practice-based teaching/research model. This chapter also reflects on the goals, experiences, and lessons learned in order to sketch out a roadmap for the future of landscape architecture education, challenging traditional assumptions about the roles of academics and practitioners.

Background

Global warming, climate uncertainty, political unrest, refugee crises, food security, and the impact of humans on the environment are just some of the wicked problems landscape architects face (Brian 2008; Rittel and Webber 1973). Today's landscapes are asked to perform a multitude of functions related to liveability, biodiversity, storm water management, public health, and community identity (Beatley 2011; Hester 2006; Hou and Rios 2003; Sander and Putnam 2010; Gehl 2013; Southworth 2003; Southworth and Ruggeri 2011; Steiner 2011). Their long-term success may depend on landscape architects' ability to involve all stakeholders in the design and planning process to co-create future scenarios that can foster greater stewardship and resilience (Bose et al. 2015; Manzo and Perkins 2006; Ruggeri 2014a, 2018).

In 2000, the European Landscape Convention (ELC) has shed light on the importance of engaging communities in decisions to preserve and transform community-recognized landscape assets for the benefit of all (Dejeant-Pons 2004). The ELC's clear mandate for a participatory approach to landscape design and planning has brought to light tensions between professional rhetoric advocating for the rights of underserved communities and the reality of consultants being financed by and operating in the interest of private clients (Butler 2018; Hester 2005; Eaton 2017). Many landscape architects are limiting their engagement with community members for consultation and information sharing, and see participation as either too costly or too constrictive of their creativity. National and local policies have done little to fill this gap, remaining vague as to the quality and extent of citizens' engagement (Arnstein 1969; Jørgensen et. al. 2015).

The uneasiness and skepticism of landscape architecture practitioners toward participation may have its roots in the current models of professional education. Modelled after the beaux-arts apprenticeship-based pedagogy (Ledewitz 1985), which emphasizes individual creativity and originality, landscape architectural education encourages competition between professionals, and is focused on training problem-solvers rather than facilitators and partners in bottom-up processes (Milburn and Brown 2003; Ruggeri 2014a). While the introduction of service learning and design-build opportunities into the design curricula of many schools has offered some students and instructors opportunities to practice civic responsibility (Battistoni 1997; Monson 2003; Salama and Wilkinson 2006), it has failed to challenge traditional expert–layperson relationships. Participatory Action Research (PAR) is slowly contributing to the re-framing of these relationships as partnership in co-creation (Park 2006; Schneidewind et al. 2016). Participatory Action Research is also changing the way experts and communities communicate. Action research is increasingly employing storytelling as a tool for conflict-resolution, goals setting, and visioning. Through the authoring of new compelling stories, planners and designers are seeking to motivate people to take direct action in fighting disinvestment and degradation (Ganz 2011; Ponterotto 2006; Sandercock 2003)

The gap: toward a transformative landscape architecture education

Contemporary landscape architecture has come a long way from its beaux-arts roots. It has become a science with a unique body of knowledge, theories and methods for the "systematic study of landscapes themselves, and of processes of landscape-making" aimed at the promotion of "substantial social, economic, and environmental benefits" (Davis and Oles 2014: 6). In landscape architecture as in many other fields, there is a renewed interest in erasing obsolete boundaries between expert knowledge and the knowledge held by communities (Eaton 2017; Schneidewind et al. 2016). Within this shifting paradigm, participation emerges not as a technique for soliciting feedback but as a new philosophical approach to promoting systemic, resilient change processes from the bottom up. Conceiving landscape architecture as a transformative science calls for the integration of rigorous scientific inquiry across a multitude of domains. Together with traditional training in the nuts and bolts of the profession—topography, creativity, aesthetics, and ecology—future agents of democratic landscape change should become skilled researchers, able to identify key research questions and choose the most appropriate methods for gathering data and analysing problems. Becoming active agents of transformative landscape should also involve strategic thinking, collaboration and collective visioning (Ruggeri 2014a).

Landscape architects and planners should be keenly aware of the political and ethical dimensions of their work, and reflective of their positioning in the world and within the established power structures (Makhzoumi Egoz and Pungetti 2011; Schneidewind et al. 2016). This includes re-thinking each field's positioning within the larger scientific discourse. "Transdisciplinarity [. . .] requires an ethic of resolute openness, tolerance, and respect for perspectives different from one's own and a commitment to mutual learning and mediational processes in which contrasting values and conflicts of interest are negotiated" (Stokols 2006: 68). Education can play a key role in the re-framing of traditional professional models by providing opportunities to become aware and critically challenge deeply rooted professional biases that may prevent landscape architects from become true agents of transformative change (Figure 8.1; Van Hulst 2012). Educator Paulo Freire sees pedagogy as an instrument for overcoming barriers between experts through shared reflectivity and awareness raising. He writes:

> education either functions as an instrument that is used to facilitate the integration of the younger generation into the logic of the present system and bring about conformity to it, or it becomes "the practice of freedom," the means by which men and women deal critically and creatively with reality and discover how to participate in the transformation of their world.
>
> *(1996: 34)*

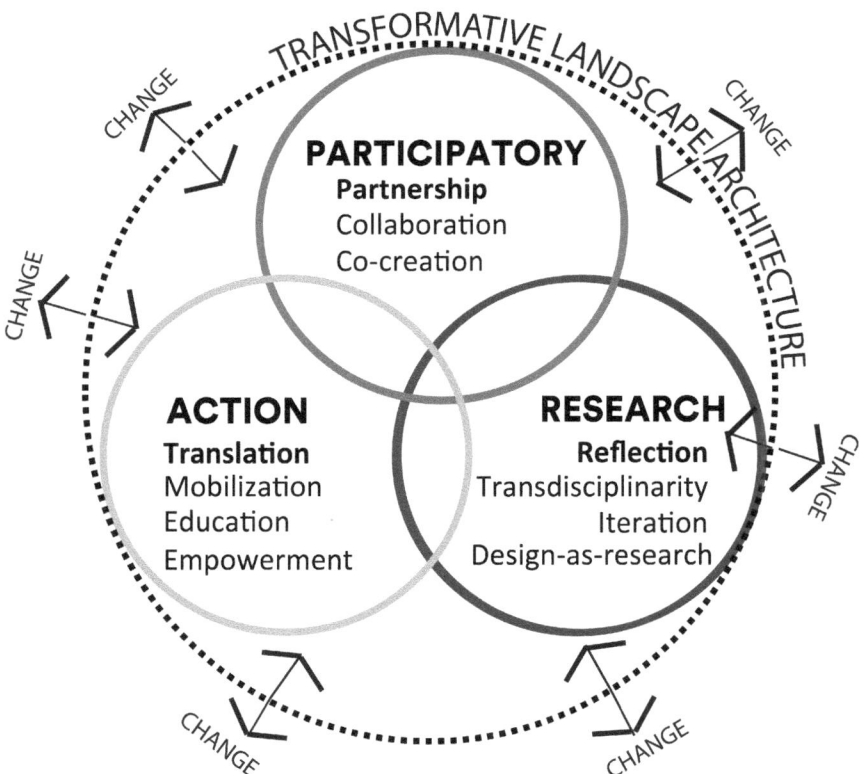

Figure 8.1 Diagram illustrating the goals and theories encompassed by a transformative landscape architectural education (diagram by the author)

The pedagogical model described in this chapter is inspired by the Participant Action Research belief that addressing the complex challenges of our contemporary landscape requires collaborations and partnerships between experts and citizen scientists, a rigorous research-based approach, and a deep commitment to democratic values. It may be operationalized into the following pedagogical goals:

- the need to raise awareness of the political nature of design and planning decisions and its ethical implications for practitioners;
- the integration of theory into the design process;
- the incorporation of new knowledge and robust research into the creative process;
- collaborations across all partners, peer teaching, student-run theory seminars;
- the ability to coalesce around a collective co-created vision;
- reflection as a foundation for assessing personal and professional growth; and
- an emphasis on storytelling, thick description and communication.

In the fall of 2014 these objectives formed the foundations for the LAA341 course "The Urban Landscape as a Social Arena" offered at the Norwegian University of Life Science, which serves here as a case study for illustrating an emergent new pedagogy to prepare young professionals for the practice of transformative landscape architecture (Figure 8.2).

Case study: the design studio as a training ground for transformative landscape change

During the fall of 2015 the LAA341 course "The Urban Landscape as a Social Arena" offered 38 master's students in landscape architecture and planning the opportunity to activate theories, methods, and processes of participation in the context of real-life planning and design processes in the Norwegian communities of Ski, Giske, and Bodø. The three pilot municipalities were chosen from over 130 municipalities that had participated in *Barnetråkk*, a research effort led by the Norwegian Design and Architecture Centre (DoGA). The project aimed to operationalize a mandate by the 1985 Norwegian Planning and Building Act, which called for the inclusion of younger generations of citizens in sustainable development decisions and their active engagement in the planning process by mapping their experiences as they walked from home to school (Aradi 2010). Yet, despite the success of this project, researchers pointed at the challenges related to the translation of these maps into concrete planning decisions. The NMBU students were asked to respond to this challenge by transforming the data gathered by the project into concrete visions and designs.

Their work began with the careful analysis of the *Barnetråkk* findings, followed by interviews and focus groups with key stakeholders aimed at identifying goals and working research questions that would guide the work of NMBU students in each pilot community. Between September 17–26 in 2015, NMBU students organized day-long participatory workshops, which were intended to give them a deeper understanding of the place and the unique challenges and visions for the future of the community as they were perceived by local youth. The participatory techniques used were inspired by the course literature on PAR, community participation, and landscape democracy, adjusted in order to fit each community's unique problems (Hester 2006; Kot and Ruggeri 2005). The workshops began with a general introduction to the theories driving their work and continued with a series of creative exercises during which children would be stimulated to discuss in an open setting their perceptions as to the issues their respective communities were facing. In Ski, children were divided into small groups and asked to use

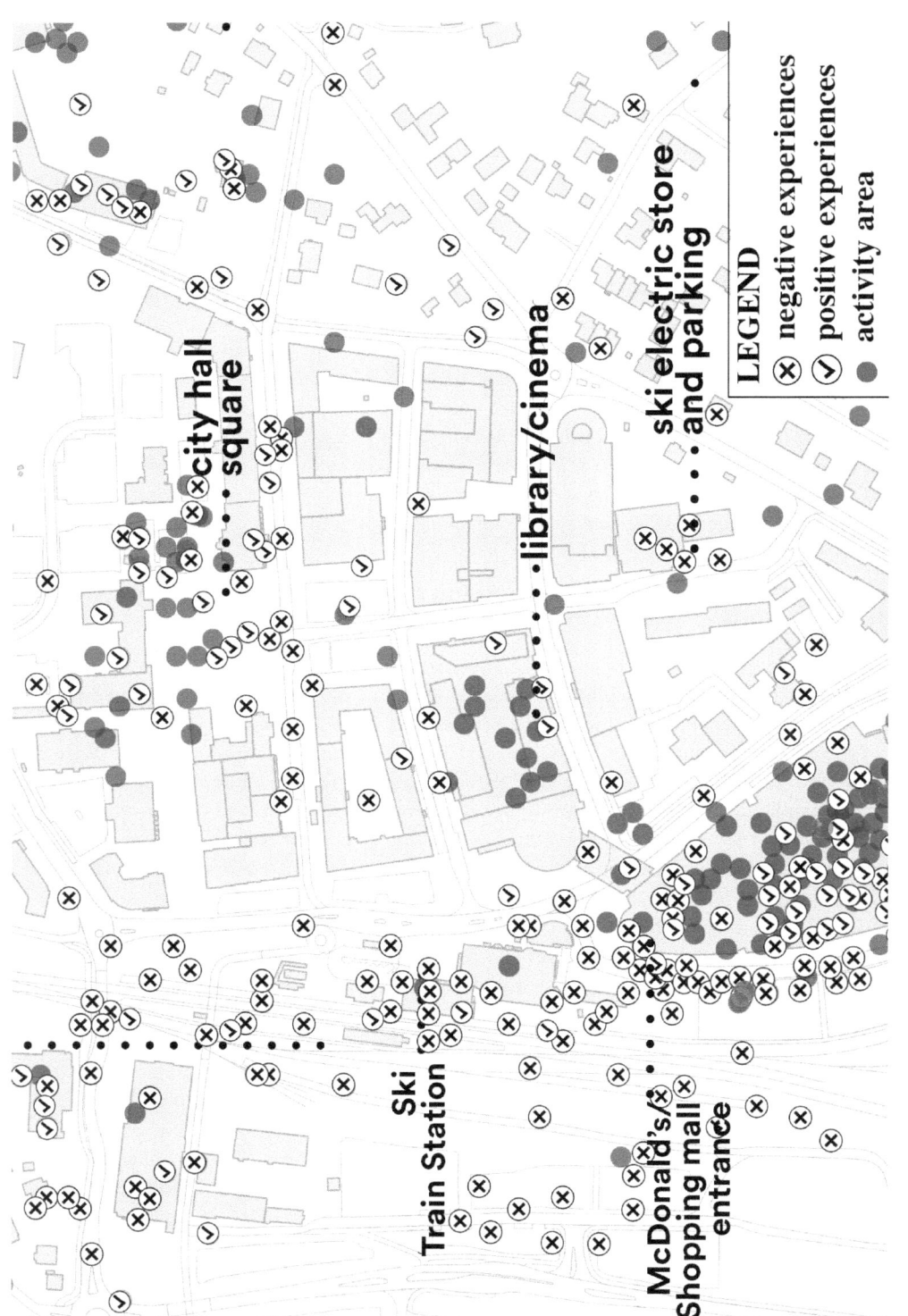

Figure 8.2 Barnetråkk maps were the starting point of the students' investigations into children's perceptions of their community's public realm (source: LAA341 students)

LEGO blocks and modelling clay to create three-dimensional models illustrating their perceptions of liveability and density. In Bodø, NMBU students took children on walking tours to clarify further the information collected through the *Barnetråkk* maps. In both Bodø and Giske, abstract images—labelled as *pictograms* (Figure 8.3)—helped them initiate a dialogue with local children about the unique sense of place, distinctiveness, and shared values. In all of the municipalities, the workshops ended with a visioning exercise that encouraged children to speculate about future scenarios through freehand drawing and collages, and to present their visions in a plenum session (Figure 8.4).

Upon returning to the NMBU campus, students were asked to negotiate and agree upon a set of shared goals, which resulted from their expert analyses and the findings from the workshops. This vision would be translated into a 'game board', a mapping and strategic design tool developed by Raoul Buntschoten to identify emergent conditions that have the greatest transformative potential, and to collectively resolve tensions between potentially conflicting strategies (Hall 2010). The game boards acted as a de-facto contract that would ensure that individual interventions would be respectful of the collective will and respond to the visions and ambitions of the children in each of the communities. In the process of developing the group game boards, group visions were constantly refined and negotiated throughout the course via peer reviews, class-wide critique sessions, interactions with guest practitioners and more traditional one-on-one dialogues with course instructors.

Figure 8.3 In Bodø, the pictogram activity helped NMBU students gain a deeper understanding of the bond between children and the city's landscape (photo by the author)

Figure 8.4   Giske junior high school students present their visions for the future of the community and their wish for more meeting places (photo by the author)

Mid-term reviews, impromptu presentations, and Instagram and Facebook posts shared with community stakeholders helped establish a continuous feedback loop that allowed the incorporation of new knowledge or insights from community members at all stages in the process. Collectively, these activities helped to strengthen and refine the students' storytelling, which would culminate in final presentations held at Oslo's Design and Architecture Centre, in front of a crowd of dozens of practitioners, stakeholders from the community and members of the general public. After the end of the course, a selected group of students were able to return to each partner municipality, where they were able to disseminate further their visions in front of citizens of all ages and backgrounds. In the words of Giske's public health coordinator, Torill Valderhaug,

> we never dreamed that the results would be so good. In the space of a very short time, the students have understood our situation and the challenges we face with impressive accuracy. And the proposals they have come up with have been well received by the politicians. In brief: it has been perfect.
>
> *(Røhnebæk Bjergene 2016)*

The PAR classroom: benefits and challenges of teaching for transformative landscape change

The following is a discussion on the performance of the pedagogical steps taken in the course to achieve the initial goals, and of the lessons learned and suggestions for promoting transformative landscape architecture education in the future.

1) A PAR-inspired pedagogy: the mutual benefits of bridging academic and local knowledge

The cooperation with DoGA and the integration of the *Barnetråkk* project findings offered the ideal opportunity to embrace a Participatory Action Research framework to envision, initiate and lead transformative landscape actions in Ski, Giske, and Bodø. While in Ski the challenge was to uncover the deeper feelings community members held about the physical changes and livability challenges resulting from densification, in Giske the students were tasked with promoting walkability, community life and re-activating people's biophilic connections to the sea. In Bodø, students responded to the need for a richer social life through the transformation of the city's streets and underutilized space into a network of public landscapes for the joyful display of community identity and pride.

The grounding of the course in Action Research resulted in a tighter than usual bond between academics and residents. The sharing of state-of-the-art theories and design strategies by landscape architecture and planning students helped inspire community members and municipalities to imagine what once seemed impossible to achieve. This resonated with the municipal staff and local leaders, who had begun to question their planning procedures and, in the case of Giske and Ski, immediately began to integrate the students' visions into their planning agendas. Through their engagement with the communities NMBU students experienced first-hand a shift from a more traditional role as outside experts and 'professional service providers' to partners and 'insiders.' All in all, participation strengthened the communities' ownership of the work of academics. The NMBU students were invited back to the respective communities after the course ended to present visions they had authored during the course to all residents. In some cases, this has led to changes in policies and plans. Giske has recently unveiled a plan for the redevelopment of an abandoned community centre in Roald inspired by the work of the students on the course (https://nb-no.facebook.com/destinasjonroald), while Bodø has enlisted some of the students to follow up on their efforts to promote participation and social capital construction.

2) The emergent, reflective nature of PAR processes

Rather than framing the course brief as a series of problem-solving challenges, LAA341 instructors challenged NMBU students to define their own scope of work, to act as 'integration actors' and become agents of transformative landscape change by identifying and prioritizing future actions strategically (Clemetsen and Støkke 2018). Each group was asked to identify a research question that would inform the workshop activities and their own goals. Throughout the fall, NMBU students reached out to community members in person, through social media, or via online surveys. Every time new information emerged, NMBU students would collectively reflect on its impact, research new theories and strategies, and tweak their visions to respond accordingly. Consistent with the PAR research framework adopted by the course, the idiosyncrasies and uniqueness of the landscape challenges faced by each pilot community resulted in a design process that was illustrated through flow diagrams synthesizing the steps and their overall influence on the project outcomes (Figure 8.5).

3) The collaborative nature of PAR-inspired education

The complexity and systemic nature of the landscape challenges that confronted LAA341 students in Ski, Giske, and Bodø required the integration of a transdisciplinary body of knowledge, theories, and methods, from planning to public health, and from community development to sociology. Many classroom activities were designed to reinforce a culture of collaboration. During the analysis phase, each LAA341 student took on the role of expert in a particular

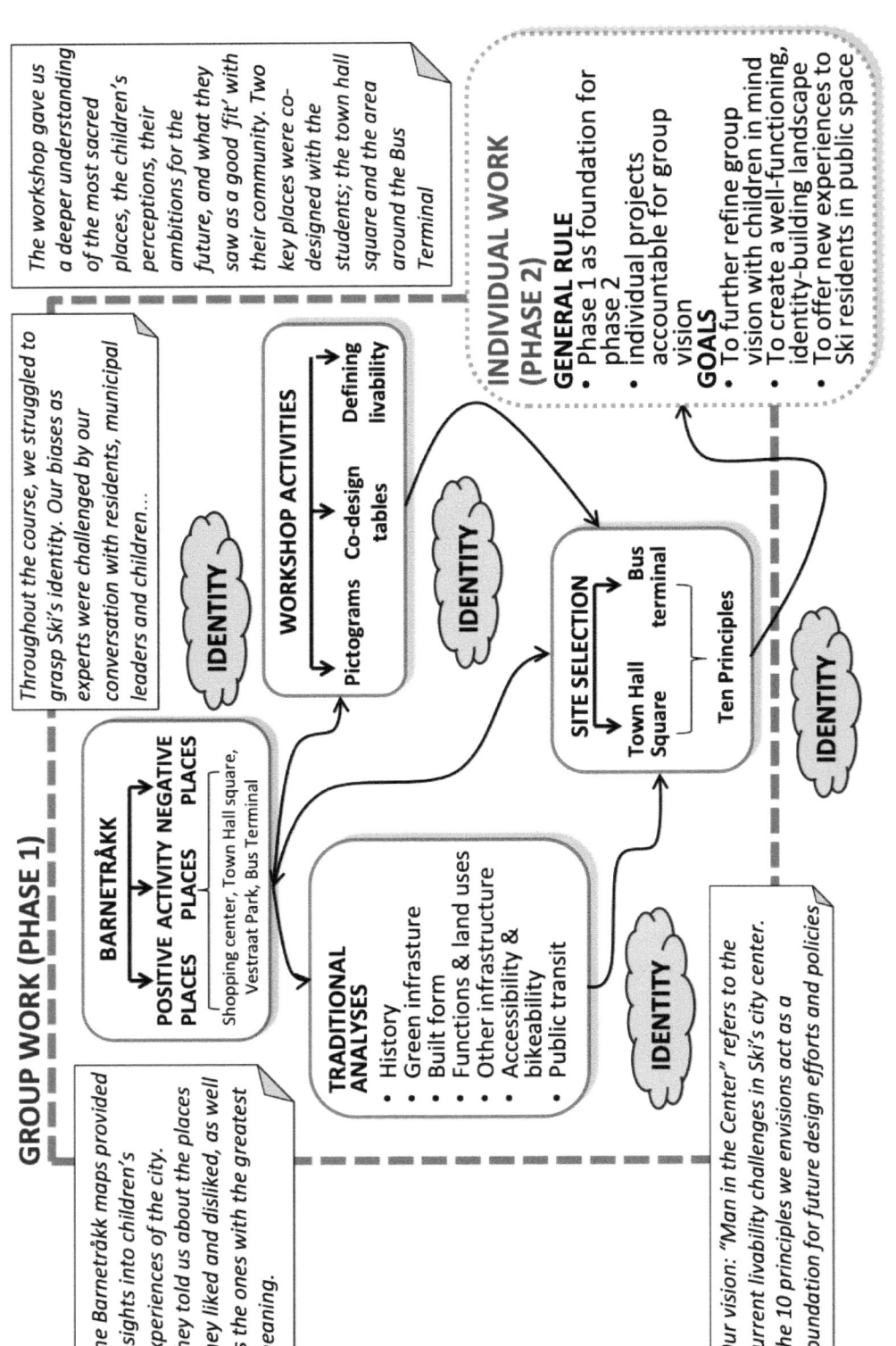

Figure 8.5 Diagram illustrating the iterative process followed by LAA341 students and their reflections on design (in italic) (adapted from an original diagram by Stine Svanemyr and Prathepa Kirubarahan)

aspect of the community relationship between people and place. Contrary to traditional studios, student-experts enrolled on the course were encouraged to share their individual analyses with fellow group members through a 'visioning chairs' exercise, in which they interviewed each other on various landscape qualities related to history and culture, physical landscape qualities worth preserving, challenges to sustainability, the role nature played in each community, their unique identity and sense of place, and any gaps in their knowledge (Figure 8.6). In order to translate the knowledge gathered during the analyses and workshops, the instructor used a Nominal Group Technique (DelBecq 1975), which revealed individual student goals and helped each team negotiate a set of shared priorities that would later be translated into visions.

4) From individual creativity to collective visions

Evidence shared during the visioning chairs and the groups' goals setting would form the foundations of strategic landscape visions that would replace the individual plans and site plans produced in traditional landscape architecture studios. Through the group game board, students were able to connect the documentary evidence gathered during their fieldwork and analyses into a group vision, which included individual landscape interventions ranging in scale from the design of a trail system to the creation of small social spaces. The game board served to promote accountability, as students were encouraged to regularly reflect on the links between individual landscape interventions and collective vision and to update the diagrams accordingly (Figure 8.7).

Figure 8.6   Through the visioning chairs exercise, students shared their individual knowledge and began to identify common goals (photo by the author)

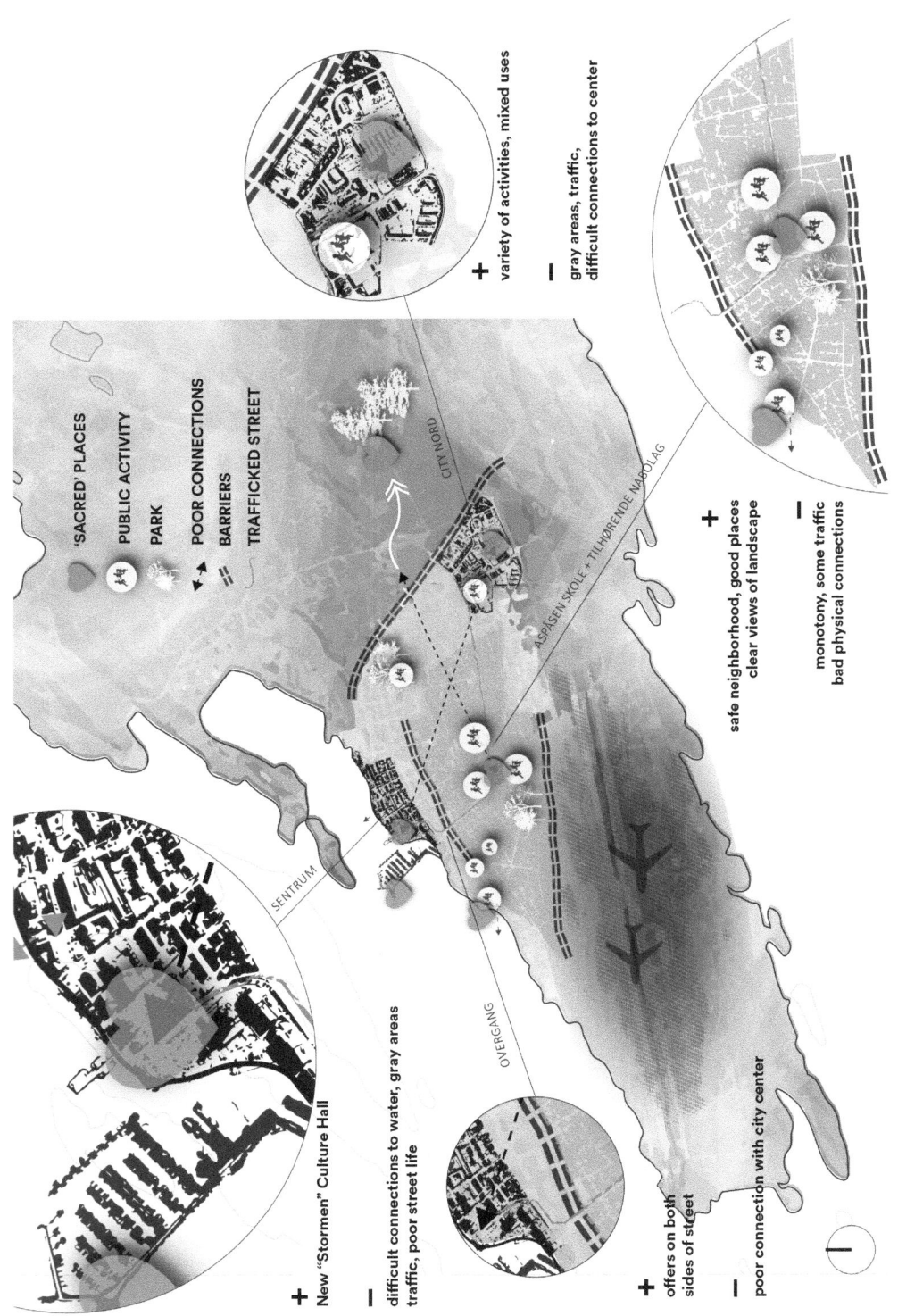

'SACRED' PLACES

PUBLIC ACTIVITY

PARK

POOR CONNECTIONS

BARRIERS

TRAFFICKED STREET

CITY NORD

SENTRUM

OVERGANG

ASPÅSEN SKOLE + TILHØRENDE NABOLAG

New "Stormen" Culture Hall

+

− difficult connections to water, gray areas
traffic, poor street life

+ offers on both
sides of street

− poor connection with city center

+ safe neighborhood, good places
clear views of landscape

− monotony, some traffic
bad physical connections

+ variety of activities, mixed uses

− gray areas, traffic,
difficult connections to center

Figure 8.7 A group game board linked the findings from the analyses to a vision and a series of landscape interventions illustrating that vision (photo by the author)

5) Reflecting on one's agency

Whether consciously or not, each landscape architect carries a vision of idealized landscapes that reflect their landscape experiences as young adults and personal biases (Francis 1995). Pre-post survey questionnaires sought to test NMBU students' personal biases, their positioning vis-à-vis issues of landscape democracy, and their personal understanding of the role as agents in the transformative visions. Grounded in the notion that landscape democracy is a culture that begins in the classroom (Ruggeri 2014a), LAA341 encouraged students to reflect, to perform self and peer evaluations, and to quantify/qualify every team member's contributions to the group. The difference between self and peer assessments gave instructors a window into the workings of each group, and helped them intervene to mitigate interpersonal conflicts or disagreements on the group visions and goals that emerged over the course of the semester. It also created a greater sense of accountability and awareness of the students' preparedness to lead and cooperate, as well as their ability to respond to challenges that emerged in the process as people were confronted with this novel educational approach.

6) The integration of innovative storytelling and representation methods

The NMBU students were encouraged to think of their visions as stories, carefully articulated to compel residents to embrace transformative landscape action (Ganz 2011). Throughout the semester, the course schedule included many opportunities to refine their verbal and visual communication skills through their constant telling and retelling of such stories. Both the mid-term presentation that took place on the NMBU campus and the December 8 final review in Oslo featured guests from participation experts, practitioners, and the partner communities. Follow-up public meetings in Giske, Ski, and Bodø offered additional feedback (Figure 8.8). Copies of the final reports describing the process and visions were made public through social media, ensuring the transfer of ownership and responsibility to the communities themselves.

Discussion: toward a transformative education for landscape architecture

The LAA341 course was an experiment in both participation and pedagogy, which wanted to test the possibility of transforming the landscape architecture studio into a collaborative setting for the performance of transformative landscape change in the context of real-life planning challenges. Students on the course learned about the theories behind participation, methods and processes for the co-production and sharing of transdisciplinary knowledge around the landscape. Their interactions with residents of Ski, Giske, and Bodø were instrumental in prioritizing future actions in the respective communities. The partnership with communities resonated in the NMBU students' responses to course post-evaluation. They concluded that while participation may require much energy and time, it is worth it: "participation can be difficult, but worth working with. Even though the result is not always how people imagine, the process can be as important as the final result." Another student highlighted the importance of participation as a place-making tool: "Participation takes time, but when done correctly and on a realistic timetable, the outcome will be better, more rooted in both the landscape and the community and it will counteract the tendencies of homogeneity we are seeing in today's landscape".

Collaborating with communities helped NMBU students gain awareness of the social agency of landscape architecture. As students wrote in their post-survey, "it will be [our] responsibility

Figure 8.8 In the spring of 2016, students returned to the communities to present their visions and receive stakeholders' feedback (photo by the author)

to encourage participation as a part of the planning processes in the future." While most NMBU students embraced participation, a few remained skeptical that it would ever become a part of mainstream landscape architecture. "I found the course very inspiring, but I worry about how much I will get to use participation when I start to work." Their work in Giske, Ski, and Bodø gave them first-hand experience of the power of participatory landscape science to "bridge between designers and the ones who lead their everyday-life in the projects we create."

Concluding lessons: toward a transformative pedagogy in landscape architecture

This chapter has outlined the vision for a new educational model for landscape architecture that aims to empower students to promote transformative landscape change. The educational experience gave students the chance to experience challenges and opportunities connected to the practice of participation, collaboration, visioning and storytelling in the pilot communities of Ski, Bodø, and Giske. In partnership with young residents, NMBU students co-designed visions that would address current needs while laying out ideal paths for sustainable landscape transformation. The following lessons learned begin to synthesize a few principles for the future of landscape architecture and its agency in addressing wicked challenges faced by real communities.

1   The wickedness of contemporary landscape problems requires adaptability.

In approaching their work with the community, LAA341 students were asked to collectively reflect and come to a consensus on the most appropriate solutions to the challenges and problems voiced by the community. Related to Randy Hester's (2006) "openness" concept, the course instructors had to embrace the notion that the final design may be more open-ended and 'visionary' than simply solving the problems. The students' work resembled the work of scientists, constantly testing their 'research question' and associated visions. The students were encouraged to develop designs that would not preclude possible adaptations and, in many cases, developed 'scenarios' for the implementation of these projects by community members.

2   Landscape architecture education and practice are not neutral. They are political and deliberate about the nature of the change they seek to generate.

This principle suggests that design should be up front in terms of values and beliefs. Students in the class were encouraged to precede their presentation with a few slides of their 'theoretical foundations,' thus setting the stage for their approaches. Their solutions may not have been the 'best,' but they represented the collective values and ambitions of the residents they engaged, integrated with the transdisciplinary knowledge and experiences brought into the process by the partnership with academics.

3   Transformative landscape architecture must engage tactics.

Landscape time is different from human time, and it often takes decades to go from initial conception to implementation. Designers must be creative and develop new tools for envisioning transformative landscape change scenarios that can adapt and be resilient. The game board assignment helped LAA341 students synthesize their analysis and find synergies and complementarities to build upon through their visions. Students' exposure to case studies in tactical urbanism and their visit to Oslo's Waterfront promenade introduced them to the need to prioritize short, medium and long-term implementation strategies and speculate on the potential systemic consequences of their visions.

4   Participatory Action Research inspires shifts in values and actions.

In the class, a very rich discussion emerged around the ethics of design, and the role and responsibility of the groups to provide the places they had studied with real tangible steps to be taken in their future. Many of these steps responded to very specific needs that they had discussed with community members, others were based on successful examples from Scandinavia, Europe and the USA. By opening up to the world, communities were empowered to 'think big' and put their best foot forward in shaping new ambitious stories. Yet the true results of their work should be assessed in the long run. Conversations with local leaders in Giske continued to reveal a hesitation on the part of community members to take leadership in implementing the NMBU student visions, a signal that participation is a culture that needs to be practiced and nurtured in order to thrive.

5   Participatory Action Research-inspired landscape architecture education is rich in methods and resourcefulness.

Through participation, students in LAA341 were able to gain a deeper than usual understanding of the communities in which they operated. They also became deeply aware that the perspective of the pupils who participated in their workshops needed to be supplemented by other information, forcing them to engage new modes of engaging the communities. In the case of Giske, they used new technologies carefully and resourcefully, starting a social media campaign to communicate their work, adapting a well-used photovoice method to Instagram (Ruggeri 2014b) in order to solicit pictures of the community's most valuable landscape assets and their importance to the lives of the people living them, and engaging the community through online surveying platforms.

6 Participatory Action Research challenges designers' creative processes.

The complex interdisciplinary nature of our field conflicts with the simplistic analysis to synthesis educational models currently being adopted in landscape architecture curricula around the world. Training agents of transformative landscape change requires redesigning old curricula to give students the opportunity to construct systemic visions of change, rather than to solve punctual problems. Students in LAA341 were challenged to respond to the complexity of the issues at play through a partnership in co-creation with children and residents. In Ski, students felt overpowered by the inevitability of these changes, and focused instead on improving the 'here and now' by envisioning practical solutions for greater livability. In Bodø, they thought to respond to the decline of social capital by embedding new life and delight into the streetscape of the city's neighborhoods. In Giske, they speculated on how residents could be brought together through green infrastructure and create stronger bonds with nature and one another.

7 Transformative landscape architecture education is a training ground for leadership and citizenry.

Participatory Action Research calls for experts to take action to solve societal challenges by engaging those who are involved to act on their social responsibility and drive to leadership. Both of these can and should be taught in the landscape architecture classroom. Students should have the opportunity to volunteer, collaborate and take responsibility when there is a need for it. The classroom should be a community of learners where everyone is engaged in both learning and teaching. This requires the design of a dialogic space where individuals can freely share their points of view and ambitions. During the course of the Fall 2015 semester, students in LAA341 engaged in rich discussions that helped them to refine their group visions and storytelling, engendering a sense of ownership and responsibility to perform at their best.

References

Aradi, R. (2010), 'Surveys of children's use of space. Experiences from Fredrikstad, Norway'. *Kart og Plan* 70: 295–310.

Arnstein, S. R. (1969), 'A ladder of citizen participation'. *Journal of the American Institute of Planners 35*/4: 216–224.

Battistoni, R. M. (1997), 'Service learning and democratic citizenship'. *Theory into Practice 36*/3: 150–156.

Beatley, T. (2011), *Biophilic Cities: Integrating Nature into Urban Design and Planning* (New York, NY: Island Press).

Bose, M., Horrigan, P., Doble, C. and Shipp, S. C. (2015), *Community Matters: Service-Learning in Engaged Design and Planning* (Abingdon, UK: Routledge).

Brian, W. (2008), 'Wicked problems in public policy'. *Public Policy 3*/2: 101–118.

Butler, A. (2018), 'Landscape assessment as conflict and consensus'. In S. Egoz, K. Jørgensen, and D. Ruggeri, (eds) *Defining Landscape Democracy. The Search for Spatial Justice* (Gloucester, UK: Edward Elgar Publishing).

Clemetsen, M. and Støkke, K. (2018), 'Managing cherished landscapes across legal boundaries'. In S. Egoz, K. Jørgensen, and D. Ruggeri (eds) *Defining Landscape Democracy. The Search for Spatial Justice* (Gloucester, UK: Edward Elgar Publishing).

Davis, B. and Oles, T. (2014), 'From Architecture to Landscape'. *Places Journal* October, https://doi. org/10.22269/141013.

Dejeant-Pons, M. (2004), 'European Landscape Convention entered into force'. *Environmental Policy and Law 34*/2: 79.

Delbecq, A., Van De Ven, A. and Gustafson, D. (1975), *Group Techniques for Program Planning: A Guide to Nominal Group and Delphi Processes* (Glenview, IL: Scott, Foreman and Co).

Eaton, M. (2017), 'The New Landscape Declaration: a summit on landscape architecture and the future'. *Landscape Journal: Design, Planning, and Management of the Land 36*/1: 90–92.

Francis, M. (1995), 'Childhood's garden: memory and meaning of gardens'. *Children's Environments 12*/2: 1–16. Retrieved [7.5.2017] from www.colorado.edu/journals/cye.

Freire, P. (1996), *Pedagogy of the Oppressed* (New York, NY: Continuum).

Ganz, M. (2011), 'Public narrative, collective action, and power'. In S. Odugbemi and T. Lee (eds) *Accountability Through Public Opinion: From Inertia to Public Action* (Washington DC: The World Bank), 273–289.

Gehl, J. (2013), *Cities for People* (New York, NY: Island press).

Hall, P. (2010), 'Diagrams and their future in urban design'. In M. Garcia (ed.) *The Diagrams of Architecture: AD Reader* (New York, NY: Wiley), 163–169.

Hester, R. T. (2005). 'Whose politics do you style? If all designers serve some political will, whose will do you serve?' *Landscape Architecture 95*(12): 72–+.

Hester, R. T. (2006), *Design for Ecological Democracy* (Cambridge, MA: MIT Press).

Horrigan, P. (2015), 'Rust to green: praxis as university-community placemaking'. *Partnerships: A Journal of Service-Learning and Civic Engagement 6*/3: 8–28.

Hou, J. and Rios, M. (2003), 'Community-driven place making'. *Journal of Architectural Education 57*/1: 19–27.

Jorgensen, K., Clemetsen, M., Thoren, A. K. H. and Richardson, T. (2015), *Mainstreaming Landscape through the European Landscape Convention* (Abingdon, UK: Routledge).

Kot, D. and Ruggeri, D. (2005), 'Crafting Westport. How one small community shaped its own future'. In J. Hou, M. Francis and N. Brightbill (eds) *Reconstructing Communities: Design Participation in the Face of Change* (Davis, CA: Center for Design Research), 107–116.

Ledewitz, S. (1985), 'Models of design in studio teaching'. *Journal of Architectural Education 38*/2: 2–8.

Makhzoumi, J., Egoz, S. and Pungetti, G. (2011), *The Right to Landscape: Contesting Landscape and Human Rights* (Farnham, UK: Ashgate Publishing).

Manzo, L. C. and Perkins, D. D. (2006), 'Finding common ground: the importance of place attachment to community participation and planning'. *Journal of Planning Literature 20*/4: 335–350.

Milburn, L. and Brown, R. (2003), 'The relationship between research and design in landscape architecture'. *Landscape and Urban Planning 64*/1: 47–66.

Monson, C. (2003), 'Practical discourse: learning and the ethical construction of environmental design practice'. *Ethics, Place and Environment 8*/2: 181–200.

Park, P. (2006), 'Knowledge and participatory research'. In P. Reason and H. Bradbury (eds) *Handbook of Action Research*, First Edition (London: Sage), 83–93.

Ponterotto, J. G. (2006), 'Brief note on the origins, evolution, and meaning of the qualitative research concept thick description'. *The Qualitative Report 11*/3: 538–549.

Rittel, W. and Webber, M. (1973), 'Dilemmas in a general theory of planning'. *Policy Sciences 4*/2: 155–169.

Røhnebæk Bjergene, L. (2016, October 31) 'NMBU students designing the communities of the future'. Retrieved from www.nmbu.no/en/news/node/26777.

Ruggeri, D. (2018), 'Storytelling as a catalyst for democratic landscape change in Modernist utopia'. In S. Egoz, K. Jørgensen and D. Ruggeri (eds) *Defining Landscape Democracy. The Search for Spatial Justice* (Gloucester, UK: Edward Elgar Publishing).

Ruggeri, D. (2014a), 'Democracy matters, beginning in the classroom'. In M. Bose, P. Horrigan, S. Doble and S. Shipp (eds) *Community Matters: Service-Learning in Engaged Design and Planning* (New York, NY: Routledge), 189–209.

Ruggeri, D. (2014b), 'My Mission Viejo. A photovoice investigation of place identity and attachment in master planned suburbia'. *Journal of Urban Design 19*/1: 119–139.

Salama, A. and Wilkinson, N. (2006), *Design Studio Pedagogy: Horizons for the Future* (Gateshead, UK: Urban International Press).

Sander, T. and Putnam, R. (2010), 'Still bowling alone? The post-9/11 split'. *Journal of Democracy 21*/1: 9–16.

Sandercock, L. (2003), 'Out of the closet: the importance of stories and storytelling in planning practice'. *Planning Theory & Practice 4*/1: 11–28.

Schneidewind, U. et al. (2016), 'Pledge for a transformative science: a conceptual framework'. *Wuppertal Papers* (191). Wuppertal: Wuppertal Institute for Climate, Environment and Energy. Retrieved from http://nbn-resolving.de/urn:nbn:de:bsz:wup4-opus-64142.

Southworth, M. (2003), 'Measuring the livable city'. *Built Environment 29*/4: 343–354.

Southworth, M. and Ruggeri, D. (2011), 'Place identity and the global city'. In T. Banerjee and A. Loukaitou-Sideris (eds) *Urban Design: Roots, Influences, and Trends. The Routledge Companion to Urban Design* (Abingdon, UK: Routledge), 495–510.

Steiner, F. (2011), 'Landscape ecological urbanism: Origins and trajectories'. *Landscape and Urban Planning, 100*(4): 333–337.

Stokols, D. (2006), 'Toward a science of transdisciplinary action research'. *American Journal of Community Psychology 38*/1–2: 63–77.

Van Hulst, M. (2012), 'Storytelling, a model of and a model for planning'. *Planning Theory 11*/3: 299–318.

9

Studying landscape as a cinematic space

Irina Pața and Ana Opriș

Introduction

What is the landscape's reason for existence? Starting from the landscape's definition (as stated by the Council of Europe in the European Landscape Convention, Florence, 2000) as part of a territory perceived as such by man, characterised by the result of action and interaction between human and/or natural factors, the scientific approach targets the second part of this definition: man's presence which shapes the landscape. The procedural process of perceiving and understanding the urban landscape is related once more to the presence of man – a superior and social being characterised by thought, intelligence and articulate speech, but particularly unpredictable.

Although the word landscape still holds extremely different perspectives and meanings, in its definitions (regardless of the source) an observer is mandatory; the contemplation/judging of the space as landscape is needed. The subjectivity of the landscape as a perceptive phenomenon is a defining trait, regardless of the observer's education and his/her relation to the urban/natural environment (professionals/residents/unexperienced individuals/decision-making factors). Thus, selective perception also depends on the relation between each individual's objective reality and his or her imaginary world. This way, every landscape unit becomes vulnerable in relation to the ratio between its role in the territory and the manner in which it is perceived.

The process of understanding and recognising a landscape is a complex and subjective one and the elements that compose it are much more than we imagine at first glance. Perception is not limited to buildings, streets, squares or means of transportation – it includes monuments, live neighbourhoods, events, street life, noise, creative tension, the feeling of belonging to a place, public access, belonging to a community.

Man acts as the receiver of the messages sent by the (built and social) landscape and perception is a complex phenomenon that goes beyond the limitations of primary sensations. 'By subsuming and extracting from various theories, we are able to determine that there are two main groups of perception mechanisms: sensory mechanisms and intellectual mechanisms' (Stan 2012: 56). When discussing the perception of landscape, Stan[1] explains that sensory mechanisms belong to the realm of visual and non-visual perception, by using the observer's senses; the process of exploration is a gradual one, usually based on a single sense that acts as a leader.

Intellectual mechanisms belong to the pre-emptive realm, involving meanings that change the impact of immediate perception.

The analytical approach to a landscape, characteristic of a specialist, includes both the objective and the subjective component. The objective component is that 'in terms of which we come to understand in depth a certain landscape' and the subjective component is 'the first one that occurs in the study of a landscape, determining the manner in which we position ourselves on the first level of approach – the intuitive one' (Stan 2009: 210).

Assuming that the objective component already has a universal set of analyses, our chapter aims to describe two exercises designed to facilitate different types of re-presentation for the subjective component using video images. This is also because the objective analysis can often be incomplete, on one hand, or can overlook relevant symbolic elements related to the perception mechanism, on the other hand. Andrei Pleşu (2009) emphasises that the classical landscape analysis still relies heavily on fragments rather than the whole, on levels it is selective, abstracted, isolated or searches differences and can often pass the overall view, its fluidity, complexity, subjectivity that is, in fact, the landscape: 'stopping cannot be a good way to understand the fundamental movement of nature, its coherence, lacking in fragments' (Pleşu 2009: 46). Landscape representation in architectural drawings has not the power to grasp its complexity due to three attributes that landscape incorporates: landscape spatiality ('places, like things, conjure up a wealth of images and ideas'), landscape temporality, and landscape materiality ('landscape experience belongs to the sensorium of the tactile, the poetry of material and touch') (Corner 2002: 147–149). This is why we consider it important to use film and cinematography in landscape studies.

Landscape can essentially be considered a cinematic space by the way it is organised in time, through the dynamics of space, and movement in space. Even though, in the last ten years, the theoretical and practical research related to film and landscape in the field of landscape architecture was more debated, it is still a segment that deserves to be and has many resources that can be explored. More, in the Romanian context this area is not so much analysed and almost not at all used in the teaching process at our university.[2] Among other assignments, the case studies presented aim to introduce and test film as a tool of analysis and representation, taking into consideration the concept of space in relation to landscape and cinematography: composing and deconstructing a space, manipulating perception, travelling through space.

Analysing the interaction between the fields of architecture, landscape and cinema, we can identify how concepts migrate, considering the fields of semiology (sequence, scenario, storyboard, cinematic path), signification (the influence that movies have on our way of perceiving and understanding architecture, cities, and landscapes), and process. Different authors have underlined the strong relations between cinema and architecture concerning the design process: 'producing images and making a whole out of fragments are common goal in cinema and architecture' (Akçay 2008: 3). Pallasmaa (2000) pointed out that not only does the 'temporal and spatial structure' make architecture and cinema similar, but 'both architecture and cinema articulate lived space. These two art forms create and mediate comprehensive images of life'. Talking about the similarities between the architectural experience and the cinematic experience, Jean Nouvel states that

> architecture exists, like cinema, in the dimension of time and movement. One conceives and reads a building in terms of sequences. To erect a building is to predict and seek effects of contrast and linkage through which one passes. [. . .] In the continuous shot/sequence that a building is, the architect works with cuts and edits, framing and openings [. . .] screen, planes legible from obligatory points of passage.
>
> *(Nouvel quoted in Bruno 2002: 69)*

A colloquium was organised in Bucharest to debate 'Space in film' (Fundația Culturală Secolul 21, 2001) by different experts such as architects, film directors, screen players, art critics and image directors. The answers to the question of defining the relation between architectural space and cinematic space were sometimes similar and sometimes different: space as adjutant to the narrative, architecture as film setting and film as architectural background, architecture as a film character, space as a modified reality through the lens of the camera, film setting as pure décor, cinematic space as virtual reality and an abstract notion. Other schools or programmes also debate this relation using different approaches: in a theoretical approach, in the teaching process or as workshops and film festivals.[3]

Case study

As teaching assistants in the Department of Landscape Design and Planning at Ion Mincu University of Architecture and Urbanism, we proposed two assignments designed to test ways in which moving images can be used as tools in the process of reading and representing the landscape. The student's training programme includes, as early as the first years of study, two manners of reading the landscape – the analytical reading (which focusses on a correct and objective analysis of the urban layers) and the sensible reading (a subjective one that focusses on one's own sensitive interpretation, based on the exchange of information that occurs between the landscape and the student).

One of the most debated questions regarding film and architecture is the relation between cinematic space and architectural space. Giuliana Bruno affirms that one of the first who raised this matter was Sergei Eisenstein.[4] He was considering space to be the element that, on one hand, is the connection point between the two fields, but, on the other hand, is the element of dissociation. If the first reduces to the way we organise space and associate meanings with the proposed structures, the latter refers to the way we can experiment with the space, one being a real continuous travelling, the other being a virtual one, a discontinuous and fragmented understanding of space. The student assignments were designed to explore these connections and differences. Also relevant for the research is the permanent reference of one field to another and the exercises were built to mirror this relation, starting with Hopkins's approach of cinematic landscape (Hopkins 1994), which identifies two different processes that add value and signification to the way we read a landscape through the question of 'representation in and interpretation of film', and Penz and Lu's approach of Urban Cinematics (Penz and Lu 2011), how cinematography uses the urban space in parallel with how urban studies uses cinema.

The first assignment investigates the relationship between real and virtual as a two-way approach to the relationship between the landscape and the image. Thus, it is useful to perform research approached from two complementary directions: the manner in which the real landscape is captured on video or in a photograph and then the manner in which it is perceived and represented through screenings (video montages) submitted by the students.

Why would such a tool be useful? Because an integrated approach to urban landscape that can help us better understand the ideas and events present in an urban landscape requires work instruments.

Even as early as their first exercise of this type, counting on the possibility of enriching the process of interpreting the landscape, the students are guided to develop their own mechanisms for understanding the landscape (decomposition and re-composition) by using video montage.

The first exercise that uses video as a tool for the sensible haptic reading of the urban landscape consists of a path/urban space and creating a video montage in order to capture and render the recorded landscape sequence in a personal manner by using one's own storyline (Figure 9.1).

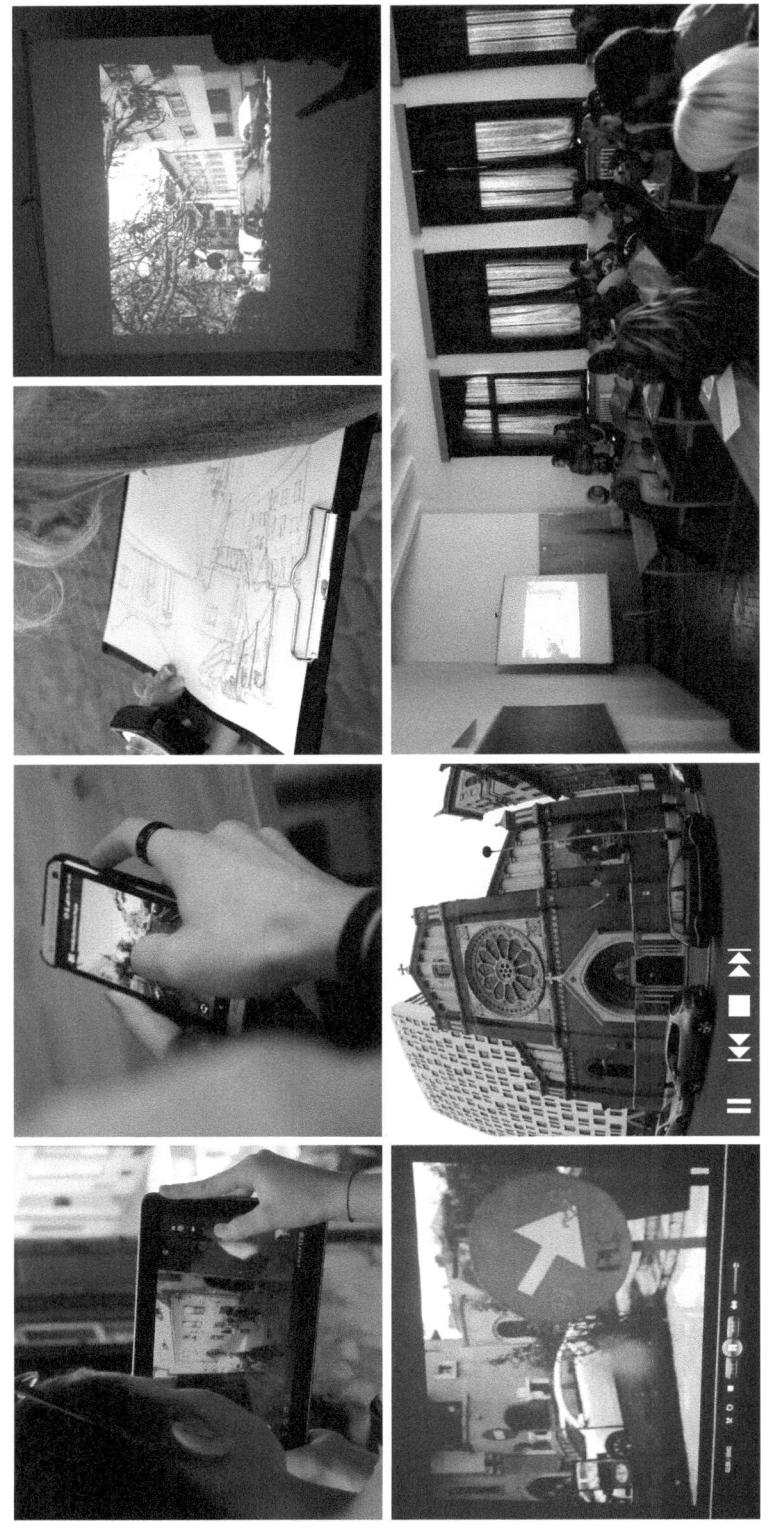

Figure 9.1 Images extracted from the video-montage assignment: recording the video file, creating a narrative sketch and the presentation of final results. First year Bachelor studio: Landscape Representation, 2014. Source: personal collection of the authors

A borderline definition of the landscape sequence can be that of an anthropic framework within which the individual discovers certain meanings, which he then reads, analyses and organises.

Both the creation of the video montages (using the bike, the portable camera, the soundtrack, the colour effects, distortion filters, etc.) and their screening are two project phases that are visibly useful for students – 'in a city, the moving elements, particularly the inhabitants and their activities are as important as the physical, stationary components', 'light changes everything in a city', 'noise changes the perception of a landscape but also the manner in which one covers a space' – are just some of the topics discussed while viewing the video montages created by the students.

Such an exercise tends to show that the urban landscape, in terms of its structure, is the result of the continuous stratification of major urban projects and minor individual projects that, in time, supersede and overlap. The urban landscape features the visible traits of a territory, combining both physical and cultural elements, indicating the human presence, thus becoming a living human–space synthesis.

Because urban vulnerability studies performed today focus on the built component and its predisposition to natural or anthropic dangers, and focus less on the discrete urban phenomena specific to the landscape,[5] the aim of our video-montage exercise is to observe the urban reactions, how the subjects position themselves in relation to the urban landscape within which they interact with the public space, the way in which they participate in shaping this space or are concerned with its evolution; this is a starting point in studying the vulnerability of the urban landscape.

Assuming that professionals (architects, landscapers, urban planners) possess complex deduction instruments and are capable of quickly and practically understanding the characteristics of their urban landscape, we considered it necessary to develop a set of 'reading the urban landscape' student exercises capable of verifying the diagnosis offered by specialists and translating it into the reality of common individuals who are also observers of the urban landscape and users of the public urban space.

The second assignment considers the reverse process, the capacity to read a space presented in a film sequence.[6] This task focusses on creating connection between classic forms of representation and the dynamic medium of film. The students had to use graphic elements that they already comprehend like points, lines, surfaces and hatches to present a virtual space as the setting for the film. They had to be able to read and recompose an imaginary urban landscape, to identify landmarks, to set them in space and construct the missing elements of connection, to translate the dynamic perception of the space into design representation specific for urban and landscape projects.

The selected sequence for the second assignment was from the movie *The Third Man*, 1949, directed by Carol Reed, 1:05:20–1:07:30. It consists of thirty-six different frames, most of them concentrated on architectural elements and urban space (Figure 9.2). The movie sequence was filmed in three different areas of Wien old city centre and combined as a montage to create the illusion of one single urban space (Figure 9.3). This is a technique used by many directors in order to be able to create a more complex and expressive décor.

This was a 45-minute exercise. The students watched the short clip multiple times and were asked to draw in plan and section (or 3D sketch) the space that the main character is transiting. Most of them were able to decode the elements presented by the film director and compose them in a continuous way, forming a logic space. They identified the possible path that the character had taken and made coherent drawings of the spatial structures from the clip (Figure 9.4 and Figure 9.5).

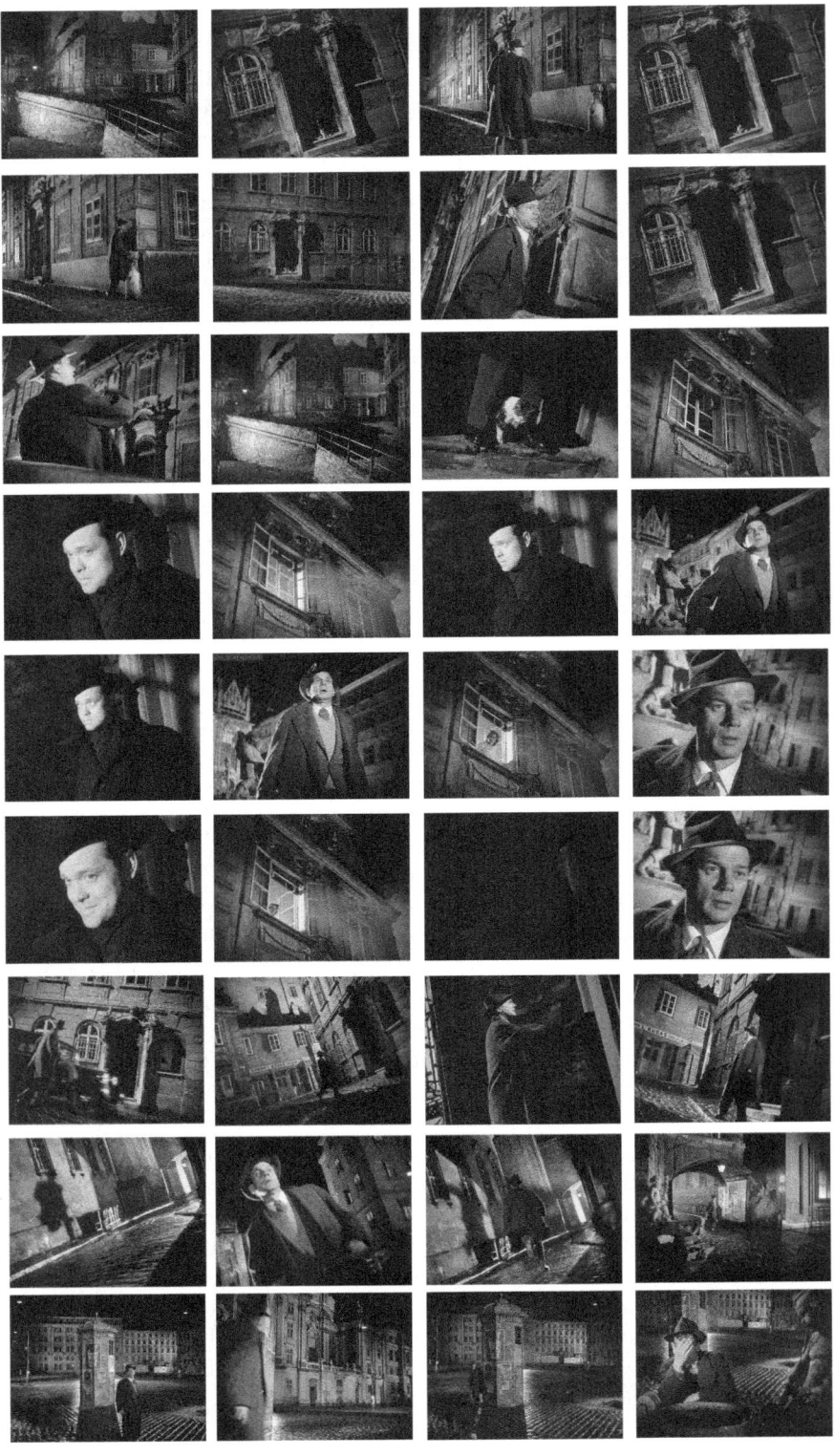

Figure 9.2 Images extracted from the movie sequence. Source: personal collection of the author

Am Hof

Schulhof

Moelker Steig + Maria am Gestade

Figure 9.3 Presenting the relation between Wien real landscape and the virtual composed landscape in the presented sequence of *The Third Man*. The images were correlated by identifying the landmarks that define the urban landscape. Source: image captures from the movie, personal collection, and Google street view.

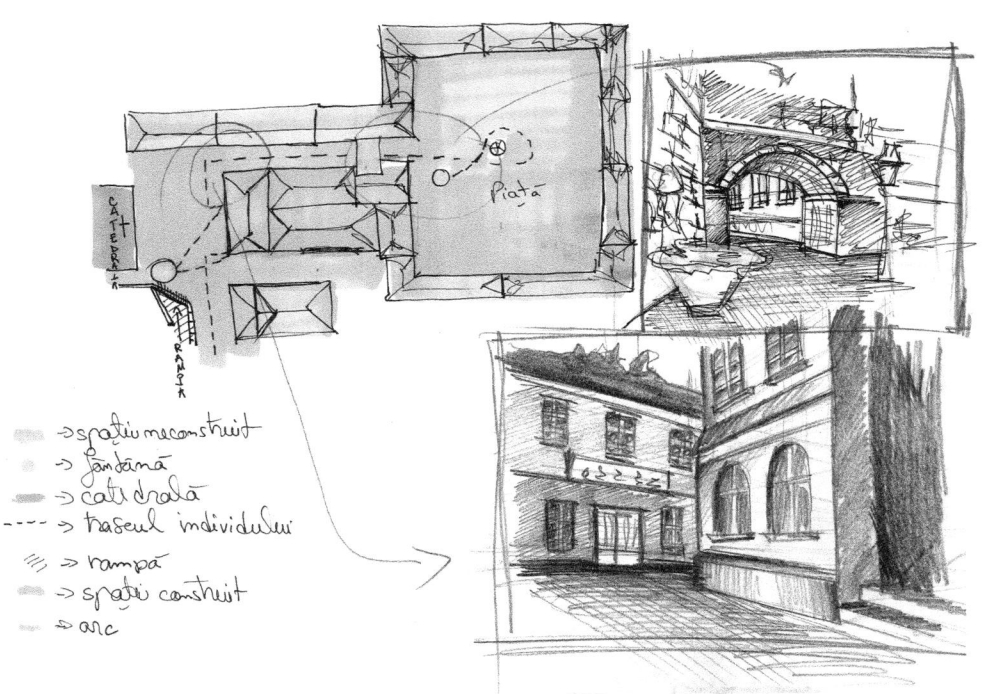

Legend (handwritten):
→ spațiu neconstruit
→ fântână
→ catedrală
---- → traseul individului
→ rampă
→ spații construit
→ arc

Figure 9.4 Assignment 2 result. Source: Carata Adina Livia, first year Bachelor student, 2016

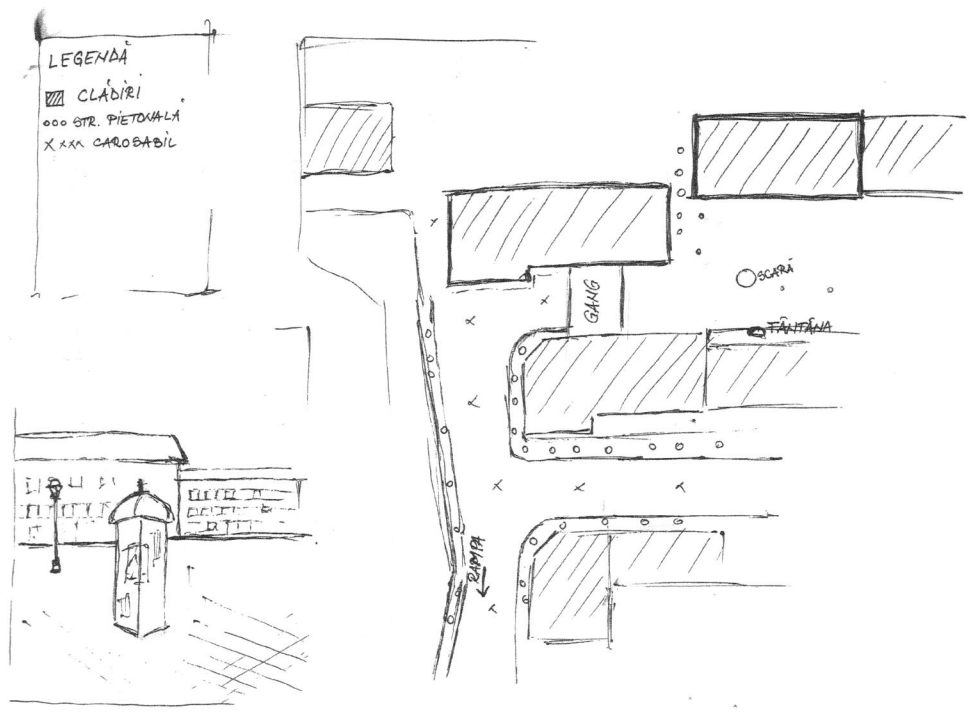

LEGENDĂ
CLĂDIRI
ooo STR. PIETONALĂ
x xxx CAROSABIL

Figure 9.5 Assignment 2 result. Source: Secăreanu Andreea, first year Bachelor student, 2016

Analysing the thirty-two delivered assignments we can conclude that students described a logic system of spaces using the technique of decomposing the moving images and recomposing an urban landscape using the identified landmarks (Figure 9.6). The elements can be grouped in three categories, relating to space and identity: urban spatial structures, landmarks and elements of décor. If 90% of the students deciphered the main structures that define the space, the streets, the square, the gangway with the arch, only 40% considered the level difference in space as a site specific. From the percentage results, we can clearly see that different elements that help in framing the space were chosen by the film director and emphasised by different cinematographic qualities. In this way, the round kiosk, which will become a trigger point in the narrative, is much more visible than the church (Maria am Gestade), a building that is more representative in an urban landscape.

Conclusions

Landscape as a cinematic space is best approached from two perspectives: how the landscape is captured in film and the interpretation of landscape through film.[7] The two assignments presented above were designed to test the way film captures different landscape features and the way we can intercept the sensitive characteristics of the landscape otherwise omitted by classic analysing tools.

Students have responded well to the assignments and were dedicated to exploring other tools of investigation and representation. We think that these types of exercises in the early years of

Number of students	32	Percentage
Identified urban spatial structures		
public square	29	90%
connection street	29	90%
the level difference in space	13	40%
narrative path	27	84%
Identified landmarks		
gangway / arch	29	90%
1st statue (near the church)	9	28%
2nd statue + fountain (in the square)	26	81%
circular kiosk	30	94%
church	8	25%
Identified elements of decor		
light pole	2	6%
the cat	2	6%
handrail	1	3%
stone walkway	3	9%
door/s	3	9%
window/s	3	9%

Figure 9.6 The table presents the inventory of the second assignment results. Source: the authors

studies are very important in the landscape teaching process because they include the haptic dimension of the landscape. Also, studying space through cinematographic means can help students hierarchize elements in a landscape design plan or an urban development plan. At the same time, it can raise questions about perception, sequential reading of space and the role that landmarks have in reading a space.

Notes

1 Mechanisms developed by Angelica Stan in the chapter referring to landscape perception included in the book *Devenirea Peisajului* [The Landscape Becoming] (2012).
2 Department of Landscape Design and Planning, Ion Mincu University of Architecture and Urbanism, Bucharest.
3 'Sergei Eisenstein bids them (the filmmakers, a.n.) to look back over architectural history, from Kallikrates to Bernini, so they may understand how to arrange space to best convey meaning'. The consumer of architecture becomes what Giuliana Bruno refers to as the 'prototype of the film spectator'. 'Eisenstein is quick to point out that architecture is the body traveling a contiguous path, as opposed to film, which is the mind traversing a stream of spatially and temporally discontinuous images'. This is the manifestation of Bruno's 'site-seeing' vs 'sight-seeing' (Newman 2012).
4 Some examples that we consider relevant for the three categories are:

- in theoretical approach: Hopkins 1994; Bruno 2002; Girot 2006, 2014; Truniger 2013; Penz and Lu 2011.
- in the teaching process: Cristophe Girot – *Institute for Landscape Architecture (ILA)*, ETH Zürich; Michael Szivos – *Reading New York Urbanism (RNYU)*, Graduate School of Architecture, Planning and Preservation, Columbia University; Mads Farsø and Rikke Munck Petersen – *Landscape Film Studio*, University of Copenhagen;
- in workshops and film festivals: *CINECITY Architectural Film Project Architectural Ideas Explored in 60 Second Films* (2009–2015); *Urbaneye Film Festival* (2014–2017); Copenhagen Architecture Festival x Summer School (2015–2017).

5 Although no generally accepted formula for the quantitative description of vulnerability is in place, while approaching the notion of urban landscape vulnerability we can refer to two key notions: acute vulnerability and latent vulnerability; the first is characterised by the susceptibility of the elements of urban landscape to be adversely affected by certain dangers (with rapid evolution); the second is characterised by their disturbance as a result of urban phenomena (with discrete evolution) such as landscape deterioration, loss of identity, cultural decline, etc.
6 This assignment is part of the 'Visual Aesthetics' course, coordinated by Liviu Veluda, teaching assistant, PhD: www.uauim.ro/en/faculties/urbanism/landscape/compulsory-courses/up-146.
7 'This cinematic landscape is not, consequently, a neutral place of entertainment or an objective documentation or mirror of the "real", but an ideologically charged cultural creation whereby meanings of place and society are made, legitimized, contested, and obscured' (Hopkins 1994: 47).

References

Akçay, A. (2008), 'The Architectural City Images in Cinema: The Representation of City in Renaissance as a Case Study', Master's thesis, School of Natural and Applied Sciences of Middle East Technical University [website] http://etd.lib.metu.edu.tr/upload/12609699/index.pdf, accessed 10 March 2017.
Bruno, G. (2002), *Atlas of Emotion: Journeys in Art, Architecture, and Film* (New York: Verso).
Corner, J. (2002), 'Representation and Landscape' in Swaffield, S. (ed.) *Theory in Landscape Architecture: A Reader* (Philadelphia, PA: University of Pennsylvania Press), 144–165.
Council of Europe (2000), *European Landscape Convention*, Florence [website], https://rm.coe.int/1680080621, accessed 10 March 2017.
Farsø, M. and Petersen, R. M. (2015), 'Conceiving Landscape through Film: Filmic Explorations in Design Studio Teaching'. *Architecture and Culture*, 3/1: 65–86.
Fundaţia Culturală Secolul 21 [21st Century Cultural Foundation] (2001), 'Spaţiul în Film' [Space in Film] in Doinaş, Şt. Aug. (ed.) *Filmul. Secolul 21* [21st Century Film] (Braşov: Editura Fundaţiei Pro), 13–42.

Girot, C. (2006), 'Vision in Motion: Representing Landscape in Time' in Waldheim, C. (ed.) *The Landscape Urbanism* (Princeton, NJ: Princeton Architectural Press), 87–103.

Girot, C. (2014), 'Landscape: Beyond the Margin of Vision', in Deriu, D., Kamvasinou, K. and Shinkle, E. (eds) *Emerging Landscapes: Between Production and Representation* (Farnham, UK: Ashgate Publishing), 81–94.

Hopkins, J. (1994), 'A Mapping of Cinematic Places: Icons, Ideology, and the Power of (Mis)representation' in Aitken, S. C. and Zonn, L. E. (eds) *Place, Power, Situation, and Spectacle: A Geography of Film* (Lanham, MD: Rowman and Littlefield), 47–66.

Newman, D. (2012), 'Polite Resistance and the Dimensions of Narrative – Week Six Response' [blog] 27 February. *Media and Architecture: A Graduate Seminar in Media Studies at The New School.* Taught by Shannon Mattern [website] www.wordsinspace.net/media-architecture/2012-spring/?p=482, accessed 10 March 2017.

Opriș, A. and Paţa, I. (2012), 'Urban Landscape: Dynamics in its Perception and Self-Adjustment'. *Journal of Singapore Institute of Planners. Planning Studies & Practice*, 4/1: 57–68.

Pallasmaa, J. (2000), 'Lived Space in Architecture and Cinema', Chapter 1 in *The Architecture of Image: Existential Space in Cinema* (Helsinki: Rakennustieto) [website] www.ucalgary.ca/ev/designresearch/publications/insitu/copy/volume2/imprintable_architecture/Juhani_Pallasmaa/index.html, accessed 10 March 2017.

Penz, F. and Lu, A. (eds) (2011), *Urban Cinematics: Understanding Urban Phenomena through the Moving Image* (Bristol, UK: Intellect).

Pleşu, A. (2009), *Pitoresc și Melancolie* [Picturesque and Melancholy], third edition (Bucharest: Humanitas).

Stan, A. (2009), *Peisajul Periferiilor Urbane – Revitalizare Peisageră a Zonelor Periferice* [Urban Periphery Landscape – Landscape Revitalization of Peripheral Areas] (Bucharest: Ion Mincu University Press).

Stan, A. (2012), *Devenirea Peisajului* [The Landscape Becoming] (Bucharest: Ion Mincu University Press).

Truniger, F. (2013), *Landscript 2 – Filmic Mapping – Documentary Film and the Visual Culture of Landscape Architecture* (Berlin: Jovis Verlag).

Filmography

Reed, Carol (dir.). 1949. *The Third Man*. UK.

10
Attention and devotion

Thomas Oles

The problem of teaching a subject is always the problem of learning it. I begin therefore with the memory, still vivid, of my own education in landscape architecture.

I entered this discipline in the middle of my career, after a decade studying language and literature. I suddenly found myself a complete novice in another field, without the luxury of years to regain the competence I had lost. I had to start all over again.

The result was considerable psychic discomfort. I remember my first weeks in the landscape architecture programme as a fog of mystery and dread. What was 'design', and how would I acquire the skills I would need to practice it? Everything about design education seemed so vaporous that my peers and I often found ourselves suspecting our teachers' motives. Was there something they were concealing out of *Schadenfreude*, the sheer pleasure of seeing us writhe as they had done? I remember running into one of my instructors, a genial man near retirement, on the way home after a long and (it seemed to me) humiliating day in studio. 'It's hard at first', he offered. 'Still—there's no need to walk around with a dark cloud over you!' I nodded agreement, but inside I seethed. 'All very well for him to say that', I thought. 'He's the teacher! This is *easy* for him!'

And yet the design studio was a forgiving environment. There the noise of the world was muted, there the beginner could fail in safety. Not so real landscapes. These were decidedly *unforgiving*, and that is probably why students were kept well clear of them for the first weeks of class. Our time was spent, instead, on studio tasks that many readers of these words will recognize. We translated abstract ideas ('tension', 'rupture', 'calm', etc.) into two-dimensional compositions, made diagrams representing conceptual opposites ('symmetry/asymmetry', 'balance/imbalance', etc.), built models of the 'volumes' that 'created space'. The fruits of these labours were then judged according to standards we understood only hazily, if we understood them at all.

These exercises, legacy of the Beaux Arts and Bauhaus, were no doubt well-intentioned. In their own way they were valuable, perhaps even essential. I remember them with some fondness. But they had an unfortunate result. They established, at the very start of my landscape architecture education, a rigid separation between *abstract concepts* and *real places*. This separation was also a chronology, and, implicitly, a hierarchy: 'site analysis' came *before* sites, 'landscape volumes' *before* landscapes. The message to the beginning student was subtle but unmistakable.

First (it went), learn these categories for representing the world; then you will be ready to manage complexity in it. If we somehow missed it, this point was driven home several weeks later. Presented with our first real-world site, a small parking lot behind the architecture building, we were instructed to use the very same techniques we had employed when there was no site at all. The world, it seemed, was so much blank paper—and we held the pen.

The field tasks we performed in that first studio are, alas, long lost. I therefore cannot say with certainty what students were told to look for, how to look for it, or how to communicate what they found. Still, it is not hard to imagine the lineaments:

> Produce a 1:500 analysis plan of current site conditions, using the graphic conventions you have been learning in this class and in studio. At a minimum this plan should show: ground cover (asphalt, grass, dirt, etc.); light and atmospheric conditions (sun, shadow, wind, noise, etc.); location of physical infrastructure (buildings, roads, paths, wires, lamps, seating, etc.); vegetation location and types (tree trunks, tree canopy, shrubs, flower beds, etc.); flows across or around the site (water, animals, vehicles, etc.); and patterns of human movement and interaction (pedestrian traffic, gathering places, etc.).

Writing this hypothetical exercise just now was a 'System 1' task (Kahneman 2011). It required almost no active mental effort on my part; unlike the passages that precede and follow it, these words practically wrote themselves. It would be easy, too, to reproduce the 'graphic conventions' to which this assignment refers (Figure 10.1).

Why should these tasks, the writing and the drawing, be so easy? They are, after all, conceptually and technically difficult: most adults would be unable to perform them at all, to say nothing of performing them well. So why does this way of 'reading the landscape' remain so automatic, so *comfortable* for me, even after many years spent outside active design practice?

The answer is simple. This ease is not a function of personality, or intellect, or talent. It is a function of training. It is the result of having learned a very particular set of practices for making sense of the world. It is a product of having read and heard, over many years, words like these:

> The kinds of information collected for our contextual analysis basically involve an inventory of existing and projected site conditions. We are not concerned with design responses to the site at this stage but rather with finding out all we can about the site. We are interested in facts.

> *(White 1983: 16)*

This passage, taken from one among dozens of similar 'site analysis' textbooks, describes the principal method for 'reading the landscape' taught to several generations of environmental design professionals, and I am no exception. It is the single most important legacy I carry from my first months as a landscape architecture student, and it remains an essential part of all my teaching.

At the same time, I remember well my own frustration with this method as a student. That 'site analysis' and 'landscape design' were treated in separate courses suggested that they were somehow different. But (like many things in design education) the nature of their relationship was never stated precisely. The passage above is an example of this reticence. It describes a process of collecting 'information' that is explicitly non-normative ('we are not concerned with design responses at this stage') and includes 'projected site conditions' alongside 'existing' ones. Yet what is a 'projected site condition' if not a 'design response'? There would seem to run, somewhere, a boundary dividing facts from values, 'inventory' from 'analysis'. But the course of this boundary, if it exists at all, is left off the map.

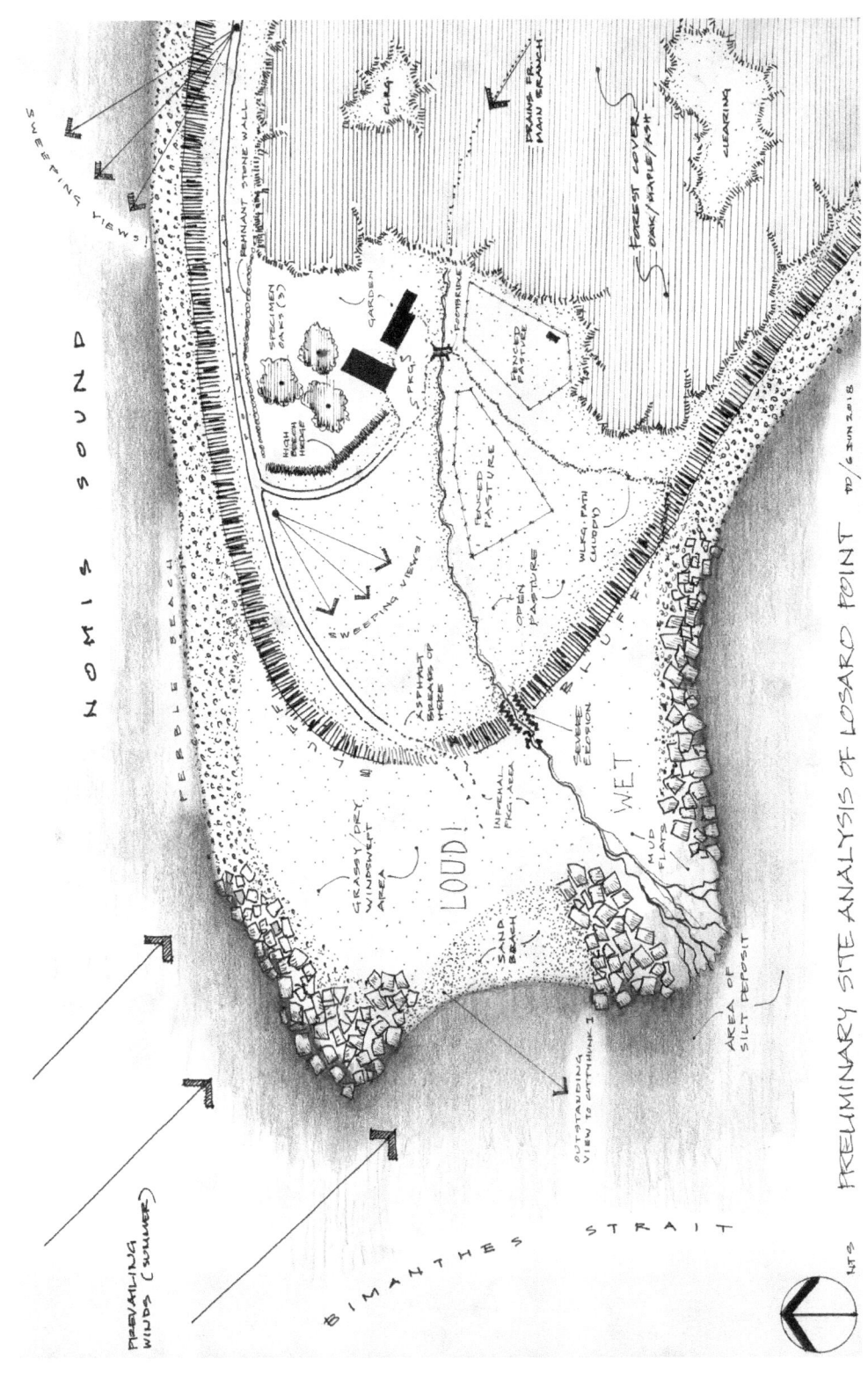

Figure 10.1 Preliminary site analysis of Losaro Point

My education was marked by similar mystification about the role of site analysis in the design process. But I remember getting the message loud and clear. Aggregation and representation of 'site conditions' and 'facts' ('finding out all we can about the site') would yield, eventually, a key insight about the particular place being observed. This insight, somehow, would then propel one forward: first to a 'concept' (in those days still called a 'parti'), then to 'schematic design' (as though a design could ever be anything other than a schema), and finally to a fully 'resolved' proposal. By following scrupulously what were, really, quite simple rules, we would learn, in time, to turn the base metal of private experience into design gold, a proposition with general validity. For a beginning student looking for some handhold, any piece of unshifting ground, it was a comforting account—not despite its naiveté but because of it.

These memories still cut close to the bone, but I am well aware that they date from another age. No doubt all teachers sooner or later begin to sense a gulf between themselves and their students. But I cannot help thinking that this particular gulf is wider, and widening faster, than ever before. After several centuries of constancy, the tools and methods of design education have changed so dramatically in the last twenty years that I sometimes doubt my ability to teach them at all. My students' experiences are increasingly remote to me, like the vast unexplored regions on a medieval globe. What fantastic monsters, I wonder, dwell in that *terra incognita*? What stories will my students—equipped with space-age kit, unlimited information, and far more savvy (I assume) about the world than I ever had—tell their students of their voyages? Are the loads they carry heavier, the risks they face greater, than those I did?

In recent years I have tried to map this territory by increments, asking landscape architecture students to reflect more or less systematically on their own learning. No doubt due to my own experience, these exercises have tended to focus on students' first encounters with new or unfamiliar landscapes, and the role that this early 'reading' plays in their subsequent design process. The responses are illuminating. 'We arrived in the morning at the site for a three-hour visit', one advanced student in a respected American MLA programme writes:

> Armed with digital cameras, clipboards, and sketchpads, we had barely stepped off the bus before we were clicking away in every direction, digital photos of anything and everything . . . Though I hate to admit it, I was one of those shutter-happy people. Without even really looking or taking the time to land, pause—breathe—I was busy filling up my memory card and frantically taking notes, as if somehow all of this recording of information would capture the 'essence' of the place. After a three-hour flurry of running around, snapping photos and, in moments of distraction, being nearly flattened by traffic and cyclists, we piled back on the bus, did a windshield tour of the neighborhood, and went back home. Three months later, we presented a complete urban design for the site, having only visited it once that morning.
>
> *(Poulin 2013: 4)*

Taken alone, this anecdote would have little weight. But a first-year undergraduate student from the same programme tells a nearly identical story:

> We got out of the car, blew up the helium balloon, and started the trip . . . At first, people were wandering around and 'bombarded by sensations,' from rippling cattails to massive flocks of birds. Then, cameras were out and everyone started to observe (or: pretend to observe) things that were worth an artistic shot . . . In that moment, this special site was not anything special to me. I kept taking pictures just like a machine . . . The camera literally

replaced my memory and I was acting like a zombie. This morning, as I was looking at the photos I took, I honestly was not sure of the purpose of this field trip. My reflection, simply put: *I was there, so what?* I talked to the other students I went with and got back almost the same feelings.

(Peng 2016: 1)

These accounts of 'reading the landscape' describe an experience very different from my own. This is not the frustration of wanting to say something but lacking the technical skill to say it. It is not self-doubt while looking down a long hard road of 'failing again, failing better'. It is not clinging to a single method of site analysis. This is the burden of *not knowing how to pay attention in the first place*. The hyperactivity these students describe ('clicking away in every direction', 'taking pictures just like a machine') is their automatic response to a world of signal where they dare not classify anything as noise. The result, they realize later, is not insight but 'distraction' or, worse, anaesthesia.

What can explain this difference? It is tempting to cite tools these students carry but that I lacked, particularly the miniature networked computer ('smartphone'). But this is too easy. These students' confusion while 'reading the landscape' is not just a matter of technics, of failure to master a new and powerful tool. Something deeper runs beneath these accounts. This is the ontological relationship between subject and object. Both assume that something called 'the landscape' actually exists in the world, something available for 'reading'. This thing is very complicated, of course. It is composed of many parts (buildings, trees, roads, sunlight, water, human bodies) related to one another in complex and unpredictable ways. But it is nevertheless *knowable*. With proper training and the right tools, one may take its true measure. The aim, in both, is to 'find out all we can' about this complicated thing—not for its own sake, not for the pleasure of finding out, but rather because we want to do something to it. The landscape is not only knowable, it is *deficient*.

In one form or another, these assumptions underwrite every practice of landscape architecture. The profession could hardly exist, let alone be taught, without them. But such objectification entails a paradox. This paradox has been formulated in many ways across time and place; in Western philosophy it reaches back at least to Plato and the pre-Socratics. Here is Sigmund Freud:

[T]he first step towards the intellectual mastery of the world in which we live is the discovery of general principles, rules and laws which bring order into chaos. By such mental operations we simplify the world of phenomena, but we cannot avoid falsifying it in doing so, especially when we are dealing with processes of development and change.

(Freud 1937: 373)

To understand the world one must represent it, but to represent it is, inevitably, to impoverish it. This tension is especially salient for landscape, a word whose meanings span 'subjective' perception and 'objective' reality. Every landscape is both 'in here' and 'out there' (Wylie 2007: 5), the world *as I see it now* and the world *were I never to see it*. No single account of the second can ever lay claim to authority, for any such account inevitably depends on the first. This surplus is embedded in the very idea of landscape.

No landscape serves itself up in ready-made digestible chunks. A real place will eventually slip every conceptual net people try to throw around it. This will not surprise anyone with experience of environmental design practice and its 'wicked problems' (Rittel and Webber 1973). It is a problem for such practice only if *not* slipping nets is made the precondition for envisioning change.

Something like this seems to be at work in the passages above. Despite their best efforts, these students cannot make the world yield in the way they expect; the result is panic and paralysis. This, however, is a diagnosis; it does not explain the source of the expectation itself. An encounter with a new place might just as well have been a liberation and delight for these students. It might have given them license to feed their curiosity or wonder, follow their intuitions and passions, exchange questions they assumed they *should answer* for questions they found they *must ask*. That they do not experience it as such reveals something profound and, I think, disturbing about their education. They cannot take pleasure in 'reading the landscape' because they have not been prepared to do so.

These students are typical. Their experience is echoed widely in my own classrooms, which span institutions, countries, and every level of the landscape architecture curriculum. Indeed, the confusion and discomfort they describe is growing more acute. The causes of this trend, as well as its character and extent, need further exploration, and this is not the place to advance formal hypotheses about them. But I believe they relate to a change in landscape architecture discourse. The humanities dimension of the field has been less and less emphasized in recent years, as landscape architects attempt to assert their status as 'researchers'. This has entailed what might be called a 'quantitative drift' in which *measurement* of landscape phenomena is privileged over 'subjective' accounts when it comes to motivating and justifying design decisions (Lenzholzer et al. 2013). This drift is, of course, part of a much wider change in the modern era, the process by which 'the qualitative was reduced to the subjective: the subjective was dismissed as unreal, and the unseen and unmeasurable non-existent' (Mumford 1934: 49). It will surely form one chapter in some future history of the profession, and it probably has more to do with current requirements of academic advancement than with any inherent properties of landscape architecture. But, whatever its nature and cause, it can hardly have failed to affect students' understanding of the discipline. If the stories above are any indication, landscape architecture programmes today are increasingly stalked by what Kant called 'the ghost of totality': the phantom of complete knowledge about the world (Kant 1997 [1781]: 237).

I do not want to be misunderstood. I am not arguing against collection of 'data'. I am not making a case against 'evidence' in landscape architecture practice (one can hardly imagine a modern profession being built on claims of ignorance). It is altogether reasonable that students should gather and analyze information about the places whose futures they are asked to envision. The impulse to do so, indeed, attests to a nascent professional ethics. The problem is that, when given free rein too soon, *this same impulse* will almost certainly inhibit the design creativity students expect it to enable.

This might strike some as a paradox, but will come as no surprise to those with practical experience in design. Generally speaking, the early insights that fuel design exploration—so-called primary generators—can be neither deduced from *a priori* categories (a 'brief') nor derived directly from any particular set of 'facts' about the world (Darke 1979; Eckert 2000). Design ideas emerge, almost always, through sustained, even obsessive *attention to phenomena*. This is important for the eventual 'solving' of particular design 'problems', of course. But ultimately the designer's attention is not a function of any single task, a reliable vehicle turned on and off at will. It is, rather, a voracious and unruly harvester of impressions, powered by the mighty twin engines of curiosity and awe. It drives the designer at least as often as the designer drives it.

Cultivating the habits of such attention is quite simply the main task of landscape architecture (and probably all design) education. And yet: it *cannot* be so simple, for otherwise these habits would already occupy a central place at every level of the professional curriculum. They do not; indeed, if my own training is any guide, they have never been. In a sense this is not surprising. Few things are more difficult to explain than the mental operations by which a *private* experience of landscape becomes a *public* argument about how that landscape should change. It is easy to understand why landscape architecture teachers might choose to rely on established methods, however great their distortions.

This is not to say there are no alternatives. Perhaps in reaction to the drift noted above, recent years have also seen attempts to formulate more qualitative approaches to 'site analysis' (see Lanfranco 2009). A few of these have made their way into professional curricula. One is the following account of 'landing' in an unfamiliar landscape at the start of the design process:

> Landing requires a particular state of mind, one where intuitions and impressions prevail, where one feels before one thinks, where one moves across and stalks around before seeking full disclosure and understanding . . . During landing, nothing is allowed to remain obvious or neutral to the designer; rather everything is apprehended with wonderment and curiosity, with subjective and interpretative eyes.
>
> *(Girot 1999: 62)*

I find this passage suggestive. I recognize the state of mind it describes, and it chimes closely with my own experience. In many ways it is a more honest account of 'reading the landscape' than those that underwrote my own education. The problem is that this is not, I have found, a very good index of its utility for landscape architecture teaching. I have discussed this text with many students and none has ever reported being helped by it to arrive at a design idea. In its own way, it is as mystifying as the ubiquitous but non-operational idea of a 'genius loci', primarily as advanced by Norberg-Schulz (1979) in his book of the same title. A few paragraphs of that work (or the Heideggerian phenomenology on which it is based) are enough to leave students and their teachers pining for the clarity of traditional site analysis, however great its distortions.

When it comes to 'reading the landscape', then, landscape architecture education appears torn between facts and feelings, the ghost of totality and the spirit of the place. Neither alone will suffice for developing design skill. We might do better, I have often thought, following the poets.

Consider the following passage, by the American poet Stanley Kunitz, on the writing of Anton Chekhov:

> He was particularly hard on examples of stylistic inflation, such effusions as "the setting sun, bathing in the waves of darkening sea, poured its purple gold, etc." Brevity, relevance, and specification were his criteria of excellence. A passage in one of his letters reads: "In descriptions of nature you should seize upon the little particulars, grouping them in such a way that they make a clear picture, even for eyes that are closed. For instance, you can convey the full effect of a moonlit night if you write that on the mill-dam a little glowing star-point flashed from the neck of a broken bottle. . . ." An entry in Chekhov's notebook, dated several years later, reveals how faithful he was to his own instructions: "A bedroom. The light of the moon shines so brightly through the window that even the buttons on his nightshirt are visible."
>
> *(Kunitz 1985: 122–123)*

Here poetry is less genre or sensibility than habit and method. The poet Kunitz reveres the prosist Chekhov's disciplined attention to particulars, to the 'intricate texture and density' (Sadoff 2010, 109) of the phenomenal world. These particulars come, first and foremost, from lived experience, but there is no straight line that leads to them, no 'rational' method for determining *which* particular to notice and when. Poems arise not from the neocortex, 'the organ of philosophy and the sciences', but rather 'out of the swamps of the hindbrain, "the old brain," dragging their amphibian memories behind them' (Kunitz 1985: 51). Indeed, to go into the world looking for a poetic image is to insure that one will not find it. The poet's skill lies, rather, in maintaining a constant state of expectancy, a readiness to accept with reverence and gratitude those 'gifts' the world might bestow at any moment (Hyde 1983).

One would need many books to explore the ways that poets from Virgil to Wordsworth to Kunitz have evoked real or imagined landscapes, to say nothing of poems, 'about' landscape as such (for example, Richard Payne Knight's 'The Landscape' [1794]). But this would miss the point. I want to open a different discussion, about how *being and thinking like a poet* might help students of landscape architecture become better practitioners. This is a question of methodology, and the way one answers it would seem to have clear implications for pedagogy. Yet virtually nothing has been said about the ways that poetry might stimulate landscape architecture learning, despite the many affinities between design methods and those of the fine arts more generally (McDonnell 2011).

Practical readers will already see the challenges. Perhaps greatest of these is the aura of mystery (or mystification) that persists around poetry and poets, yoked by Romanticism to the notions of 'inspiration' and 'genius'. As has been widely noted, these concepts are at best unhelpful when attempting to develop creativity in oneself or liberate it in others (Nachmanovitch 1990). It is safe to assume, then, that at least some landscape architecture students will harbour assumptions about poetry that, unaddressed, will sow doubt rather than instil confidence.

The best antidote for these assumptions is etymology. It is worth recalling that the Greek *poesis* denotes, quite simply, 'making' of any kind. The Middle English 'maker' was synonymous with the modern English 'poet'; Scots English used the older word well into the nineteenth century. This sense of poetry as making (or forging) is preserved in the colloquial 'wordsmith', a term that originally referred to poets but later came to denote any person who has acquired writing skill. Like those of any other craft, then, poetic skills are a function of disposition and discipline, not 'talent'. They can be learned, practiced, *improved*.

One could do worse, then, than to consider how poetry *itself* is taught. Doing this suggests many affinities with landscape architecture pedagogy, particularly when it comes to developing skill in the 'art of paying attention'. Take the following exercise, developed by poet and educator Roger Mitchell:

> Write a poem that is simply a list of things.

The 'simply', again, gives the game away. Compiling such a list is not nearly as easy as it seems, and that is the point. Mitchell assigns this task to beginning students to help them

> look outside their minds to or towards the objective world . . . The exercise can help with a number of lessons all writers need to learn: that language is a more plastic medium than we are taught, that the objective world is our best (perhaps only) source of images, and that poems are made things.
>
> *(Behn and Twichell 1992: 37)*

Its most important lesson, however, is *subjectivity*. Unlike conventional 'site analysis', this task furnishes no normative categories or hierarchies whatsoever, no understanding of the world that precedes the experience of the person who completes it. It is silent on what this list should include, where the things in it should come from, or indeed what constitutes a 'thing'. In the absence of such handholds (or crutches), students are forced to follow their own intuitions, memories, or obsessions.

And yet, like other professionals, landscape architects must eventually organize and classify such intuitions in order to understand how (and indeed whether) they relate to the design problem at hand. In practice this almost always means organizing one's particular experience of a given landscape (the 'things' in one's 'list') according to general categories. This process is almost always unavoidable, but it is also among the most dangerous moments of the design process, since it is here that one's early design insights are at greatest risk of being lost. Poetic methods might help students understand and navigate this difficult phase. Another exercise used by Mitchell, aimed at mastering metaphor, asks students to 'describe an object or scene that particularly interests you without making any comparisons of one thing to another' (Behn and Twichell 1992: 46). Or consider this task, by the poet Michael Pettit, on distinguishing individuals and classes, the particular and the general:

> Write a description of two or three paragraphs in which you describe one *particular* member or element of a set:
>
> - one sparrow in a flock of sparrows
> - one baby in a nursery of babies
> - one fish in a barrel of fish
> - one scream in a stadium of screams
> - one somersault in a series of somersaults
> - one Rockette in a chorus line of Rockettes
>
> *(Behn and Twichell 1992: 33)*

The habits that these exercises aim to instil are, it seems to me, very close to practices essential for creative design practice. I can therefore see many ways that versions of these tasks might be introduced in landscape architecture education, particularly during those stages of the design process, usually early, where 'primary generators' are most commonly discovered (Darke 1979).

At the same time, one must acknowledge the possibility of undesired effects and take steps to avoid them. For example, integrating poetic methods into landscape architecture teaching would necessarily involve disentangling poetry from writing. As Kunitz notes, the first is far older than the second, the words of a poem going back 'to the beginnings of the human adventure when the first symbols were not spoken but sung or chanted or danced' (Kunitz 1985: 50). This distinction is critical for design students, many of whom would likely see poetry as having little to do with technical skills they seek to acquire. It would be important to confront such scepticism directly, and demonstrate how poetic methods extend to many other kinds of making. It is also possible that using poetry to teach landscape architecture would call into question the very idea of 'landscape' as the domain of a specialized discipline. Rather like landscape, poetry is a *general* art that sits uneasily at best within modern notions of professionalism. If poetry is to be at all helpful as a teaching method in professional landscape architecture programmes, it will require open discussion of what are, to poets at least, quite obvious tensions between poetry and 'the crowds streaming onto the professional or business highways' (Kunitz 1985: 52) pursuing the twin gods of money and status.

And yet, despite the difficulties, something tells me that landscape architecture programmes would, on balance, end up with more creative and, quite possibly, more radical practitioners were the curriculum opened to poetic methods. Of course, one's view on these methods will depend on one's vision for landscape architecture itself. The question of poetry's place in landscape architecture education (if it is to have one) only makes sense when asked as part of wider discussions about landscape architecture in its third century. What is the nature of landscape architecture knowledge? What will constitute LA expertise? How will landscape architecture position itself with respect to other disciplines, particularly those (like civil engineering) with which it will need to compete for 'territory'? How will the profession retain its identity while committing itself to the idea of landscape as a fundamentally collective enterprise?

Many people are now reflecting on such questions, and the answers they give vary widely. But it is not at all clear that, on balance, the people who design, administer, and accredit programmes in landscape architecture are ready for them. The extent and causes of such resistance are beyond my capacity to explore in any satisfactory way here. But various hypotheses might be advanced. One concerns the nature of landscape architecture as a profession. The idea that poetic methods are relevant to teaching landscape architecture inevitably rests on a proposition, however implicit, of *identity* between the two. That proposition may simply be too destabilizing for a young and still relatively weak profession. At the same time, professions are not monoliths; there are as many modes of landscape architecture practice as there are landscape architects. I often find myself wondering, then, whether the real problem might lie instead with *teachers*. Most landscape architecture educators today work not in professional design practice but in academia, a different profession altogether, and a growing share of these never worked in design practice at all. Could it be that they—*we*—are quite a bit less skilled in the 'art of paying attention' than we should be, yet resist ceding an inch of turf to those who practice it better, whatever the lessons they might impart? Do we fear the poets when it is we ourselves who are 'a danger' (Pound 1934: 83)?

I give the last word to Mary Oliver (2016: 8):

> Teach the children. We don't matter so much, but the children do. Show them daisies and the pale hepatica. Teach them the taste of sassafras and wintergreen. The lives of the blue sailors, mallow, sunbursts, the moccasin flowers. And the frisky ones—inkberry, lamb's quarters, blueberries. And the aromatic ones—rosemary, oregano. Give them peppermint to put in their pockets as they go to school. Give them the fields and the woods and the possibility of the world salvaged from the lords of profit. Stand them in the stream, head them upstream, rejoice as they learn to love this green space they live in, its sticks and leaves and then the silent, beautiful blossoms.
>
> Attention is the beginning of devotion.

I can think of no better ethic for landscape architecture learning. Will we, the teachers, listen?

References

Behn, R. and Twichell, C. (eds) (1992), *The Practice of Poetry: Writing Exercises from Poets Who Teach* (New York: HarperPerennial).

Darke, J. (1979), 'The Primary Generator and the Design Process,' *Design Studies* 1 (1): 36–44.

Eckert, C. (2000), 'Sources of Inspiration: A Language of Design,' *Design Studies* 21 (5): 523–538.

Freud, S. (1937), 'Analysis Terminable and Interminable,' *The International Journal of Psycho-Analysis* 18: 373.

Girot, C. (1999), 'Four Trace Concepts in Landscape Architecture,' in J. Corner (ed.) *Recovering Landscape. Essays in Contemporary Landscape Architecture* (New York: Princeton Architectural Press), 59–69.

Hyde, L. (1983), *The Gift* (New York: Vintage).

Kahneman, D. (2011), *Thinking, Fast and Slow* (New York: Farrar, Straus and Giroux).

Kant, I. (1997 [1781]), *Critique of Pure Reason*, trans. P. Guyer and A. Wood (Cambridge: Cambridge University Press).

Kunitz, S. (1985), *Next-to-Last Things* (Boston, MA: The Atlantic Monthly Press).

Lanfranco, C. (2009), *Site Divine: An Alternative Method of Site Analysis* (San Francisco, CA: Montag Press).

Lenzholzer, S. et al. (2013), '"Research through Designing" in Landscape Architecture,' *Landscape and Urban Planning* 113: 120–127.

McDonnell, J. (2011), 'Impositions of Order,' *Design Studies* 32 (6): 557–572.

Mumford, L. (1934), *Technics and Civilization* (New York: Harcourt, Brace and Company).

Nachmanovitch, S. (1990), *Free Play: Improvisation in Life and Art* (Los Angeles, CA: J. P. Tarcher, Inc.).

Norberg-Schulz, C. (1979), *Genius Loci: Towards a Phenomenology of Architecture* (New York: Rizzoli).

Oliver, M. (2016), *Upstream* (New York: Penguin Press).

Peng, F. (2016), 'Reflection' (unpublished essay).

Poulin, M. (2013), 'Learning to Land' (unpublished thesis), Cornell University, Ithaca, USA.

Pound, E. (1934), *ABC of Reading* (New York: New Directions).

Rittel, H. and Webber, M. (1973), 'Dilemmas in a General Theory of Planning,' *Policy Sciences* 4 (2): 155–169.

Sadoff, I. (2010), 'Poetic Memory, Poetic Design,' *New England Review* 31 (3): 109–117.

White, E. (1983), *Site Analysis: Diagramming Information for Architectural Design* (Tucson: Architectural Media Publishers).

Wylie, J. (2007), *Landscape* (New York: Routledge).

11

Time out!

Thirty years of experiences from outdoor landscape teaching

Roland Gustavsson, Allan Gunnarsson and Björn Wiström

Educational motives for outdoor and action-based teaching

Time is not out, rather the opposite, when it comes to landscape teaching. Based on thirty years of experience in outdoor teaching and a course in the Blekinge archipelago, we know that in landscape teaching it is essential to leave the lecture theatre occasionally, join the 'real world' and experience the particular essence of outdoor learning by action or 'doing'. As one student said:

> After the days out in the archipelago with hands-on work, it will be very hard to go back indoors at the university. But that is the way it is. We need the theory to be able to do professional practical work.

Despite the benefits of outdoor teaching and learning, indoor teaching is becoming increasingly dominant in European schools of landscape. Our intention is not to challenge indoor teaching as the main approach, but to call for a balance. From its current marginalised position, professional training and acting in practical outdoor situations, mixing theory with practice and conceptual thinking with physical skills, deserves a renaissance. It is time for a paradigm shift and a new era of education.

In the Blekinge course, we have identified the following crucial perspectives and approaches for our profession.

- *Contextual training* in an era when abstract conceptual thinking is favoured.
- *Outdoor training* to help students acclimatise and become professional in outdoor situations.
- *Conceptual thinking based on concrete action* as a professional strength, through leaving lecture halls to enter reality, learning to understand local contexts and the concrete world and complement computer-based skills.
- *Experts as outsiders and actors in top-down models*, and how alternatives can be linked to dialogue integrating *local knowledge among skilled local stakeholders*.
- *Multiskilling and the ability to cover many aspects, while able to see a place and a landscape as a whole with spatial qualities.* Covering the whole may be impossible, but it should be the ambition

in university teaching to counteract today's fragmented world, in which multiple experts partition the landscape based on their perspective.

- *Grasping and handling complexity* at a time when urban vegetation and landscape architecture systems generally prize simplicity.
- *Describing and developing the unique and special* at a time when standardisation is practically the rule.

The course and Blekinge islands

The outdoor course involves undergraduates from two degree programmes, landscape architecture and landscape engineering, at the Swedish University of Agricultural Sciences, Alnarp. It was established in co-operation with the local authorities in Ronneby and Karlshamn and with Blekinge regional council. Building trust between these parties through dialogue was essential for success. The students themselves and their engagement and creativity are important complements to the management plan. Instead of being experts in one aspect, we apply a wider scope covering cultural heritage, aesthetic-tourist aspects and biological and management aspects. Day one involves getting to know the landscape, local history and the actual intervention sites, followed by development of ideas. Day two involves implementing ideas in hands-on work and day three covers presentation of concepts and results, and documentation. Students (and teaching staff) leave the classroom and immerse themselves in reality on a very attractive island, where there are no interruptions. All become acquainted with a different landscape with many

Figure 11.1   Just arriving at the island of Tjärö. Over the thirty years of the course, we have had thousands of students out in the archipelago. Reaching the island is always special, leaving the mainland and the problems behind, knowing how valuable it is to be able to concentrate a hundred per cent.

surprises and more well-known cultural and natural roots. We socialise, work, try to think theoretically and creatively, swim, eat, drink and sleep in a three-day break from the world.

The two islands

A course is held on the islands Tjärö and Karön in the Blekinge archipelago in the southern Baltic Sea, a two-hour drive from the university campus. An island visit is always special. Leaving the mainland by boat brings a feeling of leaving problems behind and just enjoying the moment. The two islands, with their small-scale landscape, provide opportunities to use a range of places and situations within walking distance. They also represent a manageable scale, where the island as a whole can be grasped in a few days. The islands comprise holistic unbroken microcosms of a Scandinavian landscape with characteristic elements such as granite rocks and slabs, boulders and stone walls, spontaneous vegetation in many biotopes, and water, elements lacking in the landscape around the Alnarp campus.

The islands of Tjärö and Karön share many similarities. Tourism is critical for both, and they are both of national interest for nature conservation and cultural heritage.

The course in detail

Every year we bring about ninety students to the islands, divided into two main groups and several smaller groups of six to seven, all comprising a mixture of landscape architecture and landscape engineering students. Each group stays two and a half days, while preparations and follow-up

Figure 11.2 Contextual training. Working actively to make places look natural using existing characteristics and the surrounding atmosphere and, in parallel, always getting a sense of a human touch.

require a further three to four days. Both Tjärö and Karön have good self-catering accommodation. Interest among local businesses, house owners, experts, stakeholders, the media and many leading politicians has contributed to the success of the course.

The days on the islands offer training in discovering different landscape identities and potential and how these can be respected and utilised in interaction with social, cultural, historical, aesthetic and biodiversity aspects. In particular, students receive training in transforming analysis and evaluations into necessary management actions, which they then implement. The students thereby experience the before and after and examine short-term and long-term dynamics. The landscape and different stakeholder views are discussed as a basis for the working process. In parallel, the course demonstrates how the university can co-operate with local forces, local authorities and regional councils.

Visits comprise the following main stages.

- An initial teacher-led introduction to the islands as part of the archipelago and region, which aims to give an overview and understanding of the local and regional context and the sites chosen for study.
- An interpretation of the sites allocated to the groups concerning materiality, history and future potential, in dialogue with supervisors. Problems and restoration possibilities concerning hidden patterns and collapsing elements from the periods of historical land use are discussed, as are needs and potential qualities related to biodiversity and visitors and users of today.
- Discussions and decisions about main goals and visions, considering multiple views and knowledge from the outside and from inside the group with its landscape architecture and landscape engineering students.
- Practical work using tools like pruning shears and handsaws to reduce and emphasise vegetation, and spades and crowbars to restore paths and stone walls. The work is explained step-by-step, with landscape managers from the region and the local authorities providing practical knowhow and chainsaw and brush-cutter manpower. Concepts are sometimes renewed and refined through discussions within the group and in communication with local experts and supervisors. All actions and results are documented.
- Presentations during a field seminar, which acts as a grand finale, awakening interest and stimulating students to communicate their visions and plans to other groups, teaching staff and invited guests, making the presentation day special. The field seminar also gives the opportunity to step back and reflect about different scales, from detailed to general, and to compare the situation before and after. The field seminar is always followed by time to relax and say goodbye to the island and new and old friends.
- In a follow-up step at the university, all students must complete a personal diary and draw conclusions with sketches, photos and text. This provides feedback on the teaching outcomes.

Teaching platform

Some basic thoughts

In a changing society, landscape architects and landscape engineers could well occupy a position between human science and natural science, between art and landscape science, and between the intuitive, experienced and the conceptual, observed landscape. However, this will require training in how landscape design, planning and management can be linked to pedagogic and philosophic theories, combined with hard practice in the essentials. Below, we summarise some theories and thoughts that serve as motivating background, permeate the didactics and determine the spirit in which we organise the course.

The pedagogy and philosophy literature, which is the foundation for our way of thinking, has two main scientific traditions. On one hand, there is a design tradition based on engagement and personal experience (e.g., Norberg Schultz 1980; Pallasmaa 2005), in which knowledge is personalised through experience and is action-led, creativity and testing are stressed and the spontaneous, naïve and intuitive are used to gain a sense of 'green fingers' (e.g., Dewey 1916; Streeck, Goodwin & LeBaron 2011). On the other hand is the logic analysis tradition that involves circling around the object to observe it from outside without affecting it and searching for knowledge through different pre-fabric perspectives, theories and analysis. Rediscovering the thinking of Bergson (1889/1992, 1928a, 1928b), which combines the intuition-personal experience and logic-analysis traditions, was essential for our thinking and actions.

Both these basic approaches are important in landscape degree programmes in Scandinavia and have remained stable for a long period. We consider them to be complementary approaches that are both essential for didactics and learning during the Blekinge course. Analytic-directed knowledge is used by teachers and students in interpretation, concept development and management plans. The course would be very vague without such knowledge; absolute knowledge about the landscape, based on what has been, what species are present today, its current state and its physical elements with links to history (e.g., Gunnarsson 2011), is essential in performing the practical work in the course.

Teaching and learning in the course is highly dependent on personalising knowledge by experience and embodying it, using all the senses (Johnson 2007; Streeck, Goodwin & LeBaron 2011; Scarinzi 2015). 'Embodying' may often be hard work. Many students are not accustomed to hard physical work, and to challenge them we say: the more blisters on your hands, the better your theoretical thinking. Besides outdoor learning, learning-by-doing and the contextual dimensions, the course uses teaching and learning based on dialogue and interaction with practitioners and local people as connoisseurs, in co-operation with organisations such as local authorities and regional councils (Mellqvist 2017). The course is short and this concentrated form can create problems but can also bring benefits, as noted by this student:

> The days on the island were a fantastic experience. To be given an area in a nature reserve in such a beautiful landscape was really great and a true privilege. To go from introduction to discussion and planning, and further to realisation and the end result, in such a short time was intense and challenging, but pretty cool.

Keys to success: examples from the islands

Based on reflective actions from the work, where we strive to make theory practical (Dewey 1916, 1925; Schön 1987), some particular keys to success have emerged. To showcase some aspects of the course, ten examples are presented below.

Development of 'place-directed creative management' as a concept

As a practical necessity in the compact course approach, an overall concept of creative management has emerged. This means that, in addition to overall rational vegetation management, such as grazing and conventional thinning performed by the local manager, specific places, paths and

Figures 11.3 and 11.4 Understanding time and changes. Reading the landscape and its historical roots. Two photos from the 1950s are illustrating common characteristics on northern Tjärö: topography, almost micro-scaled fields, stone walls, and a grazed juniper landscape as a remnant from the nineteenth century.

Figures 11.5 and 11.6   Discussing the danger of transforming everything within the open, tree-rich and semi-open into an attractive pastoral landscape with unreflected associations to the English landscape park as a garden ideal. Neglecting important structures such as wooded meadow and mosaic-type landscape with shrubs like junipers. This should be compared with the state in 'our time', if we make time elastic. Photo in the late 1990s: students of this year show how many more trees there were growing in the former field, before they were cut, and they show where each tree was positioned. In 2016, the last photo shows a situation in which a new student group has cut down the small trees and shrubs but left the large oak to help visitors understand different time layers in the landscape.

objects are selected for more detailed 'design-oriented and place-related management'. These 'places' are allocated to the students as 'spatial areas and passages' to be defined and given identity (e.g., Norberg Schultz 1980). The students can thereby really make a difference and the landscape is given a new articulation. Depending on the situation, the work changes or strengthens the existing identity, variation, contrast and complexity without exceeding the management budget. At the same time, the students get a feeling of being needed and capable. The landscape architect students understand that management involves design and the landscape engineering students understand that design involves management (Gunnarsson 2011). This concept, called 'creative management', has also been implemented in Alnarp's landscape laboratory and has attracted international interest (e.g., Duinker et al. 2017).

Discovering and discussing environmental aesthetics

One central aspect is to relate discussions on aesthetics to a wider and deeper context, following the ideas of environmental aesthetics of, e.g., Berleant (1997) and Eton (2001). While on the island, students become aware of the strong identity and beauty of the landscape, which is enhanced by memories and knowledge among local stakeholders and their stories, raising a strong feeling of belonging and engagement. This is enhanced by the responsibility given to the groups for transforming a weakly articulated site into a place with a specific identity.

Involving stakeholders and local connoisseurs

A university education prepares students to take the role of top-down outsider. The strength of local knowledge is often a surprise during the days on the island, and students begin to think about how to include it in their work. Stakeholders and connoisseurs (local experts, often

Figures 11.7, 11.8 and 11.9 Karön. The island became important for its spa, which, in the 1880s, was the most popular in Sweden for sea bathing, views of the archipelago and walks in nature. All villas were constructed at that time, based on spa ideals and the Jugend style (Art Nouveau). The Annexet villa, Karön, with its beautiful blueberry-fern 'garden'. The photos show how we have restored the view to the sea but in a sustainable way by not opening up the canopy but by lifting tree crowns. This reveals boulders in the understory and emphasises the blueberry carpet, and the path that crosses the view. Karön, 2004 and 2014.

leaders of local associations) are identified and take part in the presentations. Connoisseurs are also stakeholders, but not always (Mellqvist 2017), and if not recognised and invited to participate their contributions could be overlooked. It felt good when local owners/stakeholders said of the students' work on Tjärö in 2016: 'What they have done is magic!' 'We expected much, but what they did is almost unbelievable!'

It is easy to recommend, but much more difficult to achieve, successful co-operation between the university and local organisations, including key individuals. For success, it is an advantage to return to the same area for several years, but perhaps not too often. After twelve years visiting Karön we felt the need for a break and moved to Tjärö. There may be a threshold above which a sense of expectation and positive surprise no longer materialises and there may also be a risk of becoming routine and taken for granted.

Pedagogically, the work must be appreciated and seen as special, and a mixture of insiders, outsiders and 'bridgers' is probably important for success. Moreover, to get a strong sense of a 'real' project, much time and effort is needed before the course days to prepare, establish dialogue with local people and become recognised and trusted. In our case, local stakeholders were hesitant at the start, but meeting outdoors on 'introductory walks' in the actual landscape during the preparation phase, and viewing the students' work, overcame their reservations.

Select appropriate working areas

Sites must have a need, but also potential, for transformation or restoration. Sites where the discussion centres on future actions that are not achievable within the course often fail to activate the group. Moreover, the site needs to offer some kind of resistance or complexity to challenge the group. Change needs to be balanced against stability over time. If a certain aspect or stakeholder becomes too dominant, the work becomes less of a process and more of a workload, and thus less stimulating. Thus supervision of the group has to counterbalance, through dialogue, the different stakeholders and different parts of the student group. An idea must be something that can be created by group members; see also Bergson (1928b) and his thinking about creativity and free will.

Select representative working areas

No single site can exemplify and showcase all aspects of a landscape. However, a wide range of different sites (e.g., Flyvbjerg 2001) can enable much wider discussions and examples in final presentations. All participants should get the impression that the examples cover many cases of high interest. Consequently, teaching staff have to check all groups to make sure their ideas are pedagogically wide and cover situations in which the contextual and high-interest principles meet and can be discussed. Students often find it more stimulating to be the first to discover a place and to find brand new solutions. Alternatively, to stress dynamic thinking, a group can be allocated a previously managed site, to consider how much could be gained if the site and the work were taken one step further.

Start with the obvious

Before starting the physical work, the groups should have an overall idea of what they want to do, but not a detailed plan. This is because many decisions should be taken during the process, when group members are more acquainted with the landscape and the situation, have met their supervisors to consider alternative perspectives and have had time for repeated reflection. The best solutions are often not found until the practical work has advanced some steps. By starting with the obvious, unclear matters are often resolved and new issues emerge. The whole working process thereby becomes an act which develops step by step. When the groups are hesitant or have differing opinions, starting by doing the obvious gets the process going. For example, it is difficult to see at the start how many trees should be felled. Taking it step by step lets students see whether to continue or to stop.

Enforce dynamic thinking

One of the most important aspects is to support the students with ideas about what will happen with the vegetation and landscape over time, through both longer and shorter perspectives, in order to avoid the snapshot experience the students could easily get by staying just a few days on the island (Foster 2000; Gustavsson 2009). This can be done by explaining how the vegetation structure and species balance have changed in the past hundred years and will most likely change ten, twenty and fifty years from now. Returning to the same islands is valuable in this regard, since previous examples can be shown to the students and teaching staff to deepen their understanding of place-specific dynamics. The thirty years of the course is a short period, but long enough to learn something from experience.

Supervise when needed, give a helping hand when needed

Dialogue between the groups and supervisors is critical. It is important to balance between challenging the students' ideas, examining them from different angles and perspectives, and supporting them by suggesting other alternatives, trying to expand their thinking, making them think less conventionally. This must be done in a way that makes students creative and eager to progress, rather than hesitant. When practical work in the groups is underway, supervision becomes more about checking and showing how ideas could be improved and sometimes also helping out physically in the work. It always adds to team spirit when the teachers roll up their sleeves and put work gloves on.

Supervise with diversity

Supervision by a pair of teachers, or a teacher and a local connoisseur, has proven to be successful, as this enforces communication between facts, views and ideas in different fields and how to integrate these in different solutions. This has also been a good way of integrating new staff into the teaching model.

Presentations in the field, a stimulating content-rich show

Much energy is devoted to the presentations in the field on the last day. Developing language is important for communication and particularly for presentation. To attract attention, students need to be articulate and professional, so the presentation task is good training. The presentations are almost always well prepared and organised in the groups on the day before the final presentation, and everyone in the group has a role. The students have a chance to present their work and to learn about the work of other groups. Local issues are combined with general issues in the discussions. The presentation event creates great interest and trains the students in how to draw attention, how to keep it and how to communicate their main intentions and work in a short time.

Teachers, other students and visitors, stakeholders and connoisseurs, all contribute to the discussion after each presentation, leading to a wider dialogue. The presentation sometimes takes the form of an outdoor performance, drama or comedy. However, it has to be well organised with respect to the audience position, direct sun and wind exposure, so that all can concentrate on the content, hear what they should hear and feel that they all are well entertained.

Happy and educated locals and practitioners

Our overall experience is that students gain a lot, but also give a lot to local practitioners and residents. Many locals have reported how much they enjoy the eager eyes of the students, their positive approach, their wish to increase knowledge and their bravery in diving into something new and difficult. Their presence is stimulating and they often bring solutions outside the conventional that are much more site-appropriate and place-related. The course also works as a seminar for non-academic managers: 'It stimulates, brings new knowledge, and is talked about for the rest of the year' (Jonny, a local worker in Ronneby parks department, in Mellqvist 2017).

Figure 11.10 An example of the presentations on the last day. The students communicated with locals who have owned villas for generations, working and communicating in the garden and surrounding nature. Discussion during the field seminar on how privacy can be combined with overall accessibility and attractiveness. Villa Siesta, Karön, 2014.

Many thanks!

Over the thirty years of the course, we have had thousands of students. It has been fantastic to work with these young people and we feel honoured. We are grateful also to island owners and organisations, such as Blekinge regional council, Ronneby and Karlshamn local authorities and local associations, who play important roles. Being out, communicating and co-operating with local people and the landscape itself have convinced us that the university campus is not always the centre of the world.

References

Bergson, H. (1928a), *Den skapande utvecklingen* (L´Évolution Créative). (Stockholm: Wahlström & Widstrand).

Bergson, H. (1928b), *Intuition och intelligens*. (Stockholm: Wahlström & Widstrand).

Bergson, H. (1889/1992), *Tiden och den fria viljan* (Essai sur les donées immédiates de la conscience). (Nora: Nya Doxa).

Berleant, A. (1997), *Living in the Landscape. Toward an Aesthetics of Environment*. (Lawrence: University Press of Kansas).

Dewey, J. (1916), *Democracy and Education: An Introduction to the Philosophy of Education*. (New York: Macmillan).

Dewey, J. (1925), *Experience and Nature*. (Chicago, IL: Open Court Publishing Company).

Duinker, P.N. Lehvävirta, S. Nielsen, A.B. and Toni, S. (2017), 'Urban Woodlands and their Management', in: F. Ferrini, C.C. Konijnendijk and A. Fini (eds) *Routledge Handbook of Urban Forestry* (New York: Routledge), 515–528.

Eton, M. (2001), *Merit, Aesthetic and Ethical*. (Oxford: Oxford University Press).

Flyvbjerg, B. (2001), *Making Social Science Matter. Why Social Inquiry Fails and How it Can Succeed Again*. (Cambridge: Cambridge University Press).

Foster, C. (2000), 'Restoring Nature in American Culture: An Environmental Aesthetic Perspective', in: P.H. Gobster and R.B. Hull (eds), *Restoring Nature. Perspectives from the Social Sciences and Humanities*. (Washington DC: Island Press), 71–97.

Gunnarsson, A. (2011), 'Om landskapsvårdens och trädgårdens hantverk' [About the landscape and garden crafts], in: G. Almevik, L. Bergström and E. Löfgren (eds) *Hantverkslaboratorium*. (Gothenburg: University of Gothenburg), 51–60.

Gustavsson, R. (2009), 'The touch of the world: dynamic vegetation and embodied knowledge', *Journal of Landscape Architecture*, 4(1): 42–57.

Johnson, M. (2007), *The Meaning of the Body. Aesthetics of Human Understanding*. (Chicago, IL: The University of Chicago Press).

Mellqvist, H. (2017), 'The connoisseur method – a study on long-term participation in landscape planning'. PhD thesis (Swedish University of Agricultural Sciences).

Norberg Schultz, C. (1980), *Genius Loci: Towards a Phenomenology of Architecture*. (New York: Rizzoly).

Pallasmaa, J. (2005), *The Eyes of the Skin. Architecture and the Senses*. (Chichester, UK: John Wiley & Sons).

Scarinzi, A. (2015), *Aesthetics and the Embodied Mind. Beyond Art Theory and the Cartesian Mind-Body Dichotomy*. (Dordrecht: Springer).

Schön, D. (1987), *Educating the Reflective Practitioner. Towards a New Design for Teaching and Learning in the Professions*. (San Francisco, CA: Jossey-Bass Publishers).

Streeck, J., Goodwin, C. and LeBaron, C. (eds) (2011), *Embodied Interaction: Language and Body in the Material World*. (New York: Cambridge University Press).

12

Caring for Arctic and Subarctic landscapes

Janike Kampevold Larsen

Arctic and Subarctic landscapes have historically formed the margins of Europe and North America, both physically and symbolically. Difficult to access, they were mythologized and conceived of as outer territories that conveyed food and fur resources to administration centres. Today, landscapes in the Arctic and Subarctic are under intense pressure, and moving to the centre of political and cultural attention.[1] The temperature is rising more than twice as quickly here as in areas on more southern latitudes, causing ice loss, thawing of permafrost, changing weather patterns, and alternating flooding and freezing in the winter. Communities may be forced to migrate or relocate infrastructure due to heavy erosion and increased geo-hazards such as landslides. The end of 2017 saw evacuations in Longyearbyen, the northernmost community in the world, as new wind patterns and wet snow threatened nearly seventy homes for the third year in a row, demonstrating that settlements and cities are left in need of new methods for geo-hazard management, and for urban infrastructures that are adaptable to changing climate.[2] Communities in the Arctic face shrinking or rapid expansion due to shifting interests from development forces, industrial actors, and speculative promises of potential income from tourism, new sea routes and resource extraction. The small Norwegian town of Vardø, off the northeast Barents coast, is shrinking as a result of fishing policies, while Longyearbyen at Svalbard is expanding due to increased tourism and scientific research and institutions.

This chapter has been written from the perspective of researching and teaching landscape at The Tromsø Academy of Landscape and Territorial Studies, a new Master's programme in landscape architecture developed by The Oslo School of Architecture and Design (AHO) and UiT, The Arctic University of Norway. The programme was inaugurated in 2013 and, as of 2018, it will become an integrated programme in landscape architecture, set up as a joint programme between AHO and UiT. The first three years will be based at AHO and will focus on design solutions for northern landscapes, infrastructures and urbanism, with a persistent partial focus on Subarctic ecologies. The last two years will be based in Tromsø and will refine the profile that has been developed so far. This chapter takes a few studio courses as examples of how the programme aims at building knowledge on specific Subarctic conditions. It emphasizes, however, the work performed in remote Arctic territories, as these are the most challenging pedagogically.

Figure 12.1 Svalbard Satellite Station (SvalSat) contributes to making Longyearbyen, Svalbard, a hyper-networked place. It is the world's largest commercial ground station for the download of satellite data.

Located at a latitude of 70° north, the programme was conceived of with the explicit aim of training landscape architects to meet the specific challenges in Arctic and Subarctic climates, urban centres and territories, and with a strong premise for regional development of knowledge within the discipline. The north of Norway is the primary geographical context, while Iceland, Greenland, Svalbard and Bear Island have served as extended fields of operation. The areas of operation are far beyond the traditional domains for landscape architecture, and sometimes perform as laboratories for testing methodologies, tools and approaches. The main landscape medium, the ground, is subject to extreme conditions in terms of climate, climate change, industrial exploitation and local pollution. The programme's approach to places and territories requires specific measures as to travel, length of field trips and preparation for often extreme conditions. Students at times operate as ambulatory nodes of designers, with a very good student–teacher ratio.

The programme works at all scales: with scientists at UiT, it develops design for ecosystem-based infrastructure and circulation in winter cities. It further bases urbanism in a cultural geography based on understanding of place as both globally networked and founded in local identity and practices. It engages with local actors and stakeholders in order to propose new solutions for social and infrastructural development. This includes strategies for social empowerment and development of social networks for food culture or preservation though use. It cultivates strong bottom-up approaches through long-term cooperation with municipalities and populations. Most importantly, perhaps, the programme is developing, in a research-based approach, a concept of territorial management as a means to negotiate between regional and traditional land-use practices, new industrial development and national management strategies.

The extremity of place forces an attention to preservation issues, place development, situated learning and mapping, and has fostered an unusually strong relationship between research and teaching, a model that AHO has exercised for some years.

City ecologies/Arctic ecologies

To our knowledge, Arctic ecology is not being consistently taught in any other landscape architecture programme. There are, however, examples of landscape architectural site-specific work in Arctic territories, probably best demonstrated by Cornelia Hahn Oberland's work in Inuvik, at 2 degrees above the Arctic Circle, a region of extreme permafrost thaw and flooding since the 1950s. Her landscape plan for the East Three School in Inuvik carefully considers the semi-materiality of snow gusts, the 'arctic amplification on soil regimes in the far north', and particularly the active agency of the permafrost's active layer, the top layer that thaws in the summer month, creating pools, saturating soil, and hence causing the sinking of buildings as well as inundation of plant groups (Herrington 2013: 50). Testing of snow gusts, longer shadows and scaling circulation to snow-moving machinery are among the basic material factors that call for specific expertise in landscape and urban design in the Arctic, and are part of the teaching approach in Tromsø.

First-semester studio courses in Tromsø are developed under the generic title 'Elemental concepts for the Subarctic city'. They explore how urban surface water, snow beds, sea level

Figure 12.2   Rain garden built by students and teachers in collaboration with Tromsø municipal officials on top of Tromsø island.

rise, watercourses, running water on frozen ground and snow deposits may be considered spatial resources in a Subarctic city. Answers involve multi-functional blue-green-white infrastructure, allowing for a broader array of professional knowledge involved in problem solving. The courses also engage students in hands-on building of Arctic-specific rain gardens, modified to enhance performativity in cold conditions.[3]

Under the rubric SNOWHOW the courses delve into exploration of snow management and city ecologies shaped by abiotic circumstances such as latitude, temperature and precipitation. Arctic species adapting to urban niches and forming new anthropogenic habitats must be analysed and understood as spatial potential. 'Inner city glaciers', as Chris Neal MilNeil labels snow deposits, will usually not melt until June, leaving a sedimentary record of whatever road and air pollution has been deposited during the winter. Snow dumps in Tromsø represent not only 'a condensed record of a city's air and water pollution' (MilNeil 2013: 79, 81), they are also the areas of highest biological diversity in Tromsø. By systematic exploration and analysis of snow beds and floras and interviews with snow-moving personnel, the studio courses have isolated small-scale dross-scapes as particularly promising for the establishment of a sustainable green infrastructure. Model making and digital representation are classical tools. Designs are developed in close collaboration with biologists, and not least municipal bodies, res ting on the premise that the academy helps develop knowledge in the region.

Bottom-up strategies and participatory project work

It seems to be a prevailing conception that settlements in the Arctic are remote autonomous nodes, withdrawn from global networks. Yet they are entangled in a complex global web of trade and migration, and have been for centuries. The academy works to give students an awareness of Arctic places as both local and globally networked, debunking the myth of the secluded Arctic, in line with Doreen Massey's thinking on place:

> It is a sense of place, an understanding of 'its character', which can only be constructed by linking that place to places beyond. A progressive sense of place would recognize that, without being threatened by it. What we need, it seems to me, is a global sense of the local, a global sense of place.
>
> *(Massey 2005: 156)*

The challenge for many smaller Arctic places is how to plan for a future that is approaching quickly, while still maintaining their character of being places that include people, as Lucy Lippard puts it.[4]

What may be the role for landscape architecture in communities where physical design solutions are not what is most needed? This was one of the challenges for the design studio 'Vardø Transformed – Cultural Heritage and Place-Based Development'.[5] The academy has had a long-time involvement with local agents in the Arctic city of Vardø through the research project 'Future North',[6] and the community development platform 'Vardø Restored'.[7] This former flourishing fishing community has been shrinking for forty years and harbours a large number of pre-war wooden structures, most of them in a serious state of dilapidation. Students have been part of participatory processes with property owners. They have been trained in interview techniques, and in the registration and documentation of cultural heritage (Mainsah 2018). They have provided design solutions for the city's common areas, one of them proposing a circulation plan for storage and re-use of building materials (see Figure 12.3). During an exceptionally long six-week field trip to Vardø, students had the opportunity to work closely with local actors in

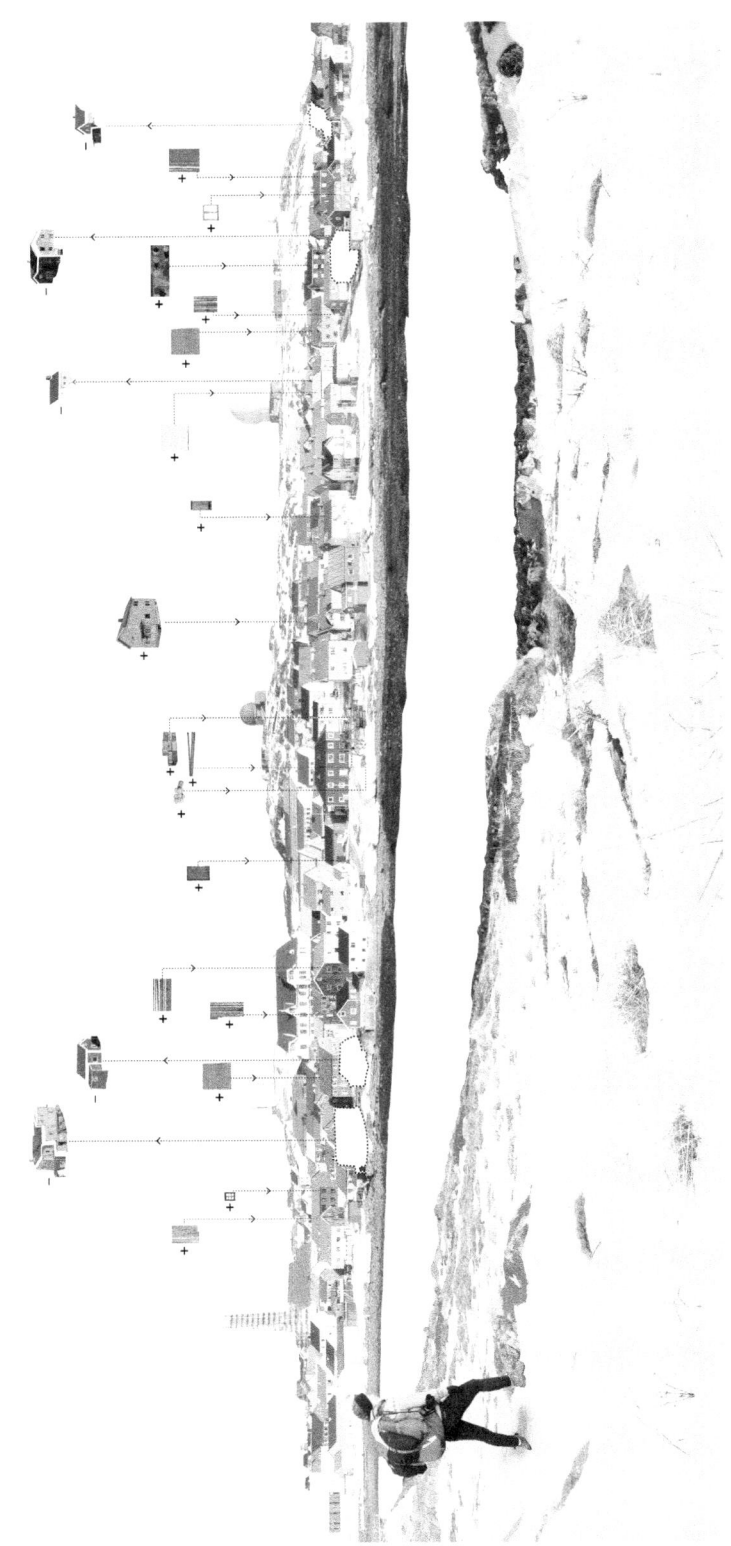

Figure 12.3 Annie Breton's project 'Unexpected Trajectories/Vardø's Transient Heritage' proposes a system for moving, storing and carefully dismantling abandoned buildings, and making building parts available for the remaining built environment.

order to project possibilities for future city development. This studio course is significant for being based on a bottom-up strategy and its in-depth collaboration with local actors that do not necessarily have strong agendas. It exemplifies a socio-cultural approach to learning, built on an inquiry-based learning more than an 'educator-provided and topic-determined learning' (Hemmersam et al. 2018). Programmes are developed with places and actors, which also means that the success of learning may be measured by the project's value for the local community. The studio course needed to carefully balance considerations of learning outcome and the needs of the external actor, and in-field lessons for future professional situations for the students.

Territorial management

The classical conception of territory is a delimited area, a physical object for a nation state to govern or exercise sovereignty in or within. We work with our students towards thinking about territories not as static objects, but as a dynamic making, where different actors and forces are contributing, interfacing, assembling – human and non-human alike. As argued by Andrew Mubi Brighenti 'territory is not defined by space, rather it defines spaces through patters of relations' (Brighenti 2010: 57), and its main characteristics are located in 'relational, processual and "eventual" perspectives' (Brighenti 2010: 57).

In the Artic today the idea of territory is still very much a question of sovereignty. Nation states claim sovereignty over land areas, continental shelfs and sea areas. The planning and building of infrastructure seems to adhere primarily to an idea of territory as manageable and administrable, even more so as one sees the possibility of new sea routes and petroleum extraction as the sea ice diminishes. How can a teaching programme negotiate what Lola Sheppard calls 'composite territories' of urbanism, infrastructure, ecology and industry (Sheppard 2013: 179)? Emblematically, 'Lateral Office' (Lola Sheppard and Mason White) has addressed composite territories by their Arctic Food Network, a spatial practice proposing a food and mobility network that sustains continued exchange between Inuit communities even when frozen waterways thaw due to climate change, offering a foundation for a sustainable independent economy. Their recent *Many Norths, Spatial Practice in a Polar Territory* charts, analyses, maps, and projects into a shifting Canadian Arctic, emphasizing spatial practice as 'the negotiation of an environment for survival or opportunity via tactical, spatially based interventions, at any scale regardless of economy' (Sheppard and White 2017: 9).

We find, too, that we need to survey the historical, contemporary and future in all areas of work. Settlements and territories in the Arctic are largely uncharted when it comes to urban, landscape and architectural qualities. Cartographic methods, hence, are invaluable and GIS-based maps often a prerequisite. Variegated territorial forces must be traced and understood by on-sight observation and by research, by classical and novel tools. Ultimately, students must be trained in questioning the current definitions of landscape and territory, and the possibility for influence and management of diverse ecologies.

How to teach at a large territorial scale

Approaching larger Arctic territories, then, raises a number of challenges, some of them associated with scale and approach. Many areas are difficult to access, have low mean temperatures even in summer time, and are extremely vulnerable.

Extended journeying and fieldwork are necessary tools in approaching large tracts of land, as much as threatened cities like Vardø. These are immersive approaches that allow students as well as researchers an understanding of the expanse of space, the vulnerability of its ecologies,

Figure 12.4   Field work by abandoned mine 5 in Adventdalen outside Longyearbyen

and not least the material component of spaces. Arctic territories are seldom covered by trees or even shrubbery. In open territories, remnants from industries, built structures, and even pollution is easily discernible. These remnants are co-constitutive of space, and often preserved. More than cultural heritage, however, they may be considered evidence of forces that have been and still are operating on the territories. They may be conceived of as assemblies of trajectories, as Doreen Massey would put it; interests and desires that contribute to the ongoing production of space (Massey 2005: 118–119).

Bjørnøya – negotiating human presence in vulnerable areas

Bjørnøya (Bear Island) forms a laboratory for understanding the idea of remoteness, vulnerability, and of strategic and industrial forces in the Arctic. Contested since its discovery in 1596, occupied both by the Russians and the British for its strategic position and resources (walrus, whales, coal, lead), exploited for coal between 1915–1925, and now preserved under the Svalbard Environmental Protection Act of 2001,[8] this small Arctic island is in many ways emblematic of a vulnerable Arctic landscape. We visited the island with a small group of students, arriving after two days at sea with the Norwegian Coast Guard. The approach itself taught students about the extreme expanse of the Arctic sea territories, and the material effort needed to build and operate a coal mining station at Tunheim, at the northeast brink of the Barents Sea.[9]

Paradoxical to the exceptionality of transport and location, the studio course 'Intervals of Neglect'[10] employed much used landscape architectural learning methods, such as walking, sketching, registration of immaterial qualities, coupled with a strong focus on approach and immersion. Mirroring the significance of the approach to the island itself, the studio teacher stressed the importance of the first arrival at the site of Tunheim. The idea of landing held extreme pedagogical importance. Students were launched into one of the most remote territories in Europe. Landing at the meteorological station was only a first step. The subsequent three-hour hike to Tunheim involved carrying supplies and equipment as well as a rifle for polar bear protection. It transitioned seamlessly into a two-hour rambling among industrial ruins and disintegrating iron, wood, glass, cords, tiles, and other materials on site. The approach was choreographed, much in line with landscape architect Christophe Girot's methodological first step in 'The Four Trace Concepts': 'The state of just-landedness is precarious, but it plays a vital role in the genesis of design. Initial landing provokes impressions and insights that often last through the entire design process' (Girot 1999: 62). What we entered was in effect a time-landscape,

Figure 12.5   Landing at Tunheim. Exploration of the train track and its time landscape

where industrial ruins had been sitting since WWII, and where large areas of ground had not been trodden for almost 70 years. Shortly after arrival, we realized the effect our steps had on the ground itself, where permafrost upheaval had reconfigured a mixed materiality, leaving frail shapes of shale and coal in patterns that are a product of human and non-human forces both, suggesting that the place is indeed not for visitors.

The raw landscape of Tunheim was a perfect training ground for founding a landscape architectural approach. Based on walking as a main immersive methodology, one in which the measured steps mapped out the extreme and unfamiliar terrain, it was a lesson in where not to walk – a lesson that is acute for large areas of the Arctic.[11] Students performed observation of material and immaterial site elements over time, recording their observations through sketches, video, photography, words and sound. The windblown Arctic location became a training ground for how to accommodate for presence in a vulnerable and hostile landscape situation, and how to identify the role of landscape architects when faced with territorial situations that are only for the very few to experience. Should an area like Tunheim be visited at all, and, if so, how to both secure the time-landscape and allow for a controlled circulation of the area? In an age of propelled exploration of Artic territories, the immersive experience of such a remote and relatively untouched landscape contributed pedagogically to raising our students' awareness of the perils of extended traffic, access and industrial presence in the Arctic as one of the globe's last unexplored nature reserves.

'Svalbard – Fluid Territory'

The studio course 'Svalbard – Fluid Territory' used an iterative logic of giving and receiving to explore the potential future of the hyper-networked space of the Spitsbergen territory.[12]

Figure 12.6 Tunheim materials bleeding into the ground

Figure 12.7 Part of the large industrial area of Tunheim at the brink of the Barents Sea

The course studied and illustrated how external influences, such as chemical pollution, tourism and resource extraction, have affected the physical terrain and climate of Svalbard, and thus the migratory and habitation patterns of its marine and terrestrial organisms. Conversely, the course explored how the territorial agency of Svalbard, as a repository of cultural heritage and through institutions devoted to science and satellite monitoring, extends outward to influence the actions of the larger global community.[13]

An understanding of the Svalbard territory as a hyper-networked space required an array of research tools and methodologies, from analogue to digital. Initial research was conducted on the field trip that involved lectures and interviews with natural scientists at the University Center in Svalbard (UNIS). Research further included immersive experiences and situated impressions during site visits, photo documentation of infrastructure and material qualities of spaces, drone flying and in-field conversations about what we were seeing, surveys of scientific literature and quantitative descriptions of the biophysical terrain, historiographical archival research, and an epistemological study of research terminology for appropriation into the project's landscape vocabulary.

The main investigative tool was the territorial scale sections that students were asked to cut across the Spitsbergen territory. Traditionally a tool for discovery, both by geology and design disciplines, the section 'serves the double purpose of reconnaissance and measure' (Desimini and Waldheim 2016: 177). Unlike Kate Orff's to-scale sections in *Petrochemical America*, our sections were not to scale, and hence served less the purpose of measure than that of reconnaissance. Similar to Orff's, however, they served as a baseline for timelined narratives that linked up to an alternating set of external forces. The line enabled students to collate historical information, material distribution and operative agencies in the territory, be they human or non-human. The final work combined panoptic cartographies delineated in plan and environmental data-scapes delineated in section (Figure 12.8).

The seven students produced analytical mapping designs of:

1   the trajectory and material transformation of coal, from tertiary deposits to present-day glacial sedimentation of by-products from coal burning;
2   geopolitical claims on the sea and land territories between mainland Norway and Canada;
3   scientific practices in the archipelago from analogue on-ground nineteenth-century measurements to digital satellite monitoring and downloads;
4   the effect of climate change on the kittiwake migration patterns;
5   the territorial agency of the coastline, and its change from a multinational hunting ground to research territory;
6   cultural heritage as memory and absence; and
7   turn-of-the-century aerial innovation in Virgohamna and the demise of the Swedish explorer Andrée's expedition to reach the North Pole.

All projects hypostasized Svalbard as a relational space, interlinking nation-state interests, human and non-human migration, industries, materials and scientific innovation (Figure 12.9).[14]

A studio course like 'Svalbard – Fluid Territory' contributes to identify and narrate the impact of forces at work in a given geographically identifiable territory. It contributed to imbue in the students a lasting sensitivity to the impact of nation state policy and industries in the Arctic, but also on the marine and terrestrial biological territories, including the human, that exist within and overlaps with the physical and spatial nation-territory. It demonstrates that territory is as much narration and fluidity as it is about area and forces. Drawing and narrating their lines, itself an act of making territory, students experienced the force of cartographic

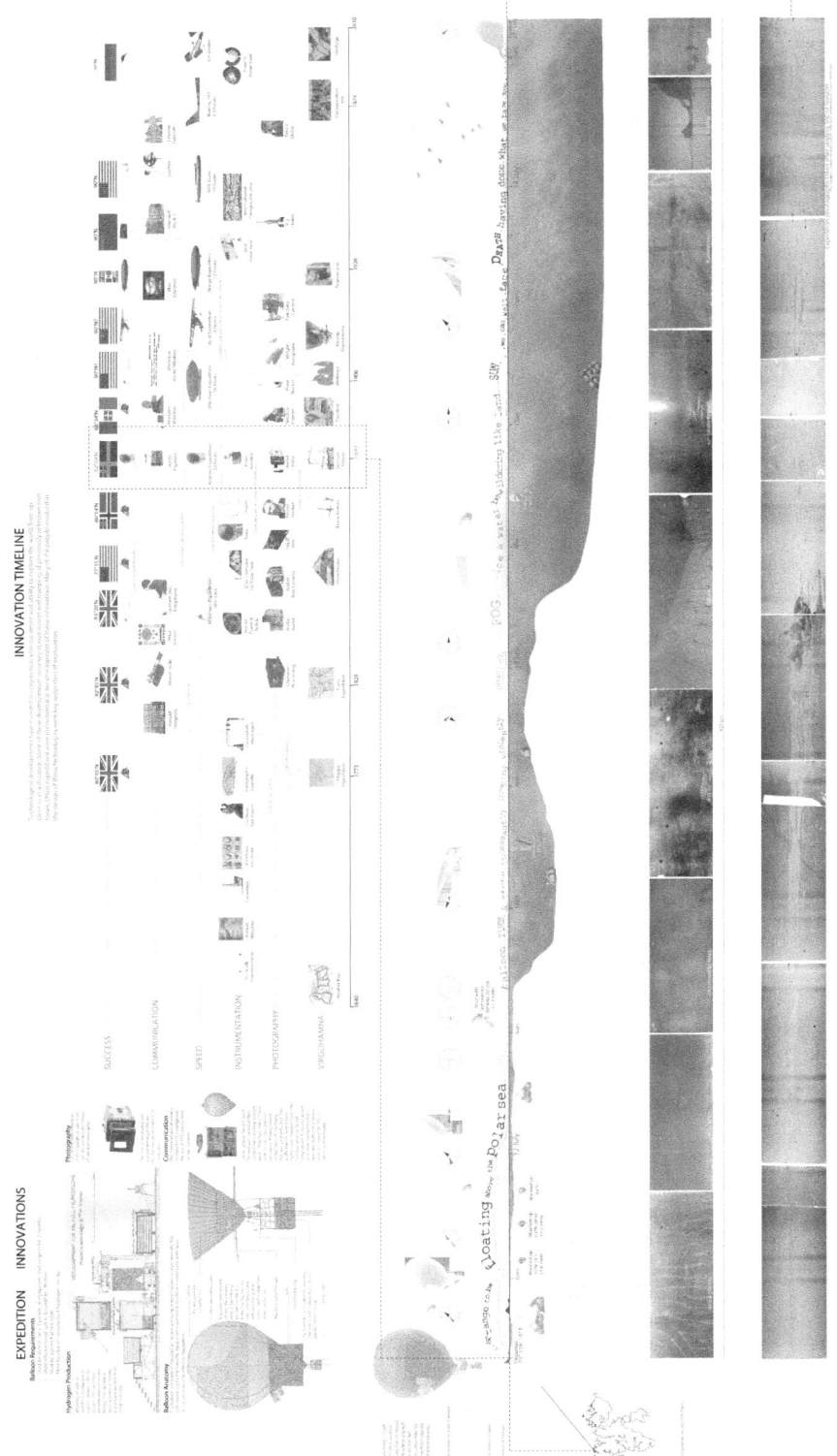

Figure 12.8   Brona Keenan's section off the coast of Virgohamna, at the northern tip of Spitsbergen, overlays the itinerary of Andrée's failed balloon journey to the North Pole. It performs as a narrative of connected forces such as innovation and documentation practices on Svalbard

This figure is shown in colour in the plate section of this book.

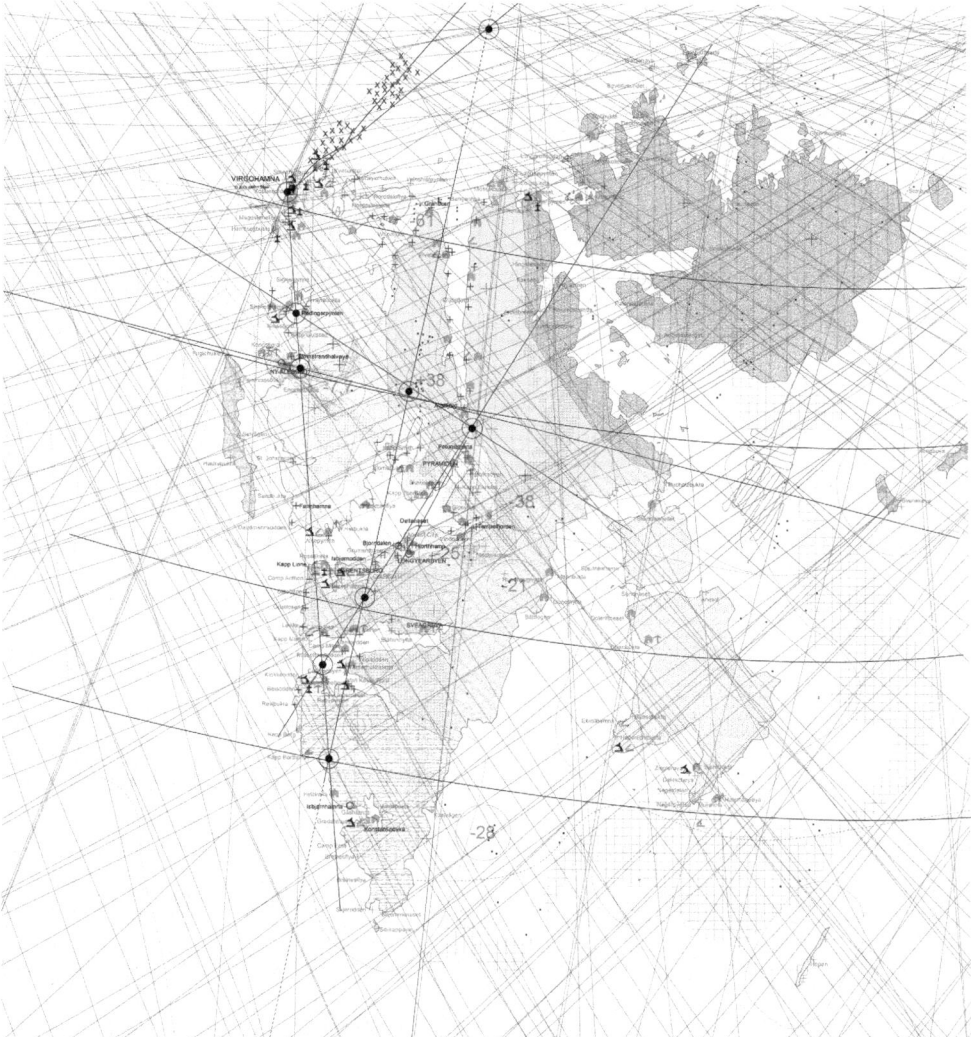

Figure 12.9 Students' sections and additional mappings were overlaid on a map of the Svalbard archipelago. The mapping includes the history of science activity, sporadic expeditions, sea ice extension, heritage sites and satellite trajectories

This figure is shown in colour in the plate section of this book.

representation as one that has indeed the capacity to unmake fixed conceptions and to remake territory as a multi-linear network of human and non-human agents. Hence, one of the pedagogical lessons of a studio course like 'Svalbard – Fluid Territory' is that the design process itself, the drawing out and making of representational territories that are not necessarily visible, contributes to the conception of territory.

Drawing on the 'landscape urbanism school', the course focussed on research and representation as critical approaches to territories rather than urban and landscape design.

The course represented a testing ground of the concept of *territorial care*, which we believe must be put forward and set in operation in the Arctic. The concept is inspired by Maria Puig della Bellacasa's investigations of matters of care as developed in dialogue with Bruno La Tour's 'matters of concern', and argues that 'the implications of care are thicker than the politics turning around matters of concern' (Puig della Bellacasa 2011: 86.) For della Bellacasa, care connotes an attention to things, issues, people, that are not necessarily equipped with or have not yet developed a human rationale or agency. A practice of care may alleviate or even substitute for a policy that is human-centred and profit-motivated. Working with the section as a device, and as a pedagogical tool, students were able to exercise care in relation to fluid territories in Svalbard by giving attention to things, not separating between the human and other materialities. For Maria Puig de la Bellacasa, the notion of 'matters of care' is a way of thinking. Our approach was to stretch this notion of care further in the context of the changing conditions and environments of climate change.

The Arctic is a region that demands critical rethinking and reconsideration of the traditional basics of landscape architecture as found in most educational programmes. Whether working on local ecologies, urbanism and community development or Arctic territories, the programme considers landscape as laboratory for extreme or composite situation that may serve to inspire work in more familiar locations. It rests on extended fieldwork and an intimate relationship with local/regional partners within planning and landscape architecture. Following the AHO model, there is a strong integration of research in studio teaching, contributing both to explore socio-cultural approaches to learning, and to integrate the acute socio-political context of Arctic communities. Students develop their briefs in collaboration with local actors and researchers, and hence are exposed to a learning process that encourages awareness of their own design agency, be this through cartographic design or landscape design.

Notes

1   To define the Arctic, we use the 10°C isotherm, which delineates the area where the middle temperature in the warmest month (July) never moves above 10°C. Hence, most of Northern Scandinavia, Iceland and Southern Greenland is defined as Subarctic. The isotherm roughly corresponds to the southern limit of the treeless tundra.
2   For an overview of what is happening in the Arctic, see Steinberg, P.E., Tasch, J. and Gerhardt, H. (2015), *Contesting the Arctic: Politics and Imaginaries in the Circumpolar North* (London: I.B Tauris & Co. Ltd); Bravo, M. and Sörlin, S. (2002), *Narrating the Arctic: A Cultural History of Nordic Scientific Practices* (Sagamore Beach: Science History Publications); Larsen, J.K. and Hemmersam, P. (2018), *Future North, the Changing Arctic Landscapes* (Abingdon, UK: Routledge).
3   See Bergset et al. (2017). Here, the studio teachers account for this studio's ongoing research on urban infrastructure in a cold climate.
4   The relevance of the concepts of site and place in Arctic communities is discussed in Hemmersam and Larsen (2017).
5   The studio was led by Thomas Juel Clemmensen, with Marianne Lucie Skuncke.
6   www.oculs.no/projects/future-north/news.
7   http://vardorestored.com/en.
8   The Svalbard Environmental Protection Act of 15 June 2001 secures, among other things, that any industrial object left in the territory older than 1949 is automatically preserved.
9   Tunheim was operative between 1915 and 1925, and partially destroyed just before WWII to keep the site from serving German Nazi forces.
10 The studio was led by Eimear Tynan.
11 For a presentation of the studio work, see Tynan and Larsen (2017).
12 Spitsbergen is the large western island in the archipelago.
13 The Global Seed Vault, Svalbard Satellite Service (SvalSat), The University Centre in Svalbard (UNIS) and numerous scientists in different research centres all provide climate, weather and scientific data to the world community.

14 The course was led by Janike Kampevold Larsen and Kathleen John-Alder with Eimear Tynan, Mats Kemppe and cartographer Riccardo Pravettoni. The research output won ASLA's 'Award of Excellence in Research 2017'. Students: Jérome Codère, Hans Stefan Eriksson, Brona Keenan, Charles Laverty, Rasmus Weitze Pedersen, Matthew Poot and Audrey Touchette. It was a parallel course to 'Arctic City: Longyearbyen', led by Peter Hemmersam and Lisbet Harboe, which took Longyearbyen as a case for Arctic Urbanism. Work from both courses is collated in the pamphlet *Future North: Svalbard*, available at the Institute of Urbanism and Landscape, AHO.

References

Bergset, M., Haukeland, A. and Skuncke M.L. (2017), 'Snøens muligheter – blå-grønn-hvit infrastruktur' [The Potential of Snow – Blue-Green-White Infrastructure], *Arkitektur N*, 2: 54–64.

Brighenti, A.M. (2010), 'On Territorology: Towards a General Science of Territory', *Theory, Culture & Society* 27/1: 52–72, https://doi.org/10.1177/0263276409350357.

Desimini, J. and Waldheim, C. (2016), *Cartographic Grounds, Projecting the Landscape Imaginary* (New York: Princeton Architectural Press).

Girot, C. (1999), 'The Four Trace Concepts', in J. Corner (ed.) *Recovering Landscape, Essays in Contemporary Landscape Architecture* (New York: Princeton Architectural Press), 59–69.

Hemmersam, P., Harboe, L. and Morrison, A. (2018), 'Building the Brief: Developing a Place-specific Urban Practice', in E. Lorentzen and K. Annabell Torp (eds) *Formation – Architectural Education in a Nordic Perspective* (Copenhagen, Denmark: Nordic Baltic Academy of Architecture/Architectural Publisher), 200–216: 201.

Hemmersam, P. and Larsen, J.K. (2017), 'Landscapes on Hold, The Norwegian and Russian Barents Sea Coast in the New North'', in K. Maier and S.J. Ray *Critical Norths, Space, Nature, Theory* (Anchorage: Alaska University Press), 171–191.

Herrington, S. (2013), 'Designing With Water Above the Arctic Circle: East Three School', *Journal of Landscape Architecture* 4/2: 44–51.

Lippard, L. (2014), *Undermining, A Wild Ride through Land Use, Politics, and Art in the New West* (New York: The New Press).

Mainsah, H. (2018), 'Visual and Sensory Methods of Knowing Place: The Case of Vardø', in J.K. Larsen and P. Hemmersam (eds) *Future North, the Changing Arctic Landscapes* (Abingdon, UK: Routledge), 106–199.

Massey, D. (2005), *For Space* (London: Sage).

MilNeil, C.N. (2013), 'Inner-City Glaciers', in E. Ellsworth and J. Kruse (eds) *Making the Geologic Now, Responses to the Material Conditions of Contemporary Life* (New York: Punctum Books), 79–81.

Puig della Bellacasa, M. (2011), 'Matters of Care in Technoscience: Assembling Neglected Things', *Social Studies of Landscape* 41/1: 85–106.

Sheppard, L. (2013), 'From Site to Territory', in N. Bhatia and L. Sheppard (eds) *Bracket 2: [Goes Soft]* (New York: Actar).

Sheppard, L. and White, M. (2017), *Many Norths: Spatial Practice in a Polar Territory* (New York: Actar).

Tynan, E. and Larsen, J.K. (2017), 'Cold Climates and Performative Territories', in Aa. Altés, A. Jara and L. Correira (eds) *The Power of Experiment* (Lisbon: Artéria Humanizing Environment), 140–155.

13

A critical approach to teaching landscape assessment

Andrew Butler

Critical thinking is considered a fundament of university education, providing a basis for questioning assumptions and revealing 'truths' (Haigh 2016; Mulnix 2012). Yet increasingly state-led agendas for universities focus on tangible goals with impetus on instrumental knowledge, frequently at the expense of critical learning (Berg and Seeber 2016; Collini 2012).

As a professional education, a central aim of landscape architecture (LA) studies is the production of future professionals to support the discipline. Landscape architecture education provides a step towards accreditation in national associations, helping form future landscape architects and landscape planners who will have defined and recognised roles in society. Consequently, the emphasis of teaching often focusses on instrumental and functional knowledge of planning and design tools and approaches that will be required in professional life. Yet in order to create a discipline that is forward thinking and capable of addressing the 'wicked problems' brought about through landscape change, landscape architecture education needs a balance between teaching instrumental knowledge and a critical approach to questioning this knowledge and the world to which it relates.

Critical thinking is the ability to think rationally and reasonably about what to believe and do. It can provide the intellectual space to question the worldview of a discipline, the approaches used to maintain this worldview, as well as revealing the assumptions and values that individual students hold. Thus, allowing students to identify cognitive biases and blindspots (Whiley et al. 2017) creates a position from which to question the world around them.

Critical thinking is often portrayed as abstract and conceptual in nature (Weissberg 2013), so we must pose the question: how can critical thinking be incorporated into landscape architecture education programmes as a complement to instrumental knowledge? In this chapter I engage with this question through my experience of teaching landscape planning and, more specifically, landscape assessment. However, the following text will touch on and have relevance for other aspects of the discipline of landscape architecture.

Educating critical landscape planners

The European Landscape Convention (ELC) recognises landscape planning as "strong forward-looking action to enhance, restore or create landscapes" (Council of Europe 2000: art. 1. f). This definition is further expanded in the 2008 guidelines for the implementation of the Convention, as concerning: "forms of change that can anticipate new social needs by taking account of ongoing developments. It should also be consistent with sustainable development and allow for the ecological and economic processes" (Council of Europe 2008a). From an academic context, Tress et al. (2006) have defined landscape planning as: 'primary attempts to influence the spatial organisation of landscape, making trade-offs between different needs, demands, values and land uses, how to solve land-use conflicts between different interest groups'. In this chapter landscape planning is a distillation of these definitions: forward-looking sustainable actions that negotiate societal and nature's needs, demands, values and aspirations in relation to the existing landscape and in light of drivers of landscape change.

Any definition of landscape planning is dependent on how landscape is conceptualised. Throughout Europe, rhetoric around landscape has increasingly come to be associated with the definition articulated through the ELC: 'an area, as perceived by people, whose character is the result of the action and interaction of natural and/or human factors' (Council of Europe 2000, 2008b). This definition places emphasis on perceptions of landscape that are formed in the context of the needs and demands of society, which are themselves in perpetual flux. Consequently, values ascribed to the landscape and aspirations for a landscape's future constantly change.

Societal needs, demands, values and aspirations also drive landscape change. Through the interrelated effects of globalisation, population growth, urbanisation, accessibility, calamities and changing cultural values societal values are met (Antrop 2005; Eiter and Potthoff 2016; Plieninger et al. 2016). It is landscape change that lies at the heart of landscape planning as both a driver for such change and as a reaction against it.

Values that planners handle and negotiate while addressing change are developed by individuals and communities in relation to landscape. These values represent social constructs, interpreted through social and cultural filters, and projected on to the landscape (Planchat-Héry 2011). The recognition of certain views privileges the holder of favoured value (Brookchin 1982; Olwig 1996). Recognising and making evident the values that are the basis for creating proposals or solving issues relating to landscape is central to the transparency of the planning processes (Butler 2016; Jones 2009). Revealing whose values are recognised illustrates who the landscape is being planned for, whose agenda is prioritised (Roe 2013; Thompson 2000).

The values advocated in landscape planning education are dependent on numerous factors. These include the prominent view promoted at the individual academic institution; be that a fine arts institution, agriculture and forestry college, technical university or 'general' university. Landscape planning in each of these institutions develops practice and procedural concepts, which help them better understand and handle specific perspectives of landscape (van der Brink and Bruns 2012). Recognised landscape values are also reliant on national contexts. The establishment of landscape planning in different countries has been reliant on the nature of neighbouring disciplines and the gap that landscape planning fills (Kidd 2013), as well as the dominant cultural understanding of landscape in each nation (Scazzosi 2004). Yet central to all landscape planning studies, irrespective of institutional and national context, is a consideration for context (relevance of place) and recognition of diverse values.

The ambiguities of the discipline across Europe, that the different contexts create, are compounded by the fact that the discipline has traditionally developed through the endeavours of practice rather than academic conceptual development (Selman 2010; van der Brink and Bruns 2012). Thus it is often difficult to discern the theoretical basis for tools and approaches used to attain knowledge of the landscape. The context in which the student of landscape planning finds themselves raises fundamental critical questions, about the discipline, the tools of the profession and the education that supports it.

Recognising landscape planning as a negotiation between diverse needs, demands and values makes it ripe for critical learning. The student of landscape planning must grapple with the complexity and ambiguity embedded in the definitions of landscape planning, landscape and landscape change, among others.

Developing critical thinking through landscape assessment

The significance of the assessment process is expressed in the ELC where it is recognised as the first step to mitigating the negative impact of development on the landscape (Council of Europe, 2000: Chapter 2, art. 6C). Assessment is listed as one of the more prescriptive aspects of the convention's text: 'assess the landscapes [. . .], taking into account the particular values assigned to them by the interested parties and the population concerned' (Council of Europe 2000, 2008). This not only lifts the importance of the assessment stage, but also the complexity and multiplicity of values held by a diversity of stakeholders (Jones 2007).

The aim of landscape analysis is to provide new knowledge and understanding in relation to past development and future change (Stahlschmidt et al. 2017). It acts as a means for making practice more systematic and has become a fundamental aspect of landscape planning (Marsh 1998; Muir 1999). The assessment stage in landscape planning can be both a standalone process for documenting, understanding and communicating the landscape or be a fundamental part of the planning and design process.

The purpose of a landscape assessment has traditionally been to provide insight and present the landscape for others to argue for its values, thus it is normally considered that an assessment reflects the landscape values. However, in practice, the assessment is dependent on how landscape is acknowledged and therefore what is accepted as a landscape value (Brunetta and Voghera 2008; Stephenson 2010). The values, whether formally recognised or informally accepted, create the basis of the scope for the assessment and consequently determine the knowledge base for decision-making.

Although the ELC advocates assessment that addresses the diversity of values recognised by multiple parties, the tools used for assessments are grounded on individual disciplinary epistemologies (Butler 2014). Most assessments are expert products, created with tools constructed on disciplinary values. A narrow, specialised assessment can develop an unquestioned impression of what a landscape is, which is then used as evidence for how a landscape should be (Muir 1999). This, contra to the rhetoric of the ELC, perpetuates the idea of landscape as a professional domain at the expense of the intimate encounters of those who inhabit the landscape.

Expressing the values on which an assessment is built provides the means to communicate and open discussion on the values present in the landscape (Brunetta and Voghera 2008). So while teaching landscape assessment may be seen as providing 'useful knowledge' for the discipline, it also provides the opportunity to explore what are often considered the 'truths' of the discipline. Providing a basis for questioning the foundations of the landscape planning in its many contexts.

A progressional approach to teaching critical thinking through landscape assessment

Landscape assessment is or can be present in many course modules throughout the landscape architecture/planning education. Teaching landscape assessment tends to be an 'add on', a way to systematically attain knowledge about an area, providing a basis for problematising issues within the landscape. Initially, the assessment process can be seen as introducing students to the idea of landscape and developing an understanding of a specific landscape, through creating an inventory of the area. This allows students to recognise what landscape means to them. At the Swedish University of Agricultural Sciences (SLU), students in their first year interpret map and text data of a landscape special to them and subsequently create a sand model of this landscape. Such an exercise requires the students to reflect over a familiar landscape; revealing the significance of familiar features; exposing new elements, and, ultimately, seeing the landscape with new eyes, reflecting critically over the taken for granted.

In the second and third years of studies I introduce specific approaches. Ecological, historical, spatial (e.g., Cullen 1961; Lynch 1960) and landscape character are presented to and used by the students in design and planning modules. The focus is as much on the benefits and drawbacks of the different approaches as on their usage. Being introduced to the theoretical and disciplinary basis for these tools helps to reveal the substantive nature of landscape they are dealing with. Consequently, this creates awareness that all assessments are partial and that a composite is needed if landscape is to be dealt with as anything approaching a holistic entity. The aim of introducing the students to multiple forms of assessment, as well as revealing the tools, is to make them critical of the tools at their disposal. Ultimately, they are provided with a smorgasbord of approaches from which to select those they feel to be suitable for specific contexts and for dealing with the issues at hand. Each of the tools and approaches they are introduced to has its own nuances and background (Fairclough and Herring 2016; Sarlöv Herlin 2016) pointing to a need to reflect over usage.

Through a landscape assessment, values are legitimised and discourses bolstered. At Master's level I critically address for whom the landscape is being planned or designed. A short exercise for getting students to question the values in an assessment is through analysing existing landscape assessments that are in the public realm. This allows the student to understand how an assessment was undertaken; what was the substantive understanding of landscape promoted; what the purpose of the assessment was; and to reveal what values are justified or excluded in the planning process. Taking point of departure from Flyvbjerg's (2004) phronetic planning, the following questions are directed towards the landscape assessment being analysed.

1   What landscape is this assessment describing?
2   Who gains and who loses, through the developing discourse?
3   Is the discourse developed through the assessment desirable?
4   What, if anything, can we do about it?

This develops the issue of power structures into the landscape education, and helps to develop a critical stance to these structures. As has already been highlighted in this chapter, landscape in line with the ELC refers to an entity perceived by people. Addressing existing assessments reveals the limitation of who the 'people' are, helping to acknowledge the relevance of public participation as an essential aspect of landscape architecture education along with the need to address power relations (Butler 2016). Consequently, the students reflect on how assessment creates or develops the public discourse on landscape, and who gets to be involved in this development.

To genuinely engage with the 'public's perceptions' of landscape would involve a focussed programme of public involvement in the studies, which is often not possible due to time restrictions and prioritisation of other aspects of the discipline. At the Master's level, to counter this and to further students' self-critical reflection, I develop a role-play scenario, where the students can learn from pretence (Nichols and Stich 2000). Students in groups take on the role of a certain actor group in the landscape (a farmer, a forager, a tourist, etc.), devise their own form of assessment suitable for their role and undertake an assessment from this perspective. The groups then split to form a composite assessment, discussing and arguing for the values from their individual roles. This allows the students to recognise the multiplicity of values in the landscape and consider who is the community who perceives what the landscape could be.

Teaching can foster a healthy critique of the tools and approaches for landscape assessment rather than taking them for granted; questioning their adequacy and capability for addressing landscape (Sarlöv Herlin 2016). It is only through engaging with these practices and revealing the values that they address that the relevance of the tool can be questioned, allowing the complexity of landscape to be recognised. A final stage in the critical study of landscape assessment in education is to question the relevance of the tools, their background and the agenda they promote. This helps build up a critical understanding of what values are recognised in a landscape and opening up for questioning the traditions and epistemological start points of these tools and whose landscape does this benefit (Sarlöv-Herlin 2016). This ultimately recognises which knowledge is seen as 'fact', and which values are legitimised in the discipline; and what landscape change they promote.

Conclusion

Critical thinking in landscape planning education provides a means for questioning the dominant discourses on the landscape as well as the tools the discipline uses for perpetuating or disrupting these discourses. As society builders, students of landscape planning need to question what society they hope to build and how individual and disciplinary values can represent this society.

A critical thinking approach can be used to question our individual preconceptions; the landscape we deal with, the tools used for engaging with the landscape, as well as the paradigmatic assumptions of the discipline and the worldviews on which the discipline is built. This opens up for addressing the prescriptive assumptions, how as both individuals and as part of a discipline we 'think' the world should work, which tools and which debates should be forwarded and whose agenda we are supporting. Are we strengthening a dominant discourse or trying to engage with subordinated values.

Pressing for critical thinking is not at the expense of other important aspects of landscape architecture education, but as a complement. It is also not only landscape assessment that is ripe for critical thinking, but all aspects of the subject can be dealt with in the same manner. Students require approaches, including landscape assessments, which provide clear instructions. Yet we also need to teach how to transfer skills to new contexts and 'mash-up' and blend approaches. As such, we as teachers must do more than just guide the learning of new concepts, ideas and tools; we need to create the space for students to be able to reflect and think critically about these approaches and the discipline itself (Whiley et al. 2017).

Education should involve risk (Biesta 2013), this is the difference between the teachers providing the student with what they want (instrumental knowledge) and offering to open up knowledge that is yet unknown to them (questioning the 'truth'). It is easy to frame problems built on taken for granted and unquestioned assumption and views, the difficult thing is identifying when the frames are questioning the norms and prejudices of the discipline and society as a whole (hooks 2010; Kumashiro 2002). Questioning what landscape is through the teaching of critical landscape assessment can provide a step towards achieving this.

References

Antrop, M. (2005), 'Why landscapes of the past are important for the future', *Landscape and Urban Planning*, 70 (1–2), 21–34.

Berg, M. and Seeber, B. (2016), *The Slow Professor* (Toronto: University of Toronto Press).

Biesta, G. (2013), *The Beautiful Risk of Education* (New York: Routledge).

Brookchin, M. (1982), *The Ecology of Freedom: The Emergence and Dissolution of Hierarchy* (Palo Alto: Cheshire Books).

Brunetta, G. and Voghera, A. (2008), 'Evaluating landscape for shared values: tools, principles, and methods', *Landscape Research*, 33 (1), 71–97.

Butler, A. (2014), 'Developing theory of public involvement in landscape planning: democratising landscape' (Uppsala: Swedish University of Agricultural Sciences).

— (2016), 'Dynamics of integrating landscape values in landscape character assessment: the hidden dominance of the objective outsider', *Landscape Research*, 41 (2), 239–52.

Collini, S. (2012), *What Are Universities For?* (London: Penguin).

Council of Europe (2000), *European Landscape Convention* (CETS No 176; Florence: Strasbourg: Council of Europe).

— (2008), *Recommendation of the Committee of Ministers to Member States on the Guidelines for the Implementation of the European Landscape Convention* (Florence, Strasbourg: Council of Europe).

Cullen, G. (1961), *Concise Townscape* (New York: Van Nostrand Reinhold).

Eiter, S. and Potthoff, K. (2016), 'Landscape changes in Norwegian mountains: increased and decreased accessibility, and their driving forces', *Land Use Policy*, 54, 235–45.

Fairclough, G. and Herring, P. (2016), 'Lens, mirror, window: interactions between Historic Landscape Characterisation and Landscape Character Assessment', *Landscape Research*, 41 (2), 186–98.

Flyvbjerg, B. (2004), 'Phronetic planning research: theoretical and methodological reflection', *Planning Theory and Practice*, 5 (3), 283–306.

Haigh, M. (2016), 'Fostering deeper critical inquiry with causal layered analysis', *Journal of Geography in Higher Education*, 40 (2), 164–81.

hooks, b. (2010), 'Critical thinking', in b. hooks (ed.) *Teaching Critical Thinking: Practical Wisdom* (New York: Routledge.), 7–11.

Jones, M. (2007), 'The European Landscape Convention and the question of public participation', *Landscape Research*, 32 (5), 613–33.

— (2009), 'Analysing landscape values expressed in planning conflicts over change in the landscape', In V. Van Eetvelde, M. Sevenant and L. Van De Velde (eds), *Re-Marc-able Landscapes: Marc-ante Landschappen* (Ghent: Academia Press), 193–205.

Kidd, S. (2013), 'Landscape planning: reflections on the past, directions for the future', in P. Howard, I. Thompson and E. Waterton (eds), *The Routledge Companion to Landscape Studies* (Abingdon, UK: Routledge), 366–82.

Kumashiro, K. (2002), *Troubling Education: Queer Activism and Antioppressive Education* (New York: Routledge).

Lynch, K. (1960), *The Image of the City* (Cambridge, MA: MIT Press).

Marsh, W. (1998), *Landscape Planning: Environmental Applications* (New York: Wiley).

Muir, R. (1999), *Approaches to Landscape* (Basingstoke, UK: Macmillan).

Mulnix, J. (2012), 'Thinking critically about critical thinking', *Educational Philosophy and Theory*, 44 (5), 464–79.

Nichols, S. and Stich, S. (2000), 'A cognitive theory of pretense', *Cognitio*, 74 (2), 115–47.

Olwig, K. (1996), 'Recovering the substantive nature of landscape', *Annals of the Association of American Geographers*, 86 (4), 630–53.

Planchat-Héry, C. (2011), 'The prospective vision: integrating the farmers' point of view into French and Belgium local planning', in M. Jones and M. Stenseke (eds), *The European Landscape Convention: Challenges of Participation* (Dordrecht: Springer), 175–98.

Plieninger, T., et al. (2016), 'The driving forces of landscape change in Europe: a systematic review of the evidence', *Land Use Policy*, 57, 204–14.

Roe, M. (2013), 'Landscape and participation', in P. Howard, I. Thompson and E. Waterson (eds), *The Routledge Companion to Landscape Studies* (Abingdon, UK: Routledge), 335–52.

Sarlöv Herlin, I. (2016), 'Exploring the national contexts and cultural ideas that preceded the Landscape Character Assessment method in England', *Landscape Research*, 41 (2), 175–85.

Scazzosi, L. (2004), 'Reading and assessing the landscape as cultural and historical heritage', *Landscape Research*, 29 (4), 335–55.

Selman, P. (2010), 'Centenary paper: Landscape planning: preservation, conservation and sustainable development', *Town Planning Review*, 81 (4), 381–406.

Stahlschmidt, P. et al. (2017), *Landscape Analysis: Investigating the Potentials of Space and Place* (Abingdon, UK: Routledge).

Stephenson, J. (2010), 'The dimensional landscape model: exploring differences in expressing and locating landscape qualities', *Landscape Research*, 35 (3), 299–318.

Thompson, I. (2000), *Ecology, Community and Delight: Sources of Values in Landscape Architecture* (London: Spon).

Tress, B., Tress, G. and Fry, G. (2006), 'Defining concepts and process of knowledge production: an integrative research', in B. Tress et al. (eds), *From Landscape Research to Landscape Planning: Aspects of Integration, Education and Application* (Dordrecht: Springer), 13–26.

van der Brink, A. and Bruns, D. (2012), 'Strategies for enhancing landscape architecture research', *Landscape Research*, 39 (1), 7–20.

Weissberg, R. (2013), 'Critical thinking about critical thinking', *Academic Questions*, 26 (3), 317–328.

Whiley, D. et al. (2017), 'Enhancing critical thinking skills in first year environmental management students: a tale of curriculum design, application and reflection', *Journal of Geography in Higher Education*, 41 (2), 166–81.

14

Teaching design critique

Jacky Bowring

Why is design critique a crucial part of teaching landscape architecture? After all, if we assume that the market will decide on the value of a built work, perhaps there is no place for design criticism. Or if we accept that 'everyone is a critic', does the practice of design critique become redundant?

On the other hand, if we recognise that design critique is an integral part of design practice, then it is a vital part of landscape architectural education. It could be said, to paraphrase A. O. Scott who was writing about art criticism, "all design is successful criticism" (2016: 22). And, if it is recognised that criticism is intertwined with the ethics of practice, then design critique is essential.

One of the challenges of teaching design critique is that 'criticism' is heavily burdened with connotations of disapproval and finding fault.[1] In an educational setting, critique's sometimes adverse associations can derive from the studio setting, where 'crits' are perceived as hostile, even attacking, and students can feel vulnerable.[2] Crucial to teaching in this area is both a recognition of the *value* of criticism within the design process, and also that critique is in itself a *creative* practice. As Blanchon puts it, "criticism is fully part of the creative process itself" (2016: 67). Stead adds to this, highlighting how it is not only part of the process, but creative in its own right: "criticism itself can be a productive and creative practice, which is literary but also specifically architectural" (2007: 79). And as Hopkins argues, it can be design's equal: "If it is done thoughtfully, criticism can be as much a creative act as design itself" (1994: 24).

This chapter focusses particularly on the critique of designed works,[3] and the perplexing questions that circle the apprehension of design as built – or as imagined. Conflicting views on the success or failure of a design can be disconcerting, and for students this can also resonate with their anxiety over receiving conflicting feedback from tutors. Debates over whether designs could or should mean anything are puzzling, and challenging students to consider the question of meaning helps to stretch their thinking about design intentions. The possibility of multiple and divergent interpretations of the same design is fraught. All of these conflicts, debates, and possibilities create a mystifying fog around design critique.

Teaching Design Critique is a form of demystification, a revealing of the often hidden assumptions and ideals that underlie criticism. Becoming aware of diverse positions and the messiness of conflicting theories reveals how criticism is not merely an act of passing judgement, but rather it is about becoming part of the conversation about a design. It involves developing a nimbleness with theory, and of a heightened awareness of one's own position in relation to theory. There are strong parallels between teaching Design Critique and teaching Design Theory. While theory can be taught as a foundation for a generative process, criticism explores how theory is also an analytical and critical tool.

Design criticism involves a multiplicity of decision making, which is usually invisible. Through pointing to these decisions, students are able to be explicit in their own choices, about why and how they are undertaking criticism. A careful consideration of the motivations and framing of design critique also heightens their awareness of the work of critics. Decisions might include taking a position on whether or not to be concerned with a designer's intentions. For some critics it is important to know how the designer intended their work to be read or experienced.

Exposure to a range of different forums in which critique takes place helps to reveal the influence of criticism, including a range of print and online media ranging from short-form critique to substantive pieces, popular and academic writing, settings such as an Urban Design Panel,[4] and design activism that uses landscape interventions to express critical points of view (for example, Battista et al. 2005).

This chapter explores some ways of working with students to expand their understanding of design critique, and to develop skills in the art of criticism. The first section suggests an approach for students to survey bodies of criticism, to appreciate the ways in which design criticism is part of the broad design discourse. Second, the chapter goes on to explore the motivations for criticism, which are important in underpinning the kind of criticism that is undertaken. The following section emphasises the importance of having a clear theoretical position from which to critique a work. Also, as well as having rigour in terms of a position, it is important to be clear about the nature of the work being criticised, as discussed in the fourth section. Finally, the chapter suggests a range of ways in which critique might be communicated or expressed, from writing to designing. The intention is not to propose a 'formula' for criticism, but to identify useful ways of challenging students to think about criticism. The types of examples, and the expectations about the depth of theory, will depend on the level at which criticism is being taught. Complex and theoretically dense criticism is expected of postgraduate students, but even students in the early stages of their undergraduate study can be challenged to think explicitly about their position as critics. Being a critic can be daunting, especially in small countries where the profession of landscape architecture has a limited number of practitioners and there can be sensitivities over critique. Developing a strong background in design critique as part of learning to be a landscape architect can contribute to the establishment of more robust debate within the profession, and through this to continue to challenge landscape architecture as a profession.

Anatomising: what is design critique and how does it work?

One of the most vivid means of demystification is an investigation into how criticism operates in the design profession, and what this reveals about the dynamic between the critic and the design. Juan Pablo Bonta's seminal work from 1975 provides a model for students to explore how criticism has accumulated for a built work. Bonta based his study on Mies van der Rohe's Barcelona Pavilion and examined the evolution of the criticism about the building, including where it was or was not written about. He structured all of the criticism that he could locate into nine categories: blindness; pre-canonical responses; official interpretation; canonical interpretation; class identification; dissemination; grammaticalisation and oblivion; meta-linguistic analysis; and re-interpretation.

Bonta termed his study an 'anatomy', bringing to mind a dissection of the body of work, cutting, examining, identifying. Analysing this criticism provides insight into the ways in which a critic can be influential in the design realm, from choosing to ignore a work, through to advocating for it, or using the work itself as the foundation for their own creative act.

The exercise of anatomising requires students to look carefully at the body of critique for a particular work. It challenges them to think about what a critique is. Should they, for example, look at everything that has been written about the work they have chosen to explore? Examples such as Parc de la Villette generate a massive volume of writing about the design, and there is therefore a need to think of some kind of limiting factor in order to tackle it (for example, for a particular time period, or only work published in refereed journals). Focussing on a particular example also reveals how, over time, perspectives on a work can change dramatically, as Bonta showed in his study of the Barcelona Pavilion. It also reveals how some critics may celebrate a work at the same time as others condemn it. Again, the Parc de la Villette is a vivid example, as on one hand it is listed as one of the "World's Worst Parks" by Project for Public Spaces (Project for Public Spaces 2004), but on the other hand is considered to exemplify "an analytical and conceptual approach to the way a human feels within a larger urban setting" (Souza 2011).

Bonta's work dates from the 1970s and covers four decades of criticism. At the time Bonta was writing, the Barcelona Pavilion was yet to be rebuilt, and its reconstruction in 1986 added further layers to the criticism of the building, accumulating further meaning and significance. Another important post-Bonta addition to the body of criticism is Caroline Constant's article from 1990, which is, in Bonta's terms, a 'reinterpretation' of the building (Constant 1990). Bonta's framework gives insight into the dynamics of criticism in relation to a built work, and is a snapshot of a particular era, when criticism was all in print form and critical conversations about built works took place at a much slower pace. As students exploring their own examples quickly discover, the internet has dramatically changed the body of criticism to be anatomised, and an important part of the Bonta-based project is the critiquing of Bonta's original framework.

The Barcelona Pavilion also raises a further significant question relating to the body of criticism. Can rebuilding a building be a form of criticism? Can the destruction of a design (for example the sequence of destruction and rebuilding at Jacob Javits Plaza in New York) be a type of critique? Another question raised by this exercise is about what the limits of criticism are and who is a critic? Is the opinion of a layperson as much an act of criticism as the considered words of a theoretically informed critic? As part of the anatomising of a body of criticism, the requirement to define what is and is not a critique, and why, demands reflection on the nature of published critical discourse. How are critics critiquing? On what basis are they making evaluations? Or, if they are not evaluating, in what other ways are they enhancing our understanding or interpretation of a work? These final questions are vital in learning design criticism, the core question of the *purpose* of a critique.

Purpose: why are you critiquing?

What reason is there for undertaking the criticism? What motivates the critic? As Dennis Sharp once asked of the intentions of architectural critics, "Are they altruistic, practical or propagandist? Are they international, nationalistic or regionalistic?" (1989: 8). Being explicit about the intentions of critique is one dimension of positioning the critic within the field of possibilities. Again, this is often implicit rather than explicit in written criticism, and teaching design critique involves pointing to this aspect and revealing how critics have a range of motivations.

The mystery of the intellectually dense work of theoretically informed criticism on one hand, and the brutal frankness of the "I don't like it" critique on the other hand, represent very different intentions for criticism. In his book *Architecture and Critical Imagination*, Wayne Attoe articulates three distinct motivations for critique: normative, interpretive and descriptive (1978). Like Bonta, Attoe's seminal text is several decades old, yet it remains an important and enduring approach to the orientation of critique. Attoe's three categories direct the ways in which a critic approaches a work, highlighting how the purpose of the critique necessitates different frameworks and approaches.

Normative

Making judgements is often an assumed part of what a critic does. In Attoe's category of normative critique, judgements are explicitly intended. The frame of a normative critique demands a set of norms or standards to which the designed work is compared and evaluated. These might be actual measurements that indicate performance (emissions, water quality), or more general rules of thumb. Familiar evaluative tools such as sustainability rating systems or post occupancy evaluations fall into this category. (And a critique of the criteria on which such tools are based can also be revealing). In design critique, the normative approach represents an important aspect of professional practice, where evaluations of functionality and efficacy provide tangible feedback on built works.

Interpretive

Criticism is not always about evaluation, and Attoe's category of interpretive critique recognises how the critic can be creative in their response to a work. An interpretive critique provides an evocative frame for reading a design, and can create a new impression of it. As Attoe points out, one effective way of re-framing a work is through shifting metaphors, perhaps encouraging the understanding of a street as a city's 'lounge', rather than its 'driveway'. Interpretive critique also encourages emotional and impressionistic responses to a work. Recent work on ficto-criticism enriches the possibilities of interpretive critique, where the critic adopts a fictional role, or creates a fictional site, amplifying the imaginary dimensions of critical practice (for example, Rendell 2007).

Descriptive

The final category of criticism situates the work within its context. Echoing practices of attribution, where the provenance of an artwork is determined, a descriptive critique can connect a design to the broader oeuvre of a designer, or to a particular type of design. Descriptive criticism can also focus on the social and cultural context of a work, adding depth to the understanding of a work. Mona Domosh's study of the New York World Building is a comprehensive exemplar of a descriptive critique (1989).

The range of Attoe's categories for the intentions of critique also highlight how a wide range of theory might be enlisted by a critic, highlighting the need for a critic to *position* themselves.

Positioning: how are you critiquing?

A critic's position is rarely strongly signalled, and students can be encouraged to be detectives when looking at criticism, finding forensic evidence of underpinning theories. As Jane Rendell stated, "I suggest a critic always takes up a position, and that this needs to be made explicit through the process of writing criticism" (2007: 180). Theory can often be 'naturalised' or invisible within a critique, and part of teaching design critique is to actively reveal the theoretical position(s).

Different theoretical positions can reveal different aspects of a built work. A feminist critique, for example, can vividly demonstrate how a design can disadvantage or empower particular groups of users (see, for example, Weisman 1992). Queer theory offers a position for critiquing the built environment from the perspective of gender, particularly in terms of affect and emotion, for example the work of Sedgwick (2003). That very same site seen through the lens of formalism would offer a critique of its visual properties of colour, line, tone, and texture, and reveal nothing about its accessibility, danger, or emotional charge.

Finding a position for critiquing requires a degree of dexterity with theory. Developing this nimbleness necessitates opportunities to be exposed to a range of theories, to see how they work, to try them out. Contrasting positions, such as feminism, queer theory, and formalism as above, can shine a very bright light on the significance of theory for critique. How can that same site be seen so differently? There is also an intertwining of the intention and the theory, where the intention can demand some particular types of theory. For example, the intention to *evaluate* a main street from a *normative* approach will require the identification of relevant norms within the theory of urban design.

Teaching design critique is an opportunity to unpack theory, and to explore how it can often become abridged or conflated. Aesthetic theory, in particular, is significant for design criticism, but is often abbreviated to be merely about beauty. Through a more thorough attention to aesthetics, its full multi-sensory dimensions can be brought into the critical apparatus. Further, the expanded understanding of aesthetics encourages an embracing of the theoretical position of phenomenology. While this can be a daunting body of theory, it also has possible avatars to which students can relate strongly – for example, a heightened sensory critique of the city from a skateboarding perspective (Borden 2001), or 'parkour eyes' as a critical approach to landscape (Ameel and Tani 2012). Other positions might be drawn from the theories like semiotics where the critique deals with the ways in which a design is an act of signifying, and the critic becomes the reader and interpreter. Critique can also draw on the complexities of psychoanalysis (Vidler 1992) and emotion (Bowring 2016).

Theory is central to positioning, determining the frames that are the foundation for making the critique. The 'how' of critique (the position) needs to be considered in tandem with the 'why' of critique (the intention). An effective means of teaching how these two factors work together is through using a matrix (Figure 14.1). Although the matrix can over-articulate the position and the intention – making it more obvious than it would tend to be in a design critique – it instils the discipline of being aware of how these factors influence the act of critique. This approach also encourages careful consideration of the ambition of a critic. For example, a critic may be seeking to explore the designed work as a means of gaining insight into the

		Theoretical position (What perspective is the designed work being viewed from?)						
		Formalism	Feminism	Semiotics	Phenomenology	Aesthetic conventions	Emotion	Psychoanalysis
	Normative				A normative phenomeno-logical critique?			
	Interpretive						An interpretive emotional critique?	
Intention	Descriptive	A descriptive formalist critique?						

Figure 14.1   A matrix that combines the intention of a critique with the position of the critic

design's foundations, and therefore needing to evaluate it from the perspective of ecological function or a cultural frame such as semiotics. Or, the goal may be to use the critique of a designed work to explore alternative ways of understanding the built environment, perhaps through interpretations through the eyes of more-than-human subjects such as animals (see, for example, Tom Turner's DIY Critique in Turner 1996).

The matrix plots Attoe's three categories of intention against a range of theoretical positions, for example visual aesthetics, semiotics, phenomenology, formalism, emotion, feminism, and psychoanalysis (see Figure 14.1). Locating an approach to critique within the matrix foregrounds the parameters the critic is working with; while on one hand the matrix perhaps oversimplifies the practice of criticism, the intention is to encourage an explicitness of approach. The matrix can be an analytical tool, as a means of identifying critics' positions and intentions. It can also be a generative tool for criticism, for example asking questions such as 'what would a critique from the perspective of an *interpretive emotional* position look like?' or 'how would a *normative phenomenological* critique be performed?' While some possible combinations might at first seem absurd, they also demand exploration of how such an approach might be possible; can there be *norms* for the *experience* of landscape, for example? And, for the more immediately graspable combinations (such as a *descriptive formalist* critique), this is also a prompt to be explicit in the ways in which the critique is undertaken. If the intention is to be descriptive in a critique, there is a discipline in following that intention, rather than slipping into a vague position of making a casual and unsupported judgement about the work. The range of combinations also presents the possibility of adopting two contrasting approaches for the same site, therefore exploring the influence of theory and intention. A heightened awareness of the need to consider the intention of a critique, and the theoretical position, also emphasises a need for critics to be self-aware of their ethical stance on undertaking the critique – i.e., their ambition or aim for the criticism. On one hand, the goal may be to contribute to best practice in the profession, through the evaluation of the performance of designs. In this case, an interpretive critique is unlikely to be useful, whereas a normative approach immediately reminds a critic of the need to be transparent and explicit about the norms they are using. On the other hand, the motivation may be one of enriching theory, and deepening our understanding of design and landscape, prompting an interpretive intention, perhaps with a variety of positions.

Selecting: what are you critiquing?

It might seem odd to come to the selection of the work to be critiqued after first discussing the why and the how. However, in learning about design critique, having confidence in how to approach it can influence the choice of subject. A sustained and in-depth critique is an exercise that draws together the questions about *why* and *how*, and highlights the need to define the *what*. Some of the parameters for further defining the subject include: is it built? Or is the critique of a representation of the design? If you are looking at a site in a city, are you including context? Are you considering the temporal aspects of a site? Looking at each of these in turn provides further dimensions to consider in the teaching of design critique.

Built or unbuilt?

The subject of criticism might be presumed to be a built or established work. Yet, in many situations, critics are considering unbuilt works. The critique of representation draws in further layers of complexity. The evolution of software, hardware, and skills sees increasingly realistic renderings being presented. In some cases it is difficult to know whether an image is a computer rendering or an actual photograph of a completed project. There is a tendency to believe that the more realistic an image is, the 'better' it is. A critique of renderings is an opportunity to carefully consider the consequences these seemingly real images may have.

Katy Kingery-Page and Howard Hahn warn that hyper-realism can have a dulling effect on perception, and can even suggest ideals to the audience that are simply not possible, and are therefore misleading (2012). The potential to deceive is supported by the increasing ability of digital rendering to manufacture atmosphere and mood. Renderers are able to carry out a sleight of hand, where otherwise dull imagery is given a virtual makeover. Jacob van Rijs of Dutch architecture practice MVRDV suggests there is a kind of 'happy filter' that can be put over images, referring to a firm of architectural renderers who claimed they could make any project look good (Van Rijs 2015). Tuning into the influences that increasingly real and seductive imagery has on the reading of unbuilt works can sharpen the critic's insight into the design in question.

Contextualising

Teaching criticism is also about the art of contextualising. There are strong parallels with site design, where the setting for designing is a vital consideration. For critique, the consideration of context can influence the critic's apprehension of a work. Considered in isolation from its context, a work might satisfy the criteria that underpin a normative critique. However, when the context is taken into account, the work's performance may be compromised – in terms of safety, beauty, access, and so on. How large is the frame for a design critique? How far beyond the site itself should the critic be looking?

Temporality

The questions of how long to spend on site, how often to visit it, and over what time period, are questions for student critics as well as they are for student designers. Site visits seem often driven by expediency rather than experience. A quick visit and a hundred rapid digital photographs might completely miss the nuance of the site for a design project, or a design being critiqued. The encouragement to think carefully about the temporal aspects around visiting a site can empower students towards a much more profound relationship with landscape. Bernard Lassus, for example, suggests site visits should be for as long as "boredom sets in, or almost" (1998: 57). The temporal considerations need to include how a site changes over time – daily (including at night), seasonally, yearly, and over its life as a design? How might design critique be different for historic sites?

Doing: critiquing, writing, designing

Having negotiated the questions of *why, how* and *what*, a final consideration is the form in which a critique is communicated. While the familiar approach might be a written critique, this too brings questions of audience, language, and length. Audiences might vary from the populist, to the professional, to the academic, and length often varies accordingly. Intriguingly *Archinect*, an international website for architects, ran a competition for haiku critique, a pared-down form that demands evocative and powerful words (Taylor-Hochberg 2016). Alternatively, written critiques can be lengthy, sustained articles, or even books. Useful sources for the craft of written criticism include Huxtable (1976) and Lange (2012).

Alongside writing, critique can be communicated – and activated – through designing. Design as a mode of critique might respond to designed works, and the evaluation or interpretation of those works. It can also respond to other aspects of the environment of design, for example policies or practices – or even the profession itself. For example, landscape architect Ken Smith's design for the Camouflage Garden, which sits on the roof of MoMA in New York, was a critical commentary on how the profession of landscape architecture was often involved in the art of camouflage, attempting to conceal elements considered undesirable in the landscape. And, in turn, Smith reflected on how such camouflaging can stand out very strongly in the landscape, where a planted screen might raise a question about what is being screened, in the same way that the Camouflage Garden ironically is very obvious when viewed from skyscrapers surrounding the gallery (Amidon 2006).

Further, designing-as-critique can recast criticism within the design process. Rather than as the respondent to a finished product, the critic-as-designer becomes an instigator, an innovator. Instilling the possibility of this role for critique empowers students to seize the opportunities of design to respond critically to the world in which they find themselves.

Overview

Teaching design critique foregrounds the importance of criticism within landscape architecture, including how it can influence the understanding of designed works, and how in itself critique can be part of our creative practice. In summary, an approach to teaching design critique directs students to have a heightened awareness of the intention for their critiquing, their theoretical position, their choice of site or subject, and the form of communicating the critique.

Teaching design critique resonates with the teaching of design theory, and also with designing. Students can become empowered to critique their own work, a crucial skill to take into practice. The demystifying and unpacking of the critical apparatus encourages an active engagement with critique. This also enhances awareness of the quality of the critique that they encounter. And, ultimately, to be more aware of their own designing, and the kinds of questions that might be asked of it. Does it matter whether or not their work has meaning? Does it matter if visitors to a site 'get' the meaning or not? What would a feminist critique of their design reveal? How would they decide if their design was successful in the short, medium and long term? In what ways might their rendering of their design influence how others respond to it? The imbricated and nested relationships between designing, design theory and design critique are all related in the teaching of landscape architecture.

Notes

1 The words 'critique' and 'criticism' can in general be used interchangeably, both referring to the practice of criticising. While the word 'criticism' can evoke the negative sense of the practice more immediately, it also has useful connotations of literary criticism and criticising as a creative act.
2 While this chapter does not focus specifically on studio crits, Parnell and Sara's (2007) book *The Crit* offers useful advice for students, especially in terms of recognising how criticism should be a constructive part of the design process and design learning.
3 Critique in landscape architecture can also be usefully extended to the built environment in general, much of which is not designed. And further to the cultural landscape at large, which might include criticism of policies and designations, as much as of the physical form of a place. However, this chapter is of necessity focussed on one dimension of critical practice in order to meet the size limitations. For further thinking about the critique of landscape see, for example, Cresswell (2003).
4 In New Zealand, the Urban Design Panel is a panel of design experts who assess design proposals. It provides independent and free advice, and follows in the tradition of examples such as CABE, the Commission for Architecture and the Built Environment, in the United Kingdom.

References

Ameel, L. and Tani, S. (2012), 'Everyday aesthetics in action: parkour eyes and the beauty of concrete walls', *Emotion, Space and Society*, 5/3: 164–173.

Amidon, J. (2006), *Ken Smith, Landscape Architect: Urban Projects* (New York: Princeton Architectural Press).

Attoe, W. (1978), *Architecture and Critical Imagination* (Chichester, UK: John Wiley & Sons Ltd).

Battista, K., LaBelle, B., Penner, B., Pile S. and Rendell, J. (2005). 'Exploring 'an area of outstanding natural beauty': a treasure hunt around King's Cross London', *Cultural Geographies*, 12/4: 429–462.

Blanchon, B. (2016), 'Criticism: the potential of the scholarly reading of constructed landscapes. Or the difficult art of interpretation', *Journal of Landscape Architecture*, 11/2: 66–71.

Bonta, J. P. (1975), *An Anatomy of Architectural Interpretation: A Semiotic Review of the Criticism of Mies van der Rohe's Barcelona Pavilion* (Barcelona: Gustavo Gili).

Borden, I. (2001), 'Another pavement, another beach: skateboarding and the performative critique of architecture', in I. Borden, J. Kerr and J. Rendell (eds) *The Unknown City: Contesting Architecture and Social Space* (Cambridge, MA: MIT Press), 180–199.

Bowring, J. (2016), *Melancholy and the Landscape: Locating Sadness, Memory and Reflection in the Landscape* (Abingdon, UK: Routledge).

Constant, C, (1990), 'The Barcelona Pavilion as landscape garden. Modernity and the picturesque', *AA Files*, 20: 46–54.

Cresswell, T. (2003), 'Landscape and the obliteration of practice', in K. Anderson, M. Domosh, S. Pile and N. Thrift (eds) *Handbook of Cultural Geography* (London: Sage Publications), 269–281.

Domosh, M. (1989), 'A method for interpreting landscape: a case study of the New York World Building', *Area*, 21/4: 347–355.

Hopkins, J. (1994), 'Critics' Forum', *Landscape Design*, February: 24–25.

Huxtable, A. L. (1976), *Kicked a Building Lately?* (New York: New York Times Company).

Kingery-Page, K. and Hahn, H. (2012), 'The aesthetics of digital representation: realism, abstraction and kitsch', *Journal of Landscape Architecture*, 7/2: 68–75.

Lange, A. (2012), *Writing about Architecture: Mastering the Language of Buildings and Cities* (New York: Princeton Architectural Press).

Lassus, B. (1998), *The Landscape Approach* (Philadelphia, PA: University of Pennsylvania Press).

Parnell, R. and Sara, R. (2007), *The Crit: An Architecture Student's Handbook* (2nd edition) (Abingdon, UK: Routledge).

Project for Public Spaces (2004), *The World's Best and Worst Parks* [website], www.pps.org/reference/september2004bestworst, accessed 1 November 2017.

Rendell, J. (2007), 'Site-writing: she is walking about in a town which she does not know', *Home Cultures*, 4/2: 177–199.

Scott, A. O. (2016), *Better Living Through Criticism: How to Think About Art, Pleasure, Beauty, and Truth* (New York: Penguin).

Sedgwick, E. K. (2003), *Touching Feeling: Affect, Pedagogy, Performativity* (Durham, NC: Duke University Press).

Sharp, D. (1989), 'Criticism in architecture', in R. Powell (ed.) *Exploring Architecture in Islamic Cultures 3: Criticism in Architecture* (Singapore: Concept Media/Aga Khan Awards for Architecture), 8–15.

Souza, E. (2011), 'AD Classics: Parc de la Villette / Bernard Tschumi Architects', 9 January. *ArchDaily* [website], www.archdaily.com/92321/ad-classics-parc-de-la-villette-bernard-tschumi, accessed 23 November 2017.

Stead, N. (2007), 'Criticism in/and/of crisis' in J. Rendell, J. Hill, M. Fraser and M. Dorrian (eds) *Critical Architecture* (Abingdon, UK: Routledge).

Taylor-Hochberg, A. (2016), 'Short and not-so-sweet: a collection of architecture haiku criticism'. *Archinect* [website], http://archinect.com/features/article/149980050/short-and-not-so-sweet-a-collection-of-architecture-haiku-criticism, accessed 12 March 2017.

Turner, T. (1996), *City as Landscape: A Post-postmodern View of Design and Planning* (London: E and FN Spon).

Van Rijs, J. (2015), Lecture at Christchurch Polytechnic Institute of Technology, 24 July, Christchurch, New Zealand.

Vidler, A. (1992), *The Architectural Uncanny: Essays in the Modern Unhomely* (Cambridge, MA: The MIT Press).

Weisman, L. K. (1992), *Discrimination by Design: A Feminist Critique of the Man-Made Environment* (Urbana, IL: University of Illinois).

15

Values and transformative learning

On teaching landscape history in a community of inquiry

M. Elen Deming

The study of landscape history contributes its share . . . by reminding us, among other things, that since the beginning of history humanity has modified and scarred the environment to convey some message, and that for our own peace of mind we should learn to differentiate among those wounds inflicted by greed and destructive fury, those which serve to keep us alive, and those which are inspired by a love of order and beauty, in obedience to some divine law.

J.B. *Jackson (1996), vii,* A Sense of Place, A Sense of Time

Introduction

In *A Land* (1951), her magnificent humanist archaeology of Britain, Jacquetta Hawkes (1910–1996) writes trenchantly of enduring landscape scars dating from the earliest phases of the industrial revolution, which (to her) continued to disfigure her homeland. The unsentimental commercial values that rose dominant in the 19th century, Hawkes felt, were exemplified particularly by "the American people, the most successful materialists in the history of the world" (Hawkes 1951, 217). Yet, to American landscape geographers such as J.B. Jackson (1909–1996) or John Stilgoe (1949–), the forms of American marketplaces, cities, spatial patterns of industrial agriculture, the infrastructure of railroad and electric transmission lines, and the scars of extraction as from oilfields and quarries, if not always beautiful, at least presented valuable material histories of American ambition, energy, and design thinking (Figures 15.1 and 15.2).

In any scale or period, "values matter precisely because they materialize" and they materialize because they guide social choices (Deming 2015, 1). Exactly *how* this happens provides the subject for numerous landmark studies by cultural landscape historians and geographers in the United States (May Thielgaard Watts, Don Meinig, J.B. Jackson), Britain (Denis Cosgrove and Stephen Daniels), Europe (Shelley Egoz, Kenneth Olwig, Don Mitchell, Tom Mels), and elsewhere (Deming 2015, 6–12). From these and other authors, we learn that historical values are deeply coded in collective social narratives, capital and spatial (dis)investments, resource (and human) exploitation, and regulatory patterns, among other things.

Figure 15.1 Developmental values—both individual and collective—encoded within America's Jeffersonian grid. Aerial view of the traditional grid of farm fields, homesteads and villages now overlain with new patterns—drain tiles, railroads, and wind turbines. Near Gibson City, Ford County, Illinois (photo by author, March 2018)

Practicing, interpreting, and even maintaining landscape design must also be understood as value-laden enterprises. Using substances from the ground itself and guided by myriad practices of human actors, landscape forms may equally encode aspirational, unjust, or merely prosaic values. And although they may be confusingly cloaked in the guise of "nature," designed landscapes may make even the most unconscious values of societies and social actors materially, socio-spatially, and aesthetically manifest. By extension then, the activities of teaching and learning the history of cultural landscapes are equally freighted.

This chapter expands upon this basic premise with special attention to values-based pedagogy in professional design education. The first section of the chapter explores the importance of foregrounding values in and for landscape architects, while the second section explains the structure and intent of an online course in landscape history developed and taught over several years at the University of Illinois, Urbana-Champaign. The course pedagogy borrows the concept of "a community of inquiry" among history students in order to focus debate about design ethics and values. Adopting this approach permits students to mutually construct a sense of historical values that makes it more alive and relevant for them. Further, the community of inquiry can safely explore techniques for opening up challenging formative, even transformative, conversations about the values—some of them problematic—inhering within contemporary landscapes.

Figure 15.2 Capitalist values. Aerial view of the urbanized grid of Chicago's north shore; the graded intensity of capital development graphs both the accrual and erosion of economic and social values. Chicago, IL (photo by author, March 2018)

The place of values in professional design education

At every stage of learning, students represent a broad cross-section of social values. They maintain a range of complicated personal and family attitudes as well as unconscious biases—possibly including sexism, racism, consumerism, technological chauvinism and/or ethical naiveté—that often remain undisclosed, unapprehended, and unprocessed in ordinary classroom settings. Left unexamined, such biases may run against the grain or even undermine the goals of the profession.

In a 2015 paper, Thomas Oles points out that although "ethical issues are at the centre of a large and growing literature on public participation and design practice," the personal ethics and values held by landscape architects often run counter to their professional practices. He characterizes landscape architecture as an "uneasy discipline," in which the drive for commercial survival often dominates other commitments or, indeed, the profession's larger aspirations to support values such as public health and well-being, ecological fitness, cultural expression, and social justice. "Virtually no attention," Oles writes, "has been given in this literature to the possibility that 'true landscape democracy' . . . might be in tension, not with particular modes of professional practice, but with the structural requirements of professionalism itself" (2015, 189).

Is there a role for addressing values in professional education? Surely there ought to be. Design educators are responsible for more than simply training up future design professionals. They also assist students toward understanding, developing, and operationalizing design values, whether personal or shared, as well as articulating ethical values-based perspectives that are relevant to the design community(s) in which they participate.

"What history should we teach and why?"

In professional degree programs, one way to open up the conversation on values is through teaching the social and humanistic history of landscape design. Because of its apparent "displacement"—both geographically and temporally—the study of landscape history may provide a lowered-risk social setting for students to work through difficult conversations about social and environmental values. Seen at an abstract distance, landscape history seems to offer a relatively safe academic context in which to tackle highly charged and debatable subjects such as religion, war, colonization and exploitation, environmental degradation, slavery and racism, political ideology, gender inequality, civil rights, memorialization, and so on.

During the mid-to-late 1990s, a vibrant debate on the purpose and quality of teaching history in landscape design programs was begun in the pages of *Landscape Journal* by then-editor Robert Riley. In a subsequent response critiquing Catherine Ward-Thompson and Peter Aspinall's focus on the instrumentality of historic monuments as a source of formal inspiration in students' design process, Dianne Harris argued that the study of history should, instead, help develop inherently humanistic values among students (Riley 1995; Ward-Thompson and Aspinall 1996; Harris 1997). A generation later, however, with notable exceptions among specific faculty and institutions, it is difficult to see broad or encouraging progress.

Might we now also consider a different dimension of the question: what history *do* we teach and why? If we see that the present is the 'child' of the past, then we can understand how society is imprinted with the DNA of its own history. But which history, or *whose* history, should we study? Why have design historians valued certain forms and practices—but not others? Such questions may be useful in provoking new scholarship as well as critiquing the development of curricula.

A recent conversation among American landscape architects challenges the ways that received notions of 'high design' seem to be "rooted in Eurocentric design principles" (McKee 2018, 12). This is hardly new. Reflecting broader changes in both the ethical principles and social composition of the profession, the argument calls for greater diversity of non-western historical subjects, social and environmental themes, research methods, regions, and case studies to be taught to budding landscape architects enrolled in professional degree programs. Critics maintain that greater diversity in the subjects and treatment of landscape history is needed for practitioners to achieve longer-term and broader-scale impact on real-world problems.

> Landscape history as it's now apprehended may indeed begin in the European garden. But that is not where it ends. To confront existential perils that are emerging globally, particularly those wrought by climate change, landscape architecture graduates will benefit from greater literacy in the ways people are challenged to live the world over, not merely how they construct leisure or beauty.
>
> (McKee 2018, 12)

Cultural critic Raymond Williams (1921–1988) long ago addressed the tangle of issues that lay at the root of the complaint. On the one hand, he examines the position that "high culture—'the best that has been thought and written in the world'—is in danger, or is indeed already 'lost,' because of widespread popular education, popular communications systems, and what is often called 'mass society'" (Williams 1974, n.p.). On the other hand, there is an alternative position that asserts

> high culture—'the tradition'—is, in the main, the product of past stages of society, that it is ineradicably associated with ruling classes and with elites, and that it is accordingly being replaced in modern democratic conditions by a popular culture.
>
> *(Williams 1974, n.p.)*

Further, Williams observes there are practical ramifications for each position in the "allocation of resources and in the political shaping of cultural institutions" (Williams 1974, n.p.).

The selective tradition

Debates over what ought to be valued (and thus preserved in the cultural patrimony) have direct effects on the ways we teach landscape design history, theory, and practice. Problems of cultural bias, sometimes unconscious, often simply dismissive, are structured into the selection, interpretation, and even consumption of past cultures. Only consider how historical accounts of the experiences or designs of enslaved or oppressed classes of people, and/or women, long considered to be beneath historical notice, have only recently started to emerge in landscape architectural scholarship. Such bias may be delicately reframed as an expression of what Williams calls "the selective tradition," in other words, a "deliberately selective and connecting process which offers a historical and cultural ratification of a contemporary order" (Williams 1977, 116). Here he explains how the selective tradition works:

> [T]he cultural tradition can be seen as a continual selection and re-selection of ancestors. Particular lines will be drawn, often for as long as a century, and then suddenly with some new stage in growth these will be cancelled or weakened, and new lines drawn. In the analysis of contemporary culture, the existing state of the selective tradition is of vital importance, for it is often true that some change in this tradition – establishing new lines with the past, breaking or re-drawing existing lines – is a radical kind of contemporary change.
>
> *(Williams 1961, n.p.)*

We need only substitute the words 'landscape design history' for 'cultural tradition' to grasp the critical relevance of historical values to the canon of landscape design history. Consider a case in point: the seminal American textbook on landscape architectural history, *An Introduction to the Study of Landscape Architecture* (1917). Compiled by Henry Vincent Hubbard and Theodora Kimball Hubbard (the first librarian for the School of Landscape Architecture at Harvard University), this book was written as landscape architecture was becoming fully professionalized in the United States (Hohmann 2006). Written to ratify the Euro-centric Beaux Arts architectural standards of the period (and the institution), this historical survey guided the "intellectualization" of American landscape architecture and established the canon of landscape history for well over half a century. In what Dix calls a "process of relative valorisation or exclusion," high design case studies illustrate and exemplify imperial aspirations, as well as principles of class and racial privilege in their many guises (Dix 2005, n.p.). Our students are increasingly aware of that.

New subjects and kinds of scholarship, accompanied by fundamental changes in the tools we have to engage students, now make it possible to teach landscape history in a more expansive way.

Contemporary teaching encourages rich and accurate historical narratives that can and should be challenging to the present; that can and should create ethical and aesthetic disturbance; that can and should tell new stories in different voices. Critical histories may analyze with equal rigor the hidden values and tangled agendas of high culture's most cherished landscapes, for example Versailles at the time of Louis XIV (Mukerji 1997), or the political significance of the engineering of national infrastructure in early 19th-century France (Mukerji 2009). Historians may tackle under-studied vernacular historic landscapes such as slave quarters in the American south (Barton 2001; Ellis and Ginsburg 2010), or foreground little-known female figures working in the design professions (Mozingo and Jewell 2012; Way 2009). Whether spotlighting shelterbelts, airports, or kitchen gardens, there is much alternative landscape scholarship to be commended—and integrated—in history courses right now.

Meeting the challenge: a role for values

While scholars can still point to a paucity of new historical work overall, they also note, with reason, that even existing scholarship remains to be digested for use in survey courses (McKee 2018, 12). Elizabeth Rogers (2001) has made an admirable attempt to expand the range of topics while maintaining the traditional Euro-centric canon. However, it is the lowering of expectations for humanistic understanding in professional education and licensure that presents the most serious concern. Rather than a catalog of forms, typology, or *parti*, landscape history should also, mainly, be taught as a humanistic discipline with ambitious environmental and architectural sweep; a broad social history of ideas as well as lived experiences (Meyer 2000, 2008, 2015).

Especially for its potential as a fertile laboratory for design students to explore alternatives (identities, experiences, values), we are therefore very interested in the generally unloved landscape history survey course. Essentially, the profession already has to hand an effective mechanism to foster values-literacy, and an opportunity to engage future practitioners in mutual construction of a framework for 'reading' the cultural values of place.

In recognizing and naming our values—both personal and collective—we take an important step towards what Abraham Maslow has called *metamotivation*, leading to self-actualization (Maslow 1970/1954). Maslow differentiates between physiological needs (physical sustenance and safety), psychological needs (belonging, love, and esteem), and higher order needs (comprising self-betterment, insight, creativity, understanding, and, ultimately, transcendence). As Maslow explains it,

> This process [of satisfying higher order needs] has been phrased by some as the search for meaning. We shall then postulate a desire to understand, to systematize, to organize, to analyze, to look for relations and meanings, *to construct a system of values*.
>
> *(Maslow 1970/1954, 50)*

From this we may infer that strategies for teaching values as instantiated historical subjects within a system of critical thinking may help contribute to transformative self-awareness.

"The History of World Landscapes": a case study

"The History of World Landscapes" is a survey course in landscape history taught over a span of 15 years (2002–2017) in two different universities.[1] Cumulative enrolment for this course has exceeded one thousand students—from freshmen to doctoral students—from a variety of fields of study.[2] Students are exposed to four broad themes including: the history of ideas; changes over time in human (perceptive, imaginative, and expressive) relationships with natural processes; historic impacts of environmental technics, industries, and infrastructures on regional landscapes; and the history of geo-political and socio-economic values. Figure 15.3 suggests how global case studies and monuments are 'mapped' onto a matrix where such themes cross standard historical periodicity.

The vast majority of sites examined are from the so-called Western world (Mediterranean region to northern Europe and North America), although featured citations and comparisons are offered from Islamic culture (northern Africa, Southwest and South Asia), as well as pre-Columbian indigenous settlements and ritual complexes in the Americas. Historical periods range from Neolithic settlements through classical, mediaeval, and Renaissance periods, up through industrial and mid-20th-century modernism. Similar to many such survey courses, it is a hugely ambitious romp across space and time.

In 2014, a decision was taken to translate the traditional course content to a new format, eschewing the twice-weekly 80-minute lectures in favor of an e-learning platform. In recent iterations, the course has been offered fully online and also as a blended/hybrid (aka "flipped") format with all content (pre-recorded audio lectures, slides, readings, YouTube links, assessments, and so on) provided online. The supposition is that students will have prepared required content before attending face-to-face discussion sections (aka "active learning sessions"). Although this supposition frequently appears to be unfounded, even unprepared students are usually able to participate and learn from instructors and peers during guided classroom activities.

"The History of World Landscapes" is described by many high-achieving undergraduates and graduate students as challenging and intellectually rewarding. However, it has also maintained a reputation for being fast-paced, overwhelmingly detailed, and occasionally quite boring.[3] One challenge therefore is how to make the online course more compelling, digestible, and relevant to students. To do that, instructors adopted the "Community of Inquiry" framework for shaping online discussion forums, specifically to allow students to explore their own social values in the context of landscape history.

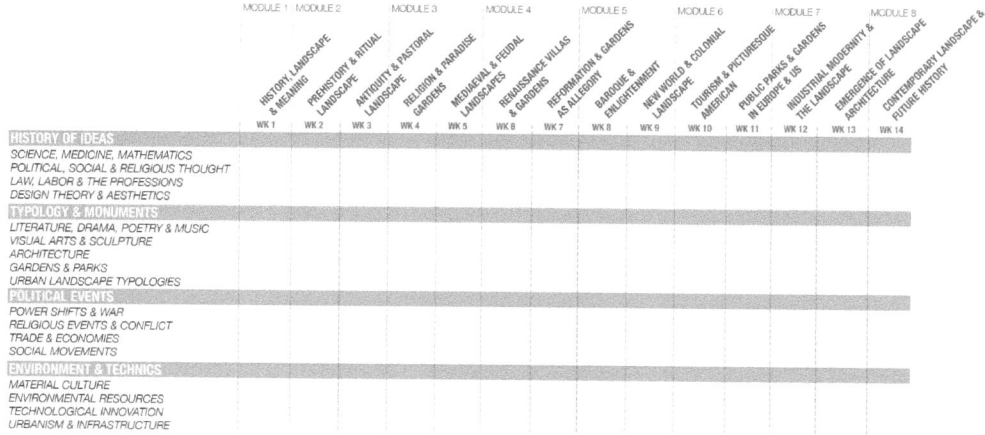

Figure 15.3 A timeline for 'mapping' interpretive themes and cases across historical periodicity. M.E. Deming (2015), History of World Landscapes (LA/ARCH 314), University of Illinois at Urbana-Champaign (USA)

Creating an online community of inquiry

Online courses typically employ a variety of techniques to present information and ideas to students who are, in turn, responsible for consuming, understanding and extending those ideas to new contexts. While distance education offers advantages (improved access and flexible pacing) for student learning, a perceived lack of social encounter (disconnection) and intensity may lead to student dissatisfaction and lower retention rates. This has challenged institutions to create online student learning experiences having "the capacity to sustain a strong sense of community that supports students both socially and cognitively" (deNoyelles, Zydney, and Chen 2014, 153).

With this in mind, instructional designers have proposed an array of strategies in support of what is termed the Community of Inquiry (CoI) framework. Community of Inquiry was originally intended to enhance asynchronous text-based discussions through engaged inquiry, similar to what might happen in a face-to-face classroom. Garrison, Anderson, and Archer propose three conceptual levels to the CoI framework: *social presence, cognitive presence*, and *teaching presence* (2000).

In brief, *social presence* indicates the ways that learners "project themselves socially and emotionally" in mediated online discussions. Online social presence may be indicated in several ways, including openness, humor, and encouragement or empathy with others. Social presence can lead to intellectual engagement and *cognitive presence*, understood as "the extent to which learners are able to construct and confirm meaning through sustained reflection and discourse." Cognitive presence announces itself in phases of learning: first a triggering prompt, then group exploration, and integration. The final phase is resolution, in which "new ideas are applied to other contexts" (deNoyelles, Zydney, and Chen 2014, 154).

The third element of the CoI is *teaching presence*, noting the critical role of the instructor in "design, facilitation, and direction of cognitive and social processes for the purpose of realizing personally meaningful and educationally worthwhile learning outcomes" (Anderson et al. 2001, 5; deNoyelles, Zydney, and Chen 2014, 154). Thus the instructor is also an active member of the community of inquiry. S/he is responsible for comprehensive design of the course including organization and content; constructing opportunities for learning; structuring prompts and assignments; and interceding and facilitating (when needed) to move discussion along.

"The History of World Landscapes" aims to foster a sense of social and cognitive community presence in online discussions, while several activities are designed to allow (or even provoke) students to co-construct knowledge along with the instructors and assistants. A variety of active learning techniques are used including peer-to-peer teaching, role-playing, debates, appreciation of historical construction methods, analysis of fictional accounts, and visual essays showing how local landscapes may encode personal values for the students. However, in the fully online version of the class, the discussion forum is the main technique for engaging students. Described below are some of the key concepts and techniques used to foster the Community of Inquiry around questions of values.

Teaching presence: linking learning objectives to historical empathy

Many learning objectives outlined for the "World Landscapes" course are normative: "To diagram the broad context of historical periods and their intellectual paradigms, religions, social forces, and geo-environmental relationships," and "to identify major stylistic movements in landscape practices (architecture/landscape architecture) and relate each to its cultural, environmental, technological, economic, and political contexts." Other objectives are crafted to help students develop historical (context and period), formal (style, typology and program), and iconological (poetic and ideological content) literacy; develop fundamental interpretive skills; and expose critical assumptions to view. For the purposes of present discussion, however, the key learning outcome is "to be able to apply methods of historical interpretation to meaning(s) and value(s) embedded in designed landscapes" (Deming 2017, UIUC).

One element that holds the course and its students together is the insistence on historical empathy. "The past," as David Lowenthal (1985) has famously written, "is a foreign country." Course materials challenge students "to engage historical 'different-ness' with curiosity and social empathy, and to be alert to historical patterns and influences in designed landscapes." Whether visible or invisible, tangible or intangible, period values motivating the design of canonical monuments are explicitly queried in the context of landscape and society. Students are then asked to explore, discuss, critique, and extend or relate those same values and motives to contemporary society, theory, and/or professional practice(s).

The course promises students that "in examining conceptual relationships between nature and society in past cultures," they will learn how designed landscapes perform meaning just as other forms of cultural expression do. More powerfully, students begin to understand that "cultural landscapes are generated by a variety of human values and motives—power struggles, values, belief systems, love and devotion, new forms of knowledge, even fear and prejudice—set within specific historical contexts" (Deming, UIUC 2017).

At the end of every online lecture, instructors develop *guiding questions* to help students summarize and synthesize lecture content. While most are open-ended and require students to interpret or explain certain historical principles (e.g., stewardship; sacredness; villa culture; romance; pleasure gardens, etc.), several are adapted or expanded as prompts for the online discussion forum, to elicit more imaginative student responses. The instructors' role in the forum thus stimulates student reflection and allows them to share their own examples of historic monuments or perceived impacts in and upon the contemporary world. By sharing their own observations and memories, historic processes and patterns tend to become more 'real' for students. And as they exchange ideas students also engage in peer-to-peer sharing, encouragement, teaching and learning, in other words *social presence*.

Social presence: telling landscape stories

The first module of the course introduces concepts and methods for understanding relationships between landscape history, form, construction and reception of intended (social) meaning. The forum prompts begin with a set of accessible questions that illustrate how social presence may be elicited in online discussions:

> Have you ever heard a story about a special place or seen a landscape in family photographs, which was a key part of your family's history? How about a historic landscape that is part of the identity of your home town or neighborhood? How are you connected to that landscape? And how does that landscape make you feel? Please share your landscape memories with the group.
>
> *(Illinois Compass website for LA/AR 314 History of World Landscapes)*

In this large (100+ students) and socially diverse (international population represents five continents; five races; ten+ native languages spoken) online class, the range of student responses to this question provides a useful basis for later discussions (*cognitive presence*) as well as a few immediate, grounded points of contact for instructor feedback (*teaching presence*). Some students are recent migrants to the United States, or international students who deeply miss their home country and their extended families. By contrast, other students may have a lineage of occupation in the same farm or homestead for three generations. The first forum alerts students that their own values and experience of the world may not necessarily be widely shared by others, let alone dominant in the class. However, their responses provide the basis for immediate connection (*social presence*), evident in expressions of sympathy, interest, and other shared points of reference between students who may never have met one another.

Telling landscape stories helps both design and non-design students notice and describe profound changes to their local landscapes, evident in just one or two generations, and to develop empathy for others. For example, one student recalls her own family's migration from Mexico to Chicago and points out that the agrarian landscape links these two places for her:

> I grew up in a small town in Guanajuato, Mexico . . . My family lived with my grandparents who owned a farm; they planted corn and had horses, chickens, pigs, etc. . . . When I was four years old my family decided to move to the United States and we settled in a town in northwestern Illinois. Although we now live in a quiet neighborhood, the town is surrounded by farms. Every time I drive out of town and see the cornfields and farms, I'm reminded of my family in Mexico. Even though we may be thousands of miles apart, I can feel a connection to them by seeing the stretch of cornfields outside my window.

In another post, an Irish-American student responds to a different classmate's story; by sharing his own experience he expresses a degree of empathy:

> Thanks for sharing that Adalberto! Also being from Chicago, I kind of know the lack of open space that comes with the city. Since I grew up toward the edge of the city, most of the ends of my neighborhood are surrounded by cemeteries, though houses are usually packed side by side with just a little bit of space in between. . . . But it's neat your parents came from Mexico with such landscapes, different than yours. My parents and relatives have been settled in Chicago for quite a while but it must have been a big change for your [parents] to come to the hustle of the city!

In a third story about place attachment and memory, a student teases out some important relationships between place, family, and self:

> In the lecture it was mentioned that sometimes history could be seen in entropy and decay; old, worn down buildings, broken off railroads, and forgotten objects can all connect us in time and space to one another in some way. However, I find that history is more embedded "deep inside the way we live and the things we dream of now" (also from lecture). This is why I am drawn to my grandfather's old house by the river. While the river, the rolling tracks, and the cornfields all bring back childhood memories, it's the way in which we lived and loved there . . . that made that place home and a huge part of who I am today.

As this student "integrates" terms and concepts from the lecture to her personal experiences and family's values ("the way we lived and loved there"), we see potential for synthetic thinking taking shape in just the first forum. "Resolution" is signaled by students' ability to "construct and confirm meaning through sustained reflection and discourse" – the highest measure of *cognitive presence* in the online forum (deNoyelles, Zydney, and Chen 2014, 153–154).

Cognitive presence: integration and resolution of historical concepts

As discussion topics track along with major themes from lesson modules, they may change direction in terms of values and the scale of their historical impact on landscapes. For instance, students might be asked to identify and discuss what are, for them, the essential properties of sacred landscapes; to compare the environmental sustainability of Imperial Rome with, say, Baroque Rome or 19th-century London; or to speculate on the interdependence of military and civil engineering in garden designs. In the final week, students are asked to imagine a future in which the landscapes made by their own generation become the subject of future historical interpretation:

> What will future historians write about the landscapes we are making today? Do you think the way they see the world we have made will depend on their values, or ours? What will be the most important factors in future historians' judgment of us (think of your own judgment of past societies and the way they treated their landscape)?

This prompt always engenders an interesting debate, revealing much about individual student capacity for empathy (and also fear). Some students go off on peculiarly apocalyptic threads with nihilistic, if poignant, visions of dystopian futures. While some defend the mistakes of past and current societies, others argue for their very right to make mistakes:

> Student 1. I think you are a little bit unfounded. When we evaluate something we cannot only look at its bad side. It is true we have destroyed the environment but because of its destruction we realize we now have to protect it. . . . [So] we will not hate the decline of Rome, the slavery of the Great Wall [of China], and the industrial development of London: they might seem to be unforgivable at that time but they had a huge impact on human development and gave us the record of human history.
>
> *(identity withheld, 2017)*

> Student 2. A few commenters possess pessimistic perspectives, some even referencing aliens and utter destruction of Earth's animals and plants after our generation. In consideration of the world's rapid consumption and detrimental practices, I understand these attitudes. This trend seems blasphemous! However, employing sustainable architectural and engineering elements is incredibly complicated with political, economic and social halts. "Being green" is not easy nor simple. But, hope is better than none. Future generations will appreciate any efforts to revert this overwhelming trend. Utilizing scare tactics or overwhelming citizens about our ultimate "doom" only further escalates the issue. Instead of lashing out at our mistakes, we need to analyze what to do next. So, please stop the negativity.
>
> *(identity withheld, 2017)*

> Student 3. Although conflict and tension have been key points for human existence for centuries, the current era is exceptionally full of political issues, conflicts and disagreements between different groups of people. I feel that the modern era reflects a certain individuality for all people and a break from the norms and constraints of past societies [and] that has allowed landscape design to take a very different turn in recent years. Unlike the uniformity of the Greeks, Romans, or virtually any other past great society, modern day society strives to stray from the norm in many different aspects of life . . . The values that I would

like future historians to apply to their analysis of our globalized culture include blatant individuality, a strive for the newer and more interesting design, and the lack of fear to try something that hasn't been done before.

(identity withheld, 2017)

Student 4. The value I'd like [historians] to apply to this culture is about "normal people." . . . Compared to the past, it is easy to see that many [contemporary] landscapes are designed for everyone equally, instead of for some wealthy families or famous persons. We have more free parks that are open to everyone and all people can have their own experiences there. For example, the Vietnam Veterans' Memorial was designed to commemorate every sacrificed soldier, instead of only the great leader. It focuses on individuals, normal people, and Lin aims to let everyone respond to the memorial, to that history. [In this] era we do not need to force ourselves to be "something" to live. We are brave enough to celebrate what is normal; and normal does not mean nothing, rather it means to be ourselves and accept who we truly are.

(identity withheld, 2017)

The online forum discussions are typically rich, sustained and lengthy. By discussing the layers of values embedded within cultural institutions and landscapes and reproduced in permanent structures—whether intellectual structures (such as laws) or physical structures (such as cities and civil war monuments)—students begin to comprehend what cultural geographer Peirce Lewis means when he writes, "Our human landscape is our unwitting autobiography, reflecting our tastes, our values, our aspirations, and even our fears, in tangible, visible form" (1979, 12). Further, they may discover their own position—either complicity or resistance—in the maintenance of landscape values and structures.

Conclusion

Teaching design history in its cultural context is a pliable, shifting, vital topic. The ability to interpret historical evidence helps students understand their own futures in the social and environmental trajectories that the past sets in motion. Analysis and debate of historical case studies in period context offer desirable intellectual outcomes. Alternatively, applying past social theories to present landscape dilemmas or, in reverse, applying current values to past landscapes may also provide rich and useful exercises for teaching and learning about one's own landscape values (Benes 1993).

Challenging students to take a conceptual distance on their own assumptions and beliefs through close, informed examination of historical case studies—canonical or not—creates an academic safe space in which they may explore, discuss, and even test their own values. For many students, history presents an abstract context that poses no apparent threat to identity. Debates over historical case studies permit students to externalize and examine personal values, to discuss theories of "goodness" or "badness," and to critique the ethical consequences of social values. And when landscape history is built upon an exploration of values it becomes possible for a variety of non-design-discipline majors to add intellectual and social diversity to the discussion.

Admittedly, these learning objectives are highly ambitious. It is true that many undergraduate students are challenged simply to understand how the basic experiences of their predecessors may differ from their own. Achieving considered historical empathy for the values and beliefs of others is considerably more difficult. Raymond Williams, again, puts it quite clearly:

We tend to underestimate the extent to which the cultural tradition is not only a selection but also an interpretation. We see most past work through our own experience, without even making the effort to see it in something like its original terms. What analysis can do is not so much to reverse this, returning a work to its period, as to make the interpretation conscious, by showing historical alternatives; to relate the interpretation to the particular contemporary values on which it rests; and, by exploring the real patterns of the work, confront us with the real nature of the choices we are making.

(Williams 1961, n.p.)

By focusing on the cultural and intellectual values latent in past design practices, history courses can be crafted to invite students' self-awareness—including reflection on personal and socially constructed values, even bias. This resolution of course begs a larger overarching question: how exactly does transformative self-awareness announce itself, and how should we attempt to measure it? While qualitative assessment of students' social and cognitive growth is beyond the scope of this chapter, it is certainly relevant to the long-term goal—to help render students' own hidden, or perhaps simply unanalyzed, values visible.

As I have written elsewhere, if the values of the profession

remain opaque . . . then at best [landscape architects] can only serve as instruments of an invisible agenda. On the other hand, once conscious of the sociocultural context of our practices . . . [students] can then begin to exercise critical agency as decision makers and landscape leaders.

(Deming 2015, 233)

In this sense, the values of professional design students may presage the world yet to come. The learning environments we design therefore ought to awaken students to the myriad ways in which social values become more than mere abstractions; they are instead actively encoded, operationalized, and received the very instant they are overlaid upon living landscapes.

Notes

1 The course was initiated at the State University of New York's College of Environmental Science and Forestry (Syracuse, NY), and then developed further for the University of Illinois (Urbana-Champaign). The Illinois course "presents a global survey of landscape philosophy and design practices as integral parts of human history."
2 At many American universities, courses in the history of landscape architecture can be framed to meet established General Education standards for all majors. At the University of Illinois, for instance, this survey meets three 'Gen Ed' standards: Western culture and civilization; philosophical and historical perspectives; and advanced writing composition, and thus attracts a broad enrollment across many majors.
3 In general, since course delivery methods were refashioned, test scores and other assessments have consistently shown normal achievement curves; the last three years show steady incremental improvement in student performance.

References

Anderson, Terry, Liam Rourke, D. Randy Garrison and Walter Archer. 2001. "Assessing Teaching Presence in a Computer Conference Environment." Journal of Asynchronous Learning Networks, vol. 5:2, 1–17.

Barton, Craig E., ed. 2001. *Sites of Memory: Perspectives on Architecture and Race* (New York: Princeton Architectural Press).

Benes, Miroslava. 1993. "Teaching History in the School of Design," *GSD News* (Summer), 25–26.

Deming, M. Elen. 2015. "Value Added: An Introduction," and "Finding Center: Design Agency and the Politics of Landscape." In *Values in Landscape Architecture and Environmental Design: Finding Center in Theory and Practice*, ed. M.E. Deming, 1–29; 220–236. Series: Reading the American Landscape. Baton Rouge, LA: Louisiana State University Press.

_____. 2017. Course Syllabus: *History of World Landscapes*. Department of Landscape Architecture, University of Illinois at Urbana-Champaign.

deNoyelles, Aimee, Janet Zydney and Baiyun Chen. 2014. "Strategies for Creating a Community of Inquiry through Online Asynchronous Discussions." *MERLOT: Journal of Online Teaching and Learning*, vol. 10:1 (March), 153–165.

Dix, Hywel. 2005. "Mark Twain: Freedom, Imperialism and Selective Tradition," *Public Resistance*, vol. 2:1, n.p., published online: http://eprints.bournemouth.ac.uk/19277, accessed June 18, 2018.

Ellis, Clifton and Rebecca Ginsburg, eds. 2010. *Cabin, Quarter, Plantation: Architecture and Landscapes of North American Slavery*. New Haven, CT: Yale University Press.

Garrison, D. Randy, Terry Anderson and Walter Archer. 2000. "Critical Inquiry in a Text-Based Environment: Computer Conferencing in Higher Education." *The Internet and Higher Education*, vol. 2:2–3, 87–105.

Harris, Dianne. 1997. "What History Should We Teach and Why? An Historian's Response." *Landscape Journal*, vol. 16:2, 191–196.

Hawkes, Jacquetta. 1951. *A Land*. New York: Random House.

Hohmann, Heidi. 2006. "Theodora Kimball Hubbard and the 'Intellectualization' of Landscape Architecture, 1911–1935." *Landscape Journal*, vol. 25:2, 169–186.

Jackson, John Brinckerhoff 1996. *A Sense of Place. A Sense of Time*. New Haven, CT: Yale University Press.

Lewis, Peirce. 1979. "Axioms for Reading the Landscape." In *The Interpretation of Ordinary Landscapes: Geographical Essays*, ed. D.W. Meinig, 11–32. New York: Oxford University Press.

Lowenthal, David. 1985. *The Past is a Foreign Country: Revisited*. New York: Cambridge University Press.

Maslow, Abraham. 1970 (1954). *Motivation and Personality*. New York: Harper & Row Publishers, Inc.

McKee, Bradford. 2018. "History in Edgewise." *Landscape Architecture Magazine* (June, 5), p. 12.

Meyer, Elizabeth K. 2000. "The Post-Earth Day Conundrum: Translating Environmental Values into Landscape Design." In *Environmentalism in Landscape Architecture*, ed. M. Conan. vol. 22, 187–244. Washington DC: Dumbarton Oaks & Harvard University.

_____. 2008. "Sustaining Beauty: The Performance of Appearance." *Journal of Landscape Architecture*, vol. 3:1, 6–23.

_____. 2015. "Beyond Sustaining Beauty: Musings on a Manifesto." In *Values in Landscape Architecture and Environmental Design: Finding Center in Theory and Practice*, ed. M.E. Deming, 30–54. Series: Reading the American Landscape. Baton Rouge, LA: Louisiana State University Press.

Mozingo, Louise and Linda Jewell, eds. 2012. *Women in Landscape Architecture: Essays on History and Practice*. North Carolina: McFarland & Company.

Mukerji, Chandra. 1997. *Territorial Ambitions and the Gardens of Versailles*. Cambridge: Cambridge University Press.

_____ 2009. *Impossible Engineering: Technology and Territoriality on the Canal du Midi*. New York: Princeton University Press.

Oles, Thomas. 2015. "Landscape Architecture: Uneasy Discipline." In *Defining Landscape Democracy: Conference Reader*, ed. Shelley Egoz. Centre for Landscape Democracy, Norwegian University of Life Sciences, 189–190.

Riley, Robert B. 1995. "What History Should We Teach and Why?" *Landscape Journal*, vol 14:2, 220–225.

Rogers, Elizabeth Barlow. 2001. *Landscape Design: A Cultural and Architectural History*. New York: Abrams.

Ward-Thompson, Catherine and Peter Aspinall. 1996. "Making the Past Present in the Future: The Design Process as Applied History." *Landscape Journal*, vol. 15:1, 36–47.

Way, Thaisa. 2009. *Unbounded Practice: Women and Landscape Architecture in the Early Twentieth Century*. Charlottesville, VA: University of Virginia Press.

Williams, Raymond. 1961. Excerpts from Chapter Two of *The Long Revolution*. First edition, London: Chatto & Windus. Online at: http://theoria.art-zoo.com/the-analysis-of-culture-raymond-williams, accessed June 18, 2018.

_____. 1974. "On High and Popular Culture," *The New Republic*, November 22, 1974. Reprinted online: https://newrepublic.com/article/79269/high-and-popular-culture, accessed June 18, 2018.

_____. 1977. *Marxism and Literature*. Oxford: Oxford University Press.

16

The landscape of landscape history

Marc Treib

Over the past decades more than a few landscape educators, like some of their colleagues in other disciplines, have questioned the role and/or value of history in design education. Those with more traditional views argue that the study of history provides a reservoir of precedents while positioning a student within a continuing tradition, others counter by claiming that, given the limited number of years needed to complete a degree in landscape architecture today, student time is better spent on more scientific and professional coursework. This latter position, a stance more commonly taken by those who are themselves more scientifically oriented, risks lacking an understanding of what currently constitutes the history of landscape architecture and the benefits resulting from its study. The question of limited study time remains an issue, however, as do the greater questions of education, content, and acquired knowledge. As I have traveled, I have found that in schools throughout the world programs of landscape architecture study have become highly compressed, at times with the resulting academic degrees somewhat inflated. What was once a professional bachelor's degree or diploma has now become a master's degree, although the actual coursework in landscape architecture may be far less than that required to earn the former professional undergraduate degree, much less what once constituted a Master of Landscape Architecture. In the text to follow I hope to establish the validity for landscape history study today and what might constitute a valid purview. Before continuing, however, I need to confess that I don't believe there is any ideal one-size-fits-all course appropriate to every country and every school. Like the subject itself, the teaching of history begins within a specific context. Another caveat: The comments to follow are not unbiased. As a historian of landscape and architecture, I feel that the study of history is crucial for educating the landscape architect for the reasons outlined below.[1] I must also note that although I have traveled widely, my views still derive from a North American perspective.

Constructing history

The construction of history is hardly a simple matter. As the canon for the histories of the design professions has broadened during the last forty years we have learned that there is not—nor has there ever been—a single "true" history of landscape architecture; instead, landscape history, like all others, exists as multiple stories comprised of many strands, preferences, and approaches. History—and arguably almost all scholarship and "scientific" study—will never be "non-fiction" but rather a story hopefully grounded in a selection of what are purported to be facts. The publication of those stories in the approved periodicals has become the goal for members of university faculties throughout the world, especially in the United Kingdom. To achieve this goal, teaching—the true education of the landscape student—has become secondary in relation to "research," much of which is not relevant to actual practice.[2] This model, which positions the reviewed journal article above teaching, actively accepts the sciences as the approved model for the humanities. This is not only unfortunate, it is a disaster for those whose primary gift is not writing, for example those more visually grounded or who conduct "research" through design. A second, and equally unfortunate, reality is that laudable research and writing in the humanities, work that in English we term "scholarly" or "academic," in many other languages is termed "scientific." I would suggest that this term creates a bias in the treatment of work in and on many subjects. Most aspects of art history, for example, are not scientific (as in science), but humanistic, perhaps interpretative, and should be treated as such. Landscape history shares this property. The architectural historian David Gebhard once quipped that "history [architectural or other] is one person's precise explanation of the way things never were."[3] Like theory, history usually follows practice rather than leads it. Given all these provisos, if history is only a branch of fiction, why should it be needed, much less required, in any landscape architecture curriculum?

Today's courses in landscape history or landscape architecture history (there is a difference between the two mind sets) have maintained their relevance by broadening their purview, cultures, and content. At the very least, a substantial course in landscape architecture history today includes environmental history—and necessarily the political, social, and institutional histories that accompany it, a study of form and space, and of course investigations into the types and roles of water, vegetation, and other aspects of the natural world that constitute any landscape. This is a complex and challenging matter and it must be admitted that such a broad course remains an ideal; rarely in any course or curriculum is history presented in such a comprehensive way. Burdened by the constraints of limited time, academic accreditation, professional licensure, and restrained budgets, most American landscape programs rarely require more than a single course, or at best a year-long survey.[4] I have been told that in Europe the situation is similar. This limitation on time compresses the investigation of content and demands that teachers must necessarily focus on selected topics. In response, some courses take form as broad surveys focused on environmental issues, others construct courses from selected case studies investigated for their formal, social, or material aspects. Others still may combine approaches, perhaps weighted toward the art of landscape design or the employ or management of plants. In addition, and quite unfortunately, the education and specialized knowledge of those teaching the courses impose a second constraint on the content and effectiveness of the courses, as it is only a rare teacher who is experienced in all these areas. A third factor bearing on programs of landscape study is the university's emphasis today on STEM: science, technology, engineering, and mathematics (one wishes that acronym could be expanded to STEAM, adding *arts* into the mix). As a result, the ideal course in the history of landscape architecture has remained, to large degree, only an aspiration. The simple solution would be a series of courses taught by those who are expert in the various subject areas, but even that approach has been difficult to implement when and where the very value of history has been questioned, especially in times when departmental financial resources have become ever more limited.

Cultural landscapes, designed landscapes

In addition to the variety of issues cited above, we must also address the distinction between histories of the cultural landscape and those of the designed landscape (that is, landscape architecture proper). If we accept that all construction and land transformation (for example by settlement, agriculture, and recreation) results in a cultural landscape, its study would provide an ideal doorway through which to enter students into the history of landscape architecture (Figure 16.1). The cultural landscape is the milieu in which students have lived and experienced, and the greater realm in which the designed landscape resides. In *The Interpretation of Ordinary Landscapes*, D. W. Meinig provided a revealing essay comparing the differing stances of the cultural landscape historian John Brinkerhoff Jackson in the United States and his contemporary, W. H. Hoskins, in Britain.[5] Jackson wrote of the American landscape as a first-hand observer who began his study with an examination of what he saw, and only thereafter interpreted the landscapes encountered through a prolonged consideration leading to an understanding of the conditions that shaped them.[6] He wrote as an unbiased witness and withheld value judgments. From specific observations he deduced generic patterns of human effort and the landscapes that have resulted. Hoskins, in contrast, was more academic, with an agenda of social and environmental continuance, decrying the continued despoiling and degradation of the English landscape and denouncing the negative forces that caused that decline.[7] Unlike Jackson, Hoskins appears to want to still time and development, especially if "progress" results in drastic changes to the English countryside. Each writer has been a keen chronicler of the thrust of landscape processes over time, and from them has established his position. Other scholars, such as David Harvey or Denis Cosgrove, have taken a more political view, characteristic of the geographer, with little discussion of the resultant qualities or aesthetics of space and form.[8]

What is to be learned from perspectives such as these? A lot. For one, the value of cultural landscape study—if not only to make students more aware of the world around them and how it came to be—situates the historical efforts of landscape architects within the greater sphere of world shaping. This knowledge provides, among other benefits, insights into the dimensions and complexity of the territory and to what degree conscious design, management, and other forms of intervention have affected its current state. The student also comes to realize how much of our world has resulted from independent and uncoordinated efforts to create human habitat and productive land. These practices, rather than conscious professional landscape design and planning, have shaped the greater part of our built environment—a sobering realization for those landscape architecture students who seek to change the world through their designs.

The history of landscape architecture per se has traditionally focused on the stylistic and ideological progression of gardens, parks, motorways, and plazas, among other landscape types. Some books and courses have exclusively restricted their content to stylistic and formal study, but today such a limited focus is rare.[9] Almost any recent landscape history course worth offering, much less attending, is far more comprehensive in its telling of stories. In addition to presenting the manner in which the sites were designed and made, and the plants, inert materials, and hydraulics used to make them, a substantial course would incorporate discussions of the clients as the instigators and administrators of the project, the people who use the landscape, the technical processes needed to transform the land, the political and economic structures within which the landscape was created, and, if relevant, colonial, racial, and economic factors. It would need be cross-cultural, perhaps enfolding diachronic and

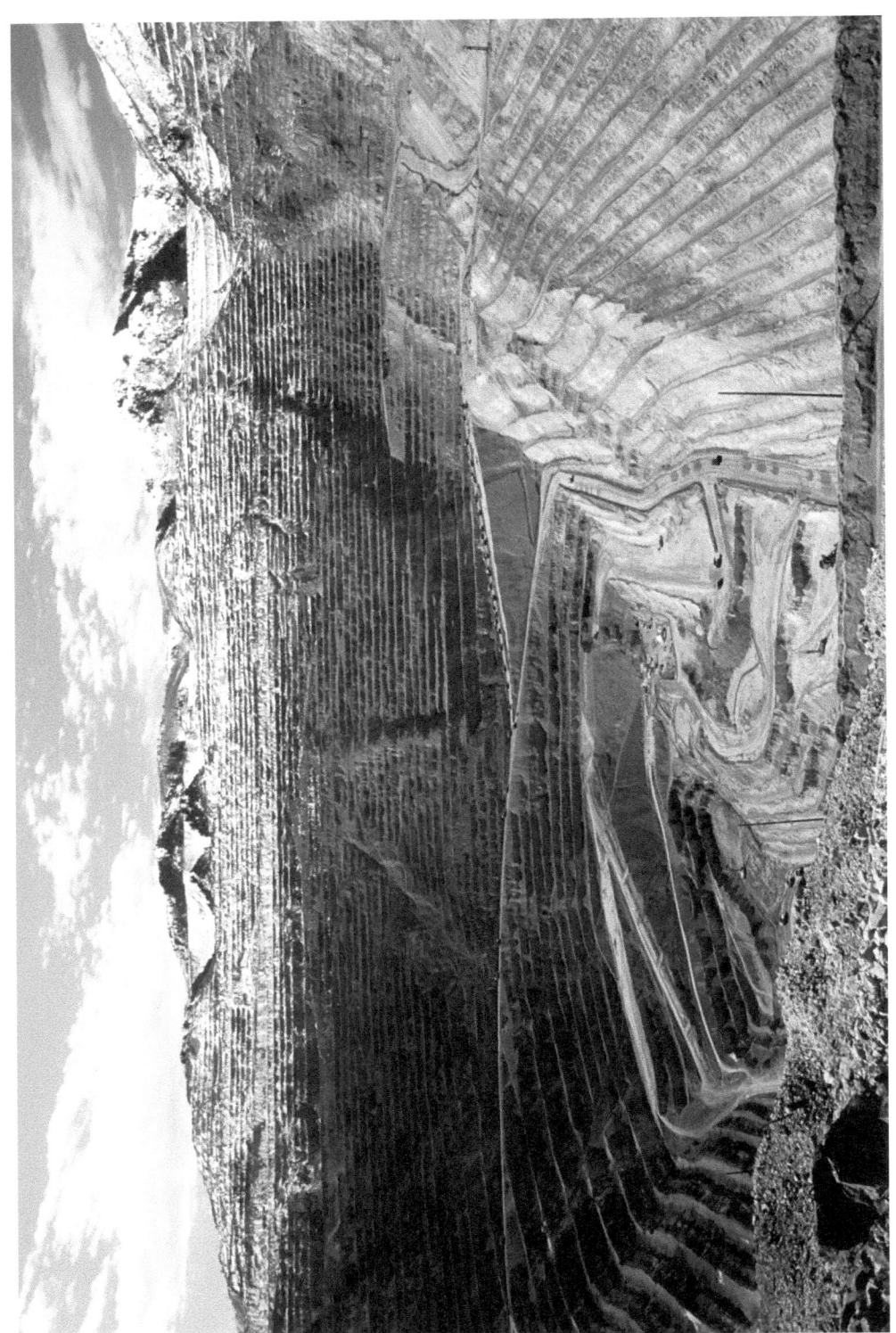

Figure 16.1 Bingham Copper Mine, outside Salt Lake City, Utah (photo: Marc Treib)

synchronic content. Additional issues might also need to be raised; perhaps other issues might be deemed irrelevant. Such a breadth of view demands much of those teaching history courses, and admittedly many courses have outlined higher goals than those ultimately realized. But beyond any other potential lesson, the landscape history course demonstrates that the making of places is dynamic and that differing social, political, and economic systems have at a minimum guided—if not directed and shaped—the form of the landscape, whether in the city, the estate, or the lowly backyard.

And what of the readings used to support such comprehensive courses? The question of the appropriate text raises several other questions. In the past such as overviews as Norman Newton's *Design on the Land* and Geoffrey and Susan Jellicoe's *The Landscape of Man* have frequently been used as course texts, although neither of them matches the depth of Marie Luise Gutheim's pioneering *A History of Garden Art*, originally published in German in 1928, although narrower in its breadth. More recently Tom Turner's *Garden History* and Elizabeth Barlow Rogers's *Landscape Design* have attempted to be more inclusive in discussing their subjects; however, physics teaches us that spreading a given volume of liquid more widely necessarily reduces its depth.[10] This axiom applies to textbooks as well. Many of those teaching history are dissatisfied with all the existing texts, either for the scope of their narrative, selection of sites, or omissions of certain cultures or times. Addressing these shortcomings they assemble "readers" of writings gleaned from journals as well as books and the internet, today commonly provided to the students as electronic pdf files. In decades past certain texts were appreciated primarily for their images, although this has become less of an issue as innumerable photographs and drawings have become available on the internet, for example the photographs of UNESCO World Heritage sites available on the institution's website.[11]

The benefits of history

History courses offer the student at least three benefits. The first is informational. History tells us about the people and events that initially shaped the site, and that under the pressure of ever-present forces the resulting landscape has continued to evolve over time; that is to say, history explains the path of the landscape's nature, and traces its various states through time, whether the content pertains to the landscapes in which we dwell today or those of the past. This knowledge contributes to a deeper understanding of the environment and, of course, to a student's general education as an cultivated human being. The second benefit is typological, that is to say, providing students with a lexicon of prototype landscapes—not to be copied, of course, but sources from which they may abstract general principles and conclusions. Although differing in specific vocabulary, the past has produced landscapes that tell us about the use and abuse of dimensions and scale, spatial ideas, the selection and application of materials, the success or failure of planting strategies, and many other lessons that contribute to an understanding in the making of places. Archaic works such as the effigy Serpent Mound in Ohio (Figure 16.2) or the Fyrkat Viking Camp in Denmark (Figure 16.3) illustrate, respectively, the reformation of topography for religious and defensive purposes. The water chain and table at the Villa Lante in Italy (Figure 16.4) tells of the potential uses of water to be gained through the force of gravity alone. The use of minimal elements in the garden of Ryoan-ji in Kyoto (Figure 16.5) demonstrates maximal effects achieved using only minimal means. In contrast, Roberto Burle Marx's Fazenda Vargem Grande in Areais, Brazil, exemplifies the informed use of vegetation used artistically (Figure 16.6). The benefits of cross-cultural relationships and exchanges are significant. Although possibly differing in plan, elements, and space than those encountered today, the lessons provided

Figure 16.2 Serpent Mound, near Locust Grove, Ohio, *c.*1000 (photo: Marc Treib)

Figure 16.3 Fyrkat Viking Fortress, Denmark, *c.*980 (photo: Marc Treib)

Figure 16.4   Jacopo Barozzi da Vignola, Villa Lante, Bagnaia, Italy, *c.*1550 (photo: Marc Treib)

Figure 16.5   Ryoan-ji, Kyoto, Japan, *c.*1500 (photo: Marc Treib)

Figure 16.6   Roberto Burle Marx, Fazenda Vargem Grande, Areais, Brazil, 1979–1990
(photo: Marc Treib)

by historical landscapes such as these may nonetheless be useful for those who will engage in the making of landscapes in the future. We might also look at the processes by which the historic landscape was planned and realized, another part of the study with relevance to contemporary practice.

No site exists without a history, whether it be its natural history or that which has been constructed at an earlier time. However, it should be stressed that no landscape history course should attempt to be directly instrumental, that is to say, that there need not be a direct connection between what is said in history class in the morning and the project in the design studio that afternoon. The lesson content should not be restricted to the design task at hand, or only rarely so. The instrumental presentation normally resides with those guiding the landscape design studios, although a historian may also contribute to these classes. The informed design instructor contributes an awareness of history in the studio, through lectures, readings, site visits, and the assignment at hand. The third lesson shows that a knowledge of landscape history, as a critical part of environmental history, is necessary for becoming an educated human being, in addition to becoming educated within your own discipline. "Think global, act local" applies equally to landscape design and political activism. And as George Santayana once claimed: "Those who cannot remember the past are condemned to repeat it."[12] We learn from our mistakes as well as successes. Yet another reason for the inclusion of history in the landscape architecture curriculum.

Manner of instruction

Over time, landscape history pedagogy has evolved from teaching focused on the progression of planning modes and styles informed by programs, clients, and power structures into a more widely conceived approach that parallels the broadening of the purview discussed above. Another question: Who teaches these courses and where do they receive their education? Until relatively recently most, perhaps almost all, landscape history courses were taught by landscape architects, i.e., those with a professional diploma or master's degree, teaching courses in other areas of the curriculum, design studios among them; more rarely by art historians specialized in the art of a certain period. While much coursework in landscape architecture history once was conceived as a progression of great works, one must credit those who have infused their courses with a growing concern for ecology, social issues, and infrastructure. How to achieve instruction in all these areas when allotted only a semester or two presents a continuing challenge to those teaching these courses.

Over time, new programs with advanced academic degrees in landscape history, paired with the current concern for issues such as sustainability, have stimulated coursework with greater breadth and depth than many of those taught in the past. Of the American programs the University of Illinois's PhD Program in Landscape Architecture and Architecture is the sole curriculum with a teaching cadre specialized in landscape history. However, it has been possible to acquire a doctorate through other paths such as within art or architectural history or under the tutelage of a single or pair of historians within a department. But faculty whose teaching lies solely in landscape history remains a rarity. My prime reason for establishing the Landscape History Chapter of the Society of Architectural Historians was to create a network for those teaching landscape history somewhat in isolation.[13] As with any academic subject, the approaches will vary and will continue to evolve. However, one would hope that today's landscape history will at least attempt to cover a wide, if not complete, range of subject areas, whether starting or ending with ecology, a historic view of the use of living materials, the design and aesthetic issues of form and space. In what order these are taught, or the balance between them, will necessarily vary with those teaching the courses.

History versus theory

Until the late 1960s one never—or only very rarely—heard of "theory" in either architecture or landscape architecture. There were courses in "history" and there were courses in "design", "plant materials", "grading and drainage", and other academic subjects and practical skills required by practice and professional registration. In my experience, theory arrived in the late 1960s when universities began to shift their mission from education to "knowledge factories," more politely and academically known as "research universities." University directives began to dictate that professional programs must no longer be focused on training practitioners within a "profession" and instead focus on educating students within a "discipline". Among several other factors, what distinguishes a profession from a discipline is its "theory." Of course there is no way that any design project with varied criteria and uncertain outcomes could ever apply a theory as it is formulated and used in the hard sciences. A scientific theory suggests that the same operation, undertaken with the same materials under the same conditions, will produce the same result (I am admittedly simplifying to some extent). And certainly, various theories for the same processes or phenomena can exist contemporaneously. But in general, this exact prediction of an outcome is rarely true of any design project, even when the task is a single room, much less a building, a park, or a region.

The design methodologist Horst Rittel once termed design as a practice characterized by "wicked problems," that is to say, projects plagued by elusive, dynamic, and hard-to-define criteria leading to solutions unpredictable to a large degree.[14] Just what constitutes theory in the design fields is still open to question, at least in my book. To some theory is the borrowing of thinking from fields such as literary criticism and applying, or misapplying, it to the design fields. To others theory constitutes a selective view that removes only those aspects of history that can be abstracted and generalized. Definitions will necessarily vary. More to the point, why do we need a theory and how can one be formulated? The two poles are staked out by literary theorist Terry Eagleton, who once wrote that those who claim they are not interested in theory are operating under an old theory. Its counterpoint: the British poet Thom Gunn, who in a poem wrote that dogs have no theory but get there all the same.[15] Experience has shown that theory may or may not be consciously applied and may lead or result from the path of design. Whether latent or overt, it is always present.

At the close of the 1960s and through the 1970s the discipline of architecture sought its theory from many places except those within the subject of architecture itself. There followed the adoption of a progression of ideas and methods derived first from operational research and its attempt to rationalize design procedures; then structuralism, whether in linguistics or anthropology; these were followed by transformational linguistics, thereafter semiotics, and more recently the many threads of post-structuralism. But, most curiously, there were few investigations of sources within architecture itself, except perhaps for the relatively brief postmodern moment in the 1980s when one group of scholars and architects attempted to resurrect an interest in classicism and to establish its place in contemporary practice. After reading Thomas Kuhn's *The Structure of Scientific Revolutions* in the early stages of theory-seeking, architects adopted the term "paradigm shift" yet never gave any paradigm broad acceptance.[16] The definition of design theory was, and remains, itself dynamic and obscure. Or so it seems to me.

Landscape, in contrast to architecture, sought a theory at exactly that moment when a concern for ecology came to the fore. Here was the basis of a theory within the discipline of landscape architecture itself. One can read history as theory, to be sure. The landscape gardeners such as Humphry Repton and later Alexander Jackson Downing, for example, were "theorists" as well as practitioners. Their writings followed the practice, however. At the close of the 1960s Ian McHarg argued for a "design with nature," a forceful contention advancing a somewhat pessimistic vision of what results from building with neither coordination nor control.[17] Paired with the innovative instrument of overlays to sort and analyze site factors, in the 1970s McHarg's vision came to dominate landscape practice—as well as the academic discipline that produced its members. Landscape architecture as the making of formally rich and intriguing, perhaps even beautiful, spaces and places was relegated to a secondary or tertiary status. Analysis trounced creation. Design did not quite die, but was relegated to a position of life support, resuscitated only in the mid-1980s when an interest in form, space, and landscape architecture as a cultural activity revived—only to be suppressed once again in the new millennium with the renewed emphasis on process, aided and abetted by the rise of the computer.

Thus we see that theory in the design professions is difficult to define or even to discern; I for one have never heard a serviceable definition of landscape theory, or, for that matter, even the need for it—as opposed to method, which often substitutes for it. But somewhere within the formulation of a theory for the design professions must lie a value system and a set of standards for evaluation. To act, to design, we require some system by which we can judge the appropriateness and efficacy of the design, and some suggestion about how it may be improved during the design process. Some may argue that theory is of little use in daily practice where other parameters govern the considerations by which designs are judged (economics and time schedules among them). Others argue that theory follows practice rather than leads it. Theory began and may ultimately end up within academia; practice will follow its own path, pursuing its own goals, and creating landscape in relation to its own values.

It has been my experience that theory is primarily the province of the academic and only borrowed and used by practitioners when needed to substantiate a design presented to a public body or individual client—or, more ubiquitously, in publications. In addition, landscape architects rooted in academia often contradict or even dismiss their former "theories" when faced with the lure of an actual commission. In their teaching many may stress process and the allowance of natural systems to determine the form of the landscape on an ever-changing basis. In actuality, regimens of maintenance by park departments paired with the citizens' desire for orderly landscapes may lead the design in a very different direction. The care and maintenance of landscapes will ultimately shape any design, whether one theoretically based or not. After several years, in fact, the landscape architect's vision becomes the gardener's province and it is the gardener who ultimately shapes its perceived form. What, then, is the benefit of theory and a knowledge of history and precedent? Is it only as an instigation to thought? If so, theory may operate as does landscape history, not instrumentally, but more generally for thinking more expansively and being open to as yet untested ideas. I believe that ultimately this provocation to thinking independent of the particular task at hand is the benefit of theoretical judgment. But we could also rightly claim that this provocation of thought is also one of the prime reasons for the study of landscape history: not just learning about the particular site and how it came to be, or how it has endured or changed thereafter, but what can be generalized and used to inform a contemporary situation.

Then just where does landscape architecture history and its teaching fit in the current university curriculum? Certainly, a regard for process should be included in any review of historical precedents. One could suggest, as well, that in any substantial history course the investigation of plants and hydraulics, among many other factors, has always been a part of the story. Are we now more aware of culture, colonialism, politics, and environmental degradation than in the past? Probably so. Like the landscape itself, the construction of its history is itself dynamic. Despite such fluidity I believe that a knowledge of history informs and deepens our understanding of our current situation and perhaps even suggests what we might do to improve our conditions now and in the future. As in the title of Paul Gauguin's celebrated painting (1897), today in the Boston Museum of Fine Arts: *Where do we come from? Where are we [now]? Where are we going?* In teaching the history of landscape architecture we need to ask related questions such as: Who made this place, how and why? Why does it appear this way and how has it been experienced in the past as well as today? Why do we regard the landscape as significant and worthy of discussion? How can we learn from the past as a way to understand the present, and perhaps even the future? The ultimate goal is not to garner factual answers to these questions but to stimulate an intellectual interest that may result in a better answer to them. The past really isn't the past, it is a key to understanding the present and the future—which makes it a rather valuable element in the education of the landscape architect.

Notes

1 Though I confess to being a critic, I do not pretend to be a theorist.
2 Informal surveys of students show that the vast majority of them do not attend university to become researchers, but, instead, are studying to become landscape architects, architects, or designers in other fields. As a result, the university's mandate for reviewed publication—often stressing quantity over quality—and the excellence of professional education are traveling in opposite directions. Within a short period of time, possible a decade, I predict that landscape faculties will be staffed only by those with absolutely little or no connection with actual landscape practice.

3 Over dinner in Berkeley, California, some time in the late 1970s.
4 "Theory" is usually treated as separate coursework, or perhaps conflated in a course with a title like "Landscape Architecture since 1945."
5 D. W. Meinig, "Reading the Landscape: An Appreciation of W. G. Hoskins and J. B. Jackson," in D. W. Meinig, ed., *The Interpretation of Ordinary Landscapes*, Oxford: Oxford University Press, 1979, pp. 195–244.
6 Among J. B. Jackson's several books are *Discovering the Vernacular Landscape* (1984) and *A Sense of Place, a Sense of Time* (1994), both New Haven: Yale University Press. The most comprehensive anthology of his essays appears in Helen Lefkowitz Horowitz, ed., *Landscape in Site, Looking at America: John Brinckerhoff Jackson*, New Haven: Yale University Press, 1997.
7 W. G. Hoskins, *The Making of the English Landscape*, 1955 (reprint, London: Penguin Books, 1970).
8 David Harvey, *The Condition of Postmodernity: An Enquiry into the Origins of Cultural Change*, Hoboken: Wiley-Blackwell, 1991; Denis Cosgrove, *Social Formation and Symbolic Landscape*, Madison: University of Wisconsin Press, 1998.
9 Although a near Bible for American landscape architecture students in the early decades of the twentieth century, Henry V. Hubbard and Theodora Kimball's *An Introduction to the Study of Landscape Design* (New York: Macmillan, 1917) is today derided, although, stylistic bias aside, it is a quite valuable text. While using historical subjects, the focus of the book is design practice rather than history alone.
10 Norman Newton, *Design on the Land: The Development of Landscape Architecture*, Cambridge, MA: Harvard University Press, 1971; Geoffrey and Susan Jellicoe, *The Landscape of Man: Shaping the Environment from Prehistory to the Present Day*, London: Thames & Hudson, 1975; Marie Luise Gothein, *A History of Garden Art*, 1928 (reprint, New York: Hacker Art Books, 1966); Tom Turner, *Garden History: Philosophy and Design, 2000 bc–2000 ad*, London: Spon, 2005; Elizabeth Barlow Rogers, *Landscape Design: A Cultural and Architectural History*, New York: Harry N. Abrahms, 2001. Christophe Girot's highly illustrated and highly idiosyncratic and spotty *The Course in Landscape Architecture: A History of Our Designs on the Natural World, from Prehistory to the Present*, (London: Thames & Hudson, 2016) is the most recent approach to the subject. The book conflates cultural with designed landscapes, however, without establishing why.
11 http://whc.unesco.org/en/list.
12 George Santayana, *The Life of Reason, Volume 1: Reason in Common Sense*, New York: Charles Scribner's Sons, 1927.
13 The chapter was founded in 2003 and, at last count, included some 200 members internationally.
14 Design methodologist Horst Rittel was a member of the architecture faculty at the University of California, Berkeley, in the 1970s and 1980s. He repeated this term numerous times in classes, meetings, and writings.
15 Terry Eagleton, *Literary Theory: An Introduction*, Minneapolis: University of Minnesota Press, 1983; Tom Gunn, unknown source.
16 Thomas Kuhn's *The Structure of Scientific Revolutions*, Chicago: University of Chicago Press, 1962.
17 Ian McHarg, *Design with Nature*, Garden City, NY: Doubleday, 1969.

References

Barlow Rogers, Elizabeth (2001) *Landscape Design: A Cultural and Architectural History*, New York: Harry N. Abrahms.

Cosgrove, Denis (1998) *Social Formation and Symbolic Landscape*, Madison, WI: University of Wisconsin Press.

Eagleton, Terry (1983) *Literary Theory: An Introduction*, Minneapolis, MN: University of Minnesota Press.

Girot, Christophe (2016) *The Course in Landscape Architecture: A History of Our Designs on the Natural World, from Prehistory to the Present*, London: Thames & Hudson.

Gothein, Marie Luise (1966) *A History of Garden Art*, New York: Hacker Art Books.

Harvey, David (1991) *The Condition of Postmodernity: An Enquiry into the Origins of Cultural Change*, Hoboken, NJ: Wiley-Blackwell.

Horowitz, Helen Lefkowitz, ed. (1997) *Landscape in Site, Looking at America: John Brinckerhoff Jackson*, New Haven, CT: Yale University Press.

Hoskins, William George (1970) *The Making of the English Landscape*, London: Penguin Books.

Hubbard, Henry V. and Theodora Kimball (1917) *An Introduction to the Study of Landscape Design*, New York: Macmillan,

Jackson, John Brinckerhoff (1984) *Discovering the Vernacular Landscape*, New Haven, CT: Yale University Press.

Jackson, John Brinckerhoff (1994) *A Sense of Place, a Sense of Time*, New Haven, CT: Yale University Press.

Jellicoe, Geoffrey and Susan (1975) *The Landscape of Man: Shaping the Environment from Prehistory to the Present Day*, London: Thames & Hudson.

Kuhn, Thomas (1962) *The Structure of Scientific Revolutions*, Chicago, IL: University of Chicago Press.

McHarg, Ian (1969) *Design with Nature*, Garden City, NY: Doubleday.

Meinig, D. W. (1979) "Reading the Landscape: An Appreciation of W. G. Hoskins and J. B. Jackson," in D.W. Meinig, ed. *The Interpretation of Ordinary Landscapes*, Oxford: Oxford University Press, pp. 195–244.

Newton, Norman (1971) *Design on the Land: The Development of Landscape Architecture*, Cambridge, MA: Harvard University Press.

Santayana, George (1927) *The Life of Reason, Volume 1: Reason in Common Sense*, New York: Charles Scribner's Sons.

Turner, Tom Garden (2005) *Garden History: Philosophy and Design, 2000 BC–2000 AD*, London: Spon.

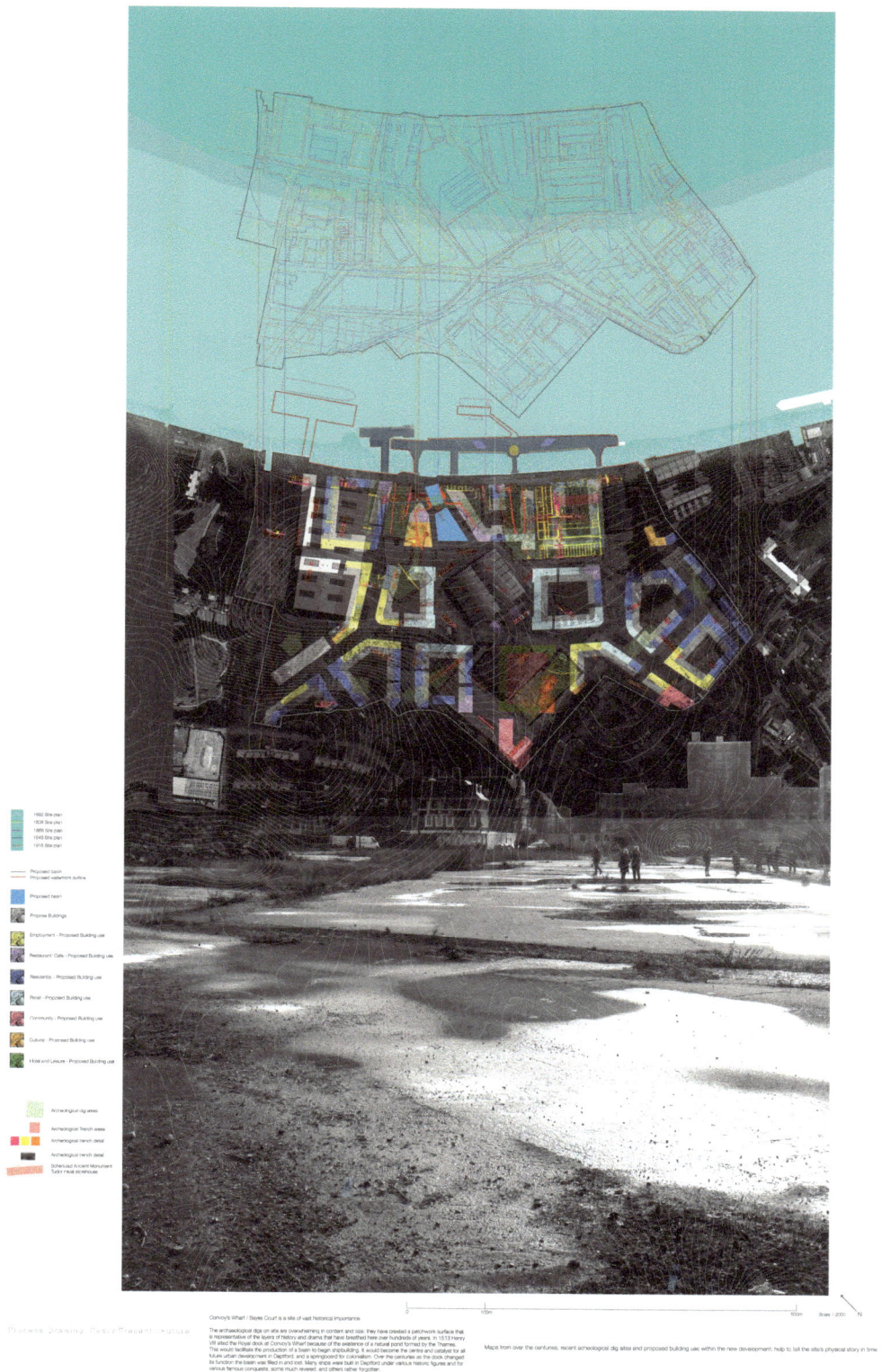

Figure 2.3   Drawing the Line/Cloud Chamber by Max Barnes, 2016

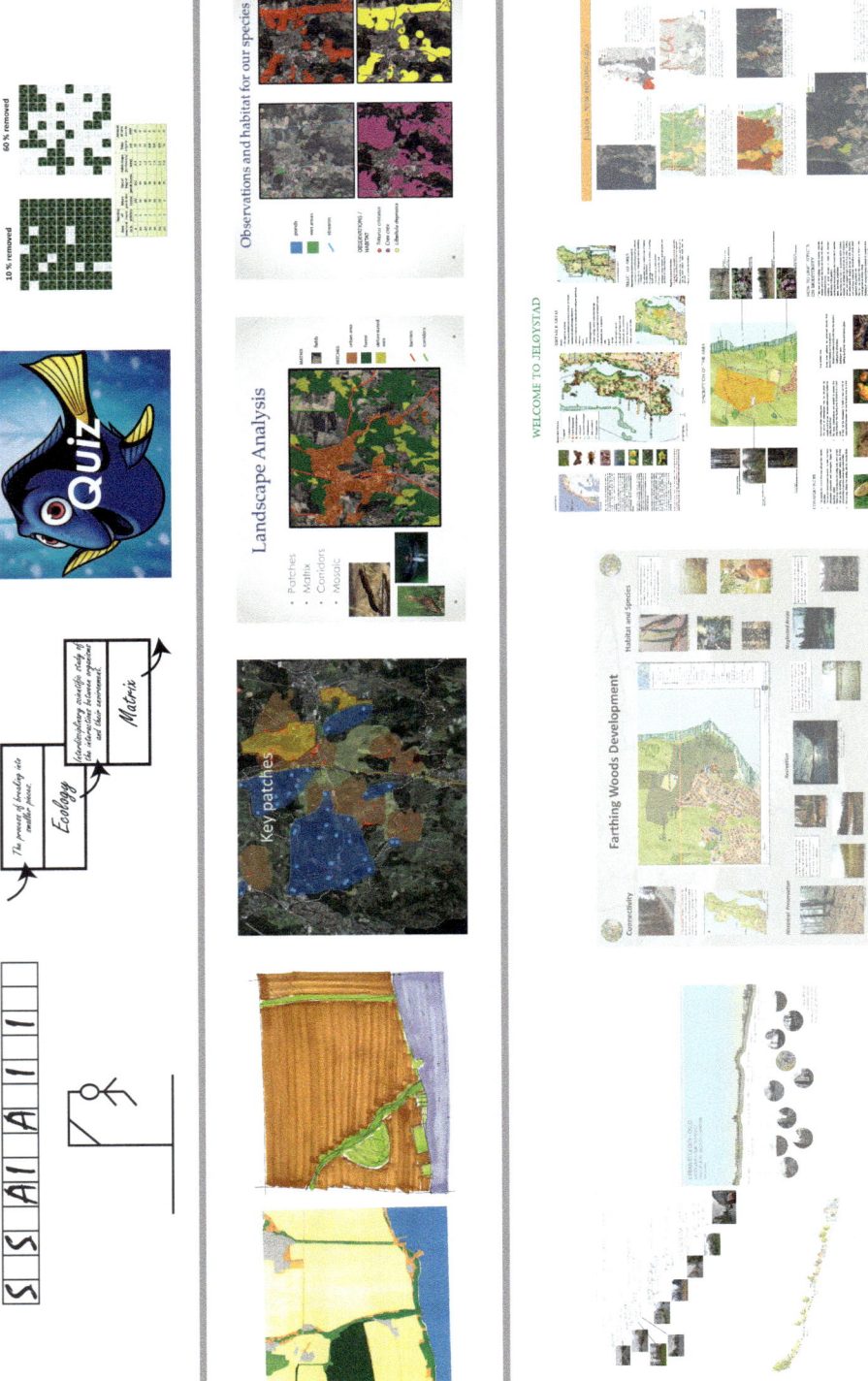

Figure 5.2   An illustration of how we see different assignments in the course representing a progression, from developing a conceptual common ground and learning tools (e.g., hangman game, definitions domino, quizzes), to doing desktop exercises on real landscapes (drawing on maps and aerial photos, adding and losing habitat assignment) to real-world exercises (plan for housing in real-world landscape, assessing urban habitat)

Colours of the season – the Vestby farmland

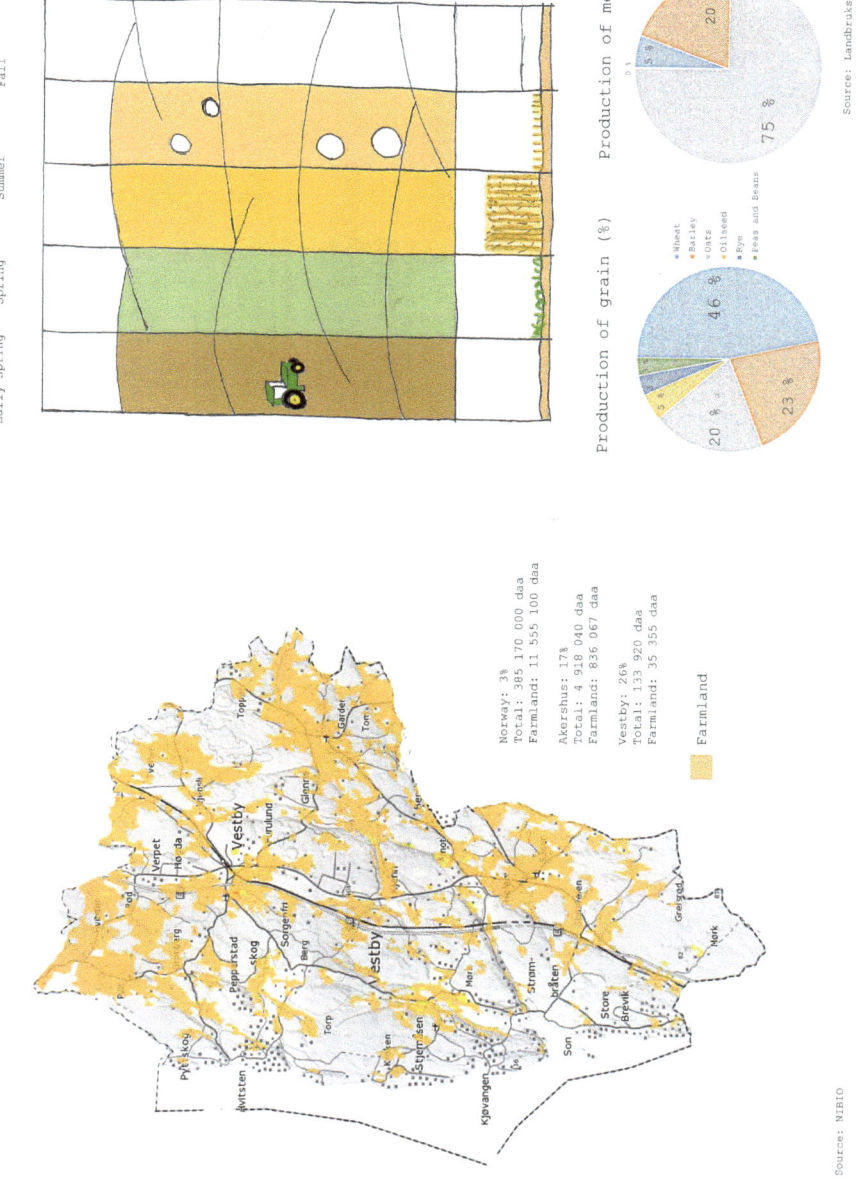

Source: NIBIO

Source: Landbruksdirektoratet

Figure 7.1 One page from the 'Edible Landscape' theme (Atlas Vestby, Akershus, 2015)

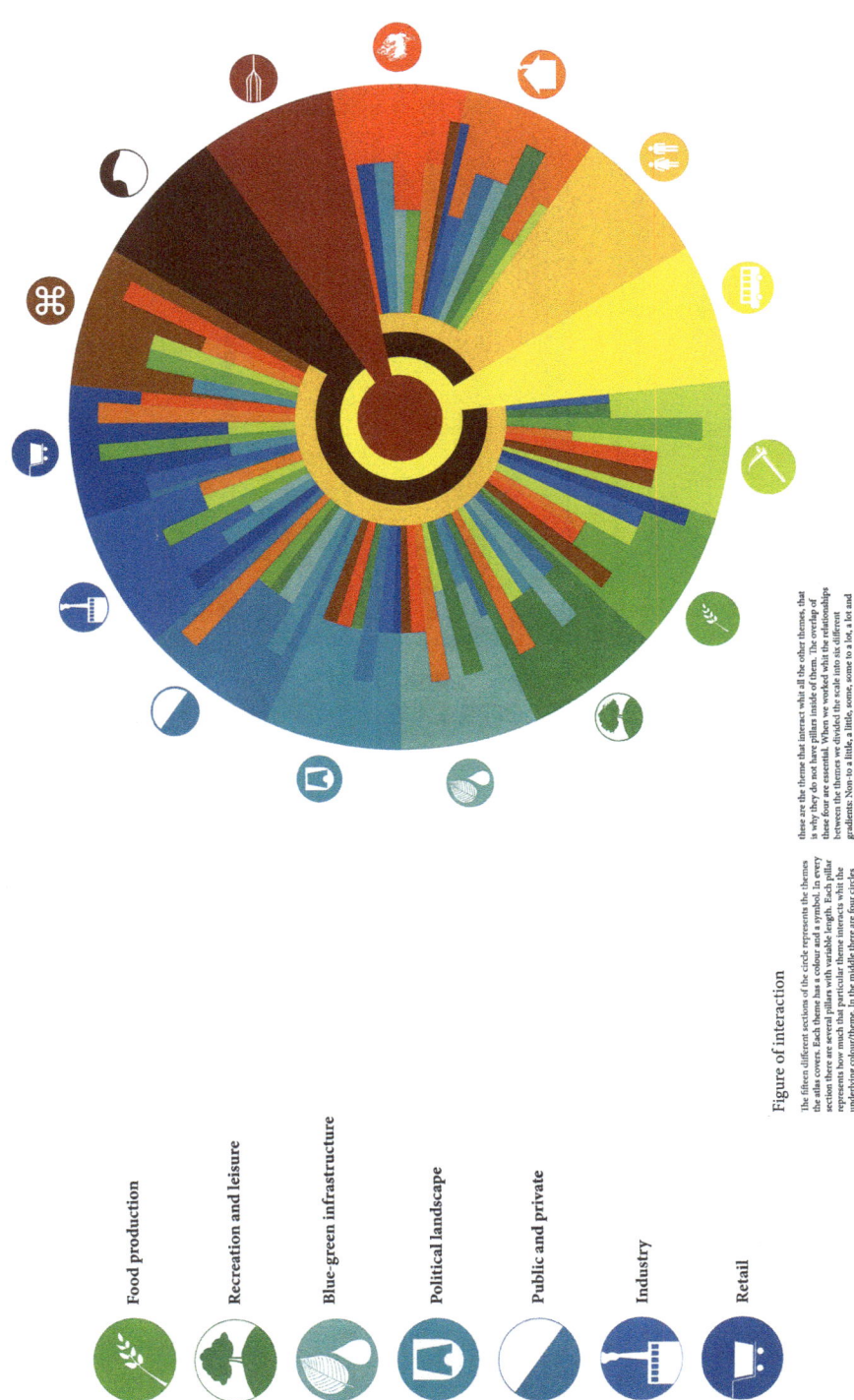

Food production

Recreation and leisure

Blue-green infrastructure

Political landscape

Public and private

Industry

Retail

Figure of interaction

The fifteen different sections of the circle represents the themes the atlas covers. Each theme has a colour and a symbol. In every section there are several pillars with variable length. Each pillar represents how much that particular theme interacts whit the underlying colour/theme. In the middle there are four circles

these are the theme that interact whit all the other themes, that is why they do not have pillars inside of them. The overlap of these four are essential. When we worked whit the relationships between the themes we divided the scale into six different gradients: Non- to a little, a little, some, some to a lot, a lot and essential for each other.

Figure 7.3   Diagram representing the relationships between themes in Bærum, Oslo (Atlas 2014)

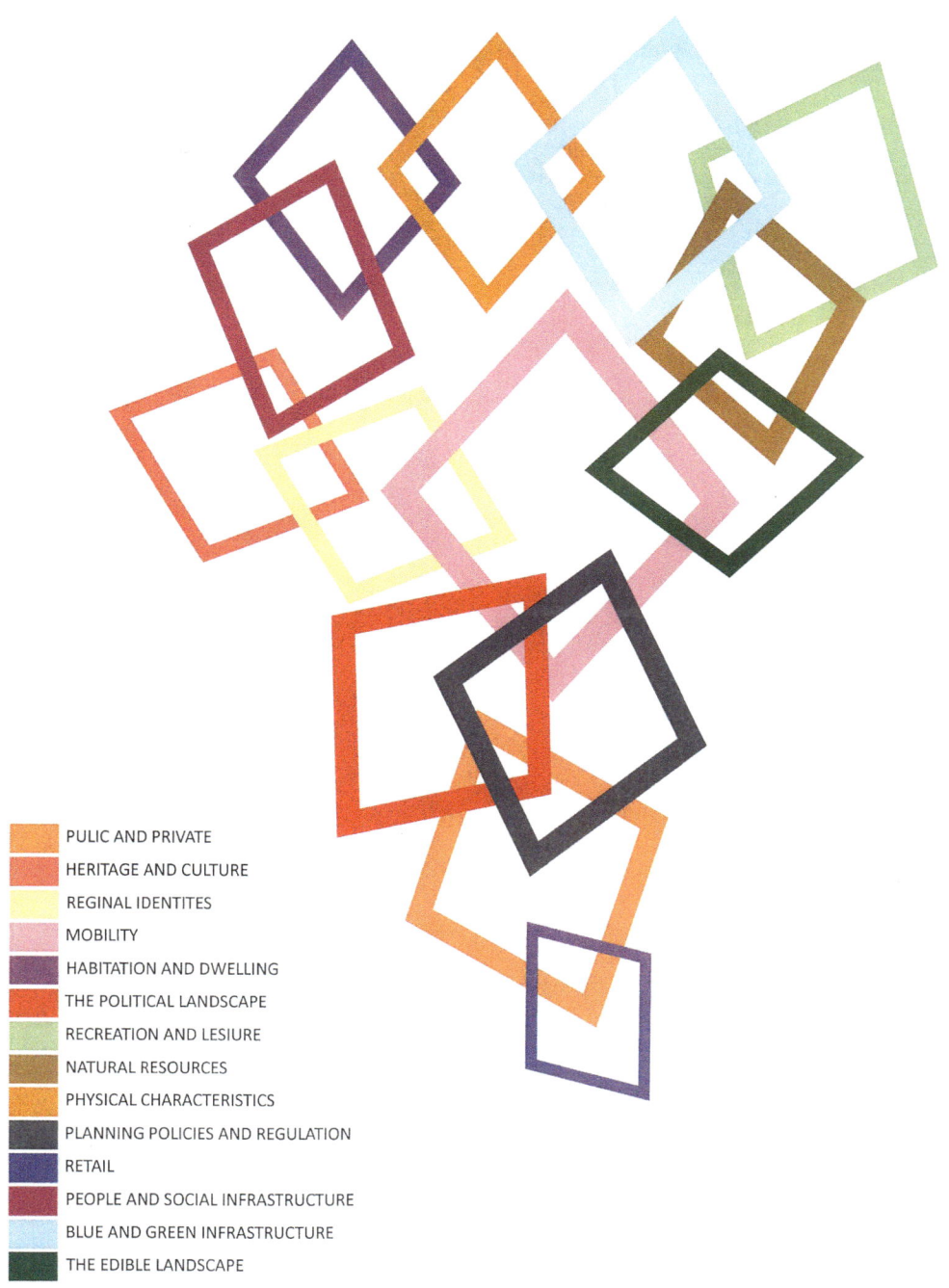

PULIC AND PRIVATE
HERITAGE AND CULTURE
REGINAL IDENTITES
MOBILITY
HABITATION AND DWELLING
THE POLITICAL LANDSCAPE
RECREATION AND LESIURE
NATURAL RESOURCES
PHYSICAL CHARACTERISTICS
PLANNING POLICIES AND REGULATION
RETAIL
PEOPLE AND SOCIAL INFRASTRUCTURE
BLUE AND GREEN INFRASTRUCTURE
THE EDIBLE LANDSCAPE

Figure 7.4　Diagram representing relationships between themes in Skedsmo, Oslo (Atlas 2016)

Figure 12.8 Brona Keenan's section off the coast of Virgohamna, at the northern tip of Spitsbergen, overlays the itinerary of Andrée's failed balloon journey to the North Pole. It performs as a narrative of connected forces such as innovation and documentation practices on Svalbard

Figure 12.9   Students' sections and additional mappings were overlaid on a map of the
Svalbard archipelago. The mapping includes the history of science activity,
sporadic expeditions, sea ice extension, heritage sites and satellite trajectories

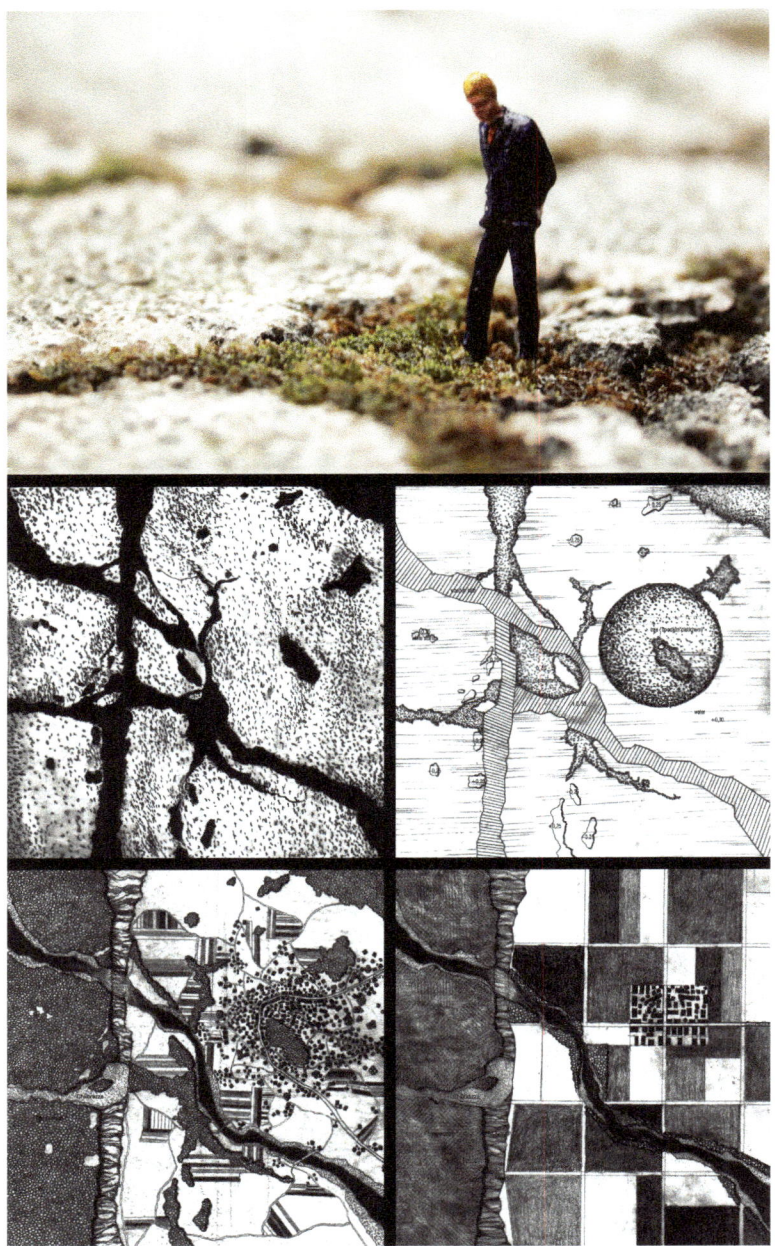

Figure 25.1   Workshop 'Out of Scale' project: The Great Discovery (Weilacher et al., 2013).
In the presented project, the fictional person is businessman Ben, who sets out
on a walk out of town to discover a new landscape that opens before him. The
photograph of the man was taken on cracked ground and illustrated to a scale
of 1:2. This was followed by a proposal of a garden to a scale of 1:100 and the
space is already set in the urban and cultural landscape to a scale of 1:5,000,
first as a presentation of the 'existing' landscape and then the envisaged spatial
changes, which on the far right of the illustration shows the de-urbanized
landscape, in which farmland dominates because of the need for food

Figure 25.2 An example from Introductory Design Studio – improving the creativity of students with the insertion of an abstract artistic exercise: colour images composed of garden elements from the first design of the house garden (Gazvoda, 2008–2014)

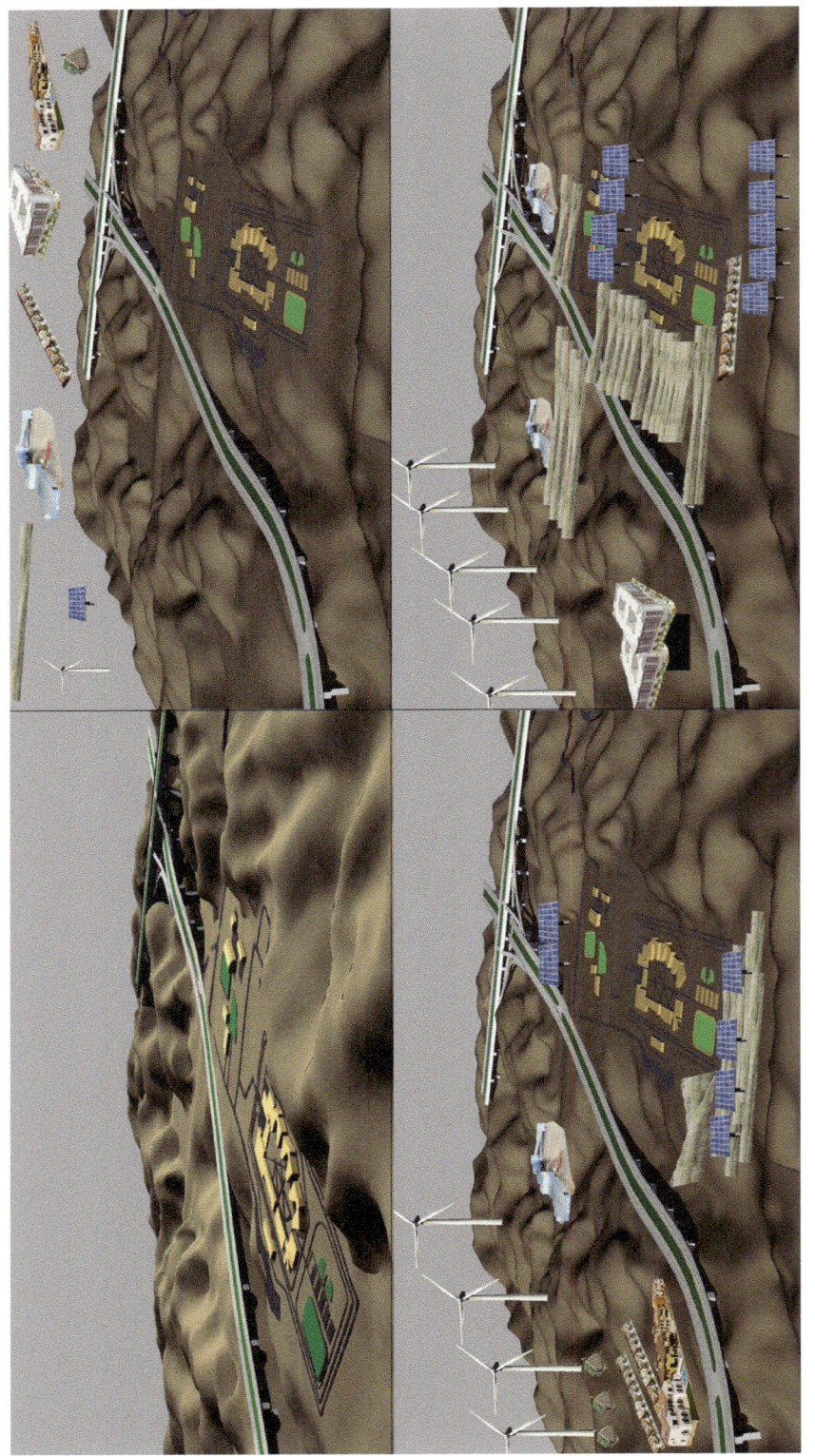

Figure 26.7 An exercise in participatory digital gaming in working with youth on the design of an alternative energy park (image credit: author)

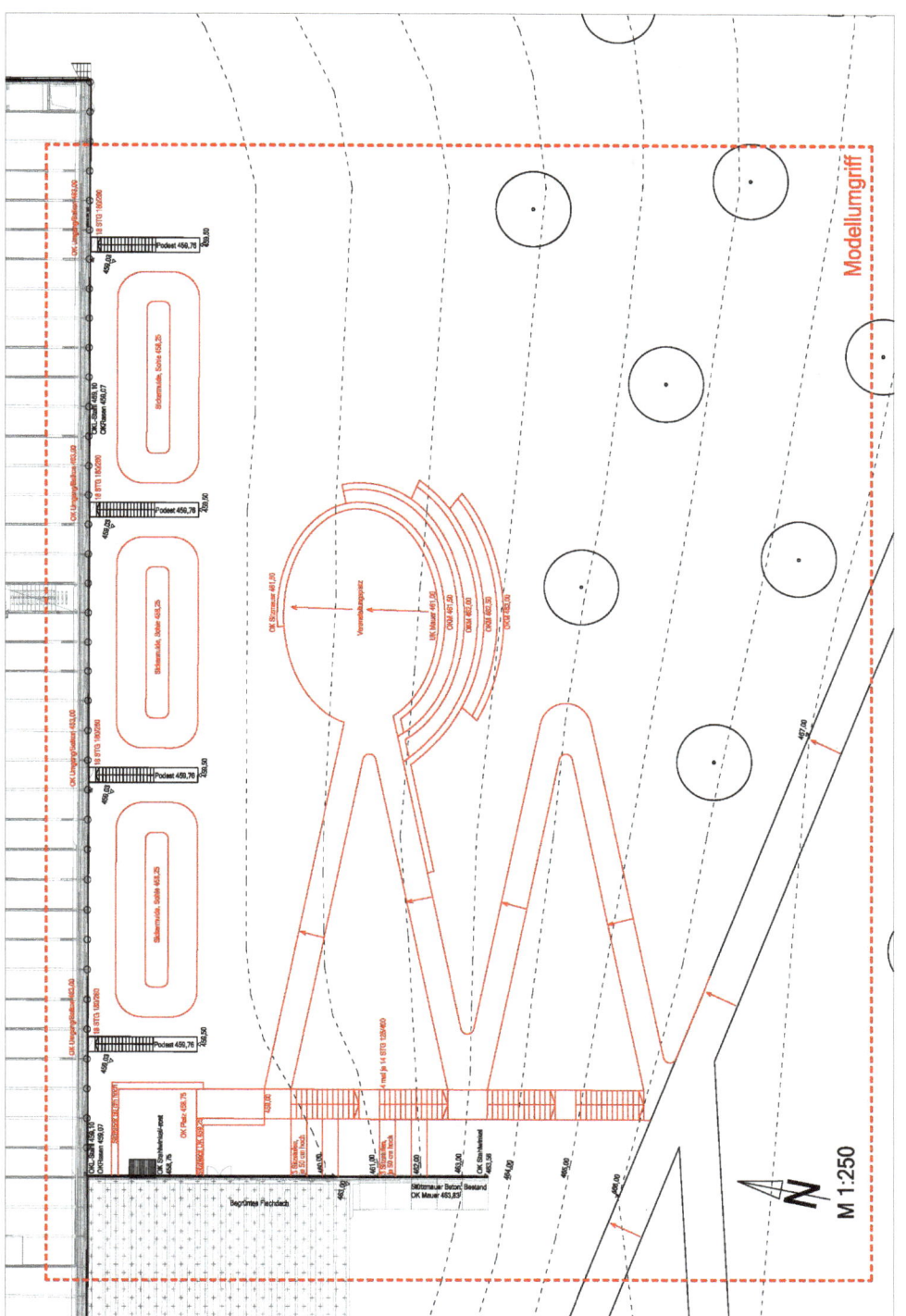

Figure 28.7 'Reflexive' terrain modelling exercise: students have to match the contour lines with the planned interventions (red lines) and build a model of the result; they have to follow certain construction rules such as steady-going slope of the paths, etc. (drawing: Ingrid Schegk; models: Saskia Schrader, Sabine Stockbauer, Kim Sander, Hanna Waschek)

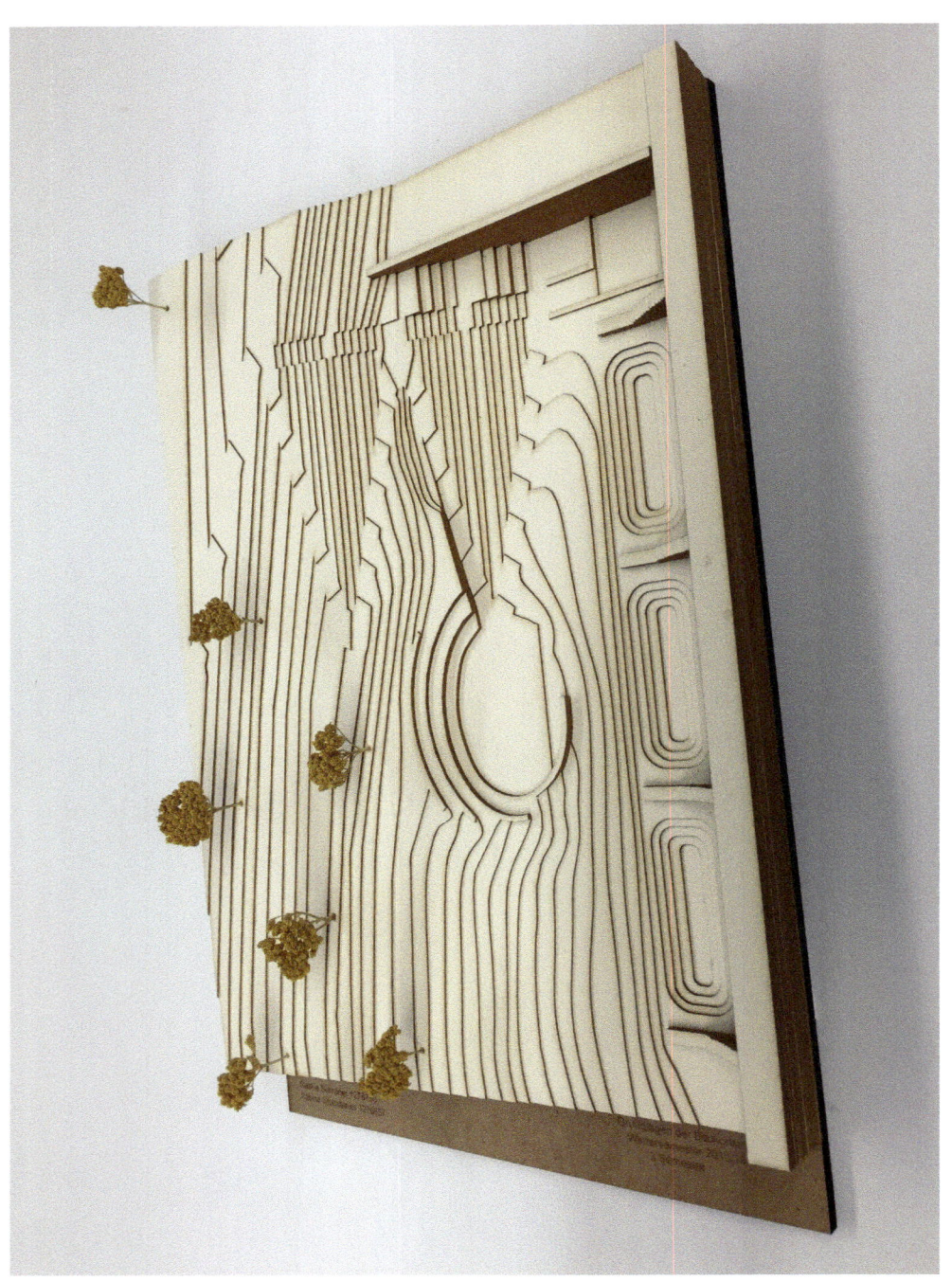

Figure 28.8 'Reflexive' terrain modelling exercise: students have to match the contour lines with the planned interventions (red lines) and build a model of the result; they have to follow certain construction rules such as steady-going slope of the paths, etc. (drawing: Ingrid Schegk; models: Saskia Schrader, Sabine Stockbauer, Kim Sander, Hanna Waschek)

Figure 30.1 Terrain and vegetation scan with a state-of-the-art handheld 3D scanner (FARO Freestyle 3D), Bukit Timah Nature Reserve, Singapore (Rekittke and Ninsalam, 2016)

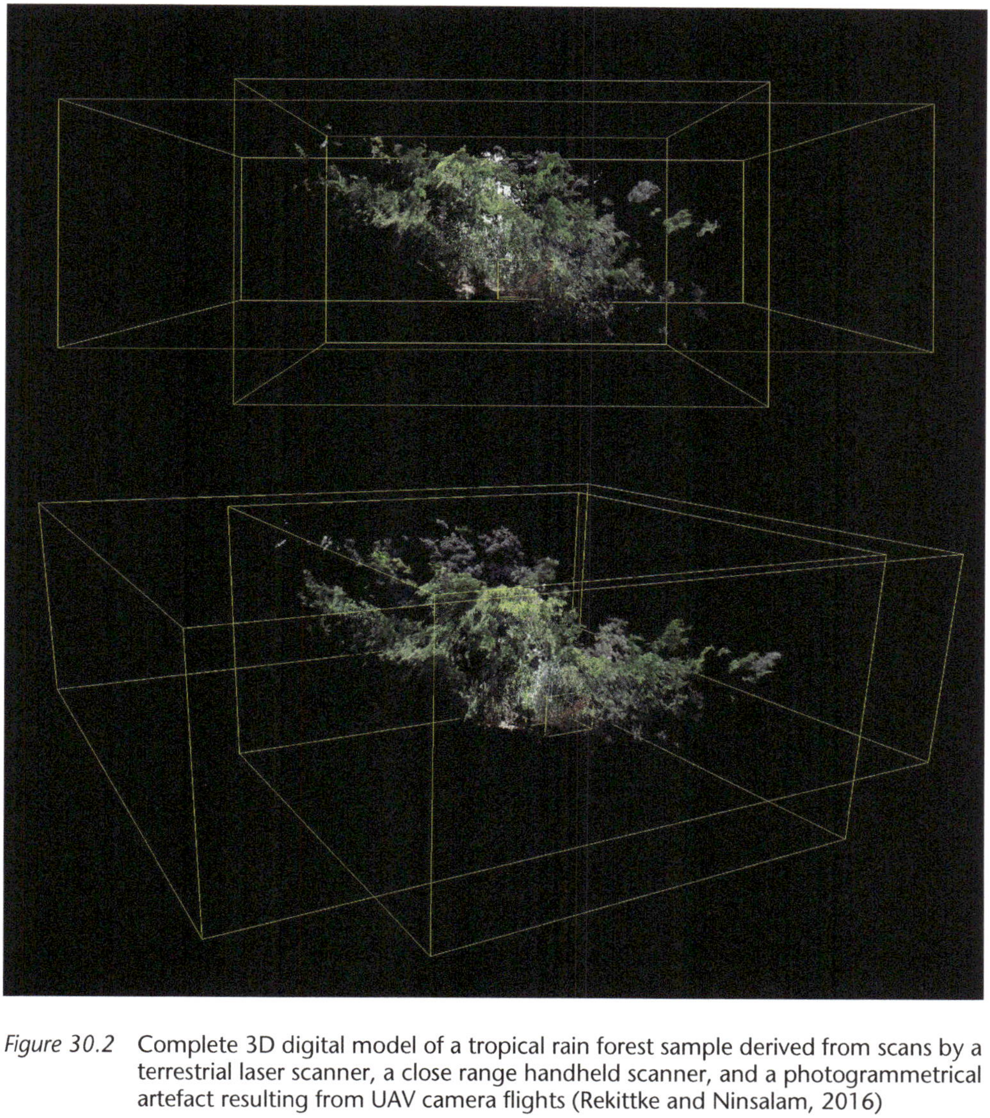

Figure 30.2 Complete 3D digital model of a tropical rain forest sample derived from scans by a terrestrial laser scanner, a close range handheld scanner, and a photogrammetrical artefact resulting from UAV camera flights (Rekittke and Ninsalam, 2016)

Figure 30.9 Photogrammetry campaign (top) with underwater camera (Olympus TG-4), in order to model the underwater aspect of a mangrove belt and seagrass meadow (bottom), Bunaken Marine National Park, North-Sulawesi (Rekittke and Ninsalam, 2016)

Figure 30.10 Pointcloud models of the above-water part of mangroves (upper part of figure) and their underwater structure (lower part of figure) on Bunaken Island, Bunaken Marine National Park. The green points and frames represent all the single photos that had been taken above and under water. The photogrammetrical artefacts—pointclouds—result from these images (Rekittke and Ninsalam, 2016)

Figure 30.11 Complete pointcloud model that combines all parts of an above-water mangrove vegetation sample with its underwater elements like stems, tillers, and roots (Rekittke and Ninsalam, 2016)

Figure 30.12 Pointcloud model sample of corals attached to a vertical reef wall in the waters of Bunaken Island, Bunaken Marine National Park (Rekittke and Ninsalam, 2016)

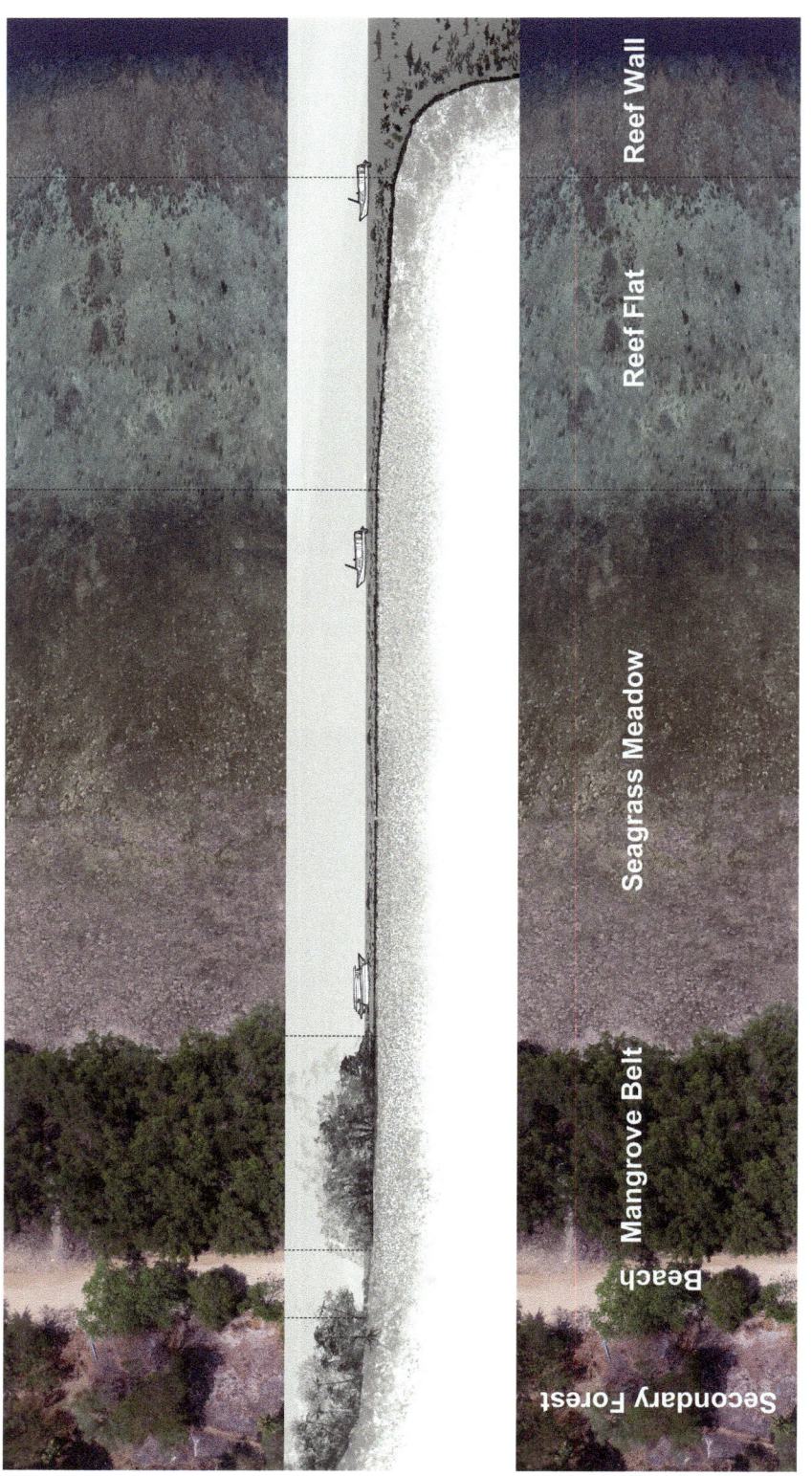

Figure 30.13 Our pointcloud models and the drived transects typify the intertidal ecosystem of the National Park, and are detailed enough for all aspects of analytical thinking, subsequent design work, and related proposals for park management (MLA Studio Rekittke, 2015)

Part II
Representing the landscape

This second part of the book is about representation of the landscape from the point of view of teaching, and forms a bridge between the didactics of landscape 'reading' and landscape 'transformation'.

According to James Corner, the primary characteristic of the term 'landscape' is a schema—a mental representation—or, as the European Landscape Convention would put it: landscape is an area "as perceived by people". Each individual observing a particular landscape perceives it differently based on their own experiences, resulting in a representation of the landscape that is unique to themselves. In teaching about landscape representation, however, there is a need to find a way of turning these unique subjective experiences into a form of representation that can find a broad enough acceptance to act as a firm foundation basis for future transformations.

At a time when digitalisation is influencing the way in which we view the world, and when it is perhaps easy to be seduced by seemingly objective photorealistic landscape representations, it is doubly important not to lose sight of the individual and the subjective as a basis for the representation process. This is perhaps the reason why the main focus of the contributions in this section tends to be on helping students to capture the personal and individual in landscape through the medium of language.

Before this, perhaps, unexpected group of three chapters about different approaches to capturing the landscape in literary terms, the section starts with a, superficially at least, more conventional consideration of the role of landscape drawing, in both representing the landscape and exploring the creative process, by Christian Montarou, who explores how we experience the landscape with our bodies, without the use of words, via a drawing and design process used to represent the landscape.

The 'writer's block' begins with Doris Gstach and Marc Kirschbaum, who describe language-based representational tools, using a design process that has the potential to be integrated into landscape architecture teaching. Kasia Gallo then focusses on the incorporation of writing into landscape architecture curricula, showcasing writing-based pedagogical efforts in design classes, and provides examples of evidence-based writing-centred activities. Finally Lake Douglas reflects on the wider importance of writing for landscape professionals and describes why writing should play a more prominent role in professional landscape curricula generally.

One characteristic that all writing-based approaches to representing the landscape have in common is the fact that they are sequential, unfolding over time, in contrast to the essentially static nature of pictorial representations. Time, and its representation in landscape teaching, is the theme of Noël van Dooren's chapter, which focusses on the function of time in landscape representation both in terms of a theoretical reflection of the role of the fourth dimension in landscape, as well as based on a set of practical design studio exercises.

Finally, the last chapter in this part moves to advocate a process through which to reconnect the many individual subjective landscape experiences into a common understanding involving co-construction of landscape knowledge in the classroom. Ellen Fetzer's contribution emphasises the positive effect of a constructivist approach, specifically regarding the three dimensions of constructivism: landscape, learning, and design. In addition, she shows how to use a digital constructivist classroom to reach out to landscape students worldwide.

Representation of the landscape as a physical and mental construct is an important bridge between reading and interpretation of the landscape, and facilitates interaction between landscape planners/designers and users, and, as such, must be a central part of teaching landscape.

17

The unarticulated dialogue in the creative process

Christian Montarou

Introduction

Teaching drawing and design in the "Landscape Architecture Studio" stands out from ordinary theory lessons in landscape architecture education. This chapter will put emphasis on the role played by body experience in the drawing and design process, when, at the interface between the physical space and the imaginary space, the act to draw lines opens up a creative potential. By stimulating tacit knowledge, this kind of intimacy with the inner landscape of the drawer can be a resource for him or her creative process. I will try to highlight how the unarticulated wordless dialogue that occurs with the phenomenal body during the drawing process can be stimulated in the studio context preliminary to a state of creative flow. It is an attempt to formulate grounds for a practice of freehand drawing beyond the technical exercises that aim at a correct representation of perspective, proportion, light and shadow.

The discussion about the relationship between landscape architecture and drawing is not new. Part of its history is linked to the discovery of the concept of space representation developed in 16th-century Europe with perspective knowledge. This mainly visual understanding of space and its representation has evolved since, both in the field of landscape architecture and in art. However, visualization techniques like two- and three-dimensional representation in plan, section, elevation, and perspective, still are originating from the Renaissance period. These rather standard representation drawings have been described, by authors like Frascati (Frascati 2011) and Catherine Dee (Dee 2004) respectively, using the concepts of "trivial and thin drawings" (visual superficial) to differentiate them from "non-trivial and thick drawings" (beyond the visual). Counter to a Rationalist understanding of man as a pure subject separated from the object it views, only linked to reality through concepts as in Renaissance times, the latter kind of drawings require a deeper relation to the subject. The attitude and method used here have similarities with an anthropological approach to the subject, using time and being immersed in the environment. The proximity of this connection is necessary for the drawer/designer to produce original ideas in expressing his or her own original feelings. Counter to digital technology, the same intimacy can be provided on the level of drawing techniques as a prosthesis of the body with pencil, brushes, pen, charcoal, chalk, etc.

Pedagogical considerations about the concept of knowledge

What makes the difference between ordinary theory lessons in landscape architecture education and the drawing/design workshop? To avoid misunderstandings it is necessary to look at the concept of knowledge generally and specifically within the field of aesthetics.

A large part of the professional education of landscape architects is sharing with the students a corpus of knowledge inherited by this field of activity. It is necessary to learn by accumulation and reproduction of existing practical and theoretical knowledge that can be used in practice later in the student's future professional life. Instead of this "nutritional" consumption of knowledge, the purpose of the aesthetics subject is more to find meaning within what one can discover by making things as "learning by doing" (Dewey 1997), like, for example, discovering existing tacit knowledge in oneself. With the help of time and maturity, this will cause a change in the way the student perceives himself or herself and the following subject of their study.

Knowledge is here understood as something emerging from experience not as reproduction of existing facts. This understanding of knowledge is similar to that sought when the designer works with the "non-trivial, thick drawings" named above. What is interesting with both Frascati and Dee's anthropological approach to the objective landscape's non-visual aspects is that, despite the concepts of eco and embodied drawings in Catherine Dee's propositions, it is all about the immersion of the physical body in nature. The question is not about embodied thinking through drawing, where the researcher, landscape architect or the drawer are an integral part of the environment, but about trying to find an inward adequacy between his or her inner landscape and the outer landscape, as expressed by Paul Cezanne: "the landscape thinks through me, I am its consciousness" (Merleau-Ponty 1948).

Embodied cognition

The immersion of the body is not only a question of immersion in the physical space of the environment. In line with phenomenology, the concept of "body" refers not only to a physical object that unfolds in space and detects the surroundings through the five senses. The French philosopher Merleau-Ponty (1908–1961) operated with two aspects of the body, the body as a physical object and the lived body as subject (Merleau-Ponty 1945). In line with those thoughts, one can denote the field between the body's presence and the "distance of the representation", understood as our parallel world of languages, symbols and images, as a potential space of creativity. Similarly, we inhabit several places simultaneously, that is, in the flow of consciousness that occurs between body, thinking and the visible. The meaning of this "gap in between" is most often ignored, whereas that cavity is understood as the inner experience of the boundary between the lived and the physical body according to the philosopher Merleau-Ponty. Embodied cognition theories (Wilson & Foglia 2011) recognize the role of the lived body in perception and thinking and do not limit perception to an interpretation and a processing of the optical image that occurs as a perception on the retina. They claim that the body shapes our recognition prior to visual perception by "an integrated set of skills that are ready to predict and incorporate the world before using concepts/formation of thoughts/assessments, a type of legitimate preparedness consisting of a kind of non-cognitive, before linguistic, a motor sensor targeting called habitus" (Merleau-Ponty 1945).

The words "body experience" refer to a dialogue with the lived body as an unexplored inner territory, named above, that we could call the drawer's inner landscape. The communication that then occurs in that dialogue is deaf, blind and dumb. Deaf, as deep concentration cannot be distracted by disturbing elements from the surroundings; blind, when registering the outer world via the visual sense, only a door opens to non-visual signals from the body's sensitivity; and dumb when the character/designer's state of concentration during work prevents temporary logical articulation of words to sentences.

The main intentions of the drawing/design course module

My teaching has evolved in line with the structural changes that took place at the university where I work, where the drawing course went from being an independent course to becoming an integral part of the LA studio course. Initially, the aim of the teaching was guided by a pedagogy focusing on the visual: teaching the students to develop basic drawing skills. The primary aim of the course is still to teach the students to see and to draw as well as learning to learn to see and draw. Later the drawing process was also used as an educational tool to give an insight into the creative process, through consciously applying creative techniques like analogy, contradiction and combination thinking for idea development in the design work. Students are challenged to think visually, to view a problem from multiple angles and to learn how to combine familiar elements in new ways. It is about exercises and tasks that also help them to learn attitudes towards the work process and to reflect on their own process. In the last five years the pedagogical goal of integrating the body more systematically into the drawing process has become a source of inspiration. The challenge is still to find strategies that facilitate access to the phenomenological body and its unarticulated knowledge so that it can become a creative resource in perception and planning of the environment.

My philosophy of teaching (presented as a hierarchy in Table 17.1), is further informed by the pedagogical philosophies of Benjamin Bloom, Mihaly Csikszentmihalyi, Nobo and Sachiko Komagata, and Zhuang Zhou (aka Zhuangzi). Zhuangzi (about 300 BC), for instance, describes the stages and transitions between different states of mind during the process of developing basic experience (Billeter 2010, pp. 41–80), while Bloom applies a target structure to the learning process. As can be seen, attaining a state of flow precedes attaining that of mindfulness—the latter of which is described by Komagata and Komagata (2010) as "a mental state of being aware of the outside and inside of oneself at present without judgment, i.e., with full acceptance."

The drawer's relationship to "own body" during different phases of the drawing process

Drawing a line on the drawing sheet opens up an imaginary space of relationships between seeing and doing, between being an actor and spectator, while the drawing is in progress. This creates a characteristic relationship with one's own body, because this activity does not just happen as head thinking in an abstract space of intellectual and mental performances. The perceptions of the task and of oneself are influenced by the flow of the action. It opens up a stream of information that comes from both the outer world, the artist's inner experience and the actual drawing.

Table 17.1 Achieving states of flow and mindfulness through mastering the task of life drawing (Christian Montarou 2018)

Continued Work **THE BODY AS A SOURCE TO PROVIDE NEW MEANING**	6) Transferring learning experiences to a new area.	**Experiencing mindfulness** This is a complex cognitive situation, "a mental state of being aware of the outside and inside of oneself at present without judgment, i.e., with full acceptance" (Komagata & Komagata 2010, p. 2). Here, the student takes a disinterested attitude by observing phenomena from a distance. The result is a paradoxical state of mind that is fully present in the situation (mindfulness) and, at the same time, completely absent (devoid of ego, mind free of thoughts). In this state, action means taking the necessary steps to adapt to any given situation.
THE BODY AS A BASIS TO PRODUCE MEANING	5) Combining learning experiences into a new and composite pattern of action.	**Flow of mental energy without the interference of ego** At this stage, intuition takes over as a creative force. Attention is now more focused on the act of drawing than on the blurred motif in the back of consciousness, and contact with the body's cognitive unconscious is established. The body's sensory-motor archive of knowledge is available as a resource and gives meaning to the input of sensory stimuli. Previous and new experiences can be combined into actions that are more complex. The drawing appears to come into existence by itself and without any effort; this stage requires discipline and perseverance.
Combinatory exercises **THE BODY AS A PARTNER**	4) Guided learning: low scaffolding. Drawing becomes a routine, intuitive and reflexive action.	**Focus of attention, i.e., 'losing oneself' and becoming one with the subject** Consciousness can be freed for other purposes—control over the process is no longer needed to the same extent as before, because the action has been partially integrated into the body. As a transitional release, other parts of consciousness take over more complicated tasks like capturing and combining fragments of information from both past and present in new ways.
THE BODY AS A MEDIUM	3) Guided learning: medium scaffolding. The focus of attention is controlled.	**Ego-centred conscious control** This stage requires discipline and determination to control the focus of attention. It entails the ability to recognize and interpret a three-dimensional motif on a two-dimensional surface, using the aesthetic repertoire of visual elements such as lines, spacing, angles, proportions and symbols. It also entails learning to exclude distracting elements while working on the drawing.
Exercises	2) Guided learning: high scaffolding. Physical, mental and emotional preparedness to act.	**Deconstruction and construction** The student learns how to draw the motif whilst breaking free from the power of perception that is governed by a conventional systematization of symbols. The transformation of the motif from a physical object to an aesthetic entity has elements of partial dissociation, but also includes the process of integrating the shapes into figurative meaning. First, the motif has to be disassembled into a puzzle of shapes, disconnected as much as possible from any symbolic significance, before it can be reassembled into new analogical meaning in the form of a figurative drawing at the end of the process.
	1) Guided learning: high scaffolding. Recognize, observe and record lines, areas, proportion.	

In other words, it links the body's inner experience to the physical action of drawing lines.

The line as a material trace of the absence of the body can be seen as a link between the physical and the imaginary phenomenological space. In lucky moments of concentration, it could give the feeling of overcoming the duality of body and mind. A door opener to creative break and discovery of tacit knowledge is the access to a state of flow and mindfulness. How can we create the necessary conditions in drawing instruction for establishing the internal dialogue to the phenomenal body in practice? This is the goal of the five following exercise proposals.

Five drawing exercises stimulating a state of flow

Topic: being present in your own body, with practical exercises in awareness using gesture drawings. The following steps of the workshop are building on each other to prepare the participants for a series of drawing exercises where the purpose is to feel the drawing tool as a prosthesis of the body. The participants will be able to experience the autonomy of drawing gestures, which, for a while, are disconnected from the control of the mind. Through working with drawing gestures connected to letters, words and breathing, the participant will investigate the forgotten connection between sound, gesture and visual expression. Using these preliminary experiences to translate different objects, the students will integrate a succession of gesture sequences to render a subject visually. The first step in this journey is to relax the body: discover hidden tensions in order to obtain a relaxed state of mind before starting to draw.

Exercise 1

Purpose: Establishing a more conscientious relation to the "inner body" and the paths to access it. *What*: Make some steps into your imaginary inner space and be aware of the frontier between outer and inner space. *Why*: We are usually separated from the body. To be creative we have to distribute our attention to the lived sensory and motoric body to access creative energy. *How*: Through different movements and focus on the breath. The participants will discover hidden tensions in their body that inhibit their awareness of their sensitive "here and now" intuitive body (Figure 17.1a and b).

The perceptual introspection is defined as relaxation because it is through this that the creative energy can flow freely and be an indispensable resource in the drawing process. The purpose of this relaxation state is to establish an attitude of non-critical, accepting and empathetic way to relate to what may come, known as "active neutrality" (Berger 2009).

Exercise 2

Purpose: Access to a flow through distributed attention focused on the outer and inner world simultaneously. *What*: The breath and the line as a bridge for bringing body and mind into an "here and now" experience. *Why*: The line opens up a dialogue with an imaginary space in the continuum of stimuli interfering with us and finds a place to be in our consciousness as a form on a background. *How*: Drawing lines horizontal, vertical, fast and slowly. Synchronize the hand movement with a free breath and working with blind drawings (Figure 17.2a and b).

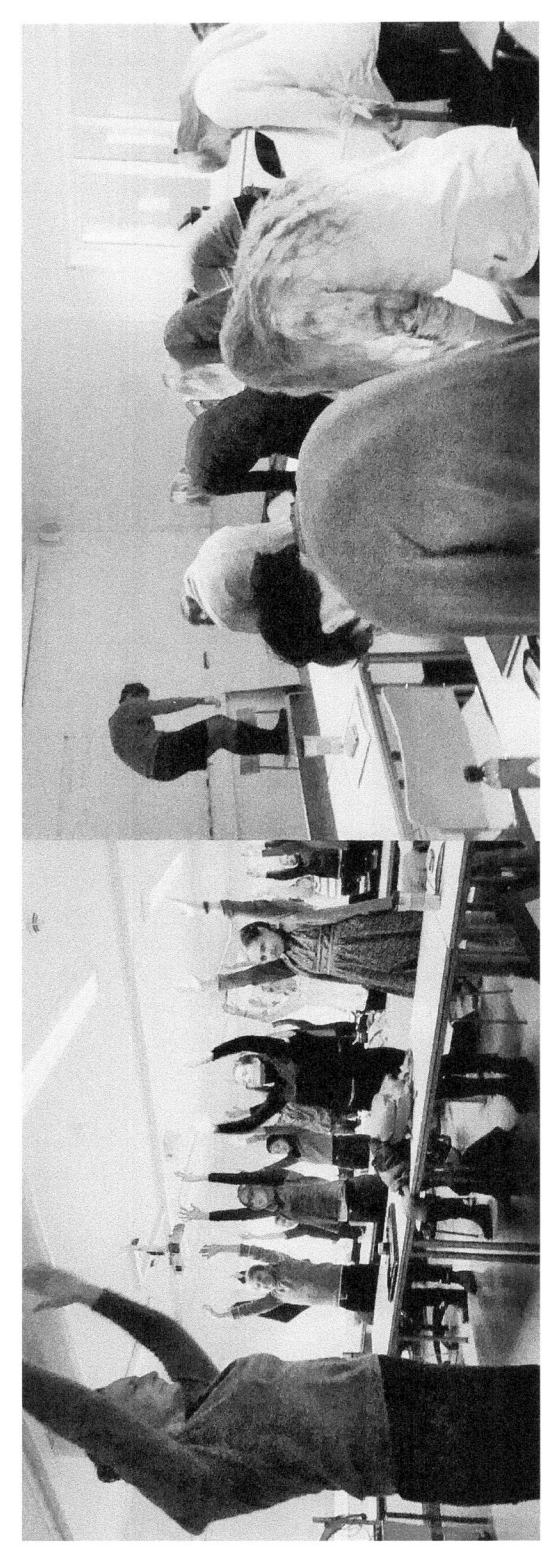

Figure 17.1a and b Relaxing the body

Figure 17.2a and b Controlling attention focus by focusing on the outer and inner world simultaneously. Synchronise movement and breath

Exercise 3

Purpose: Make your mind and body in flow, through identification with outer flow movements (ribbon) (Figure 17.3a and b). *What*: 1) The first person forms a free fluid movement with a gymnastic ribbon in 3D space. 2) The second participant creates from the first ephemera a drawing in empty space, a big scale drawing vertically on the 2D surface of the wall. 3) The third person has to create a smaller 2D drawing on the horizontal plane of a table translating the drawing made on the wall by the second participant. *Why*: The outer flow movement is including its inner counterpart in yourself. Planning a drawing action provides a visual analogy in coincidence with the original stimuli. *How*: Concentration on the rhythm of the ribbon movement and the drawing action.

Exercise 4

Purpose: Engaging the "inner body" sensations in a creative way to draw. *What*: Choose a short word like "Sun," "Love," "Bravo," etc. Take one letter at a time, pronounce the sound, and find an expression to it through serial drawings. Finally, draw the entire word within a blink of an eye. *Why*: To investigate the forgotten connection between sound, gesture and visual expression, Re-"member"-ing yourself as a kid learning to draw, write and read. *How*: Deconstructing the process in sequences with fragments of movements, before integrating them in an entire drawing of a word. Draw in different positions like sitting, standing, using a table or the wall (Figure 17.4).

Exercise 5

Purpose: Using previous experiences to imagine a succession of drawing gestures connected to the body sensory motoric memory—to render a subject visually. *What*: Capturing the drawing's lines in a gesture that begins long before any line is drawn on paper. Miming the motif in space to build a motivated answer to motivating stimuli. *Why*: Because the drawn line like the "red thread" of the "garden spider" is connected to a network. *How*: Searching through the repertoire of embodied knowledge, a corresponding schema to adapt the emotion experienced with the subject. Though this occurs on the imaginary plan, it is rooted in the body tonicity (Figures 17.5 and 17.6a and b).

The inner referent is used to catch the characteristic of the object. "It is not only tied to the visuals, but the perceptions of all senses experience" (Lakoff & Johnson 2003).

CONCENTRATION ON THE RHYTHM OF THE RIBBONs MOVEMENT and THE ACTION TO DRAW

Make mind and body in flow through identification with the physical outer flow movements of the ribbon

Figure 17.3a and b   Making mind and body in flow

Figure 17.4 "Made me aware of things that otherwise goes automatically, such as writing letters. A letter gives a clear form, but can look different if we separate shape from sound." (Student comment)

GESTURE DRAWING AND ANALOGY.

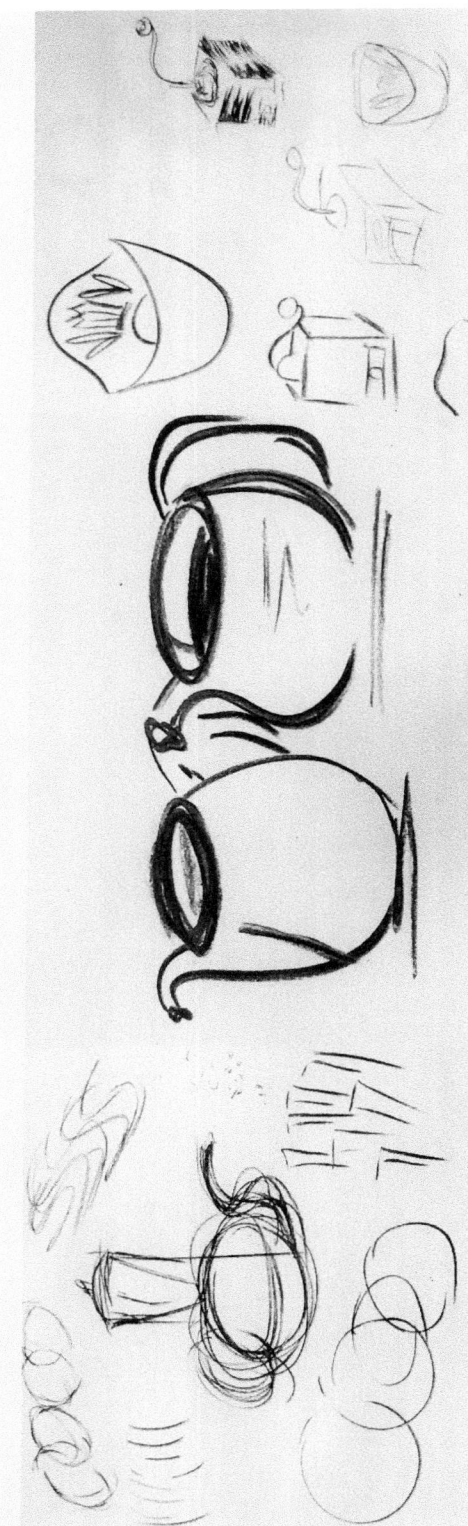

Drawing an object (teapot, tea maker, bubble bottle, coffee grinders). Creating a series of gesture drawings in a larger surface where you revisit the subject principally in different sequences. Instead of observing/copy, the subject outer contours you shall "mime" the imaginary movements its forms arouses in yourself of emotions. Create a spontaneous interpretation of the subject as an action procedure, which commemorates the necessary sequence of gestures needed to recreate the scene as a visual motoric analogy. Make several sketches of the same scene.

Figure 17.5 "This was a very exciting day. We had relaxation exercises before we started drawing. I think this helped me a lot. I thought the task where we draw the sound of letters was especially interesting." (Student comment)

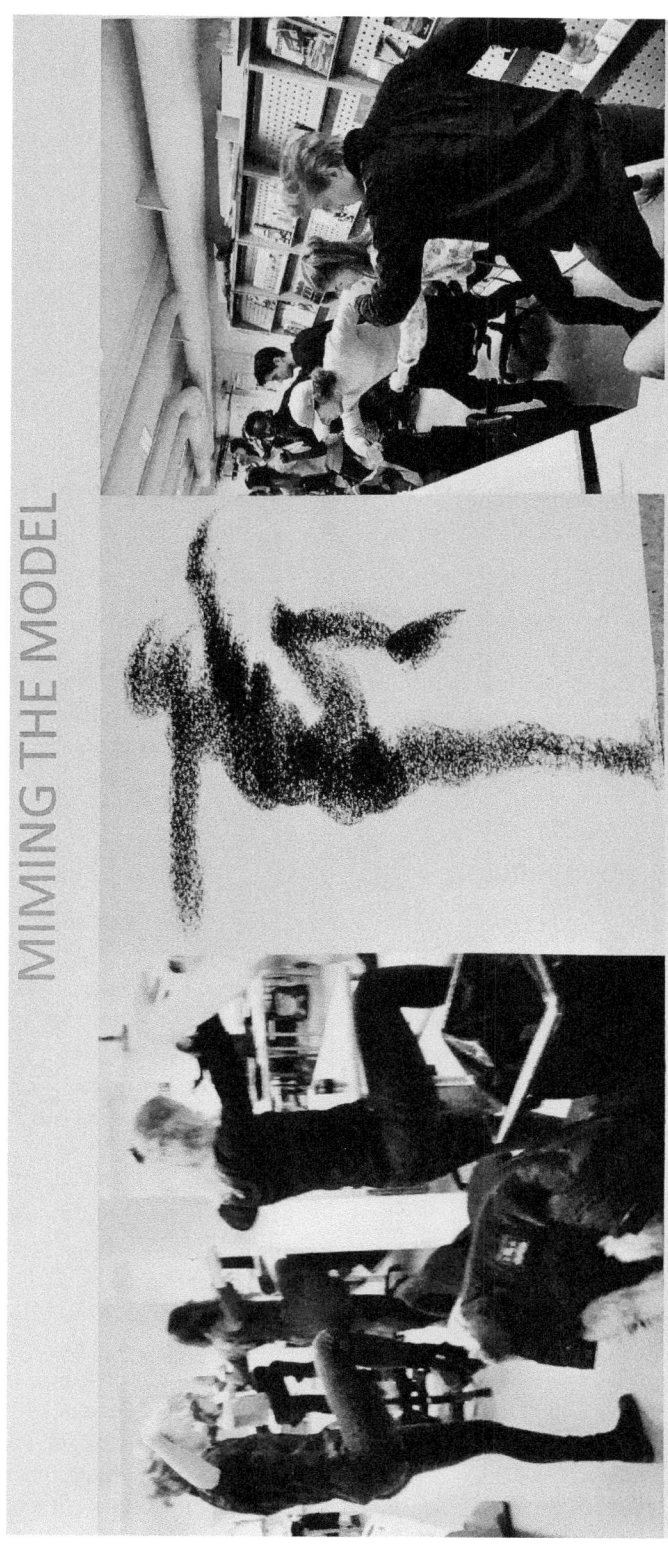

MIMING THE MODEL

To activates the designers proprioseptive schemata before drawing session

Figure 17.6a and b Miming the model to connect to the phenomenal body before starting to draw

Students' feedback to exercises

Exercises two and three:

> The task in analogy drawings with charcoal technique was nice. As with other similar exercises we have done before, we could let the line go free and interact with your body and breath, instead of the brain. It is very delicate and I see that the drawings get a very vibrant expression with a good flow. Being able to draw as natural as you write feels very good. I would like to bring this self-esteem and the flow forward. It creates character and a strong expression. In addition, I will breathe and relax more before I begin to draw to easier access the creative mode.
>
> *(2017)*

Exercise four:

> Made me aware of things that otherwise goes automatically, such as writing letters. A letter gives a clear form, but can look different if we separate shape from sound.
>
> *(2016)*

Exercise five:

> It was incredibly fun and exciting. Felt I managed to draw what I felt. You felt the stroke when you used your entire body when drawing in standing position by the wall. It was interesting to be able to draw the imaginary interpretation of movements of a teapot.
>
> *(2015)*

Embodied and localized thinking

The wider dialogue in question is also highly localized, that is, the environment's structure and the psychosocial atmosphere of the context are of utmost importance for the development of a creative potential space. I will emphasize some important aspects of that in the following sections. However, another question is how can our relationship to the body's experience contribute to cognition in the drawing process? Inspired by the French researcher Eve Berger's analysis of the role of the body in scientific research (Berger 2009), it is possible to sketch four relations or states of mind that can arise in successive sequences or simultaneously during the drawing process.

1   The body as a medium (Table 17.1, point 3). This is the first step in the discovery of another part of oneself. That is, to experience oneself in different ways in relation to the environment by exercizing control of the transitions between the right and the left hemispheres. Controlling the transition from the left to the right hemisphere can help to get you into drawing mode by seeing a change of mind. How the hand and the eye are connected along the horizon, how shapes and proportions can play a trick on the eye, how only a small change in point of view can literally change the entire image you see.
2   The body as a partner (Table 17.1, point 4) emerges when access to pre-conscious or semi-conscious activity in the brain becomes available to affect the interpretation of data flowing from the moment's action. This stimulates creativity to amodal perception where awareness of the body as a presence gets more space than it would otherwise when the

starting point for enrolment of the environment is centred on the five senses. The body also gets more fullness as it is complemented by a proximity to the "sensitive body," understood as matter and memory. In other words, it is to relate to the body in a special way by listening to its signals, attentively presenting what is happening at different levels of consciousness at the same time.

3   The body as a basis for meaning production (Table 17.1, point 5). Further development of concentration for keeping focus for a longer period can be ensured by repeating previous experiences, but with greater focus on the actual act of drawing while in progress, rather than imitating a motive. This is leading to a state of flow that is a particular state of mind where all sense of linear chronometric time is erased. This means that the notion of past, present and future temporarily no longer exists. Csikszentmihalyi's definition of flow is: "the state in which people are so involved in an activity that nothing else seems to matter" and "the experience itself is so enjoyable that people will do it even at great cost, for the sheer sake of doing it" (Csikszentmihalyi 1991). The transformation process from the information stimuli that come from both the surroundings and the artist's inner experience takes place in the mind through an embodied process. It is a partially unconscious elaboration of an intricate action plan in the form of a series of connected gestures loaded with visual intention, made before starting to draw. This happens while the mind is emptied and the ego's usual judgmental and critical sense is temporarily put out of play. By taking over the control of the hand and of the split pen's movements across the drawing sheet, the creative life energy transforms as the process tracks muscle tonus into aesthetic form. The attention focus has now been distributed to both the inner experience, the drawing and the object in the environment. The body memory bank becomes available and perception means more than collecting visual information from the outside world.

4   The body that is the source of the creation of new meaning (Table 17.1, point 6). The artist's "dialogue" and dynamic relationship to their own bodily perceptions lead to a fracture beyond the ego's will and logical thinking, and to an intuitive transformation (transmutation) of bodily feelings to a new category reflection, that is, without deduction. Through a broader focus, which emerges in consciousness, different levels of information are included as suggested above. In this undifferentiated chaos of stimuli, everything becomes potentially equal and meaningful in all directions at the same time. This means that the dualistic thinking is temporarily abolished when the boundaries between form and background, past and present, subject and object, rational and irrational, visible and invisible, time and space, body and mind are disturbed. With drawing as embodied cognition, the artist can come to the nearest symbiotic contact with an expanded amount of data (Stucke 2011). The emergence of the idea is not the result of head thinking alone but of wider grounded interaction between the body and its emotions, memory, and mind, social situations and the physical world. It happens in a creative moment in which the drawing's idea appears spontaneously and apparently unintentionally as a kind of sudden "Eureka" that emerges overwhelmingly as a result of an immersion process where you get swallowed up in work. At the same time, this alert state has been compared with that of rapid eye movement during dream work (Virilio 1984). This is a complex body-cognitive situation, "a mental state of being aware of the outside and inside of oneself without judgment, i.e. with full acceptance" (Komagata & Komagata, 2010). However, it requires competence to let this happen. It is, among other things, the ability to wonder and to be able to open up to accept the new that can come in the reciprocal and contemporary relationship that is established between the outer surroundings and the inner focus of the drafter himself or herself.

Mastering the task and experiencing flow

From an ego-centric and rational controlled state of mind, separated from an awareness of the body and the surroundings, the creative process moves to the transitional state of becoming one with the motif and the surroundings through control of focused attention. Once the ego gives up its disturbing habit of constantly producing new ideas, the intensity of attention allows intuition to take over as a creative force and to release the flow of mental energy that provides optimal conditions for action. The artist feels in top condition, wide-awake and present in the situation, mastering the task. This situation, which is experienced neither as a wholly mental nor as a wholly physical state, breaks down routine rational thinking and allows access to body knowledge through a "head-over-heels" process. This involves a new way of processing visual information by using non-focal vision to "distribute" one's attention. In contrast to focal vision, in which a single source of information is processed, distributed attention compels the artist to "scatter" his or her gaze across the scene and record more complex patterns of information. This also opens up for combining fragments of information from short- and long-term memory.

Interaction, feedback and reward

Proper feedback given at the right time can be enough to confirm an experience of flow, during which mental energy is released between the artist and his or her work. Confirmation of the newly created sketch gives students the feeling of being present and part of the here-and-now. In keeping with the students' realistic evaluation of their own skills, the drawing itself may not be exceptional, but the endorsement of an intuitive and spontaneous action represents such a strong and self-affirming reward that it overshadows the actual result and reinforces the students' confidence in their own creative potential. According to Csikszentmihalyi, one of the conditions for achieving flow is a balance between challenge and perceived mastering ability; other conditions are clarity of objective and instant feedback. The longing to recapture the first experience of flow is the best motivation to try again, free from any distracting thoughts. As a result, mindfulness allows students to direct their attention, letting mental energy flow between their own intent and the motif in question. The key to motivating an experience of flow is to foster a sense of self-development and discovery that allows the artist's mind to grow and develop levels of greater complexity (Csikszentmihalyi 1991). The greatest obstacle to creativity is often brought about by one's own thoughts and concerns.

Music rhythm as inspiration ritual

In addition to the participants' personalities, the particular space lends a distinct character to the situation, as does the location and position of the individual participants in the room—whether sitting or standing, body position plays a role in activating embodied knowledge. In addition to the primary visual focus and the way in which students are sitting or standing in the room, auditory stimuli (sounds, noise, music, etc.) may be used during the drawing session to affect the mind and stimulate the brain to concentrate on the task. I usually vary the register of sound from total silence—so that you can hear the sound of the drawing tools on paper—to jazz rhythms, didgeridoo and/or classical music at various sound levels. This can help to structure attention and dispel distracting thoughts.

Time organization as ritual

The studio work sessions are four hours each week and can be organized in different ways depending on activities. For example, they can be divided into twenty-minute drawing intervals and ten-minute breaks when working with a model. In any case, the activities that take place in the studio space require silence and concentration. The regularity of this repetitive pattern works as a positive ritual, preparing the student to settle into the situation, it calms down the left side of the brain, and helps concentration on the subject of the day, pushing away other preoccupations.

Conclusion

The educational value of working with drawing for landscape architecture students of today, born in a digitized data world, is not obvious for everybody. The extremist stance is that: "Learning to draw can develop drawing skills, but absolutely nothing of the visual skills that landscape architects need in their profession" (Moore 2009). Neither can freehand drawing develop creative abilities if we believe in those type of arguments. This does not matter given the feedback messages we receive from students on course evaluations at NMBU, nor to the starting point for this chapter. The aims of this chapter are in line with the focus of so-called "non-trivial/thick drawing" as defined at the beginning of this chapter: to formulate a practice of freehand drawing that goes beyond the technical exercises aiming at correct representations. It is also an attempt to shed light on the rarely discussed theme of body experience in the drawing and design process. Moreover, it questions how the unarticulated wordless dialogue that occurs with the phenomenal body during the drawing process can be stimulated with series of exercises, to take place in the studio context as preliminary to a state of creativity break.

This chapter is based on my own observations and experiences as a teacher of LA students over a 20-year period. Reflections are inspired by first-hand knowledge of students' behaviour in the studio, as much as by their written notes on the drawings along the way during the work process. In addition, a whole host of educators and philosophers whom I have used to illuminate and understand my own practice also influence me.

My findings about what the students learn through the drawings exercises may be summarized in six points:

1   They learn to "see" in the sense of interpreting the flow of visual stimuli that continuously interferes with our two eyes; "to see as" spaces or lines and surfaces in relation to each other, axes, rhythms, structure, textures. "Seeing as" is a visual-language act in that it makes sense in choosing and separating a difference in the field of view between a figure and its background.
2   The experience of being connected to the "whole body" in the drawing process makes the students discover its role in the creative process and what concentration means to solve a challenge.
3   Practicing the proposed exercises increases the students' feeling of mindful presence.
4   A secondary very important effect of a mindful praxis of the drawing process is a strengthening of the students' self-confidence that increases faith in their creative potential.
5   The pleasure to draw and to communicate with drawing is cultivated.
6   Finally, one must not forget the importance of a collaborative studio atmosphere and the associated dialogue in the context of the learning situation.

The landscape architect's relation to space is essential, when working with spatial planning and architecture of places. Training the relation to the lived body combined with a visual approach to space through freehand drawing leads to beneficial experiences for landscape architecture students. To see becomes not only "to look at", to register the functions of a place or other facts about the place that are also necessary to record. To "see as" is triggering imagination by interpreting landscape as space, proportion, line, foreground/ background, color, etc. It is a creative investment; by inhabiting the Euclidean space by projecting the lived body with his or her imagination, the designer sees not only what appears on the first level. By using his or her body as a perception tool, non-visual information can also be recorded.

References

Berger. E. (2009). Rapport au corps et création de sens en formation d'adultes: étude à partir du modèle somato-psychopédagogique, p. 349, accessed 29 March 2017 from: www.theses.fr/2009PA083183.

Billeter, J.F. (2010). *Leçons sur Tchouang-Tseu*. Paris: Editions Allia.

Cerap.org (2015–2016). Centre d'étude et de recherche appliquée en pédagogie perseptive, accessed 26 March 2017 from: www.cerap.org.

Csikszentmihalyi, M. (1991). *Flow: The Psychology of Optimal Experience*. New York: HarperCollins Publishers.

Dee, C. (2004). "A critical light on landscape architecture", conference proceedings 16–19 September, ECLAS Conference, Norwegian University of Life Science, Aas, Norway.

Dewey, J. (1997/1938). *Experience & Education*. New York: Simon & Schuster Editions.

Frascari, M. (2011). *Eleven Exercises in the Art of Architectural Drawing: Slow Food for the Architects Imagination*. Abingdon, UK: Routledge.

Gombrich, E.H. (1987). *L'art et l'illusion*. Paris: Gallimard.

Komagata, S. and Komagata, N. (2010). *Mindfulness and Flow Experience*, accessed 24 May 2018 from: http://nobo.komagata.net/pub/Komagata+10-MindfulnessFlow.pdf.

Lakoff, G. and Johnson, M. (2003). *Metaphors We Live By*. Chicago, IL: The University of Chicago Press.

Merleau-Ponty, M. (1945). *Phenomenology of Perception*, translated by Landes, D.A. (2012). New York: Routledge.

Merleau-Ponty, M. (1948). *Le doute de Cézanne, In Sens et non-sens*. Paris: Nagel, p. 32.

Moore, K. (2009). *Overlooking the Visual: Demystifying the Art of Design*. New York: Routledge.

Stucke, A. (2011). *Embodying Symbiosis. A Philosophy of Mind in Drawing*. AMBER PROJECTS, 2010 Vine Street, Berkeley, CA 94709, accessed 24 May 2018 from: www.academia.edu/11204484/Embodying_Symbiosis_A_Philosophy_of_Mind_in_Drawing.

Virilio, P. (1984). *L' espace critique, essais*. Paris: Christian Bourgois.

Wilson, A.R. and Foglia L. (2011). Embodied Cognition. *Stanford Encyclopedia of Philosophy*, accessed 26 March 2017 from: https://plato.stanford.edu/entries/embodied-cognition.

18

The underestimated role of language-based tools in landscape architecture

Theory, empiricism, practice

Doris Gstach and Marc Kirschbaum

It is common understanding that the sophisticated handling of different forms of visualisation belongs to the core qualifications of a design process. Language-based tools, however, seem to play an inferior role in such a process. In this chapter, the aim is to examine their true role and potentials. This is based on the understanding that designing is a highly complex process. It is also believed that language as part of the design toolbox, for intuitive as well as for cognitive parts of the process, can significantly contribute to high quality design and, therefore, has to be part of education and subsequent design skills.

In the first part of this chapter, the design process and its different phases are examined. In a second step, the role and potentials of language and language-based tools in the design process are discussed. Further, the question will be raised about how such tools are taught. In the concluding part, the potentials of language-based tools and the options to integrate them in landscape architecture teaching are critically addressed.

When dealing with the fundamental characteristics of the design process and its tools, we can find various similarities between different design disciplines, especially between landscape architecture and architecture. The following debate, therefore, will also refer to positions developed within the discipline of architecture, design and design theory (Rittel and Reuter 1992).

Characteristics of the design process

How designing works, which skills are required and which tools are applied, is a controversially discussed topic. Skills are often put into two categories; one is described by terms such as analytical, rational and cognitive whereas the other is understood as intuitive and creative. There are different viewpoints and preferences concerning their relevance among professionals. This also becomes evident in the different phases of the design process.

A design work normally starts with a *phase of preparation* in which the design task and related problems are examined and relevant information is collected and analysed (Rittel and Reuter 1992: 75). Here, analytical approaches dominate. The *phase of finding ideas*, on the other hand, is the part that is often considered as a mainly intuitive approach, expressed by descriptions such as sudden inspiration ('Gedankenblitz') or enlightenment (Seggern and Werner 2008: 34). Following Archer (1965), the creative phase needs empathy, subjective assessment and deductive thinking. Steinitz (2008: 7) distinguishes between intuitive and analytical or systematic approaches in the phase of finding ideas while pointing out that, in practice, these two are often combined.

> The first way is anticipatory and deductive. You are sitting in the middle of the night, at your table, and you have an idea [. . .] The other way is explorative and inductive. You put together a set of issues and choices — a 'scenario'. A scenario is a set of assumptions and policies that guide you to the future.
>
> *(Steinitz 2008: 7)*

Within the design process, the developed idea has then to be *tested* using criteria defined by the designer. In this phase of reflection, the idea is checked concerning its fit and it will be adapted as far as necessary or even dismissed. According to Schön (1984), this check is necessary to critically examine the consequences of the defined order for the normative and theoretical dimensions of the individual profession. He mentions criteria such as usage, local conditions, spatial organisation, form, technology, costs, character or visualisation (Prominski 2004: 96). Based on the complexity of the process, each designer has to make numerous decisions, or, as Fischer (1991) puts it: 'designing means deciding'. Decision-making is a rational process, which Rittel defines as the determination of options and the estimation of their consequences. Based on this sequence, the decision is taken (Rittel and Reuter 1992: 65). The position of Peters (2004: 30) is similar: he understands the thought process as a 'process of argumentation', which can be influenced directly by the planner.

Finally, the design has to be *communicated*, explaining the decision process as well as the substantive idea. Making a design readable and understandable in order to make communication between all groups involved possible is seen as an intrinsic part of the design task (Seggern and Werner 2008: 34; Rittel and Reuter 1992: 91 and 142).

In order to understand which role language as a tool can play in such a design process, the characteristics of language have to be examined.

Characteristics of language and its role in the design process

Language is expressed in written and spoken form with each of the two having specific characteristics. Written language in comparison with spoken language requires a stronger reflection of the content to be expressed. This reflection allows for an explication of meanings and depiction of inconsistencies, which are easily drowned out in spoken language (Goody 1999: 299). On the other hand, speaking, as Schwemmer (2011: 52) points out, stimulates imagination and creates bridges between different contents and, by doing so, generates meaning. The general influence of language on the thought process is discussed controversially (Beller and Bender 2010; Pinker 1996; Chomsky 1977). However, there is universal consensus that it is crucial for cognitive development, meaning the internal as well as the external processing of thoughts (Goody 1999: 259).

As designing has to be understood as a complex thought process and language seems to be a crucial component in such a process, language must play a role in a design process. But how do we apply language in designing?

Language seems to play a key role in analytical phases of a design, respectively in the decision-making and reflection processes and, of course, in communication and discussion. This corresponds with the common understanding that language refers to the left hemisphere of the brain and, by doing so, refers to analytical thinking, order, abstraction and hierarchy but not to comparative, associative and emotional aspects (Gänshirt 2007: 64). In terms of design tools, Gänshirt and Strübing consider visual tools important for the development of form. Such visualisation expresses the unconscious, the indescribable, based on emotional and intuitive decisions. Verbal tools, on the other hand, would describe the invisible (background, context), which would help us to construct the meaning of a design. Both approaches, they emphasise, are needed in the design process and should be employed complementary in a target-oriented way (Figure 18.1) (Gänshirt 2007: 64, 103, 226; Strübing 2009: 19).

Although this suggests certain specific potentials of language in the design process, we observe different positions concerning the perception of its relevance and use. Again, the role of design personality and individual preferences seem crucial. Krasny distinguishes two main design types. One starts with immaterial aspects such as speaking, thinking, considering, reflecting before the first line is drawn. The other type starts the thought process by sketching, where the hand is guided by imagination (Krasny 2008b: 9). However, as other statements from design practitioners show, this distinction is not just related to the initial steps of the

Figure 18.1 The comparison of right and left brain hemisphere with regard to design processes (Gänshirt 2012: 63)

design process. Own thoughts, according to visually oriented representatives, would have to be expressed visually. Words would not allow doing so.

> I wondered about the fact that when we are doing architecture, an architectural project, we are not thinking with language. I think through schemas and impressions; I draw fragments; I glue a couple of photography [sic]. [. . .] I am not in a thought that could be spoken, that needs to be spoken and then translated into a form.
>
> *(Portzamparc et al. 2008: 20f)*

A similar position is emphasised by Olin, who says that the development of spatial forms has to be developed out of forms and cannot be generated from words. 'Words can describe physical forms, but they do not (or did not) originate them; nor can they perform operations upon them' (Olin 1988: 155).

As much as language as design tool seems to be rejected by some, it is favoured by others. UN STUDIO describes their design process as communicative practice where teams design. Inspiration would rather not evolve from sketches but from communicating its content with words to others (van Berkel in Krasny 2008b: 121f). Hermann Czech (in Krasny 2008b: 34ff) describes communication in his office as being primarily with words, but also with drawings. The library, for him, is an omnipresent tool (Figure 18.2). Adolf Loos even says that drafts would not need drawings but that good architecture could be clearly described by words (Loos 1983: 85). The work of the architects Denise Scott Brown and Robert Venturi is based on thoroughly edited and carefully worded texts without professional jargon. Precise wording would contribute to clarification of the design. The clearer the design can be described by words, the clearer gets the imagination of the design and how it has to be developed (Strübing 2009: 18). Niemeyer uses a similar approach. For him, the lack of arguments to explain his design is a hint that the design is not there yet (Gänshirt 2007: 131–133).

The different positions make clear that language is used as a design tool throughout all phases of the design process. The diverse appreciation and use of language in design practice might relate to the different existing cognitive styles, with visualisers thinking with images and verbalisers thinking with words (Mayer and Massa 2003: 833). Considering that language is not just a crucial tool in the communication of a design but that, for one part of the designers,

Figure 18.2 Photos and talk of the visit in the office of Hermann Czech in Vienna by Elke Krasny (2008a)

language-based design approaches seem essential, we then have to ask how far this is considered in education. Do our curricula address language as design tool and how do they respect these different design personalities?

Language as design tool in the curriculum

Education in landscape architecture has to address manifold skills and tools needed in the discipline. Language-based skills, however, are barely addressed. Even communication skills are hardly addressed explicitly in the curriculum. Oral presentations might be trained for at some point throughout various courses, whereas writing skills are rarely taught at all. Rather, they are seen as a basic skill, which was taught at school already, enabling the students to apply writing in their later profession. Students with difficulties in written expression might be asked to take additional electives in writing (offered at basic schools or similar institutions). Further, 'WritingAcrossCurriculum' programmes, installed at various universities, especially in the US, might have raised the awareness of the relevance of writing in all disciplines, but still these courses are electives. But, as Clarke points out, translating visual experiences into spoken or written verbal statements is difficult.

> Much like translating or interpreting any one written or spoken language into another, these skills can be learned, but they must first be identified, examined and evaluated so that an appropriate choice can be made from the available options to suit the specific requirements of a particular visual experience and its verbalisation.
>
> *(Clarke 2007: 8)*

So, as much as this seems needed, teaching communication skills is a rather forgotten aspect in most design curricula. This is even more true for any language-based approaches related to the phase of developing a design idea. Language-related tools are traditionally assessed extremely low by many professionals – in teaching as well as in practice. Some instructors even assess language-based approaches as counterproductive.

> Many design instructors believe that an emphasis on written communication within the context of design instruction might hinder the development of basic design skills. [. . .] Design instructors often suggest that only when courses allow the visual and intuitive side of the brain to take over will students be able to generate truly open, creative options. Many believe that if the left brain's emphasis on verbal labelling is dominant, only traditional design solutions can emerge because the mind cannot be set free from standard, verbal definitions and rigid categories of thinking.
>
> *(Martin 1992: 1)*

Although there seems little awareness of the need for teaching language as a design tool in general, literature as well as our own teaching experiences and research give strong evidence on language potentials and how related tools can be addressed in teaching.

Some empirical findings

A well-documented research project on language as design tool was done by Roger Martin (1992), a professor in the landscape architecture programme at the University of Minnesota.

Being aware of the scepticism on that in a highly visual field of design, he wanted to find out if writing could improve the ability of students to design projects in landscape architecture. The empirical study addressed three essential phases of a design process, 1) observation and problem definition, 2) creative idea making and 3) idea communication. So far, language-based tools had not played a role in the studio work. As a literature search on available writing tools in design education showed them to be rare, Martin developed different exercises for the three phases, which were adapted from the field of composition as well as from design education. In addition to the exercises, the students were asked to keep a journal including visual as well as verbal notes.

Martin (1992) reports that although the designs based on the exercises were not superior, the targeted creative writing exercises could enhance not just the writing competences but also the design competences of students. 'We found that writing exercises, when carefully integrated into a design course project assignment can raise the quality of related design skills' (Martin 1992: 55). The study showed an increased level of observational skills, clearly attributed to the writing exercises (Martin 1992: 53). Participating students also saw a positive influence on their thought process. 'Many appreciated the workshop because it helped focus their design thinking and helped them become more open about design possibilities' (Martin 1992: 38). Applying writing in the design process could even change the perception of visually oriented participants of the study, as Martin states. 'Many students, being visually oriented and not interested in verbal communication, began to see writing as a "powerful design tool" for the first time' (Martin 1992: 19). Based on the findings from his empirical study, Martin concludes that writing could obviously play a valuable role in design process skills such as form-giving and place-making, and that verbal and visual skills would not need to be separated in order to create high quality design solutions (Martin 1992: 2).

This is supported by other teaching experiences in Europe and the US. Gulwadi (2009) reports on the application of reflective journals in teaching in order to enhance the reflection on complex design tasks such as sustainability. The results of the design studio indicated that, by means of the journals, an in-depth thought process was created as a valuable starting point for the design development. Erickson (1996) used an intensive semester-long writing exercise in a graduate landscape architecture and planning course. Based on the results, she concludes that writing is not merely a communication tool but acts as a valuable discovery technique, supporting students' pervasion of a design task. Similar experiences were gained by Van Haaren and Schmidt-Kallert (2015). In an interdisciplinary seminar between urban planning and art students, they used freestyle drawings and writings as an approach to spatial analysis. The creative use of words and drawings allowed the students to widen their perception and cognition of urban settings. Lappin et al. (2015) see writing also as a key skill in the generation of spatial design ideas. Based on the awareness of this potential, they argue for a training of writing skills throughout the whole curriculum, especially at undergraduate level. In their undergraduate 'WritingArchitectureProject', they applied a variety of assignments and methods to train writing. Several years of teaching experience with this approach showed a positive influence on the students' cognitive skills of description, analysis and synthesis.

The examples illustrate the value of language-based tools (focussing on written language) for various steps in the design process and different approaches for how to implement them in design courses.

Evidence that students actually apply language in their design processes, even if it is not explicitly taught as a design tool, is given by an assignment in a site design studio in landscape architecture co-taught by Doris Gstach.[1] The students were asked to document their design process in order to reflect on the methods they apply to solve a design problem. Most students referred to the common phases of a design process as described above. Some mentioned language-related approaches such as diagrams (functional, concept), noting keywords, brainstorming or simply the activity 'think'. Notes seem relevant to most of the students, especially in the phase of analysis where they combine them with sketches or other forms of visual notes. Some also refer to language-based approaches when they address the step of developing a design idea, such as 'jotting down phrases', 'brain dump: write down any ideas that come to mind' or the reflection of the design 'concept: what does the design mean – the message presented by the design'.

Further evidence on students' awareness of the value of language-based tools in the design process is given by an empirical study, done by the authors during winter semester 2012–2013, at four German universities (SRH University Heidelberg, University of Applied Sciences Erfurt, University of Applied Sciences Darmstadt and the University of Kassel). Although the study involved students from other disciplines besides landscape architecture, such as architecture and urban planning, the results give an idea of the students' perception in the field of spatial design.

Using a standardised questionnaire, 125 students were asked about the relevance of a variety of design tools, including the students' personal design approach and their opinion on the competences needed for designers, the media needed to understand a design concept and the tools to communicate design concepts to others. Among other possible approaches, language-based tools were regularly addressed in the survey. The results from the survey support the above discussed aspects. Language-based tools are mainly seen as relevant in the analysis (for example by making notes). Not surprisingly, sketches and drawing ranked highest. But also creative language-based tools such as brainstorming were ranked quite high (Figure 18.3). The relevance of thinking within the design processes was considered very important (mean value 1.38), whereas language was ranked quite a bit lower (Figure 18.4) (for further details on the results of the study see Gstach and Kirschbaum 2016). Altogether, the relegation of language-based tools behind visual tools was quite moderate and much less than expected. Although it becomes obvious that most students have not really reflected on their own design processes so far, they intuitively assessed the language-based tools as somewhat important and having potential for their own design work.

That this assessment can change when students are introduced to language-based design tools is shown by a second round of the survey, done after two little design tasks applying two verbally-oriented design techniques ('manipulative verbs'[2] and 'story telling',[3] according to the exercises developed by Martin 1992), done by 26 of the students in Kassel and Erfurt. These students did the questionnaire again after the exercises. The results show a significantly higher estimation of language within design processes. This result could be traced back to the phenomenon of a generally higher estimation of familiar approaches. They had not been aware of this potential because it is rarely addressed explicitly as part of the curriculum in university design studios. Maybe introducing non-language-based approaches would have had the same effect, but still, as Rittel and Reuter (1992) observed, the inner coherence of thinking and language seems to be an important facet within the complexity and heterogeneousness of design processes.

Conclusions and outlook

Spatial design is a complex process that is not just highly demanding in terms of organising the different steps but also asks for a complex thought process, which demands a fruitful combination of creative and analytical approaches. As a result of the literature review and our own empirical studies, it becomes clear that language-based tools can be a useful addition to this process and the traditionally visually focussed toolbox in landscape architecture. This might be especially needed in projects of growing complexity, as we face them today. The enlargement of the toolbox would also support the different cognitive types of personalities and their talents and ways

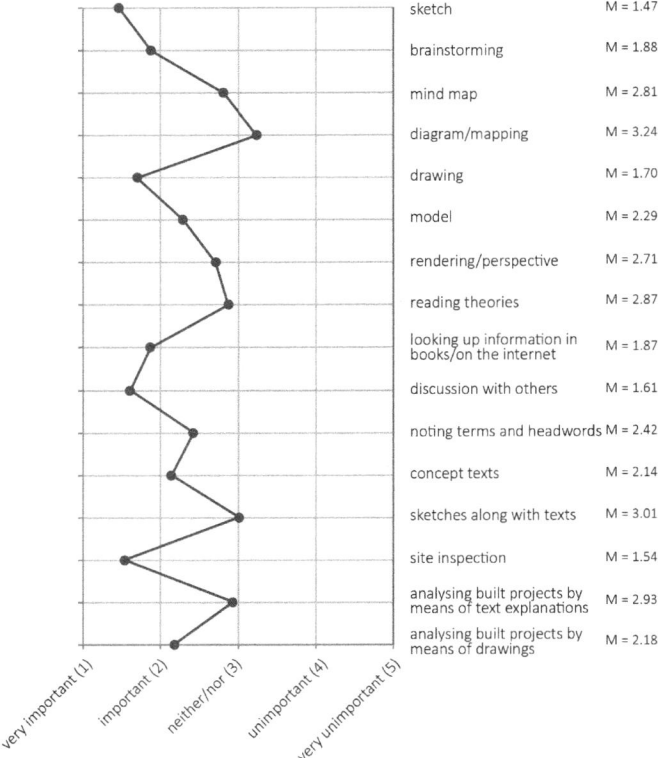

	M =
sketch	M = 1.47
brainstorming	M = 1.88
mind map	M = 2.81
diagram/mapping	M = 3.24
drawing	M = 1.70
model	M = 2.29
rendering/perspective	M = 2.71
reading theories	M = 2.87
looking up information in books/on the internet	M = 1.87
discussion with others	M = 1.61
noting terms and headwords	M = 2.42
concept texts	M = 2.14
sketches along with texts	M = 3.01
site inspection	M = 1.54
analysing built projects by means of text explanations	M = 2.93
analysing built projects by means of drawings	M = 2.18

very important (1) important (2) neither/nor (3) unimportant (4) very unimportant (5)

Figure 18.3 Student survey: Question 1: Within design processes, ideas and concepts need to be developed and different decisions need to be made. How do you assess the relevance of the following tools for your own ideation and decision-making process? (N = 125) (Gstach and Kirschbaum 2016)

of thinking, since it would offer different approaches among which the students could choose. Further, we have to recognise that language plays a more relevant role in landscape architecture than is commonly perceived or admitted. Professional reality demands even more analytical than creative skills. This would also speak for more language-based tools, even if we saw creative parts as being mainly based on visual tools. That this falls short was pointed out above.

Although the examples show that language-based tools are valued by certain colleagues in the field of design education and are, therefore, addressed in teaching, corresponding courses are still far from common in landscape architecture programmes. It seems necessary to invest in further research. More attention has to be given to the design competences taught at universities, because they constitute the toolbox the students will refer to in their professional work afterwards. This asks for further knowledge and empirical evidence on the potentials and relevance of language-based tools in the different phases of a design process and, subsequently, in the development and application of related didactic tools. Approaches such as the ones briefly addressed above might be a helpful starting point. Further research is also necessary concerning the various design approaches applied by professionals, first to make design processes more transparent and, second, to raise the awareness of further, also language-based, tools available to create a successful and reflected design process.

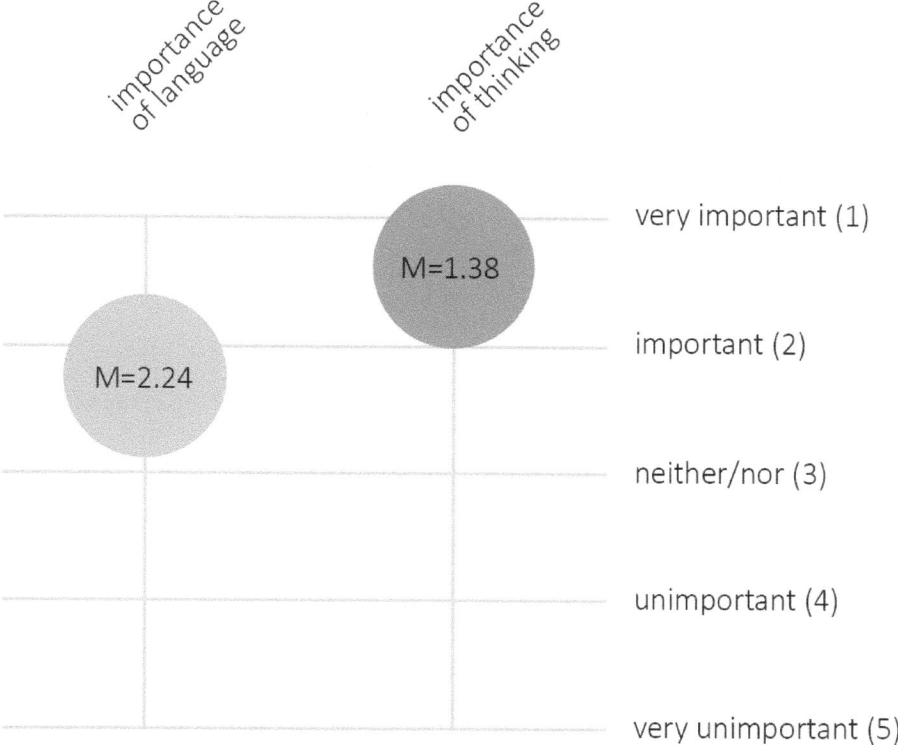

Figure 18.4 Student survey: Question 5: In your opinion, how important is thinking (in terms of pondering, reflecting) within design processes? (N = 125). Question 6: In your opinion, how important is language (spoken and written) within design processes? (N = 125) (Gstach and Kirschbaum 2016)

Notes

1 Site Design Studio Spring 2010, Master Program of Landscape Architecture, Clemson University/USA, co-taught by Doris Gstach, Mary-Beth McCubbin, Galen Newman.
2 Manipulative verbs: the students first arbitrarily select a number of active verbs, then choose a few of them. Then they write phrases that describe the effects on the design site if the verbs were applied there. In addition to this, the students develop a little sketch that visualises the outcome of the verbs' application (see Martin 1992: 19–29).
3 Storytelling: the students are asked to write a narrative in which they imagine what might happen to people who inhabit the design site. The site is described with current or imagined attributes that enhance people's activities. In this exercise students are forced to imagine people's possible needs and behaviours (see Martin 1992: 39).

References

Alexander, C. (1964), *The Synthesis of Form* (Cambridge, MA: Harvard University Press).
Archer, L. B. (1965), *Systematic Method for Designers* (London: The Design Council).
Beller, S. and Bender, A. (2010), *Allgemeine Psychologie - Denken und Sprache* (Göttingen: Hogrefe).
Chomsky, N. (1977), *Reflexionen über die Sprache*, 1st edition (Frankfurt am Main: Suhrkamp).

Clarke, M. (2007), *Verbalising the Visual: Translating Art and Design into Words* (Worthing: AVA Publishing SA).

Erickson, D. L. (1996), 'Deserving a wider audience: an interactive process for graduate student writing in landscape architecture and planning', *Journal of Planning Education and Research* 16/2: 137–144.

Fischer, G. (1991), *Architektur und Sprache: Grundlagen des architektonischen Ausdruckssystems* (Stuttgart: Krämer).

Gänshirt, C. (2012), *Tools for Ideas: Introduction to Architectural Design* (Berlin: Birkhäuser).

Gänshirt, C. (2007), *Werkzeuge für Ideen: Einführung ins architektonische Entwerfen* (Berlin: Birkhäuser).

Goody, J. (1999), *The Interface between the Written and the Oral* (Cambridge: Cambridge University Press).

Gstach, D. and Kirschbaum, M. (2016), 'Language as design tool: an empirical and design perspective in the field of architecture and planning', *Architectural Science Review* 59/6: 465–473.

Gulwadi, G. B. (2009), 'Using reflective journals in a sustainable design studio', *International Journal of Sustainability in Higher Education* 10/1: 43–53.

Krasny, E. (2008a), *The Force is in the Mind: The Making of Architecture* (Basel: Birkhäuser).

Krasny, E. (2008b), *Architektur beginnt im Kopf: The Making of Architecture* (Basel: Birkhäuser).

Lappin, S. A., Gül Kaçmaz Erk, G. and Martire, A. (2015), 'ArchiBabel: tracing the writing architecture project in architectural education', *International Journal of Art & Design Education* 34/2: 224–236.

Loos, A. (1983), *Die potemkin'sche Stadt. Verschollene Schriften 1897–1933* (Wien: Prachner).

Martin, R. (1992), *Writing in the Design Disciplines* (Minneapolis, MN: Center for Interdisciplinary Studies of Writing, University of Minnesota).

Mayer, R. E. and Massa, L. J. (2003), 'Three facets of visual and verbal learners: cognitive ability, cognitive style, and learning preference', *Journal of Educational Psychology* 95/4: 833–846.

Olin, L. (1988), 'Form, meaning, and expression in landscape architecture', *Landscape Journal* 7/2: 155–157.

Peters, S. (2004), *Modell zur Beschreibung der kreativen Prozesse im Design unter Berücksichtigung der ingenieurtechnischen Semantik* (Duisburg, Essen: Universität).

Pinker, S. (1996), *Der Sprachinstinkt. Wie der Geist die Sprache bildet* (München: Kindler).

Portzamparc, C. de, Sollers, P. and Tihanyi, C. (2008), *Writing and Seeing Architecture* (Minneapolis, MN: University of Minnesota Press).

Prominski, M. (2004), *Landschaft entwerfen. Zur Theorie aktueller Landschaftsarchitektur* (Berlin: Reimer).

Rittel, H. W. J. and Reuter, W. D. (1992), *Planen, Entwerfen, Design. Ausgewählte Schriften zu Theorie und Methodik* (Stuttgart: W. Kohlhammer).

Rittel, H. W. J. and Webber, M. M. (1973), *Dilemmas in a General Theory of Planning.* Reprint no. 86 (Berkeley, CA: Institute of Urban and Regional Development, University of California).

Schön, D. A. (1984), *The Reflective Practitioner: How Professionals Think in Action* (Aldershot, UK: Ashgate).

Schwemmer, O. (2011), *Das Ereignis der Form. Zur Analyse des sprachlichen Denkens* (Paderborn: Wilhelm Fink Verlag).

Seggern, H. von and Werner, J. (2008), 'Designing as an integrative process of creating knowledge', in H. von Seggern, J. Werner and L. Grosse-Bächle (eds), *Creating Knowledge. Innovation Strategies for Designing Urban Landscapes. Studio Urbane Landschaften* (Berlin: Jovis), 34–62.

Steinitz, C. (2008), *On Scale and Complexity and the Needs for Spatial Analysis. Proceedings, Expert Conference on Spatial Concepts in GIS and Design* (Santa Barbara, CA: University of California).

Strübing, C. C. (2009), *Sprache als Werkzeug im Entwurf* (Muttenz: MAS Design/Art+Innovation. Fachhochschule Nordwestschweiz FHNW, Hochschule für Gestaltung und Kunst).

Van Haaren, B. and Schmidt-Kallert, E. (ed.) (2015), *Schreiben und Zeichnen als Erkenntniswege im Städtebau* (Essen: klartext).

19

Writing across the landscape architecture curriculum

Kasia Gallo

While landscape architecture relies on seeing and experiencing more than on speaking or reading, no one can deny the importance of the written word to the profession. Requests for proposals or qualifications, professional correspondence throughout the design and construction phases, public relations releases, post-occupancy evaluations, theoretical analysis, movement manifestos, or peer-reviewed scientific literature focussing on the built environment all require excellent writing. As landscape architecture students must master the discipline's basics in a relatively short time, writing may be de-emphasised in school. However, incorporating writing into landscape architecture programmes in a planned and structured way offers a multitude of benefits. Writing instruction prepares the students for the day-to-day reality of professional practice, complements the means of the designer's expression, and promotes a steady voice of the profession in communicating with the public (Thompson 2006a; Thompson 2006b; Thompson 2009). Additionally, writing increases critical thinking, fosters active learning, catalyses an acquisition of professional identity and, naturally, improves one's writing ability. This chapter outlines reasons for incorporating writing into landscape architecture curricula, showcases writing-based pedagogy efforts in design programmes, and suggests evidence-based writing-centred activities for project and lecture-based classes.

Writing in landscape architecture programmes: challenges and opportunities

Design students' writing abilities vary; while some are terrific writers, others struggle with grammar, structure, or word clutter. Many design instructors may be apprehensive about focussing on writing, given the breadth of discipline-specific knowledge students must master during their schooling. Others wish to improve student writing but are unsure how to proceed as they lack background in writing instruction and assessment. Both apprehensions are reasonable; luckily, both can be addressed.

First, engaging students in writing – even without worry about grammar and punctuation – facilitates learning, thinking, and growing as designers. Such 'unofficial' writing tasks include annotating sketches, reflecting on one's design process, analysing theoretical or ethical design issues, or pondering qualities of the natural or built environment. Second, while design faculty may lack confidence to address style or grammar, many universities have facilities to help students master the basics of writing, reducing the pressure on design faculty to incorporate detailed writing instruction into their courses. These resources include university learning centres, writing centres, or tutoring/editing sessions at native language departments. Various writing assignments can be used in all class formats including lecture-based, studio-based, and design-build. A detailed description of each writing modality follows. See Table 19.1 at the end of this chapter for a summary of writing categories and their usefulness in landscape education.

Sketching and writing

A typical design studio buzzes with excitement and creative energy. Students sketch, build models, and generate design and construction drawings, often developing multiple schemes for each project. In the fervour of designing, writing is sometimes de-emphasised. However, rather than subtracting from the process, writing-based practices strengthen it.

Sketching is an essential component of all design education. It serves multiple purposes: to record existing conditions during inventory, to graphically portray idiosyncratic relationships of existing conditions during analysis, and to present ideas during design phase and beyond. Interestingly, the word 'sketch', per the Merriam-Webster Dictionary, has multiple meanings: it can refer to visual representations of 'chief features of an object or scene', and also a words-based 'tentative draft' or a 'brief description'. This duality is well reflected in research analysing various aspects of design sketching; it also appears that combining the two modalities enhances design process. This is not surprising. Writing and drawing are both iterative processes, revised for content and quality. They evolve through multiple drafts, and serve as an externalised mode of idea generation and elaboration. While some design ideas or relationships can be drawn, others need be written to be remembered.

Interestingly, the quantity of text may relate to the quality of the sketch. Sun, Xiang, Wang, and Shao (2013) studied the sketches of twenty-one industrial design students. Participants designed, for example, a seating element, and considered how it would 'act/behave' in a park. The researchers used Sketchpaint software with a pen-like stylus, and observed participants 'sketching behaviours' (i.e., aimless vs. deliberate lines, or hesitation while drawing/noting the sketch). The authors reported that the amount of text accompanying the sketches positively correlated with the overall quality of the drawing: the text-heavy images featured increased precision and a degree of deliberateness. In other words, they were less 'rough'. More importantly, two independent raters indicated that the text-heavy drawings presented more original solutions than the sketches with less text.

While Sun, Xiang, Wang, and Shao (2013) did not speculate on why the presence of verbal description coincided with deliberateness and creativity, these features can be explained through the lens of cognitive science. Writing takes more time than thinking or fast sketching. It slows the designer down, potentially promoting deeper analysis and synthesis of ideas. Such a deliberate expression of thought may result in a reduced need for cognitive closure, or 'jumping to conclusions'. Torrance (1966) defined creativity in terms of Fluency (number of responses), Flexibility (categories into which they fall), Originality (how different they are from most commonly proposed solutions), and Elaboration (the degree of detail of a response). The findings of Sun and colleagues suggest that writing in conjunction with sketching enhanced Elaboration and Originality of proposed solutions. This has been observed by others (i.e., Soygenis, Soygenis, and Erktin, 2010).

Others observed that while Originality appears to positively correlate with domain experience, both novices and experts generate more original solutions when writing prior to sketching.

Participants of Sun, Xiang, Chai, Wang, and Liu's (2013) study (ten industrial design students with about four years of sketching experience on average, and eight novices from other departments with no prior sketching experience) were randomly assigned to two groups, both tasked with designing a chair and explaining their design rationale. Group One thought about the solution prior to sketching; Group Two wrote their ideas down before commencing drawing. Novices and experts generated a comparable number of sketches, however the expert sketches featured more original elements than novice sketches. Additionally, both experts and novices generated a higher number of creative ideas when their drawings were preceded by verbal descriptions. Interestingly, while Fluency increased, the solutions preceded by text featured poorer spatial relationships than solutions drawn without verbal descriptions. This may indicate that while the 'quick' style of writing helps with idea generation, another writing modality (perhaps analytical or reflective writing) may be more appropriate for subsequent design development.

To summarise, writing and sketching correlates with a higher quality of sketching. It appears to stimulate idea generation, and to increase the creativity of generated solutions, as expressed by number and quality of unique features. This suggests that faculty should consider incorporating some writing before design idea generation. However, investigating and expressing ideas relating to spacial qualities of proposed solutions may be best done in an elaborate verbal form (especially by novices).

Reflective writing

William Faulkner famously said: 'I never know what I think about something until I read what I've written on it'. Reflective writing, or writing that illuminates an individual's thinking, is an invaluable learning tool. It helps in tracking the development of expertise and professional identity in students, exposing and correcting misconceptions about a domain, and monitoring design thinking during any design endeavour. Sketchbooks provide means for in-studio reflection; entries can be shared out loud by volunteers. Alternately, short reflective essays can be completed by students on looseleaf paper or small notecards and turned in for a grade, in studio or lecture-based classes.

It is well known that project-based teaching results in higher critical thinking gains than lecture-based teaching (Bean 1996; Zoller 1993). Yet, without a written explanation, this thinking may go unnoticed or underappreciated in the case of design projects. Students sometimes write their design project statements at the last minute and fail to successfully lay out the thinking that drove their design process. Therefore, students may benefit from tracking their design's evolution in reflective project journals, perhaps in the form of a blog, shared with the classmates (Schon 1986).

Reflective writing is not without controversy. Students may hesitate to expose their lack of understanding or difficulties in learning. They worry this may demonstrate weakness, and could subconsciously bias the instructor while grading students' work. To combat this, it is essential to frame the question in a non-threatening way. For example, an instructor can ask 'what was the most important thing you learned during this exercise?' or 'what will you keep working on during the next project?' The grading of reflective writing is contentious too. Excessive emphasis on grammar, spelling, and linguistic clarity will limit the amount of reflection from poor or self-conscious writers. Assigning more than a nominal point value to each entry will bias students to write what they believe the instructor wants to hear. It appears that assigning some points to each entry helps students to take the assignment seriously; keeping the points low assures that students do not hesitate to say what is truly on their mind (Cannady and Gallo 2014).

In summary, reflective writing offers a peek into aspects of a student's thinking, which may otherwise go unnoticed (to both the student and the instructor). To serve its purpose, the writing should be completed and reviewed with an understanding that it is a helpful 'thinking tool', and not a high-stakes item to make or break one's grade in the course.

Writing to learn

Design students are accustomed to active learning, or academic activities that foster consideration of what one is doing and why (Bonwell and Eison 1991). Therefore, incorporating writing-based active learning activities presumably would not excessively challenge this population. Cannady and Gallo (2014) tested this assumption by converting an undergraduate, lecture-based, twenty-two-student Urban Planning class into a writing-focussed active learning experience, incorporating nineteen written assignments into the course. The assignments ranged from one-minute papers and analytical essays, through audience-specific letters, to short poems. Additionally, students completed regular at-home essays.

One-minute papers are short written responses to questions posed by the instructor during class. The answers, written in full sentences or bullet points, served to check class preparedness, and gave students time to gather and articulate their thoughts. These may be graded, used as an attendance check, or swapped with neighbours for comparison of ideas (Lucas 2010). Aside from educational gains, these may serve as tools for increasing immediacy, or the feeling of connection between a student and an instructor, expressed with verbal and non-verbal cues (Mehrabian 1968). While immediacy may be quickly established in the design studio because of repeated and extensive interactions, it may be absent in larger lecture-based classes.

Analytical essays written during the last fifteen-to-twenty minutes of each class by the Urban Planning class forced learners to consider and apply the content of the given day's lecture. For example, when a member of the city's Board of Aldermen (an administrative body responsible for planning and design decisions among others) presented local policy impacting design and maintenance of the built environment, students were tasked with identifying a local design or policy problem, proposing a solution, and writing their thoughts in a professional letter format. When the day's lecture centred on the impact of design on social issues, students responded to the following question: 'can planning and design foster diversity and promote racial, cultural, and class equality? If yes – how?' The latter prompt resulted in varied responses, and a lively discussion during the next class period.

Two assignments challenged the students more than others: the aforementioned letter to an alderman, and a series of haiku poems written to capture the sense of place of urban environments presented on slides. The letter assignment provided practise in expressing design thinking in a formal manner, not unlike what design professionals do daily. The haiku assignment was a bit tongue in cheek, and aimed at distilling an essence of a place without long-winded explanations. The authors suspect this perceived increased difficulty was associated with a specific form requirement, in addition to a content requirement.

When asked for their opinion on the various types of writing they had completed in the class, students commented on several things. They noticed, without prompting, that they had to generate and contemplate diverse ideas regarding real-life urban design problems, thus exercising their critical thinking abilities. They claimed their writing improved, as did their ability to quickly and succinctly formulate and express ideas (Cannady and Gallo 2014). One of the most memorable comments included on the survey was 'thank you for valuing our thoughts'. This suggests that interesting, relevant, and somewhat challenging prompts, combined with individualised feedback from the instructor, may result in enjoying writing instead of dreading it.

Writing to learn activities can be used in all class types, and serve many pedagogical purposes. Two applications warrant particular attention: mastering the design jargon, and understanding the construction of analogies and metaphors. These are addressed in greater detail below.

Writing to learn: mastering Archispeak

While many landscape architecture graduates become practitioners, some also teach and/or conduct research. An ability to read, understand, and write scientific literature benefits all. Scientific peer-reviewed literature fosters an understanding of existing work and disseminates new information. Such literature highlights the profession's best practices, including sustainable stormwater management techniques, historic landscape preservation guidelines, and others. To master scientific writing, students must be able to understand the scientific literature of their domain, learn the discipline-specific formatting style, and write clearly and grammatically. Additionally, teaching scientific writing aids the mastery of professional jargon. Instruction in research and scientific writing can take place in a stand-alone class, or be incorporated into a theory or a studio course. While writing-centre staff and research librarians can help to familiarise students with database search skills and formatting style guidance, design faculty must help design novices navigate the discipline's specialised language.

Hirst (2003) differentiated between 'good' and 'bad' professional jargons. Jargon is 'bad' when it is used without defining the terms to a lay audience, or when the terms are poorly constructed (i.e., incorrectly translated from a foreign language). Jargon is 'good' when specific and well-formed terms are used with fellow practitioners, or used with adequate explanation in communication with the discipline's outsiders. However, Hirst emphasised, even the proper terminology used with an expert audience can be 'bad' if it creates excessive brain strain in listeners. Therefore, faculty should pay attention to their jargon use around students, modelling and differentiating design lingo use when dealing with students of different levels. This can be accomplished by incorporating scientific writing into the design curriculum.

Per the mocking definition in *Urban Dictionary*, Archispeak is

> Large, made-up words that architects and designers use to make themselves sound smarter than you (you being the client or the confused observer of design). It does nothing to inform or enlighten the consumer of architecture and mostly serves to numb them into obedience or self-doubt.

(www.urbandictionary.com)

While some Archispeak expressions may create barriers to communication, many words or phrases are instances of design specialist language, for example 'urban fabric', 'light pollution', or 'acoustic glare' (Gallo 2016). Integrating writing consciously into landscape architecture design programmes allows a focus on multiple issues surrounding the design jargon: mastering the technical terms, establishing group belonging, and acquiring the judgement skills to know when to use Archispeak and when to abstain (Glock 2009).

Mastering Archispeak is important for students: it is well known to sociolinguists, cognitive scientists, and educational psychologists that professional jargons (examples of an in-group speech) serve important roles in asserting group belonging and professional identity (i.e., Morrison, Decety and Molenberghs 2012; Woodward-Kron 2008). While a substantial body of literature exists on professional jargons in various professional training programmes (i.e., medicine, education), scientific peer-reviewed literature on Archispeak is lacking. However, one can assume that the findings are universal: gaining expertise in an area correlates with increased usage of the field's jargon; improper usage of the discipline-specific language, by either practitioners or lay public, results in confusion and mistrust.

Including verbal components in inventory and analysis tasks, in addition to writing project statements/concept descriptions, provides a perfect opportunity to both introduce design jargon to students and to encourage and monitor its proper use. This could be amplified by engaging students in peer editing (but not grading) of each other's work, then revising based on peer comments. Students are less intimidated by comments penned by fellow students, and less likely to repeat the mistakes they note in the writing of others (Crossman and Kite 2012). It is helpful to teach students how to provide feedback to each other. Statements like 'good job' are not constructive as the author does not know what he or she did well. Informative feedback should include justification of both praise and criticism, for example, 'good explanation of reasoning behind your choice of materials for the courtyard'. Additionally, students need not edit each other's writing; pointing out problems is sufficient. For example, a constructive peer may say 'this sentence is too long and hard to follow; divide into smaller units and simplify'. Ideally, peer review is implemented and comments are addressed before an assignment is turned in to the instructor.

Writing to learn: Archispeak and metaphors/analogies

One aspect of design writing that deserves a special mention, and may benefit from structured instruction, is the forming of metaphors and analogies (Caballero 2003; Rodriguez 2015). The success of 'good' metaphors and analogies is rooted in cognitive science; people capitalise on analogical thinking when learning new things (for example, when learning new languages, or figuring out how to use public transport in a foreign country – based on previous experiences). Calling a building addition 'a parasite' is a result of analogical thinking (Gallo 2016). Such a metaphor may be addressing the aesthetic of the structure; perhaps the addition seems to be forcefully disrupting the original building's rhythm or clashing with the old materials. It can also indicate excessive use of the original building's resources, perhaps putting strain on a now-inadequate HV/AC system or water supply.

Analogies (of which metaphors are a special case) are made up of two parts: a source and a target. In simple terms, 'something' is like 'something else'. If the similarities pertain exclusively to how things look, the analogy is weak. If the similarities are drawn based on how things work, the analogy is much more powerful and helpful. 'Building skin' is a pretty good structural metaphor; computer mouse is a rather weak one, though we all use it quite ably (Holyoak and Thagard 1997). Maya Lin explained her concept for the Vietnam War Memorial by saying: 'I imagined taking a knife and cutting into the earth, opening it up, an initial violence and pain that in time would heal' (Lin 2000: 2). In this example, the black reflective marble form of the memorial, emerging cleanly from the ground, was the target, while the violence, pain, loss, and, in time, healing, of the Vietnam War were the source.

In some cases, a design concept may not be fully reflected in the student's final product. In extreme cases, the concept is an afterthought formulated at three in the morning, the night before the project is due. Perhaps guiding novices through the mental steps of generating analogies would help students to conceptualise and process design ideas with greater ease. Structured training in analogy and metaphor formation may be accomplished through writing exercises, and has a potential to vastly enhance formulating project concepts. The training would begin with an introduction to cognitive aspects of analogical thinking, and differentiating between analogies and metaphors (an atom is like the solar system vs. living-machine buildings). Students then could learn and practise the identification of the key terms (surface vs. structure analogies; source vs. target), and assess several analogies and metaphors presented by the instructor. Finally, the training could end with generating and discussing original analogies and metaphors by students.

While landscape architecture relies heavily on the visual to convey metaphors and analogies, learning how to form them may be achieved most easily through writing. This is so because, while new to design, university-level students have been expressing their thoughts in writing for at least a decade and a half. Forming and mastering metaphors and analogies can be demonstrated through explicit instruction and writing exercises. These could include analytical writing based on presented images ('what does this structure look like?'; 'analyse how a drawbridge is like a wrist'), reflective writing ('to you, does this space feel like a womb or like a tomb?'), or creative writing ('describe an experience of an ant moving through an anthill, from an ant's point of view'). Once the linguistic concepts are in place, the transition to understanding and exploring spacial qualities in three dimensions may be eased.

Table 19.1 A summary of the writing types used in landscape architecture education

Writing type	Description and benefits	Use
Reflective writing	Brings thinking to light; metacognitive in nature. Contains analysis and synthesis. May take the form of a journal, blog, or stand-alone assignments. Assignments can vary in length, but should be substantial enough to allow for nuanced explanation of student's thinking processes. May overlap with writing to learn by inclusion of analysis and synthesis of design/theoretical issues in addition to metacognition.	Appropriate in all class types: studio, lecture, and design/build. May be helpful in learning the design jargon. Prompt examples: What is your design philosophy? What was the hardest part of the last design project for you? What have you learned from it? How did you study for the last test? Were your strategies effective?
Sketching and writing	Compliments graphic/visual information. Varies in length from single words and bulleted lists to full sentences or longer forms. Appears to enhance creativity (esp. originality and elaboration) in design endeavours.	Appropriate in all class types: studio, lecture, and design/build. Writing may be utilised before, after, or during sketching. The order of writing and sketching, and the type of writing, may influence the characteristics of the final product.
Writing to learn	A category of active learning. Encompasses a wide range of activities, including analysis and synthesis; may overlap with reflective writing. May take many forms, from single-sentence responses, through one-minute papers, and short-answer responses to full-length scientific manuscripts.	Appropriate in all class types: studio, lecture, and design/build. Very helpful in learning the design jargon, and mastering analogies and metaphors.

Conclusions

Incorporating writing into landscape architecture classes can take many forms, from short one-sentence activities to bona fide publishable peer-reviewed research papers. Benefits are many: better communication skills, increased sense of professional identity, deeper thinking about the discipline and design process, and an increased ability to meaningfully contribute to the field's theoretical and practical knowledge base. Faculty need not hesitate to incorporate writing assignments due to lacking knowledge of writing instruction; help is readily available at the institutions of higher learning.

Good writing promotes good thinking; good thinking results in great design. Landscape architecture needs good designers and writers to promote the discipline to the world at large, in both spatial/visual and written forms.

References

Bean, J. C. (1996), *Engaging Ideas: The Professor's Guide to Integrating Writing, Critical Thinking, and Active Learning in the Classroom*. The Jossey-Bass Higher and Adult Education Series (San Francisco, CA: Jossey-Bass Inc).

Bonwell, C. C. and Eison, J. A. (1991), 'ASHE-ERIC Higher Education Report no. 1; Active Learning: Creating Excitement in the Classroom'. Washington, DC: School of Education and Human Development, George Washington University.

Caballero, R. (2003), 'Metaphor and Genre: The Presence and Role of Metaphor in the Building Review', *Applied Linguistics*, 24/2: 145–167, doi:10.1093/applin/24.2.145.

Cannady, R. E. and Gallo, K. Z. (2014), 'Write Now! Using Reflective Writing Beyond the Humanities and Social Sciences', *Journal of Further and Higher Education*, 40/2: 188–206, doi:10.1080/0309877x.2014.938266.

Crossman, J. M. and Kite, S. L. (2012), 'Facilitating Improved Writing Among Students Through Directed Peer Review', *Active Learning in Higher Education*, 13/3: 219–229, doi:10.1177/1469787412452980.

Gallo, K. (2016), 'Zrozumieć Żargon Zawodowy' ['Understanding Professional Jargon'], *Psychologia Wychowawcza*, 1/52: 164–171.

Glock, F. (2009), 'Aspects of Language Use in Design Conversation', *CoDesign*, 5/1: 5–19, doi:10.1080/15710880802492870.

Hirst, R. (2003), 'Scientific Jargon, Good and Bad', *Journal of Technical Writing and Communication*, 33/3: 201–229, doi:10.2190/j8jj-4yd0-4r00-g5n0.

Holyoak, K. J. and Thagard, P. (1997), 'The Analogical Mind', *American Psychologist*, 52/1: 35–44, doi:10.1037/0003-066x.52.1.35.

Lin, M. (2000), 'Making the Memorial', *New York Review of Books*, 47/17: 33–35.

Lucas, G. M. (2010), 'Initiating Student-Teacher Contact via Personalized Responses to One-minute Papers', *College Teaching*, 58/2: 39–42, doi:10.1080/87567550903245631.

Mehrabian, A. (1968), 'Some Referents and Measures of Nonverbal Behavior', *Behavior Research Methods & Instrumentation*, 1/6: 203–207, doi:10.3758/bf03208096.

Morrison, S., Decety, J. and Molenberghs, P. (2012), 'The Neuroscience of Group Membership', *Neuropsychologia*, 50/8: 2114–2120, doi:10.1016/j.neuropsychologia.2012.05.014.

Rodríguez, M. D. R. C. (2015), 'Thinking, Drawing and Writing Architecture Through Metaphor', *Ibérica: Revista de la Asociación Europea de Lenguas para Fines Específicos (AELFE)*, 28: 155–180.

Schon, D. A. (1986), *Educating the Reflective Practitioner: Toward a New Design for Teaching and Learning in the Professions* (San Francisco, CA: Jossey-Bass).

Soygenis, S., Soygenis, M. and Erktin, E. (2010), 'Writing as a Tool in Teaching Sketching: Implications for Architectural Design Education', *International Journal of Art & Design Education*, 29/3: 283–293, doi:10.1111/j.1476-8070.2010.01646.x.

Sun, L., Xiang, W., Chai, C., Wang, C. and Liu, Z. (2013), 'Impact of Text on Idea Generation: An Electroencephalography Study', *International Journal of Technology and Design Education*, 23/4: 1047–1062, doi:10.1007/s10798-013-9237-9.

Sun, L., Xiang, W., Wang, C. and Shao, S. (2013, March), 'The Function of Text Description in Sketch Promotion', 3rd Interdisciplinary Engineering Design Education Conference: 173–176, doi:10.1109/iedec.2013.6526782.

Thompson, W. (2006a), 'Land Matters', *Landscape Architecture Magazine*, 37/6: 13.

———. (2006b), 'Land Matters', *Landscape Architecture Magazine*, 37/7: 13.

———. (2009), 'What LAM is for', *Landscape Architecture Magazine*, 76/9: 13.

Torrance, E. P. (1966), The Torrance Tests of Creative Thinking: TTCT Manual and Scoring Guide (Lexington, KY: Ginn).

Woodward-Kron, R. (2008), 'More Than Just Jargon: The Nature and Role of Specialist Language in Learning Disciplinary Knowledge', *Journal of English for Academic Purposes*, 7/4: 234–249, doi:10.1016/j.jeap.2008.10.004.

Zoller, U. (1993), 'Are Lecture and Learning Compatible? Maybe for LOCS: Unlikely for HOCS', *Journal of Chemical Education*, 70/3: 195–197.

20

Back to basics
Writing for design professionals

Lake Douglas

Introduction

In efforts to expose students to the tools most beneficial for successful entry into professional practice, American landscape architecture undergraduate and graduate curricula emphasize certain disciplinary threads over others. Many factors contribute to curriculum content; among them are academic circumstances (e.g., resource limitations; program age, mission and traditions; faculty interests and leadership agendas; academic budgets; institutional expectations); accreditation standards; professional trends; regional traditions; and levels of engagement and symbiotic relationships between the academy and the profession. Obviously, the degree to which these factors influence the academy with regard to curriculum and pedagogy varies.

Certain knowledge bases, analytical tools, and office practices have changed with developing technologies that address the complex challenges with which the profession now grapples, and certainly contemporary curricula are different from those of decades ago. Ultimately, the character of a professional curriculum is defined by where a program sits within a spectrum of delivering the canon (teaching the accepted basics required for entry-level success) at one end and questioning the canon (expanding professional boundaries by experimenting with new technologies and theories) at the other.

The roots of the profession in America lie in the nineteenth-century philosophies, social and reform issues, and urban projects. Writings, appearing in agricultural newspapers, treatises, and books were the vehicles through which advocates for a landscape-based ethos disseminated philosophical theories, design rationales, and accounts of landscape interventions.[1] As America evolved from being predominantly rural/agricultural in the beginning of the nineteenth century to urban/industrial at the century's end, topics addressed shifted accordingly. By the early decades of the twentieth century, the profession was organized and subsequent generations of professionals continued the practices of their predecessors.

With regard to the post-WWII evolution of contemporary practice, three observations are relevant:

1   Many landscape architects were strongly influenced by a handful of books (with accompanying images) written by those in practice.[2] The works of a few professionals and their practices defined the profession, and exposure to these works, particularly for those in remote areas away from metropolitan centres, came in book form.

2   A second transformation happened at the beginning of the twentieth century's fourth quarter when new areas of exploration, and the analytical tools and methods of representation used to illustrate them, emerged and gained acceptance, largely pioneered and developed in graduate academic programs. These exponentially expanded the definition of the profession and charted new paths of engagement as new analytical tools seeped into office practice and gained acceptance in other academic programs.[3]

3   Most recently, we find practitioners/academicians whose influence on the profession comes from a combination of their academic activities, theoretical writings and office practice involving high-profile competitions and projects.[4]

All three cases, in unique ways, shape design pedagogy and professional practice. Each represents a different calculus of influence: in the first case, practice shaped what the academy taught; in the second, the academy gave new definition to the profession and propelled it in new directions; and in the third, the combination of academic grounding and practical experience has created new theories and approaches to professional practice. Often these have been tested first in academic or competition settings, and then disseminated into the profession. Notably, all three phases have involved written works publicizing projects, advocating positions, and raising professional (and personal) visibility.

Such give-and-take relationships should continue as environmental issues and professional opportunities become more complex and boundaries among design professionals continue to blur. Such interchanges are crucial for continued vitality and relevance because of the challenges and inspiration they provide for those who study and practice this broadly defined profession. Faced with new challenges and increasingly awash with innovative technologies, the profession's boundaries continue to overlap, whether with design-oriented professions or not. These professional reconfigurations, together with other technologies (including social media), have irrevocably changed office practice, professional education, and personal interactions; most would argue these changes are for the better. However, with the efficiencies of new technologies, will the profession lose critical characteristics and sacrifice basic components that define, humanize, and distinguish us and what we do from other design disciplines?

A thread throughout the profession's evolution is the use of the printed page (words and images, in printed formats) to present work, propose new ideas, and demonstrate professional activities. Print is an efficient and effective means of dissemination and a lasting means of documentation. As we advance deeper into the realm of non-print communication, we may also be losing sight of the value of written communication and the skills it requires. The result is ominous: if few are writing intelligently about (and advocating for) the profession, it will become irrelevant.

Background

In about ten years' experience in the academy, I have found that many undergraduates do not read and cannot write. While this may be an indictment of American secondary education, it is a reality with which our pedagogy (and profession) must grapple if we are to educate professionals who can, like their predecessors, write thoughtfully about new ideas. As students graduate into the profession, advance, and assume leadership roles, they will be responsible for the discourse necessary to keep the profession vital, relevant, and responsive to future challenges. We who teach today's students are responsible for preparing them to become the next generation of design thinkers, communicators, and writers, and we have work to do.

In my experience, students often respond with surprise when assigned to write about their work, leading me to conclude that, in their view, new technologies have rendered this basic form of communication irrelevant. No wonder; students today get information in abbreviated formats and from unexamined sources (often non-print), and communicating in other ways is inconceivable. Certainly, deficiencies in written communication are attributable to the means through which American students prepare for higher education. Yet similar pre-college deficiencies exist in areas of technology, representation, and other skill sets, and we surmount them successfully through concentrated focus. Let us consider that developing effective writing skills is as valuable as other design curriculum mainstays, and that neglecting this professional tool is an abdication of pedagogical responsibility.

DesignIntelligence analyzes professional and academic trends in the design fields and publishes annual rankings of American design programs.[5] A cursory examination of curricula from webpages of the 2016 five top-ranked American undergraduate landscape architecture programs reveals that none has a required (as opposed to elective) writing class *specific to the profession*.[6] While writing exercises may well appear elsewhere in the curriculum (in history, professional practice, or research seminars), I suggest such limited exposure is inadequate and that the best way to indoctrinate students into writing is to integrate its practice as a requirement *throughout* the curriculum.

Offering an elective writing class for design students, I endeavor to enhance students' writing skills to give them additional qualifications for their résumé and to acquaint them with opportunities within the design office environment where their capacity to communicate effectively via writing is relevant. My interest comes from two sources: first, much of my four-plus-decade career has involved writing about design issues in various forms, and now I edit the works of others for publication. Second, my interest in teaching writing emerges from what our School's alumni relate, anecdotally, as being a widespread deficiency among potential hires: the capacity to communicate effectively in written forms.

By coincidence, my institution emphasizes effective writing through a Communications Across the Curriculum (CxC) program, an innovative campus-wide initiative that seeks to "advance the communication skills of all LSU undergraduate students . . . [by helping] students improve their written, spoken, visual, and technological communication skills and deepen their learning of course content."[7] This program offers faculty development resources to facilitate developing communication-intensive courses that address communications skills in four areas: written, speaking, visual, and technical. Courses are certified by including specific assignments and activities involving at least two of the four focus areas, and students who complete CxC academic requirements are recognized at graduation as "Distinguished Communicators."[8] In existence for over ten years, the program is nationally recognized as a model for improving communications skills among undergraduates.[9] Students who have completed the program anecdotally report that the "Distinguished Communicator" designation was relevant in securing initial employment, and the experience gained through the program has been of professional relevance and employment benefit.

Sabbatical research

Based on these interests, I proposed for sabbatical research (fall 2016) to investigate writing as a pedagogical tool for successful professional practice to determine if interest exists in practice for better writing skills (ideally developed as part of the academic experience). If so, are demands for writing competence being met, as shown in recent hires? If not, how should undergraduate/ graduate curricula respond to meet professional demands?

The point of departure for this study was the content of my Writing for Design Professionals class, developed over time. From the class syllabus:

> Writing is integral to effective communication, and effective communication in all its forms is essential to success in the design community. This class will examine different forms of written communication useful in design practice; a secondary emphasis will be on using writing skills effectively in other means of communication (graphic, visual, oral). Writing for design professionals can be organized by categories:

- Marketing: project proposals; correspondence with prospective clients; design award submittals; exhibition boards; multi-media presentations; press releases; brochures and portfolios; promotional copy; job applications;
- Project-related: client team and in-house project communications; research and precedent studies; feasibility studies; planning documents;
- Office communications: in-house communications, communicating with contractors and clients; records of meetings and conversations; in-house memoranda;
- Publicity: getting work published and publicity for projects and office activities;
- Books and portfolios: increasingly design offices are self-publishing their work in book form to demonstrate design expertise to potential clients. Often, because of a lack of in-house skills, this work is out-sourced to those who know little (if anything) about design professions;
- Academic work: if you go into academics, writing is an essential part of what you will do ("publish or perish"); this includes writing for academic journals (research articles; reviews, etc.), and popular press, writing conference papers and case studies; and
- Grant applications: often the work of designers can increase with supplemental financial support from funding agencies that require applications through specified formats; as a design professional, how do you effectively participate in this activity?

Methodology

The methodology employed (abbreviated due to constraints of time and other resources) involved an initial survey, followed by discussions and work sessions with design office staff in two offices of a landscape architecture office with a national clientele (Design Workshop), to ascertain if what I was teaching coincides with what the profession uses in practice and seeks in new hires. This investigation was not as rigorous or scientific as would be possible with greater resources; yet it provided basic information that confirmed earlier assumptions and provides preliminary data for future, more rigorous, exploration. The investigation's design involved the following:

- A survey was sent to professional offices in the United States.[10]
- Information was solicited from the Vice President of Education, American Society of Landscape Architects, to determine which – if any – schools with accredited programs in landscape architecture are currently offering writing in their curriculum. This effort was not successful; subsequent telephone interviews might be more successful in gathering this data.

- Focused work with staff of Design Workshop (DW), followed by on-site multi-day sessions, individually and collectively in the firm's Denver (Colorado) and Austin (Texas) offices.[11] My visits were facilitated through DW's "Faculty-in-Residence" program, enabling me to work directly with staff collectively and individually.

The sabbatical's three goals were to determine specific curriculum implications (therefore content revisions to my class); to develop broader recommendations integrating writing throughout the undergraduate curriculum; and to publish results for dissemination to colleagues in academic practice.

Prior to the scheduled office visits, the survey was circulated among DW's offices for responses, and staff from DW's Denver and Austin offices forwarded writing samples of various lengths and purposes (e.g., project proposals, executive summaries, planning documents, etc.) enabling me to review the variety of writing the office generates (see below). Examples included completed documents as well as documents in progress. Before arriving in the offices I had made editorial comments and suggestions, highlighting areas that later became in-office discussions among staff, including the following:

- different written formats common in professional practice;
- consideration of writing mechanics (content, clarity of expression, specificity);
- the value of research in professional practice and various methodologies of research as components of professional writing;
- skills needed for effective writing within a design office context (organization, focus);
- identification of tools needed for writing success (outline, resources); and
- current needs for writing in offices of different sizes and practice characteristics and implications for the future.

The following writing genres were reviewed with DW staff:

- project documentation books (master plans for parks, open spaces, healthcare, etc.; development plans; design guidelines, etc.);
- marketing initiatives (proposals; DW boilerplate for project documentation; website text; etc.);
- Awards submissions (internal, external);
- Individual staff writing projects (professional publications, article/book ideas); and
- Grants/non-profit partnerships.

General tips for improving writing were given and discussed with DW staff in group discussions.

Results

The results of my investigations (survey data, discussions/follow-up with staff) and reflection on findings suggest the following.

- Professionals consider writing a critical and necessary professional skill.
- Being able to communicate in written form is a desired qualification for new hires.

- Overwhelmingly, and at all levels, those in practice say they wish they had received more/better instruction in writing prior to completing their degree.

- The writing elective's content is on-target with respect to marketplace demands/expectations; and respondents indicated course content is integral to office practice at rates between 95%–100%.

- From examples of professional written work observed, problems evident in practice-based writing are similar to problems in undergraduates' writings, leading me to conclude that if these areas are addressed in school, new graduates will be better prepared to enter practice and be productive in this area.

- Based on survey results, there may well be a market for post-graduation distance (on-line) instruction in writing: 74% of respondents said such a class would be instructive.

Increasing writing components in the curriculum could happen in two ways, either requiring it as a component in every class, or enlarging the professional practice class to incorporate more writing, thereby making the current elective writing class a curriculum requirement.

With regard to job portfolios reviewed by offices in hiring entry-level positions, 70% of respondents said that they "routinely look for skills in written modes of communication" during the interview process. Such scrutiny happens often with the first letter of enquiry: if poorly written, the writer might never be considered and any work submitted is irrelevant; on the other hand, if such a letter is thoughtful and well written, the student's portfolio will be examined and the applicant is far more likely to be considered for employment.

Forty-two percent of respondents said they "rarely (less than 10% of the time) see evidence of writing skills in potential hires" via portfolios, resumes, interview discussions, etc. Faculty should give this more emphasis in discussions with students on portfolio development and interview techniques.

Fifty-seven percent of respondents said they would look for writing "directly related to landscape architecture (a project statement of design critique)."

Finally, all respondents (100%) rated "professional writing as an educational priority" as follows: 43% assigned it a "high priority" (a requirement) while 57% assigned it a "medium priority" (integrate as possible).

Implications of study

This subject deserves a greater analysis, ideally with more significant resources than were initially available. My goal was to confirm (or dispute) assumptions (made from anecdotal information, discussions with colleagues about the relative importance of writing in professional practice, and observations from teaching a writing class in a design curriculum) by testing them in the office environment of one of the profession's leading firms. Regardless of my study's limitations, several observations are relevant.

- Writing is a valued skill in contemporary practice and one that is sought after as a distinguishing factor among entry-level hires.

- While present throughout the undergraduate curriculum, writing in a professional context is not emphasized as a separate class in undergraduate curricula structures; as a result, writing does not get academic attention commensurate with what the profession has determined as relevant.

- There are multiple genres of writing employed in professional practice; students would be well served by exposure and practice in multiple genres.

- This exercise confirmed the legitimacy of my elective course's content and my sense that this content is not addressed in traditional 'professional practice' classes and that it should remain a stand-alone class.
- Finally, curricula that are focused on providing their students with a professional edge should consider inserting writing into their curricula, either as electives or as components of required classes.

Looking ahead

Sabbatical field investigation and subsequent reflection has reinforced my initial supposition that writing is a useful – if not required – professional skill. Since my elective is not taught every academic year, I have since begun to insert writing into the other classes I teach on a more regular basis. In fall 2017, I taught two graduate seminars and an undergraduate studio, and writing requirements were part of all three classes. From these experiences, I have learned that writing improves with two things: reading (widely) and practice (often). Instructors should encourage students to read, starting first at the student's comfort level (graphic novels or consumer-oriented periodicals) but moving on to more serious thought-provoking journals, (professional and otherwise), and they should also require students to write in journal format or another format about their interests, experiences, design ideas, or anything else. Content matters more than form; initially, spelling and grammar should not count, though, of course, infelicities should be noted.

Writing numerous short papers (1–1½ pages) on an informal basis – impressions, thoughts, responses – is a valuable tool and helps to 'break the ice' with students who feel insecure in their writing. Have students exchange papers in class and read aloud. Grammatical errors, lapses in logic, and organization problems will become obvious and can lead to useful class discussions. Such informal exercises need not be graded; in fact, they are useful windows into students' minds, interests, and ways of thinking. Make notations for individual discussion with students outside of class.

Make note of consistent student errors (grammar, spelling, punctuation, etc.) and compile a list for distribution to students. Such 'cheat sheets' are helpful for students and handy for quick desk reference. My list, compiled over several semesters, now has about 50 mistakes regularly seen, with instructions on how to correct them.

When students do not know how to start, suggest they develop an outline first, regardless of what is being written; assign relative weight/word length to each section in order to evaluate for consistency among sections.

Give students guidance on where to find answers to writing, grammar, and style questions; I use *Chicago Manual of Style* (latest edition) but there are others. Encourage students to seek out university resources for help with basics.

Assign at least one long-form analytical paper and require it to be well organized, researched, and properly referenced, per accepted standards (university, *Chicago Manual*, etc.). Require students to develop an outline, then an abstract, then a draft prior to submission of the paper.

Spend time outside of class where it will give the most return for students. Reading and talking with students about the short-form papers works well; marking and returning longer papers is helpful, if time permits.

Finally, encourage students to view writing as another means of communicating design thinking, akin to technical skills, representation, and oral presentation, and, as such, it is a critical tool for professional success.

The results of this study suggest areas for future conversations with regard to writing and curriculum development among a wider range of offices, professionals, organizations, editors, academic colleagues, and others.

Notes

1 Nineteenth-century examples of influential writings include the works of Ralph Waldo Emerson, Walt Whitman, Andrew Jackson Downing, Frederick Law Olmsted, Thomas Affleck, Charles Eliot, and numerous others. I examined agricultural literature in "To Improve the Soil and the Mind: Using Context and Content of Nineteenth-century Agricultural Literature for Environmental Research," *Landscape Journal* Vol. 25, No. 1, 2006.

2 For many (including this author), the introduction to the profession came from Garrett Eckbo's *Landscape For Living* (1950), Thomas D. Church's *Gardens Are For People* (1955), James Rose's *Creative Gardens* (1958) and John O. Simonds' *Landscape Architecture* (1961). All authors were practicing professionals and their books, by default, often became textbooks for academic programs in landscape architecture. In addition, Eckbo, Church, and Rose's published commentary and their works in popular consumer magazines, exposing the public to the profession of landscape architecture in general and their ideas in particular, thereby increased their professional visibility and opportunities for themselves and other landscape architects.

3 Examples include Ian McHarg (University of Pennsylvania, 1960s), Carl Steinitz (Harvard GSD, 1970s), and Charles Waldheim (Harvard GSD, 2000s).

4 James Corner (Field Operations) and Michael Van Valkenburgh (MVVA) come to mind, with academic associations at University of Pennsylvania and Harvard Graduate School of Design, respectively.

5 *DesignIntelligence* is a publication of the Design Futures Council, a membership organization composed of "an interdisciplinary network of design, product, and construction leaders exploring global trends, challenges, and opportunities to advance innovation and shape the future of the industry and environment." [See: www.di.net/articles/the_design_futures_council accessed 15 January 2018]. "*DesignIntelligence* is the Design Futures Council's bi-monthly report on the future, delivering original research, insightful commentary, and instructive best practices. Design leaders rely on *DesignIntelligence* to deliver insight about emerging trends and management practices, allowing them to make their organization a better managed, more financially successful enterprise." [See: www.di.net/about accessed 15 January 2018].

6 As of this writing (2017), the top five undergraduate programs (2016) were: #1: Louisiana State University; #2: Pennsylvania State University; #3: Cornell University; #4: University of Georgia, and #5: Texas Agricultural and Mechanical University. All offer required or elective classes that include writing (basic English, "composition and rhetoric," "technical and business writing," or "written or oral expression") but these seem to be general education classes taught elsewhere in the university.

7 www.lsu.edu/academicaffairs/cxc/ accessed 15 January 2018.

8 www.lsu.edu/academicaffairs/cxc/distinguished-communicators.php accessed 15 January 2018.

9 In 2010, LSU's CxC Program was the only academic program in America recognized by the Conference on College Composition and Communication (CCCC) through its "Writing Program Certificate of Excellence" program. The CCCC is the "world's largest professional organization for researching and teaching composition, from writing to new media." See: www.ncte.org/cccc accessed 15 January 2018.

10 The survey was sent to those with whom I had professional and/or personal relationships; recipients were encouraged to distribute the questionnaire to others (in their offices as well as to other colleagues).

11 See: www.designworkshop.com accessed 15 January 2018.

References

Church, Thomas D. (1955) *Gardens Are For People*, San Francisco, CA: McGraw-Hill Book Co.

Douglas, Lake (2006) "To Improve the Soil and the Mind: Using Context and Content of Nineteenth-Century Agricultural Literature for Environmental Research," *Landscape Journal* Vol. 25, No. 1.

Eckbo, Garrett (1950) *Landscape For Living*, New York: Duell, Sloan & Pearce.

Rose, James (1958) *Creative Gardens*, New York: Reinhold Publishing Corporation.

Simonds, John O. (1961) *Landscape Architecture*, New York: McGraw-Hill Professional.

21

Exercising drawing time

Noël van Dooren

Introduction

Formally, time can be seen as the fourth dimension of landscape, and hence of designs in landscape architecture. Time has many manifestations – it can be considered as a container that includes growth, change or dynamics, and so on. Speaking about time in landscape refers to cyclical (for example, the seasons) as well as progressive phenomena – think of the growth of trees. It implies very short durations (hours) as well as extremely long durations (centuries); repetitive and predictable happenings (the weekly market) as well as irregular events, such as floods. Thus, many aspects of landscape and its performance are touched by, or subject to, these different manifestations of time. In fact, speaking about time can be seen as speaking about what essentially distinguishes landscape architecture from adjacent disciplines, such as architecture, as the very material of landscape is subject to permanent change.

Are students in landscape architecture confronted with this aspect of time in their education? Partly: most practitioners and teachers would state that pointing out aspects of time *obviously* is part of their work and that it pervades their entire teaching, even without mentioning it. At the same time, even if the aspect of time is generally understood as inherent to landscape architecture, it has hardly any role in education, and is rarely found in representations of landscape architects. This chapter aims to promote a more explicit position for time and representations of time.

Background

Are aspects of time indeed often absent from landscape architecture drawings, and, if so, why? To answer such questions, I have studied the current production in professional practice[1] (Figure 21.1). Obviously, such a study reacts to concrete drawings. But drawings often do not speak for themselves. Anthropology teaches us that interviews can help to not only derive conclusions from drawings as such, but also to reveal the considerations behind those drawings: *why* is time drawn, or not? In order to provide a framework, such questions were also explored via theory and the literature. This showed that time was addressed more explicitly in early phases of the history of landscape architecture. Think of

Figure 21.1 The office of karres + brands, Hilversum, 2015. Photo Betul Ellialtiogliu

the remarkable written work of Humphrey Repton, which in a striking way connects an idea of landscape and landscape design with time and representation (see Loudon 1988). Also garden handbooks of the eighteenth century and drawing experiments as done at the Napoleonic engineering school *Ecole des Ponts et des Chaussées* are very informative if we want to understand this problematique (see Picon 1992). And, after a long silence, the issue is once more being discussed in the literature, thus providing support for the point this chapter aims to make.

The word representation can be understood in many ways. Here, a practical perspective is taken, following Levine (2009), and Fraser and Henmi (1994). Representation evokes the landscape that does not exist yet, by means of text and a system of drawing types. Drawings in this view include various types such as plans, sections, diagrams and visualizations on screen or on paper, but also three-dimensional ones such as models or mock-ups. New digital technology can help to open up new opportunities, or to speed up drawing processes. Drawings can be regarded as artefacts; artefacts that we can look at, appreciate and understand as vehicles that transport design intelligence. But drawing also is an activity, and, in that sense, drawing is parallel to exploring and understanding landscape architectural ideas. Juhani Pallasmaa coined the phrase 'thinking hand', suggesting that drawing is also a way of discovery. This idea obviously is very relevant for landscape architecture education programmes, as it implies that not only does drawing have to be learned, but that it is a way of learning as such. Therefore, this chapter, and the exercises in it, accentuate hand drawing. Software tends to draw attention because of its own technical difficulties, whereas hand drawing in general allows a focus on the landscape problem in question – and that was the aim. This learning experience is important in general and, in this specific case, in dealing with the issue of time: are students trained in representing time, are they inspired to develop their thinking related to it, and to enrich their landscape knowledge via drawing time? At the time of study, only in a few programmes, such as the one at Harvard, was this explicitly the case. Today it is more common to find time aspects and its representation explicitly addressed in studio teaching, or lectures. Nevertheless, it is still far from self-evident. This chapter argues that the position of time, the representation of time in professional practice, and the theory on the representation of time in landscape architecture, lags behind.

Some of the most pressing problems of today have a strong time dimension, such as climate change and sea level rise. To be relevant for society, landscape architects must respond to that. That underpins the actuality of this chapter. At the same time, the argument is timeless, as this observed absence of time confronts teachers and professionals with a theoretical flaw that simply has to be repaired. It is part of a necessary process of developing into a mature design discipline. It would also help to distinguish landscape architecture from adjacent disciplines. Although architecture also deals with time – think of day and night rhythms, or weathering of materials – in the case of landscape architecture the issue is fundamental (Mostafavi and Leatherbarrow 1993). Today's challenging design problems ask for an awareness of uncertainty, flexibility, and incompleteness, as well as for careful processes of realization, new ideas on management and a response to changing circumstances. One could say that, both within and beyond landscape architecture, today's design challenges are often landscape-oriented. Thus, there is a huge opportunity for strengthening the position of landscape architecture and its performers. However, if one would want to take that opportunity, landscape architects themselves should be clearer, especially in their representation, on aspects of time.

The principal innovation proposed here is a new approach towards the basic system of representation in landscape architecture, creating a new partition, a domain of *spatial* and one of *temporal* representations. Most drawing types we know, such as plan and section, fit in the spatial group. The temporal domain has never been explicitly acknowledged. Drawings in this new temporal group aim to display the relevant time aspects in a design. They clarify the time scale at which the design operates and the manifestations of time at work. Typically, they explain the actions as needed to guide the dynamics in a design, and the persons or institutions responsible for that. We see in these drawings how a design grows and matures, and we are informed on the potential evolution in the further future. There are several types of drawings that fit into this group. The word drawing is taken here in its widest sense, as we would perhaps not look at an animation film or a comic as a drawing, but, in this context, they are certainly appropriate. The *score* as proposed by landscape architect Lawrence Halprin in his 1969 *The RSVP Cycles* is a new and crucial proposal of such a temporal representation. The year 1969 may seem a long time ago, and in the last decade we have seen new solutions emerging. Some of these solutions derive from the latest digital technology. Others stem from new schools of thought, such as landscape urbanism. But what makes Halprin's work stand out is that it integrates a debate on representation, time and landscape into an innovative theoretical solution: the score. Halprin's death a few years ago prompted many new publications that may help to reintroduce his proposal, as it did not get enough attention (see Hirsch 2006). Halprin's score, derived from choreography, is essentially a drawing showing *who* is doing *what, where* and *when*. In that sense the score is fundamentally different from most drawings in landscape architecture, as common drawings such as plan and section focus on *what* and *where* only.

Put in series, and given a time tag, plan and section are perfectly capable of discussing time aspects. This confirms that drawing time does not have to be difficult – it mainly requires a change in attitude, and an awareness of the mechanisms operating in landscape. Other options, such as the score, need to be exercised, and perhaps yet unknown solutions for drawing time can be revealed. This was the point of departure for 'drawing experiments' in existing educational programmes: to examine whether inventions in representation can expand on this.

James Corner's essay 'Representation and landscape' (1992) is essential for the theoretical foundation of this thinking about temporal representations, and also for the corresponding role of landscape urbanism. Currently, if we want to look for both drawings that display time and a thinking that includes time, landscape urbanism offers interesting material. *The Landscape Urbanism Reader* (Waldheim 2006) is a key reference. Corner, who contributes to that reader, puts it like this: it is about sowing the seeds of 'future possibility, staging the ground for both uncertainty and promise'. This shifts the focus from 'compositional design' to 'operational logic', which in turn leads to another much-used word in the rhetoric of landscape urbanism, 'performativity' (Waldheim 2006: 31). For me, as a Dutchman, this kind of thinking, and related logic such as scenario planning, has formed a significant part of my work experience. Dutch urban planner Frits Palmboom summarizes it like this: 'In our drawings we also seek to make the operation of time visible. They visualize strategies in which time and uncertainty play a role. We practise the art of determining things minimally, leaving as much as possible open' (Palmboom 2010: 34). Despite this progression in theory, in landscape architecture practice in general and in landscape architecture programmes, it is still not common to represent time aspects fully. Interviews with practitioners and a study of the literature suggest that the nature of practice (for instance, how will the client understand this?), the lack of a well-established 'best practice' of drawing, and the absence of a clear theoretical frame impede a rapid application. Recent literature, such as Diane Balmori's *Drawing and Reinventing Landscape* (2014), confirms this, but stresses that it is time to definitively integrate temporal aspects in theory and drawing. A recent essay by Duempelmann and Herrington (2014) summarizes the state of things very well, but now practice and education have to follow. Therefore, in as much as the representation of time is not yet an explicit part of the training, this chapter calls for making it explicit and for providing it with the basis to do so.

Exercises

Teaching experience at different schools has made clear that practising how to represent time indeed raises fruitful conversations on the nature of landscape and the role of time. In fact, it even seems to strengthen important general qualities in design processes. Expanding on this, in collaboration with TU Delft, a series of exercises was developed that address both the representation of time and certain general skills, as far as they are considered to be helpful in the representation of time. What this chapter wants to convey is the *idea* of such exercises, as the options are by no means limited to this proposal. One exercise is given as a whole as an invitation to use it.

It is important to learn to think about landscape and designed landscape as a series of *moments in time*. The first exercise, very well suited for the Bachelor's degree, aims to have students trying out representational solutions, and considering more precisely how landscape evolves over time. See the entire exercise at the end of the chapter. A second exercise, *exploring animation*, comes with some technical difficulties, as students have to understand the process of creating an animation film. Obviously, today's software can help to make quick animation films, but, as argued, the challenge is to understand precisely the development of a landscape over time. The representation of time shifts the attention from what and where to who and when. Animation film is very suitable for this purpose. Animation film is sequential, and, because animation film almost unavoidably introduces a narrative, the *who* in the development of landscape comes to the forefront. Our experience is that creating animation films can be very time-consuming and is therefore a risky road to take. But, if taken successfully, it is very helpful in raising awareness regarding the necessary actions in the making or managing of landscape. The social context in which such actions are done is also better understood: planting a tree is done *by* somebody, and (hopefully) *for* somebody – a reality that is easily forgotten in the studio (Figure 21.2).

Drawing the score is the third exercise. The system of types of representation in landscape architecture does not mention the score. It would be a new type – that is to say, new from the perspective of landscape architecture. If we look at Nelson Goodman's work (1976), which discusses notational systems in different arts, the score is an accepted type of notation in dance and music. This is useful to realize, as it explains the focus on who and when. A choreographer 'designing' a dance performance thinks of dancers acting at a certain place at a certain time and thus what, where, who and when are connected. From Halprin's work or examples in current practice such as the GROSS.MAX diagram for the development of Tempelhof, Berlin, we can deduce that scores in fact strongly resemble diagrams – but not all diagrams are scores. By their very nature, scores are complex, as they must cope with several layers of information. In terms of graphical quality, they are difficult to handle. However, as Tufte argues, we must accept that some drawings are complex due to the complexity of the information they have to display (Tufte 1990: 50–51). I consider, therefore, the score to be well suited for advanced work at Master's level (Figure 21.3).

Figure 21.2 Creating animation films is time-consuming and a relevant learning experience. Photo: the author

Figure 21.3 GROSS.MAX, Parklandschaft Tempelhof, competition entry 2011. This timeline could be categorized as score

Also suited for advanced Master's level is an exercise called *Speculative reconstruction*. This exercise was developed in collaboration with Copenhagen University. This exercise asks students to re-invent how an existing piece of landscape design may have developed over time to become what it is now. We applied this to a C. Th. Sørensen park design. The remarkable narrative of Højstrup Parken in Odense is that about 900 oak whips were planted in circular beds in 1954. These beds were thinned over the decades, and, today, we find 29 huge oaks, about one per bed. Although there is some archival material, it is impossible to really trace the development over time. But that creates the room for students to indeed re-invent the development. We consider the student products as very inventive. In how far they come close to reality is irrelevant; what is asked for is a convincing story, and the exercise proves that many fundamentally different and convincing narratives are possible. This challenges the entire register of representation, from plan to animation film to comic. One of the most striking outcomes addresses the future. Sørensen in 1954 had in mind the mature situation of about 40 years later, and in his 1975 autobiography he confirms his vision. Now we see the reality of 29 oaks and realize that the process of thinning could go on, and in fact *must* go on to strengthen the spatial composition (Figures 21.4 and 21.5).

Of a more general nature is an exercise inviting the students to experiment with drawing materials. Annelies Bloemendaal's use of cress provides an excellent example. In a design for a forest, she placed cress seeds on moist paper and photographed the sprouting seeds over seven days. This not only generates a symbolic representation of the growing forest, but it also alludes to the very process of growth and decay. Before sprouting, the seeds act as a two-dimensional drawing. While sprouting, a third dimension emerges. After two weeks, the young plants die,

Figure 21.4   Højstrup Parken, Odense, in 2012. Park design by C. Th. Sørensen, 1954

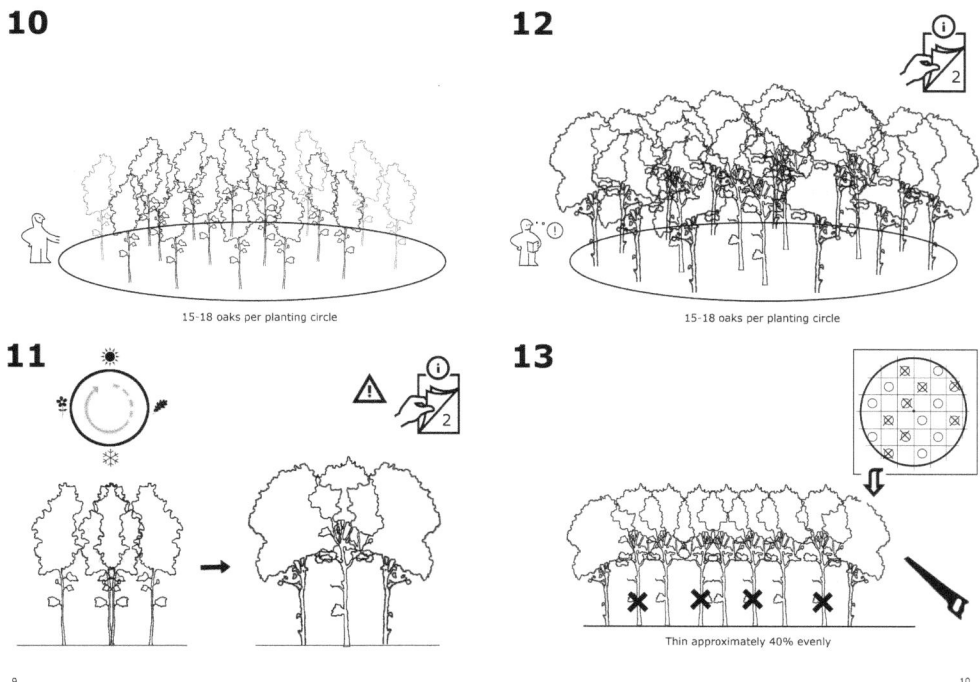

Figure 21.5   Højstrup Parken revisited. Do-it-yourself management of time in the style of a well-known Swedish firm. Copenhagen University, 2015, by Janka Bulath, Julie Skajaa, Mia Nordow and Veronika Haas

and the third dimension fades (Figure 21.6a–e). In the exercise, *Wachsen Lassen* [Let it grow] at TU Stuttgart, we proposed dough as material to make models with. Yeast and sugar provoke a rapid growth of the dough. Registering and designing by means of this growth inspired fascinating drawing experiments. Such an exercise using unusual drawing materials supports one's thoughts about time, but it also stimulates a creative approach towards drawing in general. Every profession has its conventions. It can be very helpful for students to realize that such conventions need to be challenged, and that many unconventional ways of drawing in certain circumstances may be extremely appropriate (Figure 21.7). The following exercise plays with time itself, and requires the creation of extremely quick and extremely slow drawings. Drawing as an activity takes time. Theories on creativity comment on the restrictions certain means of drawing pose, as they require a substantial amount of time. Our creative thoughts are incredibly quick, and may require a very direct response in drawing. For some, this indicates good old pencil or chalk – drawing means that allow quick working. Others argue that recent developments in computers and software also enable very intuitive sketching. Being forced to produce drawings very rapidly necessitates very fundamental choices with regard to drawing. No time for detail, no time for doubt. At the other extreme of the palette we ask for extremely slow drawing. This can suggest using very elaborate drawing means, but very large or very detailed drawings also come to mind. The entire point is this: a drawing that requires a lot of time to be made also offers plenty of time to contemplate design ideas *while drawing*, and to some extent to physically experience the extension of the landscape (Figure 21.8). The most radical idea in this exercise is that not only does the drawing represent the future landscape, but that also the drawing process in some way stands for the making, maturing and managing of the designed landscape. Again an exercise of more general nature focusses on *presentation* instead of representation. A theoretical starting point for this is James Corner's thought that drawings are incapable of transporting the essence of landscape (see Corner 1992). To start with, landscape is generally *around* us, whereas drawings are often seen as *in front of* us. In the same way, the perception of drawings suggests an instant overview, whereas landscape generally can only be known by moving around, and may require coming back more often, in different conditions. In several design experiments this was researched. The *panopticum* provides an elegant solution: what if the drawing is around us? A book form was also tried out – in some ways an old-fashioned concept, but when used in landscape architecture presentations it can become an innovation as it may help to control the rhythm and the velocity of studying the design, and to direct our concentration towards certain aspects (Figure 21.9).

Further steps

Above, a general idea of a series of exercises for different levels is given – obviously, more exercises are possible, and we will continue to work on that. But how could this function in education in its daily practice? The architecture faculty at TU Delft developed a standard-ized model to provide exercises for a range of topics in architecture. We followed this model

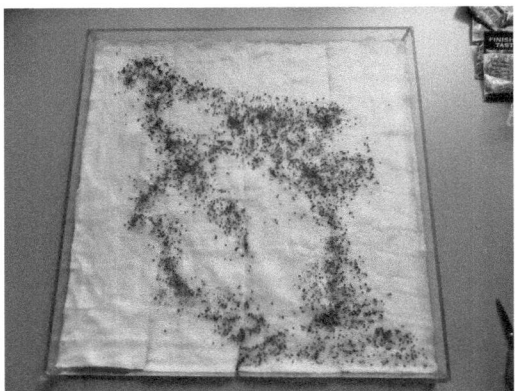

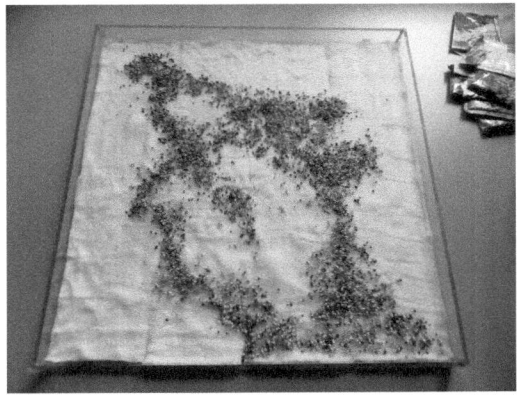

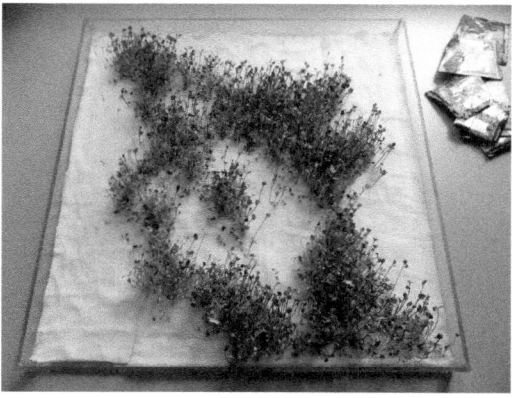

Figure 21.6a–e Sprouting cress seeds representing the growth of a forest. Academy of Architecture Amsterdam, 2011, by Annelies Bloemendaal

to transform one of the exercises into an autonomous product that can be used in teaching. Obviously, the other ones as described above can and will be handled in the same way. This chapter primarily wants to stress the benefit of such exercises, and to raise the awareness of aspects of time in landscape and representation. We will continue to test exercises and try to improve the descriptions so they are of use to other students, teachers and schools.

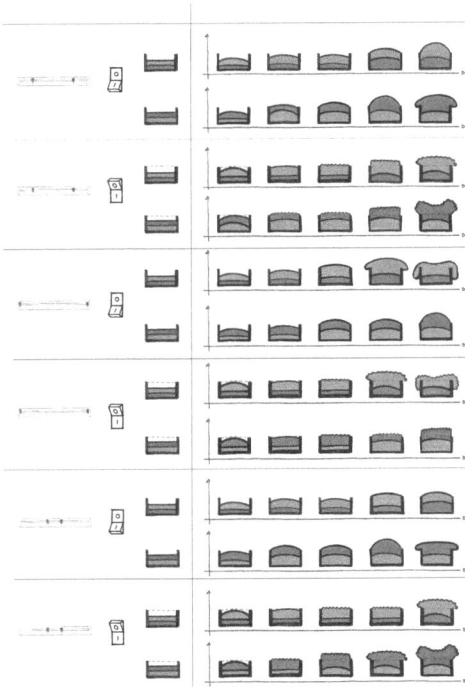

Figure 21.7 *Wachsen Lassen.* Notation or score of designed experiment with dough. TU Stuttgart, 2011, by Ina Neusch, Lu Yi and Nina Bruns

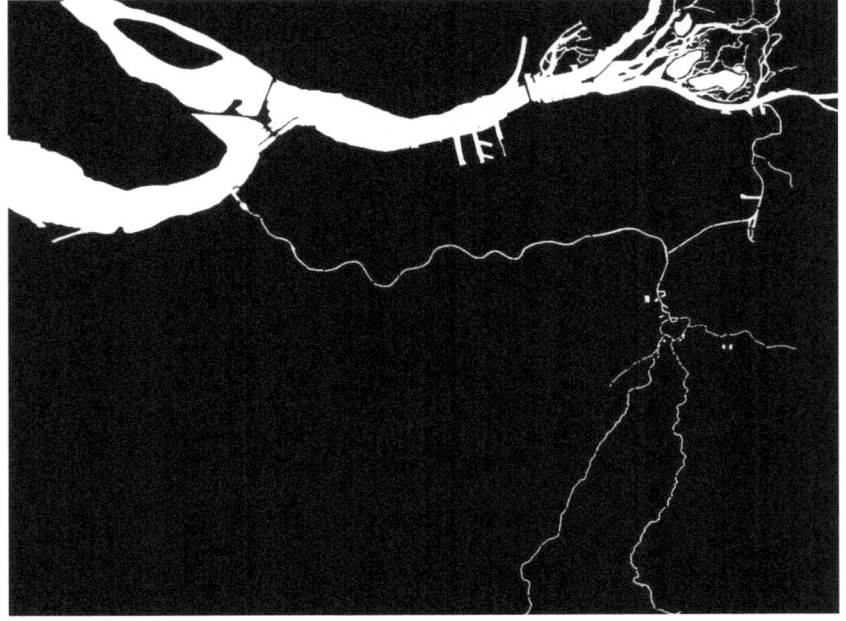

Figure 21.8 H+N+S landschapsarchitecten, De singels van Breda. A hand drawing that took three days. 1994, drawing by author

Figure 21.9   Surrounded by landscape. Form study program, 2009, Academy of Architecture
Amsterdam

Series of moments in time

When learning to draw time, it is essential initially to draw series of moments in time. As this does not require new representational techniques, it is an easy technique for getting to grips with aspects of time.

When can the method be used?

In landscape especially, designs are subject to time in many ways – think of maturing, seasonality, phasing, insecurity, and usage at different moments. If this applies to your design, in any situation in which you would normally produce *one* plan drawing, section or visualization, consider *multiplying* these drawings. You can assemble these drawings into one composite drawing – Edward Tufte speaks of 'small multiples' – or on different pages or slides. Note that a series displayed on separate slides comes close to an animation film (Figure 21.10a–f).

How to use the method?

A series of moments in time requires primarily a definition of the appropriate moments to choose. To make this choice meaningful, the moments must relate to the time mechanisms at work. To communicate the qualities of a planting plan, seasons or months may be the useful moments in time, whereas a river system should be shown in low, normal, high and perhaps extreme water levels. These examples show that moments can be defined very exactly (as in five years, or ten years) or rather qualitatively (as in 'during festivals' versus 'on rainy Sundays', if you want to show how your public space design is responsive to different intensities of use).

Possible procedure

A series of moments in time can be produced by means of analogue and digital drawing techniques. Just as in regular situations, this depends on your own preferences, and on the phase you are in: are you exploring design options for yourself, communicating a design idea to your client or regulating future management? If copies are to be made rapidly, digital drawing serves best.

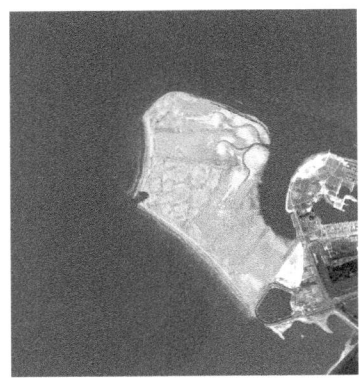

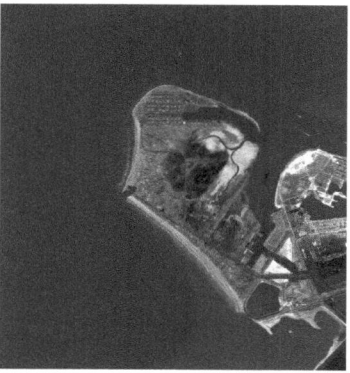

Figure 21.10a–f Vista landschapsarchitectuur en stedenbouw. Development of Maasvlakte Rotterdam. Study, 1993

Step 1: define the time mechanisms at work.

Step 2: define the moments that, following these mechanisms, are relevant to the design.

Step 3: decide, or inform yourself, on how exactly the design is being influenced at that particular moment.

Step 4: evaluate the coherence of your steps, and, if necessary, adapt the design so that it responds in a convincing way to the relevant time mechanisms.

Limitations of the method

In fact, this exercise has no limitations. It poses intellectual challenges though: it requires thinking in an analytical and creative way on how a design may behave through time. Both experienced and inexperienced designers will meet situations where they are not capable of predicting the design's response to a certain happening. One option is: find out! Following this strategy, drawing series of moments of time is an excellent way of gaining more knowledge on landscape. Sometimes you are confronted with a more fundamental lack of knowledge. In such cases, you can choose to explore scenarios based on assumptions that seem appropriate. Observe that also your client or the public has to face this lack of knowledge, and that your drawn scenarios may inspire a fruitful exchange on these uncertainties.

Tips and concerns

Be aware that none of the possible time mechanisms or the possible distribution of moments in time apply to all design situations as a general rule. Try to use this exercise to enhance your skills and to educate your public in an understanding of the *specificity* of landscape for this precise design situation.

Future perspective

With this chapter, I aimed to address two separate areas. One is a contribution to the debate on 'Teaching Landscape': how can we transform ideas on landscape architecture into practical exercises that can be implemented in the daily reality of teaching landscape? Bottom-line, these exercises are instrumental to a better understanding of landscape. The second aim of this chapter is to plead for adapting the theory of landscape representation, and to implement temporal in addition to spatial types of representation. And this would serve the larger goal of distinguishing landscape architecture as the prime discipline that is capable of handling time. Think of the young landscape architect, confronted with climate change in reality: how to make urban open space liveable and sustainable in the context of rising sea levels, more extreme droughts and heat, or very intense precipitation? Knowing how to draw time and how to inform the public on the performance of landscape in time is one of the tasks this young landscape architect has to face, but *by drawing* he or she also explores its functionality over time. The future of landscape, and thus the future of landscape architecture, urges us to innovate teaching landscape with regard to the representation of time, and this chapter indicates a route for doing so, based on an emerging theoretical foundation.

Note

1 This and other comparable remarks refer to the PhD research *Drawing Time* (Dooren 2017).

References

Amoroso, N. (ed.) (2012), *Representing Landscapes: A Visual Collection of Landscape Architectural Drawings* (Abingdon, UK: Routledge).
Balmori, D. (2014), *Drawing and Reinventing Landscape* (Chichester, UK: Wiley).
Corner, J. (1992), 'Representation and landscape. Drawing and making in the landscape medium', *Word and Image* 8/3: 243–275.
Cross, N. (2011), *Design Thinking* (New York: Berg Publishers).
Dooren, N. van (2012), 'Speaking about drawings', *Topos* 80/3: 43–54.
Dooren, N. van (2013), 'Reflexiones sobra representacion/Thoughts about drawings', *Paisea* 27/1: 4–12.
Dooren, N. van (2017), *Drawing Time. The Representation of Growth, Change and Dynamics in Dutch Landscape Architectural Practice after 1985*. PhD diss. (University of Amsterdam, 2017).
Duempelmann, S. and Herrington, S. (2014), 'Plotting time in landscape architecture', *Studies in the History of Gardens & Designed Landscapes* 34/1: 1–14.
Dunnet, N. and Hitchmough, J. (eds) (2008), *The Dynamic Landscape* (Abingdon, UK: Taylor and Francis).
Fraser, I. and Hemni, R. (1994), *Envisioning Architecture: An Analysis of Drawing* (New York: Wiley).
Goodman, N. (1976), *Languages of Art: An Approach to a Theory of Symbols* (Indianapolis, IN: Hackett Publishing Company).
Halprin, L. (1969), *The RSVP Cycles: Creative Processes in the Human Environment* (New York: George Braziller).
Hamilton Thompson, I. (2012), 'Ten tenets and six questions for landscape urbanism', *Landscape Research* 37/1: 7–26.

Hirsch, A. (2006), 'Lawrence Halprin's public spaces: design, experience and recovery. Three case studies', *Studies in the History of Gardens & Designed Landscapes* 26/1: 1–4.

Levine, N. (2009), *Modern Architecture: Representation and Reality* (New Haven: Yale University Press).

Loudon, J. C. (ed.) (1988), *The Landscape Gardening and Landscape Architecture of the late Humphrey Repton* (London: Forgotten Books. Original work published in 1840, Edinburgh: Longman & Co.).

Lynch, K. (1972), *What Time is this Place?* (Cambridge, MA: The MIT Press).

Mertens, E. (2010), *Visualizing Landscape Architecture: Functions-Concepts-Strategies* (Basel: Birkhäuser Verlag).

Mostafavi, M. and Leatherbarrow, D. (1993), *On Weathering: The Life of Buildings in Time* (Cambridge, MA: MIT Press).

Pallasmaa, J. (2009), *The Thinking Hand. Essential and Embodied Wisdom in Architecture* (Chichester, UK: Wiley).

Palmboom, F. (ed.) (2010), *Drawing the Ground: Landscape Urbanism Today* (Basel: Birkhäuser Verlag).

Picon, A. (1992), *French Architects and Engineers in the Age of the Enlightenment* (Cambridge: Cambridge University Press).

Reid, J. (1988), *The Scots Gard'ner. Published for the Climate of Scotland by John Reid Gard'ner* (Edinburgh: Mainstream Publishing Company. Original work published in 1683).

Schön, D. A. (1983), *The Reflective Practitioner: How Professionals Think in Action* (New York: Basic Books).

Sørensen, C. Th. (1975), *Haver: Tanker og arbejder* (København: Christian Ejlers' Forlag).

Steiner, F. (2011), 'Landscape ecological urbanism: origins and trajectories', *Landscape and Urban Planning* 100/4: 333–337.

Torres, C. (2009), 'Crisis in landscape representation', *Kerb* 17/1: 53–59.

Treib, M. (ed.) (2008), *Representing Landscape Architecture* (Abingdon, UK: Taylor and Francis).

Tufte, E. (1990), *Envisioning Information* (Cheshire, CT: Graphics Press).

Waldheim, C. (ed.) (2006), *The Landscape Urbanism Reader* (New York: Princeton Architectural Press).

22

Landscapes as co-construction of knowledge

Implications on the classroom

Ellen Fetzer

Introduction

In the past decades the emergence of constructivist schools of thought has influenced various scientific domains, in particular sociology, psychology, neurosciences and philosophy. All of these domains are involved in shaping our concept and understanding of landscape, learning and designing. This chapter is about how to rethink landscape education in the light of this 'constructivist turn'. Constructivist thinking does not intend to replace established epistemological approaches. It rather promotes a shift in emphasis. The suggested focus is not on the definition, transfer and adoption of concepts from a sender to a receiver – which we could understand as a common idea of 'teaching' and 'learning'. Constructivism aims to make the constructed nature of any concept, which is its key axiom, the core of any observation, discourse and – if translated to pedagogy – any learning process. All of this is currently accelerated by the overall digitalization of various domains of human life, the pluralization of our societies and the growing need for lifelong learning. Promoting lifelong learning has been an EU policy goal for more than a decade now. The most recent European funding programme (2013–2020) goes further by promoting the idea of opening up education through new technologies. This goal is based on the European Union's Strategic Framework for Education and Training 2020 (ET, 2020). Therefore, if learning, including landscape learning, is to happen increasingly across age groups, sectors, disciplines, cultures and institutions, we need an inclusive pedagogy. Constructivism can be one possible approach, not excluding the value of various other learning theories. In the following, I will elaborate on what we might understand by a constructivist approach to landscape, learning and designing. I am trying to outline which requirements for landscape education we might derive if we follow a constructivist line of thinking. I will conclude with an example that shows how both the digital and the constructivist turn can come together and open up new possibilities for landscape education.

Learning as construction

We still cannot say that neurosciences have been able to decipher how our brain actually works. But a largely supported idea is that of our brain being a largely self-referential system.

It seems that the human brain is much more concerned with a constant internal reorganisation of information than with the perception of new external stimuli (Roth, 2001: 212). Consequently, constructivist learning theory assumes that our brain develops reality by itself. This approach further suggests that reality does not exist outside of the human brain. This builds a dialectic relationship to the common notion of teaching as conveying an agreed representation of reality from a sender to a receiver. In this model, it is assumed that learners can recognize and internalize an external representation of knowledge. Again, constructivism does not deny the relative validity of these knowledge representations. However, it asks for a shift of emphasis from external knowledge representation to the process of individual and internal knowledge construction. This is particularly relevant for higher and continuing education in which all learners are adults. Adults enter the learning process with their own learning biographies, value schemes, experiences and expectations and, in our increasingly pluralistic societies, also with their specific cultural lenses. Learning can only happen if an individual learner takes the decision to link new concepts to his/her existing knowledge schemes. This decision requires that these new concepts be considered to be relevant and viable with respect to the prior knowledge and the expectations of the learner. The decision on the viability is only taken by the learner him/herself and teachers have in fact no influence on this, regardless of their competence and enthusiasm for the subject (Siebert, 2012: 27). Constructivist theory does not give an answer to the question of how we can enhance the viability of knowledge for each individual learner. But another relevant criterion is offered by the social dimension of constructivism: learning is also considered to be a social interaction and the notion, recognition and discourse of differences within groups are important factors in any learning process (Vygotski, 1986). Constructivist pedagogy therefore tries to design learning as a discourse, which eventually turns the teacher partly into a facilitator and partly into a coach, responsible for an active and critical dialogue. Such active and critical engagement lies, not at least, in the foundation for our democratic culture. 'Democracy is more than a form of government, it is primarily a mode of associated living, of conjoint communicated experience' (Dewey, 1930: 101).

Luckily, landscape education already provides various foundations for integrating a constructivist dimension. The idea of the 'reflective practitioner' (Schön, 1983) is widespread in the discipline and the design studio is the common translation of this principle into educational practice. The design studio deals with a real-world context, which allows learners to contextualize their experience and values. The learners' process is iterative, largely self-organised and the tasks are usually to be solved in teams, which mimics the professional reality for which the students are being prepared. The design studio process allows teachers to observe and accompany the learners while they are creating, reflecting, synthesizing and evaluating ideas. Alternative solutions are possible and desired and ideally verified against a set of criteria.

However, the landscape studio could even go further by integrating more dimensions of educational constructivism. This might imply the following.

- Offer learning environments with open agendas in order to address landscapes and urban environments for which there are no defined programmes yet. This way, learners can train the capability to uncover the needs and potentials of a place, by observation or dialogue, and to specify a design programme iteratively.
- Confront students with as many different actors as possible. This needs to be done with great care of course in order not to raise false expectations from the public. But if a university

builds community partnerships, ideally over many years, there are various opportunities for mutual benefits and learning, creating an authentic and motivating learning environment.

- Make the development of evaluation criteria part of the learning process. This means that learners are asked to develop evaluation criteria based on a comprehensive reflection of a project's objectives and values. This can even go further and include peer evaluation, i.e., students are evaluating each other. This does not necessarily mean replacing teacher-centred evaluation, but it can be an important additional dimension fostering students' critical thinking skills, self-awareness and social responsibility.

In the following, I will further elaborate on the implications of the constructivist turn on the concept of landscape and the concept of design, as both understandings have a strong influence on the pedagogy of landscape architecture and related disciplines.

Landscape as construction

I explained that, from a constructivist point of view, reality would not exist outside of our individual consciousness. Consequently, this principle also applies to the concept of landscape: any landscape is an individual and unique construction build by our consciousness. These constructions are shaped by our experiences and value schemes (Kühne et al., 2015: 20; Gailing and Leibenath, 2013). Similar to any knowledge field, landscapes also need to be viable and linked to people's experience patterns. Some of these landscape value schemes are shared with a social group at large, some only with a local community and some are unique and very personal. The European Landscape Convention (ELC) clearly supports this way of thinking by defining landscape 'as an area as perceived by people' (Council of Europe, 2000). Landscapes have certainly also physical, socio-economic and spatial dimensions, which we may study with established research methods, and geographical information systems provide us with ever more data for understanding this dimension. However, this knowledge leads us simultaneously into the 'expert trap': we see what we know in the landscape and automatically take this knowledge for granted. It is impossible to 'un-learn' what we have once internalized and linked to our knowledge schemes. This would not pose any serious problem if landscape planning, design and management involved only a cooperation of likeminded experts. But this is not the case in practice for at least two reasons: first, landscape constitutes the everyday environment of people and forms the foundation for their daily action, community experience and identity building. Based on people's multifaceted life experiences, values are constructed over time and associated with landscape structures and elements. This eventually develops into a symbolic dimension and a landscape of 'sacred spaces' (Hester, 2006). A landscape expert does not know any of this in the first place. But it is an acceptable condition as long as s/he is aware of the fact that there is an 'unknown' dimension in the landscape. Second, if we follow the ELC consequently and assume that landscape is the everyday environment of people, then the values, knowledge and needs of these people need to be included in the assessment, planning, design and management of landscapes. Academic education in the field of landscape needs to prepare students for these realities so that their work can bring the European Landscape Convention to life. Educational programmes can foster this by equipping students with methods and tools that help in uncovering the hidden layers of landscapes. This might involve, on the one hand, observer-based approaches such as behaviour mapping (Moore and Cosco, 2010), sociotope mapping (Ståhle, 2006), interviews or the compilation of landscape biographies (Kolen et al., 2017). On the other hand, there are also interactive methods in which the community and its observers are equal partners. These participatory action research methods include, amongst others, open space workshops, storytelling, photovoice, world cafés, community design events and digital or analogue collaborative mapping (Wates, 2000). These methods are highly practical for planners and designers and, above all, they are not restricted to planning disciplines. They can be very well applied in interdisciplinary contexts where different landscape-related professions collaborate.

Planning and designing as co-construction

Planning and designing can also be viewed through constructivist lenses if we assume that both are processes with multiple actors in which learning occurs at various levels.

> Planning, and specifically environmental planning, is a process for collectively, and interactively, addressing and working out how to act with respect to shared concerns about how far to go and how to "manage" environmental change. . . . We cannot therefore predefine a set of tasks which planning must address, since these must be specifically discovered, learned about, and understood through intercommunicative processes.
>
> *(Healey, 1992)*

Such thoughts came to be known as the *communicative turn* in planning. This new perspective extended the professional profile of the planning professions to a new set of tasks such as process design and community visioning facilitation. Even though Patsy Healey's article was published 25 years ago its implications are more than relevant today, especially with regard to the New Urban Agenda adopted in October 2016 at the United Nations Conference on Housing and Sustainable Urban Development (Habitat III). This agenda aims to lay out a new framework for how cities should be planned and managed to best promote sustainable urbanization. Article 92 of the New Urban Agenda suggests that:

> We will promote participatory age- and gender-responsive approaches at all stages of the urban and territorial policy and planning processes, from conceptualization to design, budgeting, implementation, evaluation, and review, rooted in new forms of direct partnership between governments at all levels and civil society, including through broad-based and well-resourced permanent mechanisms and platforms for cooperation and consultation open to all, using information and communication technologies and accessible data solutions.
>
> *(United Nations, 2016)*

Clearly, the New Urban Agenda asks for integrated and participatory approaches to which also landscape-related planning and design professions need to contribute. Coming back to the concepts of landscape and learning from which this chapter has departed, it can be concluded that planning and designing are complex learning processes in which the landscape serves us as a discourse framework. The framework itself evolves during this learning process, which is why understandings, visions and values of landscapes can be created and changed by intercommunicative processes. It seems relevant that landscape pedagogy is aware of this context. We operate with, at least, three constructed domains: the learner, the landscape and the design process. Maybe we should not call them construct*ed*; they are, rather, continuously construct*ing*.

I will now move on to the role of digitalization in this context. Information and communication technologies, as mentioned throughout the New Urban Agenda, will open up new opportunities for sharing landscape knowledge, shaping new discourse dimensions and envisioning alternative futures. In the following, I will elaborate on how this dimension could be integrated into our landscape classrooms based on constructivist pedagogy.

Facilitating intercultural landscape discourse with online seminars

Digitalization is certainly shaping various aspects of landscape planning and design, particularly since drawing and visualization tools have become not only standardized ways of representation but also driving forces that are changing the planning and design process itself. Another dimension of digitalization is connectivity of people and systems enabled by the internet. The idea of cross-institutional teaching and learning for the academic community of landscape disciplines has emerged within the course of the three subsequent LE:NOTRE ERASMUS projects, which eventually led to the foundation of the LE:NOTRE Institute. The project coordinator, Professor Richard Stiles, presented this idea together with the author for the first time in September 2008 at the ECLAS conference in Alnarp, Sweden. The authors emphasized that:

> the internet in fact offers academics new opportunities to collaborate internationally in teaching and research. International research collaboration is nothing new, although it has much potential to develop both quantitatively and qualitatively in new ways, but so far international teaching collaboration has received little discussion. This surely is an important part of the real potential of the application of the internet in universities – the development of what we might call e-teaching. International e-teaching is far more than an incremental development of what was already done within the existing university structures. Instead it exploits a completely new potential, which has previously not existed at all. As such it can indeed provide not just an important impetus for the further development of disciplines but also for the re-structuring of the higher education sector as a whole.
>
> *(Fetzer and Stiles 2009: 235)*

Curriculum innovation by cross-institutional and cross-cultural collaboration facilitated by web-based formats started in 2007 as part of the LE:NOTRE project and the process continues to the present as part of the core activities of the LE:NOTRE Institute, which has been established as a self-standing institution after the EU funding for academic networks had ended. European Union funding for higher education continues to support these activities, most recently by the establishment of two strategic partnerships, 'Landscape Education for Democracy' and 'Inclusive Coastal Landscapes', in which the LE:NOTRE Institute is a partner. The pedagogical, organizational and technical development of an educational format suitable for the needs of landscape architects and related professions was done in the form of a pedagogical action research study and published in 2014 (Fetzer, 2014). This open access publication presents the educational framework in great detail and can be consulted by anyone wishing to dig deeper into the subject of e-pedagogy and landscape. One core motivation shaping the design of these online seminars was the idea of enabling a virtual learning environment to which learners could link their personal knowledge, contextualize it internationally and participate actively in a collaborative process of reflection and co-creation.

The most recent example of this kind is the collaborative online seminar 'Landscape Education for Democracy'. This seminar is offered jointly by a partnership of five European universities and the LE:NOTRE Institute. It combines formal education of enrolled students with open participation modes involving learners from outside the university. The participants are introduced, amongst others, to the theory of landscape as a social construction. This is accompanied by an exercise called 'Landscape Symbols'.

Learners are invited to identify features in their every-day surroundings that have symbolic meaning for them and to interpret that symbolism. The task employs a method called *photovoice*,

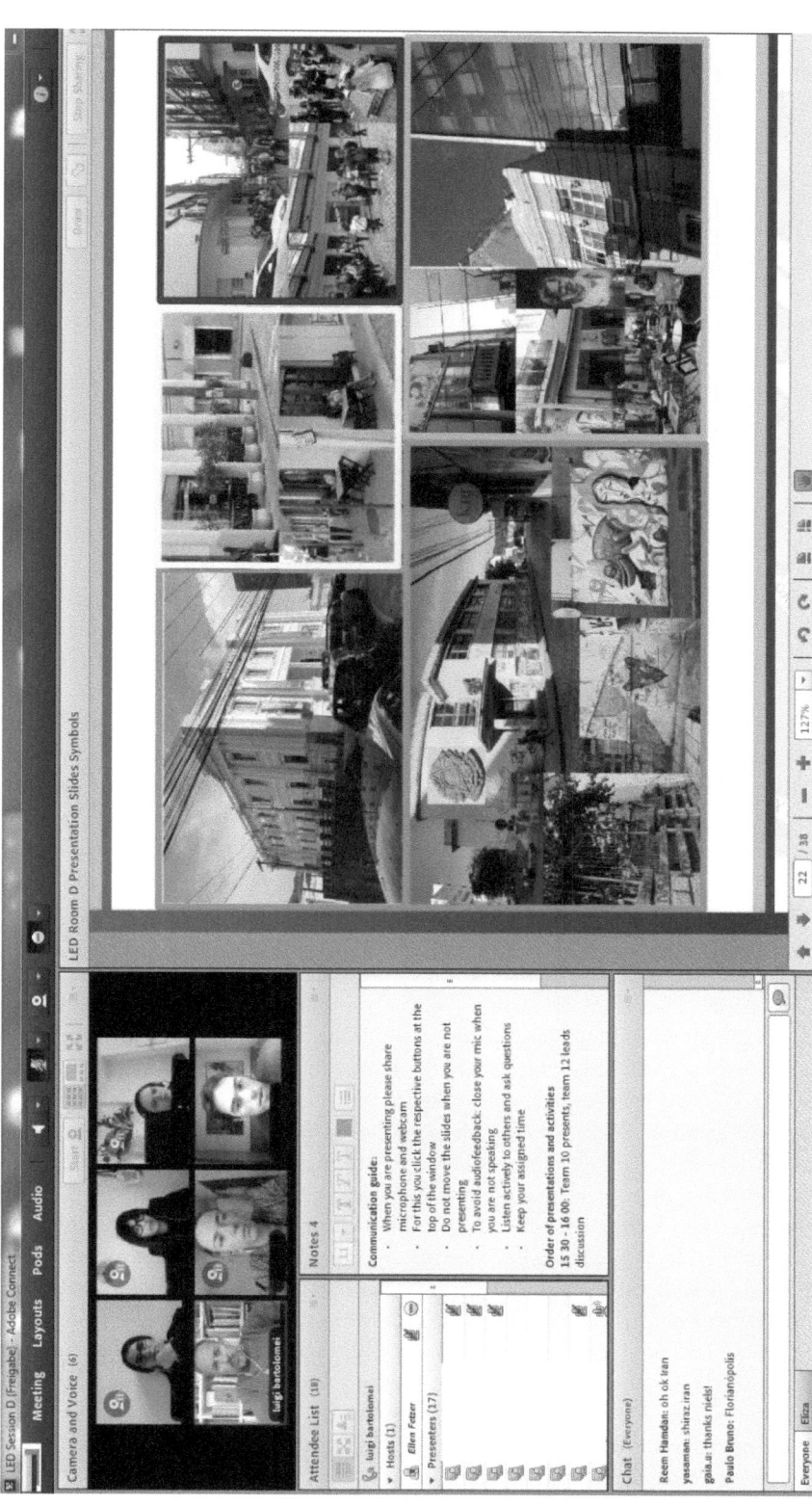

Figure 22.1 Landscape Education for Democracy: screenshot from a breakout room session held in April 2017. A virtual team with learners from Brazil, Kazakhstan, Italy, Iran and the US presents reflections on landscape symbols in their respective local environments. On the screen there are some scenes from Florianopolis in Brazil. The group findings were commented on by project partner Luigi Bartolomei from the University of Bologna

which uses pictures to identify particular landscapes and their symbolic nature. The participants not only share these landscape symbols on a public wiki, they also reflect in intercultural teams about their meanings and present the outcomes of their reflection in the virtual classroom to an international plenary. This exercise addresses various dimensions that seem relevant for contemporary landscape education: 1) to reflect on what is a relevant symbol in one's own landscape, 2) to understand that symbolic meanings are key to perceiving, memorizing and validating our landscapes, 3) to recognize that symbols are not perceived in the same way by different social and cultural groups and that their meaning changes through history, 4) that everyone is able to identify symbols in a landscape, 5) that the diversity of landscape symbols and their relativity is a basic condition for planning and designing landscapes, of which a future professional needs to be aware.

This idea of linking the specific local landscape issues to an international discussion and reflection is a common pattern in the design of the exercises applied in these online seminars. It is complemented by the possibility to invite speakers from across the world to talk about their experiences in the online classroom. In the 'Landscape Education for Democracy Seminar' the learners' journey continues with an identification of a local landscape democracy challenge, a live role play in the online classroom and the development of a democratic change scenario with the help of their international virtual team.

This online learning format is meant to allow for open access from anywhere in the world, thus providing educational opportunities also for those who have only limited access to landscape education. This global openness of the European landscape community is a relevant future vision given the many environmental challenges across the world, especially in the fast-growing metropolitan areas of the global south, places that are largely underserved with regard to landscape education. The learning activities are always designed to allow for a discussion of the learners' specific local landscape conditions. Next to this vision of opening up education, learners are automatically practicing virtual teamwork as part of the seminar process. This form of teamwork is already prevalent in the professional world of landscape practitioners as well as in research teams, and is likely to become even more relevant in the future in the light of globalization on the one hand, and the need for energy efficiency on the other. Altogether, the added value of this form of learning crystallizes in the opportunity for sharing local landscape knowledge with an international community and thus learning what is relevant in one's own landscape while creating an understanding for other landscape challenges and potentials from anywhere in the world. The technical facilitators are wikis for collaborative writing and virtual classrooms for synchronous knowledge exchange and discourse. Since 2007, we have conducted 16 seminars involving in total approximately 400 learners and 80 lecturers from 40 different countries in the world. Subjects are always selected in such a way that they allow for a reflection from various international perspectives for which different experts from the European and international communities are being invited to the online classroom. Participants link to the seminar by contributing knowledge on the subject from their local contexts. For example, in the seminar on Green Infrastructure in 2014, participants looked for green infrastructure potential in their everyday environment, documented these cases on the seminar wiki and reflected on them in international teams. This way, team members could learn about the local conditions in this specific landscape and contribute their knowledge to creating new ideas that would enhance the green infrastructure potential of this place. Another example is a seminar on landscape concepts that asked participants to present the understandings and meaning layers of the concept of landscape in their national languages and to discuss similarities and differences across cultures (Faurest and Fetzer, 2015).

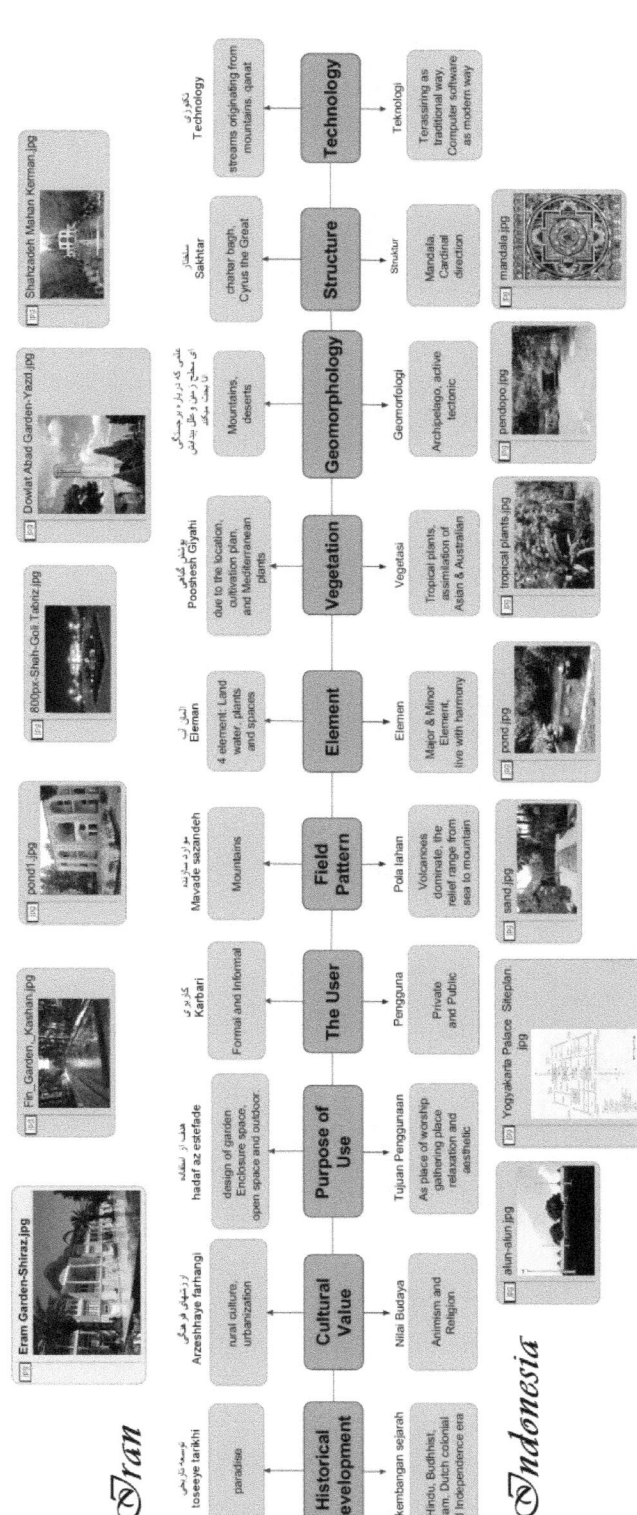

Figure 22.2 Example of a concept map comparing the concept of landscape architecture in Iran and Indonesia. Outcome of an online seminar with international participants held in 2012. Authors: Masoumeh Rajabi (Iran) and Melissa Abas (Indonesia)

Outlook: the global landscape classroom

Opening up education by means of the internet is certainly not the only development direction for landscape-related pedagogy. Disciplines like landscape architecture will always be taught in relation to very site-specific conditions and addressing a unique local community. But, while landscape features are always site-specific, the discourse on their values, meanings and futures is not. All of these are human constructions that can be documented, exchanged, reflected and developed by any possible means. The unique opportunity of online courses on landscape issues lies in the possibility to bring multiple perspectives into the classroom. This can be done also in a campus-based setting, but substantially more time, energy and money are required if people from different countries want to meet physically, especially if the audience comes from outside Europe. One could argue here that such an effort could reinforce a Eurocentric worldview. Is it a virtual mission spreading landscape architecture into the remotest corner of the world without listening to local solutions that might already be in place? Do we need anything like a global landscape mission? Or shouldn't we rather try to protect the manifold local interpretations and the linguistic richness that luckily still exists around issues of place, environment and identity (which some people in the world call landscape, but certainly not everyone)? This is exactly the reason why a constructivist perspective on pedagogy, which is after all a culture of listening, is more relevant than ever. Classroom activities can be designed in a way that allows for this diversity to take shape and become the actual core of the learning process.

However, major challenges remain if we want to mainstream ICT-based teaching across institutional boundaries in Europe and maybe even beyond. There are still many obstacles, the most relevant being the prevalent difficulties regarding curricular integration and full academic recognition of international online seminars. During past implementations this has been observed as the most relevant source of problems for the participants. Another challenge is the required capacity building among the staff members, which is part of the ongoing EU-funded projects, but certainly reaches only a limited number of teachers at the moment. The LE:NOTRE Institute will continue step by step on this path and gradually involve more and more learners and educators in this experience. Systemic change is always a learning process. Ideally, this development will be supported in the near future by strategic objectives of the landscape-related study programmes in Europe and beyond. I want to end with the vision of an 'international landscape hour', which all study programmes across Europe and maybe even the world would share in their schedules and curricula. In this hour, learners, educators, PhD students, practitioners and the interested public will meet online to share knowledge on our landscapes and become sensitive for what it means to be a global citizen. It could be a good contribution to the New Urban Agenda on behalf of academia.

Project websites to which this article refers

LE:NOTRE Institute: www.le-notre.org (last access: 24.04.2017).
ERASMUS+ Strategic Partnership Landscape Education for Democracy: www.led-project.org (last access: 24.04.2017).
Online Seminar WIKI: http://fluswikien.hfwu.de (last access: 24.04.2017).

References

Council of Europe (2000), *The European Landscape Convention*, www.coe.int/en/web/landscape (accessed on 10.12.2018).
Dewey, John (1930), *Democracy and Education: An Introduction to the Philosophy of Education*, New York: The Maximilian Company.

European Union (2015), *ET 2020 – Strategic Framework – Education & Training 2020*, http://ec.europa.eu/education/policy/strategic-framework_en (accessed on 03.12.2017).

Faurest, Kristin and Fetzer, Ellen (2015), A Condition of the Spirit: Mapping Landscape, Language and Culture in: Bruns, Kühne and Schönwald, Theile (Hrsg.) *Landscape Culture – Culturing Landscapes – The Differentiated Construction of Landscapes*, Wiesbaden: Springer VS.

Fetzer, Ellen (2014), *Knowledge Building in Landscape Architecture. A Pedagogical Action Research Study with International Online Seminars*, dissertation, Kassel University Press: https://kobra.bibliothek.uni-kassel.de/handle/urn:nbn:de:hebis:34-2014091846030.

Fetzer, Ellen and Stiles, Richard (2009), Broadening the Basis of Landscape Architecture Education. Computer-Supported Collaborative Learning: A Strategic Potential for Landscape Architecture Education in Europe. In: Ingrid Sarlov-Herlin (Ed.), *Proceedings of the ECLAS Conference Alnarp, 2008. New Landscapes, New Lives and New Challenges in Landscape Planning, Design and Management*, Faculty of Landscape Planning, Horticulture and Agricultural Science, Swedish University of Agricultural Sciences, 2008, pp. 233–242.

Gailing, Ludger and Leibenath, Markus (2013), The Social Construction of Landscapes: Two Theoretical Lenses and Their Empirical Applications, *Landscape Research*, 40(2): 123–138, doi: 10.1080/01426397.2013.775233.

Healey, Patsy (1992), Planning Through Debate: The Communicative Turn in Planning Theory, *The Town Planning Review* 62(2): 143–162.

Hester, Randolph (2006), *Design for Ecological Democracy: Sacredness*, Cambridge, MA: The MIT Press.

Kolen, Jan, Renes, Hans and Bosma, Koos (2017), Landscape Biography in: Brink, Adri van den, Bruns, Diedrich, Tobi, Hilde and Bell, Simon, *Research in Landscape Architecture. Methods and Methodology*

Kühne, Olaf, Gawroński, Krzysztof and Hernik, Józef (2015), *Transformation und Landschaft. Die Folgen sozialer Wandlungsprozesse auf Landschaft*, Wiesbaden: Springer VS.

Piaget, Jean (1977), *The Development of Thought: Equilibration of Cognitive Structures*, Oxford: Blackwell.

Pörksen, Bernhard (2015), *Schlüsselwerke des Konstruktivismus*, Wiesbaden: Springer VS.

Robin C. Moore and Nilda G. Cosco (2010), Using Behaviour Mapping to Investigate Healthy Outdoor Environments for Children and Families: Conceptual Framework, Procedures and Applications, in: Ward-Thompson, Catherine, Aspinall, Peter and Bell, Simon, *Open Space, People Space 2* Abingdon, UK: Routledge.

Roth, Gerald (2001), *Fühlen, Denken, Handeln*, Frankfurt am Main: Suhrkamp.

Schön, Donald (1983), *The Reflective Practitioner. How Professionals Think in Action*, New York: Basic Books.

Siebert, Horst (2012, 7th edition), *Didaktisches Handeln in der Erwachsenenbildung. Didaktik aus konstruktivistischer Sicht*, Augsburg: ZIEL Verlag.

Ståhle, Alexander (2006), Sociotope Mapping: Exploring Public Open Space and its Multiple Use Values in Urban and Landscape Planning Practice, in: *Nordic Journal of Architectural Research*, 19(4): 59–71.

United Nations (2016), *New Urban Agenda*, resolution adopted by the General Assembly on 23 December, http://habitat3.org/wp-content/uploads/New-Urban-Agenda-GA-Adopted-68th-Plenary-N1646655-E.pdf, accessed on 21.04.2017.

Vygotski, Lev (1986), *Thought and Language*, Revised and Expanded Edition, Cambridge, MA: MIT Press.

Wates, Nick (2000), *The Community Planning Handbook: How People can Shape Their Cities, Towns & Villages in any Part of the World*, Abingdon, UK: Routledge.

Part III
Transforming the landscape

This final part of the book recognises the dynamic and continually changing nature of the landscape. Throughout the year, it develops naturally with the different seasons and the cyclical development of plant life; over the years, the scenery changes with the growth and maturation of vegetation. With this kind of natural evolution, for example in its appearance, its possibilities for use and ecological function vary. The biggest and most abrupt changes happen through human intervention with the construction of new built developments and infrastructure projects, as a result of which landscape may be said to undergo a 'phase change' from rural to urban. Rural or peri–urban landscapes may mutate into urban or industrial landscapes or they may become the locations for new built infrastructure. Furthermore, sites formerly used by industry, military or infrastructure may be reused and redesigned for another purpose. It is therefore an indispensable part of any landscape education programme to consider the transformation of landscape in one form or another.

After analysing, or 'reading', a landscape and having found a suitable way to represent it, transforming the landscape is one of the most important tasks of landscape architects, planners and any other disciplines with a focus on the development of landscape. The transformation should aim to result in a better landscape, maybe a more useful or more sustainable landscape, maybe to an ecologically richer area, and maybe to a more beautiful landscape. In most curricula, we therefore find course units aimed at teaching the students to develop ideas and methodologies for transformation, in the form of planning, design and site construction classes.

Project-based teaching of planning and design is widely used in landscape programmes. The studio setting aims to simulate the real-life situations students may find themselves in after graduation. In this part, different forms and phases of both studio-based and real-life simulating teaching are discussed, as well as other types of classes dealing with aspects of the transformation of landscape.

In the first chapter of this part, Pinar Köylü introduces Inquiry-Based-Learning (IBL) as a strategy that engages students to develop their skills and appropriate values, attitudes, and habits and encourages self-initiative and lifelong-learning. She also explains Kolb's theory of experiential learning (ELT) and different learning styles. Tutors of design studios should focus on teaching the students how to make use of their knowledge and evaluate their progress in terms of skills they develop throughout the studio.

Following this, Jacky Bowring and Mick Abbott introduce what they call 'the Design Lab method' in which the design process is the key method for research and solving 'wicked' complex problems. The term laboratory is intended to demonstrate the experimental character of a methodology based on testing and collaboration. The strategies here are questioning, collaboration, designing, grounding and communication.

Teaching a workshop-studio approach in the creative phase of student's work requires appropriate teaching. Referring to the ECLAS Guidance Paper (Bruns et al. 2010), Davorin Gazvoda focusses on the role and abilities of the teachers in his chapter. The involvement of special experts, local laypersons as well as professionals in the field of landscape is welcome in a studio to formulate realistic approaches to solve existing problems. For the responsible teacher, involving experts means being flexible and open to react to unforeseen developments. The focus very much lies on the mastering of procedures and on the development and application of students' creativity.

By taking an interdisciplinary approach, a landscape architecture studio can also be supportive for a community service-learning facility within university centres that supports community engagement. In his chapter, Peter M. Butler describes this opportunity where extended transdisciplinary teams of faculty, professionals and students, together with community members, create the possibility for change and positive effects in many ways for the benefit of both the university (members) and the community. It defines broad learning outcomes for courses as activities and desired impacts for communities and for students within the classroom.

Karl Kullmann, in his chapter, points out the significance of a landscape approach in the form of landscape urbanism in landscape architecture education. In addition to the different methods to understand the urban landscape, this approach addresses such different aspects as urban decline, infrastructure, topography, etc. This spectrum of approaches, it is contended, should ideally infiltrate throughout the whole landscape architecture education.

A crucial part of landscape teaching involves teaching landscape construction. Construction is often regarded as separate from the design phase. In her chapter, Ingrid Schegk describes methods for teaching construction as part of a holistic design process.

Simon Colwill's chapter also focusses on teaching landscape construction but using built landscape projects as a basis for on-site learning activities. This teaching method aims to improve learning by involving students in on-site critique, analysis and evaluation of how construction details support the design.

Finally, Jörg Rekittke and Yazid Ninsalam introduce their particular form of field research in which teaching and research coincide. In the described case study, the landscape was explored by land, from the air and by the sea, using different devices and software tools for the different parts of the site, which, according to the authors, gives a sound and necessary knowledge and understanding of the landscape in focus.

Teaching landscape architecture often aims at transforming the landscape. All serious planning is not just about initiating, but also taking responsibility for changes. Students need to learn about the consequences of their ideas and plans if they were put into reality. Many different methods and approaches for finding the best way of dealing with a special site in focus as part of the complex and diverse landscape are described in this chapter and in the whole book. We hope that, as a result of consulting this book, teachers as well as students will be motivated to choose the best ways for their teaching and learning to help find specific solutions in response to the multiplicity and complexity of landscape contexts. If so, it will have proved itself both valuable for education and beneficial for the transformation of the landscape. Hope, utopian aspirations and the best teaching methods will lead to a better landscape in the future.

Reference

Bruns, D., Ortacesme, V., Stiles, R., de Vries, J., Holden, R. and Jorgensen, K. (2010), *ECLAS Guidance on Landscape Architecture Education. Tuning Landscape Architecture Education in Europe, Version 26*. The Tuning Project ECLAS–LE:NOTRE www.tuningacademy.org/wp-content/uploads/2014/02/elcas_tuning.pdf.

23

An overview of the landscape design studio in the context of experiential learning theory

Pinar Köylü

Introduction

Landscape architecture, both as a field of professional activity and an academic discipline, aims 'to create, enhance, maintain, and protect places so as to be functional, aesthetically pleasing, meaningful and sustainable and appropriate to diverse human needs and goals' (Bruns *et al.* 2010: 11). In order to achieve these purposes, it essentially necessitates acquisition of knowledge from the fields of natural and social sciences, as well as arts and humanities. Hence, the landscape architecture programmes in a variety of schools encompass courses in the history of landscape architecture, planning and design of urban and rural areas, ecology, plant materials and planting design, as well as site construction and engineering, which involve domain material incorporated into the design studio learning experience.

Studio-based learning takes up 40 to 60% of a student's workload (Bruns *et al.* 2010). Accordingly, the design studio, as with other disciplines like architecture, interior architecture, and industrial design, stands at the heart of most of the landscape architecture curricula. Given that the design studio holds a pivotal position in landscape architecture pedagogy, it becomes vital for the studio tutors to understand how students learn in order to enhance the structure of the course. Besides, considering students' learning styles would help instructors apply appropriate teaching techniques that would match their students' preferences.

Learning, as defined by the United Nations Educational, Scientific and Cultural Organization (UNESCO n.d.), is 'a process that brings together personal and environmental experiences and influences for acquiring, enriching or modifying one's knowledge, skills, values, attitudes, behaviour and world views'. In view of this definition, the key issue for learning to arise in a design studio calls for direct involvement of students throughout studio sessions. Such an approach, as compared with traditional learning methods that put learners as passive recipients of information, relies on the active engagement of students with landscape design so that they can both explore how to design and develop an understanding of becoming a landscape designer. This type of learning strategy by which students construct their own knowledge points to inquiry-based learning (IBL).

Inquiry-based learning is a type of holistic learning strategy, which engages students in their own knowledge production in such a meaningful, purposeful, and self-regulated way that they can develop requisite skills as well as appropriate values, attitudes, and habits that are essential for building up self-initiative, and for encouraging higher-order thinking and lifelong learning (Blessinger and Carfora 2014).

Inquiry-based learning involves two sub-forms of interactive learning mechanisms—active and experiential learning—which, despite their difference in terminology, both aim at increasing students' motivation and at developing skills, attitudes and values (Salama 2010). Hence, owing to these similarities in their intentions, although active and experiential learning relate dually, they differ in terms of the emphasis they put on the importance of 'being active' and of 'experiences' in learning. Accordingly, while active learning draws our attention to the active involvement of the learner throughout the learning process, experiential learning focusses on the importance of experiences in constructing knowledge. Thus, within this context, this chapter aims to interpret the landscape design studio in its associations with experiential learning, particularly with frameworks developed within experiential learning theory (ELT), as well as with the learning styles of students including Kolb's nine-style typology.

Experiential learning and Kolb's ELT

Experiential learning, which considers experience as the crucial basis for learning, traces its roots back to early times. In his quote from 400 BC, the ancient Greek philosopher Sophocles called attention to the importance of experience for learning by stating that 'One must learn by doing the thing, for though you think you know it—you have no certainty, until you try' (quoted in Gentry 1990: 9). From that time onwards, this thought of Sophocles, which addressed doing and trying as the key points of a guaranteed learning, continued to be acknowledged as a means for learning by the educators. Indeed, as mentioned by Lackney (1999), in the guilds of the Middle Ages, which actually formed the origins of studio-based learning, master craftsmen trained young apprentices primarily in the arts and crafts by exposing them to the adult world and letting them work on real products under their supervision in the ateliers. Subsequently, by the foundation of the Ecole des Beaux Arts in Paris in the 19th century, learning by doing as a teaching strategy became embedded in an educational institution (Holgate 2008). Moreover, having been established by Walter Gropius in Weimar, Germany, in 1919, learning by doing was also employed by the Bauhaus, which adopted an instruction mainly of a practical nature (Lackney 1999).

Although the importance of experience for the training of apprentices and education of students has been appreciated for a long time, educationalists, as expressed by Weinstein (2016), have commonly agreed on the matter that it was John Dewey who, in his 1938 book *Experience and Education*, initiated the use of the term 'experiential learning', which assumed an increased public prominence in the 1960s. At this time, David A. Kolb, with his writings concerning the learning cycle, became one of the most important theorists working in the field of learning (Stein 2004).

Kolb developed ELT, which hinged its intellectual origins on the works of well-known western scholars (e.g., William James, John Dewey, Kurt Lewin, Jean Piaget, Mary Parker Follett, Lev Vygotsky, Carl Jung, Carl Rogers, Paulo Freire) who placed experience at the centre of their theories of human learning and development (Kolb 1984; Kolb and Kolb 2005, 2013). Among these scholars from varied professions and cultural perspectives, Dewey's pragmatism, Lewin's social psychology and Piaget's cognitive development, which appeared different in practice but shared common characteristics—principally in the nature of the learning process upon which they were built—stood as the major traditions of experiential learning from which Kolb was primarily inspired (Kolb 1984, 2015).

Similar to its precedents, learning, according to ELT, occurs by proceeding successively through the four-staged cycle of the idealized learning process that involves experiencing,

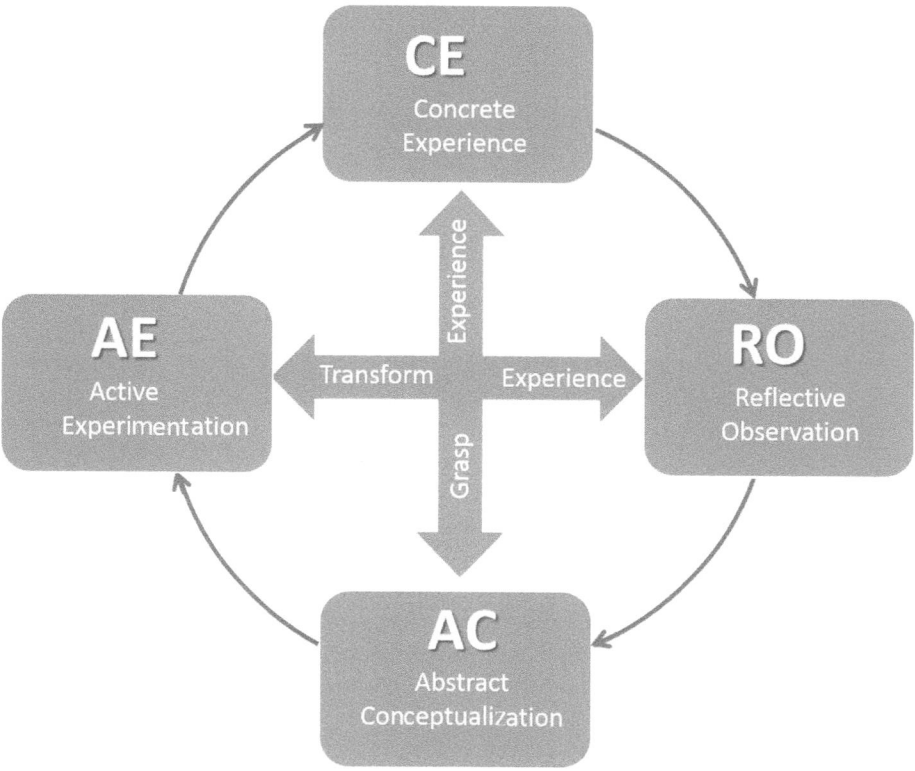

Figure 23.1   The experiential learning cycle (Kolb and Kolb 2013, p. 8)

reflecting, thinking, and acting (Figure 23.1). As the model portrays, concrete experiences form a basis for observations and reflections, which are then assimilated into abstract concepts that serve as guides for actions that, consequently, create new experiences (Kolb 1984). These learning modes that constitute the learning process are identified namely as Concrete Experience (CE), Reflective Observation (RO), Abstract Conceptualization (AC) and Active Experimentation (AE); and they essentially point to the two dimensions of learning—grasping of information and transmission of information—which occur on two intersecting axes, with two poles each. Hence, during the learning process, while the learner grasps information on a concrete to abstract scale, he/she transforms that information on a passive to active scale (Kolb and Kolb 2005, 2013).

Although Kolb affirms his theory as 'a holistic integrative perspective on learning that combines experience, perception, cognition, and behaviour', and as one which 'emphasizes the central role that experience plays in the learning process' (Kolb 1984: 20), it is at this point that his theory receives criticisms. For instance, Miettinen (2000) argues that the ingredients that Kolb gathered in his model are historically and theoretically distinct. Moreover, he points out that the overemphasis of the individual experience in ELT dismisses the analysis of cultural and social conditions, which are indispensable for change and learning in real life. Likewise, Kayes, who made a review of the criticisms of ELT, mentions that dissatisfactions with ELT generally coalesced around the idea 'that the theory's emphasis on the centrality of individual experience has come at the expense of psychodynamic, social, and institutional aspects of learning' (Kayes 2002: 142). Nevertheless, despite these criticisms, Kolb's ELT continues to attract a great deal of interest among researchers and/or educators from various disciplines.

Why refer to experiential learning and Kolb's ELT in the design studio?

The landscape design studio, along with being a physical setting, refers to the course in which the students are expected to acquire design and communication skills, learn to integrate and put theoretical knowledge into practice, and develop competencies in critical and creative thinking, as well as in problem-solving and decision-making. At this point, an experiential learning approach in the design studio plays a crucial role in accomplishing these learning outcomes.

First and for the most part, a simplistic definition of design considers it as a process that, according to Murphy (2005), involves formulation of appropriate and inspiring design ideas, and realization of these ideas as physical forms in the landscape. Hence, ELT, both by relying on the proposition that 'learning is best conceived as a process' and by suggesting that 'learning occurs through the course of connected experiences' (Kolb and Kolb 2013: 6), would fit well into the design pedagogy. Indeed, the stages of landscape design that typically involve problem identification, site inventory and analysis, creation of the design brief, concept development, drawing of the plans, as well as construction of the models, all find their place in the learning cycle in correspondence with the learning modes as suggested by ELT. Hence, pushing students into the realm of landscape design by the assignment of a design project would enable them to progress through the stages of landscape design, and, concurrently, employ each learning mode that constitutes the ideal learning process. To be clearer, a student, after being given an assignment, becomes directly involved in a design task and starts to gather information by making site visits. During these visits, he/she starts to discover realities by walking through the site, touching the ground, smelling the scents, feeling the breezes, talking with the users, thus, by collecting information about the physical and socio-cultural features of the project area. Hence, this concrete experience with the design problem and the site forms a basis for observation and reflection. Indeed, by relying on his/her concrete experience, the student identifies the opportunities and the constraints that particular site presents, and develops a brief, if not already given by the studio tutors. These reflections, subsequently, help him/her to develop conceptual design ideas. Finally, he/she attempts to translate his/her conceptual design ideas into the development of real world spaces via drawings, simulations, and/or models. Hence, accomplishing the given task by experiencing, reflecting, thinking, and acting, the student would learn essentially from the process, and develop an understanding of the distinct components of landscape design.

Indeed, at some stages in landscape design, specifically while developing conceptual design ideas and translating them into the formation of physical spaces via different visualization techniques—in other words, while thinking and acting—the student needs to consider different components of the curriculum that involve various issues related to landscape design, such as design principles, knowledge about plants, ecological, social, and psychological concepts, drawing techniques, etc., and seeks ways to incorporate this knowledge, which he/she gained in a number of courses, into the design task in which he/she is currently engaged. Thus, by being involved in the design task, the student would develop skills in integrating different aspects of landscape design and, as mentioned by Lewis and Williams (1994), would become familiar with relating theory to practice.

Moreover, throughout the design process, since students grasp and process information over and over again both by doing and reflecting on their design actions and decisions (Salama 2016), tackling the design task would also enable students to develop competencies in creative and critical thinking, as well as in problem-solving and decision-making. Given that role-playing in experiential learning enables students to consider various perspectives (Weinstein 2016), encouraging students to role-play while developing design alternatives would enable them to experience the design proposal from the users', developers', and clients' points of view, and hence help them critically reflect on what they are designing/have designed. In Schön's (1983) terms, this would make students 'reflect in action/reflect on action' or 'have a conversation with' their design proposals. Subsequently, relying on their critical reflections, the students would think of ways of improving their ideas and/or of generating new ones in a creative manner. Besides, as pointed out by Yatmo and Atmodiwirjo (2007), self-reflection makes students fully aware of the reasons for doing certain things or employing certain strategies in order to solve problems. Hence, the learner constructs a unique representation of knowledge by developing his/her own solution(s) and approach(es) to problems and ideas (Oxman 2004).

Indeed, this way of knowledge construction and learning that depends on the experiences of students proposes a constructivist model of learning in contrast to that of the transmission model of education in which pre-existing fixed thoughts are transmitted to the learner (Kolb and Kolb 2005, 2013). By promoting a learner-centred model that involves active participation of students, experiential approaches, as Lewis and Williams (1994) point out, appear to be more effective in developing communication skills and the ability to work in teams. Within this context, students, when engaged directly in a design task, would enhance their design proposals not only by self-reflecting on what they did, but also by observing each other's drawings, simulations, and models that provide a reference point for discussions, as well as by receiving feedback from the studio tutors and peers. Hence, this interactive atmosphere of the design studio would prepare students for their future careers by mimicking the ambiance of a design office where professional designers usually work in collaboration with each other.

Learning styles and Kolb's Learning Style Inventory

Although idealized learning occurs by means of the dual dialectics of experience and abstraction on the one hand, and of action and reflection on the other, and effective learners rely on the four learning modes, the relative emphasis given to each mode differs among individuals owing to their personality types, genetic makeup, educational experiences, cultural factors, and the specific task or problem he/she is currently working on (Kolb 1984, 2015; Kolb and Kolb 2005, 2013). Hence, depending on these factors, while taking in information, some individuals may prefer to rely on factual evidences whereas others may grasp knowledge better by abstractions; and on the other hand, while transmitting information, some individuals may prefer active experimentation whereas others may have a preference for reflection or observation. Thus, according to Kolb and Kolb (2005, 2013), a learner's preference for the four different modes of the learning cycle identifies his/her learning style.

An individual's strengths and weaknesses in the four modes of the learning process, and accordingly his/her learning style, are identified and measured with the help of a widely used tool called the Kolb Learning Style Inventory (KLSI). The earlier versions of KLSI identified four types of learning styles—diverging, assimilating, converging, and accommodating—by combining the scores that measured one's preference for AC over CE while grasping information, with those that measured his/her preference for AE over RO while transmitting information (Kolb and Kolb 2005). Thus, the point where the two scores intersected in one of the four quadrants, which were formed as a result of the intersection of the two axes with the four learning modes on their poles, identified one's learning style. In this regard, since each person's learning style was thought of as a combination of the two scores recorded both in terms of grasping and transmitting information,

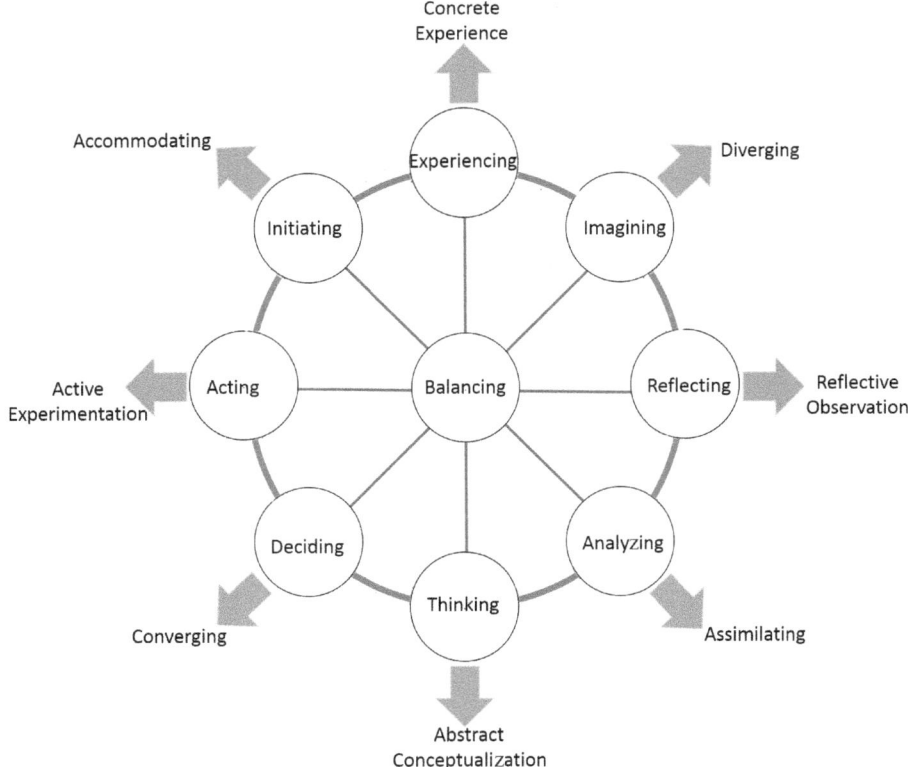

Figure 23.2   The nine learning styles and the four dialectics of the learning cycle (Kolb and Kolb 2013, p. 16)

the learning style he/she mostly preferred relied on two learning modes. Accordingly, as identified by Kolb and Kolb (2005), the dominant learning modes for a diverging style were CE and RO, for an assimilating style were AC and RO, for a converging style were AC and AE, and for an accommodating style were CE and AE.

However, research with KLSI over the years revealed, first, that there was a fifth 'balancing style' that identified individuals scoring close to the centre of the learning cycle, and, then, that there were four distinct borderline cases that identified individuals scoring close to the boundary lines that separated the learning styles from one another, and presented dominance of one of the four learning modes (Kolb and Kolb 2013). Hence, as a result of the recognition of these additional five types, the learning styles were updated to a nine-style typology (Figure 23.2).

Kolb and Kolb (2013) describe these nine learning styles as follows:

- The *Initiating* style is characterized by the ability to start action in order to manage experiences. People with this style learn by relying on AE and CE; thus, they prefer hands-on experience and real-life situations. Since they are fond of trying out new and challenging experiences, and are willing to take risks, they can initiate new projects and set goals, and put ideas into practice by experimenting with different approaches. They prefer working with teachers who act as a coach or a mentor in helping them learn from tangible experiences.

- The *Experiencing* style is characterized by the ability to involve deeply in life experiences while keeping AE and RO in balance. Rather than making logical judgements, those with this style first rely on their intuitions, and then they seek for validation through reflection and action. Since they deal with problems intuitively, they can employ innovative and unconventional approaches in solving problems. They enjoy working both in groups and individually. Since they deem it important to receive constructive feedback on their progress at work, they prefer a one-to-one relationship with their teachers.

- The *Imagining* style is characterized by the ability to create meaning both by reflecting on specific experiences and by observing. People with this style step back from experiences, and try to observe and think about what is going on. Thus, they can consider things from different viewpoints, reflect diverse opinions, and bridge the gaps between them. They are also good at conceiving of the consequences of a particular course of action, and generating alternative ways and approaches. They prefer working in groups where they can gather information from the free-flowing conversation, and receive personalized feedback. Thus, they enjoy brainstorming sessions with creative teachers who take the role of a motivator.

- The *Reflecting* style is characterized by the ability to connect experience and ideas through a continued reflection. Those with this style tend to use observation and reflection as the main ways of learning, while keeping CE and AC in balance. They have a strong sense of observation; thus, they are able to comprehend the deeper meaning that underlies events as well as interactions among people. Besides, they have the ability to generate creative ideas and come up with alternatives in solving problems, but they prefer to leave the implementation of these ideas to others. Since they are good at finding a common ground by bringing together different ideas and perspectives, they enjoy dialogues, discussions, and lectures. Besides, because of their preference for deep reflection, they also benefit from independent projects and readings during which they can reflect and make sense of their experience. Thus, they prefer working with teachers who offer opportunities both for individual and for group reflection, and who are open to exploring ideas.

- The *Analyzing* style is characterized by the ability to integrate and organize ideas through reflection. People with this style learn by relying on RO and AC. Hence, those with this style first grasp a wide range of information, and then put it into a logical form. Since they analyze and assess each step before taking action, they are able to minimize mistakes and anticipate potential problems and pitfalls. Owing to their analytical and conceptual skills, they succeed through readings, analytical models, as well as by thinking; and they also enjoy lectures in which teachers model their thinking and analysis processes, and have interactions with them.

- The *Thinking* style is characterized by the capacity to involve deeply in abstraction while keeping AE and RO in balance. People with this style are fond of carrying out mental activities that involve abstract reasoning as well as analytical thinking. They are distinguished as good goal-setters and planners as they tend to act as a result of much thought. They learn best in well-structured learning programmes with clear directions.

- The *Deciding* style is characterized by the ability to employ theories and models during decision-making. People with this style learn by relying on both AC and AE; thus, they tend to prove concepts and theories in the real world. Accordingly, they enjoy experimenting with laboratory assignments and practical applications with teachers who set clear standards and goals.

- The *Acting* style is characterized by the ability to use action as the main way for learning while keeping CE and AC in balance. Since people with this style have a strong preference for action over reflection, they are likely to put plans into practice without considering alternative solutions and caring about their consequences. On the other hand, as they have the ability to combine their experience of the immediate situation with concepts and ideas, they can work both in a practical world taking advantage of emotions and actions, and in a technical world entailing conceptual abilities. They enjoy learning on the job and succeed through discussions, and prefer working with teachers who have practical experiences.
- The *Balancing* style is characterized by the ability to navigate flexibly through the learning cycle by weighing the advantages and disadvantages of experiencing vs. thinking, and acting vs. reflecting. Thus, depending on the situation, people with this style can easily switch from one mode of learning to another. Since they have the ability to see issues from different viewpoints, they enjoy learning environments where the teachers concurrently use different teaching approaches, such as lectures, hands-on applications, brainstorming sessions, and discussion groups.

These nine learning styles, although they represent one's preferred way of learning, as Kolb and Kolb highlight, should not be thought of as 'a fixed personality trait but more like a habit of learning' (Kolb and Kolb 2013: 27). Indeed, depending on various factors, such as the nature of the matter being learnt, individuals' learning styles may show discrepancies from time to time. Nevertheless, since these styles give a general idea of an individual's preferred way of learning, educators can benefit from them in structuring their courses.

Why know the learning styles of students?

Students differ from each other in terms of their personal histories, strengths and weaknesses, ambitions and interests, senses of responsibility, levels of motivation and intellectual development, as well as learning styles, approaches to learning, and orientations to studying (Felder and Brent 2005). Relying on some evidence from literature, Kolb (1984) mentions that some individuals show a higher preference for certain modes of learning. Although one learning style is neither better nor worse than the others (Felder and Brent 2005), recent studies (e.g., Demirbas and Demirkan 2003, 2007; Kvan and Yunyan 2005; Tezel and Casakin 2010) reveal the effectiveness of different learning styles under different design conditions, as well as in different stages of the design process. Accordingly, remembering that the stages of landscape design, as mentioned earlier in this text, correspond with the learning modes of the idealized learning process, and also considering that not all students are alike, knowing the learning styles of students at the beginning of the semester would help tutors become aware of the troublesome situations that would be likely to occur during specific stages of the landscape design process, particularly for certain students. Hence, in order to minimize or even overcome those potential difficulties that a group of students would encounter as a result of the conflicts between their preferred learning styles and the learning mode essential to a certain stage of the landscape design process, tutors, in advance, can set strategies for adapting the most appropriate teaching technique(s) for meeting the needs of that particular student group; and, in turn, this effort of studio tutors would provide better communication between students and tutors. Indeed, realizing that the teaching techniques at some stage(s) of the landscape design process match his/her learning style, a student would involve voluntarily in studio courses, and thus develop a sense of belonging.

Moreover, owing to the vast range of research that indicates significantly better performance of teams having members with diverse learning styles (Kolb and Kolb 2013), knowing students' learning styles would also enable studio tutors to divide the class into teams of students with heterogeneous learning styles, and thus provide an opportunity for students to achieve successful results. Since learning styles are different from each other with distinctive strengths and weaknesses (Felder and Brent 2005), each student with his/her characteristic learning style in one project team would draw the attention of his/her teammates to some aspects of the project that, otherwise, might be missed if the group consisted of students with the same learning style. Such an approach would surely support collaboration among students, and, hence, encourage peer learning.

On the other hand, since the nature of landscape design process necessitates employment of the four learning modes that make up the idealized learning cycle, studio tutors, by knowing not only the dominating but also the non-dominating learning styles of students, and, accordingly, exposing them especially to the learning modes in which they have weaknesses—by means of employing more than one teaching method, such as lectures, discussions, desk crits, pin-ups, field trips, and hand-on activities both in a single studio session and across a project—would help their students not only enhance themselves in those modes but also gain competence in being able to shift from one learning mode to another. Indeed, such a flexibility would promote effective learning.

Effective learning in the design studio: tips for studio tutors

Effective learning, as defined by Pritchard (2014: x), 'is learning that is lasting and capable of being put to use in new and differing situations'. Accordingly, for effective learning to develop in a design studio, tutors should prepare their students with what to do when they encounter different design situations other than those introduced to them during studio sessions. This can surely be achieved both by focussing on teaching 'how to design' rather than 'what to design' and by making students experience 'how to make use of their knowledge' in various design situations. In this respect, rather than considering outcomes as the basis of assessments, studio tutors should evaluate their students' progress in terms of the skills they develop throughout the studio.

On the other hand, an increase in the overall learning effectiveness can also be accomplished when students become highly skilled in engaging all four modes of the learning cycle by developing their skills associated with each learning mode (Kolb and Kolb 2013). Indeed, owing to the correspondence of the stages of landscape design with the learning modes of the idealized learning process, a landscape architecture student, regardless of his/her preferred learning style, is supposed to employ all learning modes efficiently in the design studio. Accordingly, while progressing around the learning cycle through the stages of the design process, he/she should be encouraged to enhance his/her strengths and diminish his/her weaknesses in relation to each learning mode. Thus, when a student's preferred learning style and his/her skills contradict with the learning mode that corresponds to a particular design stage, he/she could be encouraged to focus both on the learning mode and that stage corresponding to it, and overcome his/her weaknesses in both of them by employing the skills associated with his/her preferred learning style (Table 23.1). But, above all, he/she should be ensured to feel himself/herself safe when he/she fails, and feel free to make mistakes because, as mentioned by Wurdinger and Carlson (2010), mistakes also help people learn.

Table 23.1 Suggestions for overcoming difficulties which students with specific learning styles would be likely to encounter in certain modes of learning and at certain stages of the landscape design process

Learning mode and the corresponding design stage	Learning styles that are most likely to encounter difficulties – strengths associated with them	Suggestions for meeting the needs of the learning mode/design stage by referring to the strengths associated with the students' preferred learning styles
Concrete experience – site inventory and analysis	Analyzing, thinking, deciding (collects information and compiles it in a logical form, carries out mental activities, puts abstract ideas into practice).	Make students collect information by feeling the site in terms of their visual, auditory, olfactory, tactile, and even sometimes gustatory senses; encourage them to think about what those feelings reminded them of; and let them convey via notes and/or drawings the information they compiled and upon what they thought.
Reflective observation – brief development	Deciding, acting, initiating (puts abstract ideas into practice, combines his/her experiences with concepts and ideas, talented in hands-on activities).	Assign them with a comprehensive narration of the opportunities and constraints of the project site via drawings (e.g., sketches), and encourage them to develop a brief by referring to their narrations.
Abstract conceptualization – concept development	Initiating, experiencing, imagining (talented in hands-on activities, relies on his/her intuitions, considers things from different viewpoints).	Encourage them to develop scenarios by relying on their observations of how people behave and how things occur in a specific place under specific conditions, and let them convey their observations by means of narrations and/or narrative drawings.
Active experimentation – design development	Imagining, reflecting, analyzing (considers things from different viewpoints, makes reflections by relying on his/her observations, collects information and compiles them in a logical form).	Let them make a compilation of real-world examples that best illustrate their conceptual design ideas either from design books, internet sources or by making observations and/or thinking about the examples that they had seen earlier, and encourage them to develop design alternatives by referring to these precedents.

Concluding remarks

The Chinese proverb 'I hear and I forget, I see and I remember, I do and I understand', and a more recent version of it by Benjamin Franklin 'Tell me and I forget, teach me and I remember, involve me and I will learn', still maintain their relevance today in the landscape design studio through Kolb's ELT, which leads to long-lasting learning. However, although the application of Kolb's model in the design studio provides a rich environment for students to gain higher-order thinking skills, and although the learning styles associated with it guide studio tutors to identify and plan their teaching strategies, it may still not be the key for effective learning in the studio. Other issues that influence learning, such as social, cultural, psychological and physiological factors, should also be considered in landscape design pedagogy.

References

Blessinger, P. and Carfora, J. M. (2014), 'Innovative Approaches in Teaching and Learning: An Introduction to Inquiry-Based Learning for the Arts, Humanities, and Social Sciences', in P. Blessinger and J. M. Carfora (eds), *Inquiry-Based Learning for the Arts, Humanities, and Social Sciences: A Conceptual and Practical Resource for Educators. Innovations in Higher Education Teaching and Learning, Vol: 2* (Bingley, UK: Emerald Group Publishing Ltd), 3–25.

Bruns, D., Ortacesme, V., Stiles, R., de Vries, J., Holden, R. and Jorgensen, K. (2010), *ECLAS Guidance on Landscape Architecture Education. Tuning Landscape Architecture Education in Europe, Version 26.* The Tuning Project ECLAS–LE:NOTRE www.tuningacademy.org/wp-content/uploads/2014/02/elcas_tuning.pdf.

Demirbas, O. O. and Demirkan, H. (2003), 'Focus on Architectural Design Process Through Learning Styles', Design Studies 24 (2003): 437–456.

Demirbas, O. O. and Demirkan, H. (2007), 'Learning Styles of Design Students and the Relationship of Academic Performance and Gender in Design Education', *Learning and Instruction* 17 (3): 345–359.

Felder, R. M. and Brent, R. (2005), 'Understanding Student Differences', *Journal of Engineering Education* 94 (1): 57–72.

Gentry, J. W. (1990), 'What is Experiential Learning?', in J. W. Gentry (ed.), *Guide to Business Gaming and Experiential Learning* (East Brunswick, CN: Nichols/GP Publishing), 9–20.

Holgate, P. (2008), 'Assessment for Learning in Architectural Design Programmes', *Northumbria Working Paper Series: Interdisciplinary Studies in the Built and Virtual Environment* 1 (2): 194–208.

Kayes, D. C. (2002), 'Experiential Learning and Its Critics: Preserving the Role of Experience in Management Learning and Education', *Academy of Management Learning and Education* 1 (2): 137–149.

Kolb, A. Y. and Kolb, D. A. (2005), *The Kolb Learning Style Inventory—Version 3.1: Technical Specifications* (Boston, MA: Hay Resources Direct).

Kolb, A. Y. and Kolb, D. A. (2013), *The Kolb Learning Style Inventory—Version 4.0. A Comprehensive Guide to the Theory, Psychometrics, Research on Validity and Educational Applications* (Experience Based Learning Systems, Inc.), http://learningfromexperience.com/media/2016/10/2013-KOLBS-KLSI-4.0-GUIDE.pdf.

Kolb, D. A. (1984), *Experiential Learning. Experience as the Source of Learning and Development* (Upper Saddle River, NJ: Prentice-Hall, Inc.).

Kolb, D. A. (2015), *Experiential Learning. Experience as the Source of Learning and Development*, 2nd ed. (Upper Saddle River, NJ: Pearson Education, Inc.).

Kvan, T. and Yunyan, J. (2005), 'Students' Learning Styles and Their Correlation with Performance in Architectural Design Studio', *Design Studies* 26 (1): 19–34.

Lackney, J. A. (1999), *A History of the Studio-based Learning Model* (Mississippi State, Educational Design Institute), http://edi.msstate.edu/work/pdf/history_studio_based_learning.pdf.

Lewis, L. H. and Williams, C. J. (1994), 'Experiential Learning: Past and Present', *New Directions for Adult and Continuing Education* 62: 5–16.

Miettinen, R. (2000), 'The Concept of Experiential Learning and John Dewey's Theory of Reflective Thought and Action', *International Journal of Lifelong Education* 19 (1): 54–72.

Murphy, M. D. (2005), *Landscape Architecture Theory* (Long Grove, IL: Waveland Press, Inc.).

Oxman, R. (2004), 'Think-Maps: Teaching Design Thinking in Design Education', *Design Studies* 25 (1): 63–91.

Pritchard, A. (2014), *Ways of Learning. Learning Theories and Learning Styles in the Classroom*, 3rd ed. (Abingdon, UK: Routledge).

Salama, A. M. (2010), 'Delivering Theory Courses in Architecture: Inquiry Based, Active, and Experiential Learning Integrated', *ArchNet-IJAR: International Journal of Architectural Research* 4 (2–3): 278–295.

Salama, A. M. (2016), *Spatial Design Education. New Directions for Pedagogy in Architecture and Beyond* (Abingdon, UK: Routledge).

Schön, D. (1983), *The Reflective Practitioner: How Professionals Think in Action* (New York, NY: Basic Books).

Stein, M. (2004), 'Theories of Experiential Learning and the Unconscious', in L. J. Gould, L. F. Stapley and M. Stein (eds) *Experiential Learning in Organizations* (London: Karnac), 19–36.

Tezel, E. and Casakin, H. (2010), 'Learning Styles and Students' Performance in Design Problem Solving', *Archnet-IJAR: International Journal of Architectural Research* 4 (2–3): 262–277.

UNESCO, 'Most Influential Theories of Learning', Paris: UNESCO [website], www.unesco.org/new/en/education/themes/strengthening-education-systems/quality-framework/technical-notes/influential-theories-of-learning, accessed 12 October 2017.

Weinstein, N. (2016), *Experiential Learning. Research Starters: Education* (Online Edition), http://connection.ebscohost.com/c/essays/27577767/experiential-learning.

Wurdinger, S. D. and Carlson, J. A. (2010), *Teaching for Experiential Learning. Five Approaches That Work* (Lanham, MD: Rowman and Littlefield Education).

Yatmo, Y. A. and Atmodiwirjo, P. (2007), 'Assessing Students' Learning Process in Design Studio: Assessment Design for Learner Responsibility', 29–31 May [Released under Creative Commons license] www.researchgate.net/publication/267789054_Assessing_students%27_learning_process_in_design_studio, accessed 8 February 2016.

24

The DesignLab approach to teaching landscape

Mick Abbott and Jacky Bowring

DesignLab is an interdisciplinary group of researchers, educators, landscape architects, and designers who emphasise the use of design methods to build possibility and complex grounded contexts. The DesignLab approach to teaching landscape is a counter to approaching designing as simply 'solving a problem'. Problem solving suggests that design is a linear process, and that there is a degree of inevitability in proceeding from the problem to the solution. For simple design problems this could be true, based on a 'determinate' framing of the problem. However, as Richard Buchanan cautions, 'the *wicked-problems* approach suggests that there is a fundamental *indeterminacy* in all but the most trivial design problems –problems where, as Rittel suggests, the "wickedness" has already been taken out to yield determinate or analytic problems' (Buchanan 2001: 15–16, emphasis in original). Indeterminate is not the same as being undetermined. As Buchanan emphasises, there is an openness of diverse parameters rather than an absence, which makes problems wicked.

The indeterminacy of wicked problems has echoes with a recognition that research is not about finding answers, but about deepening and enriching our understanding of the world. Rather than the limited intention of proving something, research is a journey into possibility. Bringing together this aspiration for researching as an expansive endeavour, with designing as a grappling with indeterminacy, establishes the DesignLab approach. The approach is fundamentally about increasing the imaginative scope of complex contexts in which designing produces new knowledge and possibilities. As Blanchon observes, 'Scholars agree that design activity produces knowledge as much as academic studies do' (2016: 67).

The DesignLab positions designing as a core method for researching, and actively pursues collaborative and interdisciplinary approaches. Just as the recognition of design as research is often resisted by design educators, so too are collaboration and interdisciplinary involvement in studios. Insular models of researching and designing represent particular philosophies within institutions. Discouraging collaboration, for example, is often driven by the complexity of group work situations where establishing effective teams within a class environment, ensuring evenness of contribution in both quality and effort, and meeting protocols for deriving individual grades for students create both organisational and administrative impediments. And while interdisciplinarity is a frequently voiced aspiration, design studios are usually just made up of design students who are working individually on a pre-given 'problem/brief'.

As an approach to teaching designers, the DesignLab seeks to empower students to be innovative and lateral in their transformation of landscape. While evidence-based approaches are accepted norms that favour the sciences' claims of objectivity, they can also lead to a shutting down of a full range of possibilities, based on the presumption that the evidence that exists is both definitive and encompassing. This can be due to the implicit presumption that, given the evidence, a single consistent outcome can be the only result. While it is not suggested that evidence should be ignored, it should also be critiqued and contextualised. 'Evidence' is rarely universally applicable or eternally relevant. The DesignLab approach also focusses on asking questions of evidence, of bringing different disciplinary frames to scientific data.

The terminology adopted by the DesignLab approach is significant. The term 'laboratory' is preferred over 'studio', as laboratories are associated with experimentation, testing, and collaboration. The known character of the laboratory anticipates collaboration across other approaches, for instance where findings associated with ecological sciences might meld with 'landscopic' forms in the study of environmental behaviours. Studios, by contrast, can have connotations of artistic practice – which, while in no way discouraged as part of the designing process, can also suggest insularity and egos. Like the studio, the lab suggests a restless realm of shared questions, ready experimentation and failures tempered with success. But, distinctively, the lab asserts collaboration. The studio has a tendency toward the personal where even the notion of a shared studio can have a compromised sensibility to it. The lab is oriented towards an implicit collectiveness in which groups of researchers work and collaborate over shared research questions.

As outlined below, the DesignLab approach is based around five strategies: questioning, collaborating, designing, grounding and communicating.

Questioning

The DesignLab approach draws on the discipline of design thinking's caution not to rush to answers as this can often be reflected in the oversimplification of the question. The questioning strategy is exemplified by two frameworks –'what to make?' and the 'Five Whys' – both of which are outlined below.

What to make?

Design theorist Charles Owen's (2001) simple graph (Figure 24.1) provides a potent illustration of the importance of explicitly exploring the problem. Owen's graph points to how the common leap to 'how to make it' completely overlooks the question of what should actually be made. As noted in the introduction with reference to wicked problems, this limitation is again related to the issue of determinacy. Owen explains that the first example of the one-step process is where 'an already determined concept is turned into a specification'. As previously noted, this works to limit the range of possibilities that a deeper process could generate, thereby already limiting the possibilities of the designing process (2001: 30–31).

This shifting to a two-step mindset, in which the opening task is to determine what to make, allows the task of designing to extend landscape architecture from its being constituted solely within the physicality of the bounded site to also including landscape's multisensory, temporal, distributed and dynamic dimensions. It also rejects the presumption that we already know what to make. A questioning of this approach is an opening out of possibilities and landscape's multiple dimensions. Landscape, after all, is not just constituted by a site's physical characteristics. Landscape is experiential, multi-sensory, fluid and temporal. Further, it arguably has the capacity,

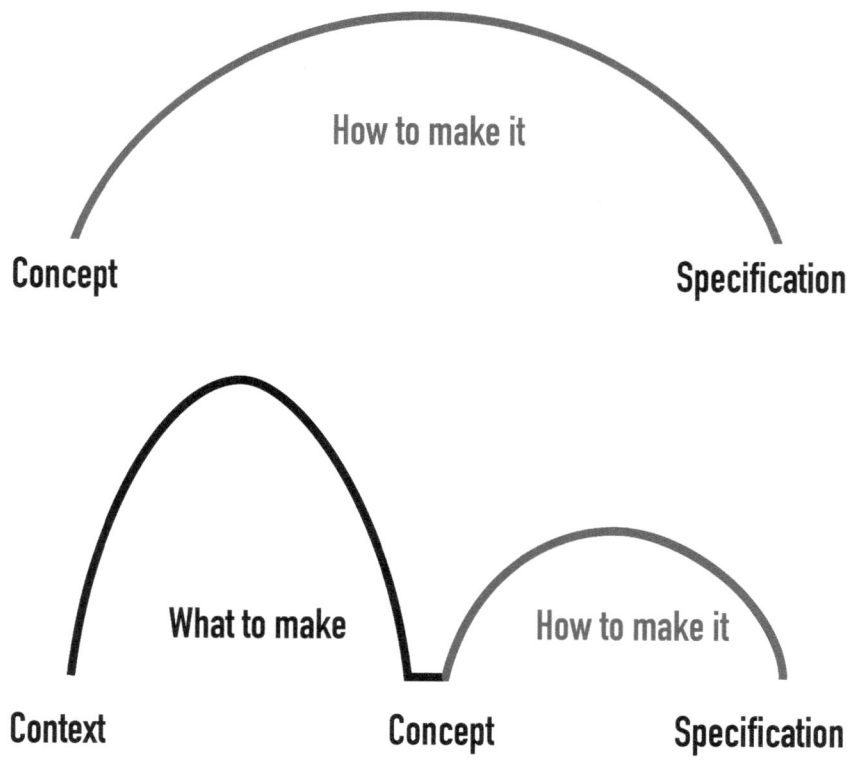

Figure 24.1   What to make before how to make it. Source: Owen 2001

the agency and instrumentality, to shape people, both as people in the landscape, and as designers of its many forms. In such a framing what to make is also reflexive, in what to make not only of landscape but also of the people within it.

Five Whys

Questioning is a process of opening out and going deep, and one of the most useful questioning tactics is the Five Whys. This approach is championed by design consultancy IDEO, and is included in their pack of 51 Method Cards. They explain that the Five Whys 'forces people to examine and express reasons for their behaviour and attitudes' (IDEO 2003, Five Whys card). The sequential asking of the question why? peels layers off assumed understandings of a situation and rapidly moves beyond the surface response to a question.

A vivid example of the effectiveness of the Five Whys approach in developing expansive questions is the situation at the Jefferson Memorial in Washington DC. As Ditkoff recounts, the memorial was becoming damaged due to the harsh cleaners needed to solve the 'problem' of excessive bird droppings dirtying the memorial (2013). The Five Whys approach unravelled the problem as follows: 1. The memorial was being damaged because of excessive cleaning; 2. Excessive cleaning was necessary because of the huge quantity of bird droppings; 3. Birds were attracted in great numbers because of numerous spiders; 4. Spiders were there eating the midges; 5. Midges were there because the lights at the Memorial were turned on one hour before dark, creating 'mood lighting' for them.

Asking the Five Whys therefore opens out the problem from being one of harsh cleaners to identifying the issue that the lights were being turned on too early. Ditkoff summarises:

> After reviewing the curious chain of events that led up to the problem, the decision was made to wait until dark before turning the lights on at the Jefferson Memorial. That one-hour delay was enough to ruin the mood lighting for the midges, who then decided to have midge sex somewhere else.
>
> No midges, no spiders. No spiders, no birds. No birds, no poop. No poop, no need to clean the Jefferson Memorial so often. Case closed.
>
> *(2013)*

This deep questioning was a way of getting to the 'what to design' that is at the core of Owen's strategy – the resulting design intervention had nothing to do with the memorial's physical structure, but was a temporal issue related to lighting. This focus on 'what to design' avoided a number of possible other responses such as encasing the memorial in Perspex to protect it, or shooting the birds.

A strategy of *questioning* expands the field of investigation. The 'what to make' and Five Whys tools are useful ways of teaching students about the power of grappling with questions, of questioning questions, and, as Thomas Oles puts it, 'Do not rush to answers: savour the asking' (2014: 109).

Collaborating

Research in the humanities does not tend to be characterised by collaboration (see, for example, Larivière et al. 2006). Design disciplines, too, are often built around this favouring of individual research endeavours. Yet given that landscape architecture, as a design discipline, has feet in both the worlds of art and science, it might also usefully look towards the sciences for valid models of collaboration. The sciences have established traditions of collaborative and collective research, with an understanding that knowledge is co-produced and co-authorship is common. Designers are adept at making the heterogeneous potent. Design as a process involves tools of synthesis and hybridisation: creating novel outcomes through melding difference. One dimension, often overlooked, is the possibility arising from the difference inherent across a number of designers. Team-based design is often understood in terms of efficiency, productivity an organisational impact. However, difference can be structured and hybridised through a deliberate configuring of the diversity creative people have. In a lab setting not only does the hybridising of forms, methods, theoretical positions and grounded contexts create innovation, but also the melding of designers working together extends the possibility space of design-directed research.

In the DesignLab teaching approach, collaboration is fostered by physical space. The form and layout of space is designed to promote interaction and exchange. Certainly, such an approach is not unusual for researchers focussed on the value that comes from reshaping environments. DesignLab seeks to reshape its research environment as a means to open up and keep opening up creative and collaborative opportunities. Large tables encourage group work, and the space allows for spontaneous exercises such as the Quattro Stagioni described below.

Arguably, collaboration should be core to all landscape architectural questions. Landscape architecture intersects with many fields, and our questions vastly exceed the capacity of any individual or any discipline alone. In our experience within a teaching setting, other disciplines are very amenable to collaboration, and find the design studio to be a place in which they can actively be part of thinking about questions.

This team-based approach to investigation, in which not only questions and design productions but also the spread of positions from which each investigator formulates them, maximises the opportunities for diverse contributions. Here teams focus efforts around common questions in ways that invite interdisciplinarity, multidisciplinarity and transdisciplinarity approaches, and that suggest different paradigms can be enlisted to arrive at a more holistic set of outcomes. Such collaboration can also enable more diverse outcomes that can range from the theoretical to the very applied.

For example, in an assessed research studio involving twelve senior students and four members of the DesignLab research team, distinctive briefs using a spread of drivers were given to two groups of two students. Following this, the combined output was 'mined' to derive distinctive drivers that could each work as a catalyst in generating a distributed conservation park that in itself was threaded through a landscape also valued for its agricultural production and tourism appeal. The source material for this work came from the collaborative 'Mackenzie Agreement', and supported an understanding that collaboration is not only interdisciplinary, or amongst the students within the studio – but also with stakeholders.

Designing

It perhaps goes without saying that designing is fundamental to a DesignLab approach to teaching landscape. However, the DesignLab approach foregrounds strategies for designing that use generative processes such as ideation, as well as analytical techniques like critique.

Quattro Stagioni

One of the most effective generative tools used in the DesignLab approach is the two by two matrix. This diagram is a powerful tool that is used by both individuals and collaborators, and that can be both analytical (retrospective) and generative (prospective). The matrix is known as *Quattro Stagioni* or 'Four Seasons' (as in the pizza topping of that name). It is a nimble technique that can be readily learnt and applied by first-time users and then just as quickly applied to other design contexts. It can also operate at a range of physical scales – from the back of an envelope, to being actually performed by students in studio, as shown in Figure 24.2. The students from Lincoln University's School of Landscape Architecture in Figure 24.2 are undertaking a concept generation activity using the Quattro Stagioni axes. Its use typically goes as follows: students are asked to develop concepts working with a range of variables, and also encouraged to bring their own perspectives to the development. Generally, we have found three to four concepts sufficient to draw out students' range of ideas before over-repetition becomes common. Then all students are asked to consider two common axes by which to locate the proposal. For instance, a national park design could have along one axis a spectrum of positions from building ecological value at one end to building social value at the other. On the other axis, the type of outcome could be teased out. This method presents a spectrum of positions ranging from site-specific solutions at one end to program-based solutions at the other end of the axis. From here, the students locate the designs according to where they best fit. To note, the axes themselves are not definitive and can also involve more pragmatic dimensions (high cost/low cost; fast build/slow build), aesthetic dimensions (green systems vs built interventions; high semiotic intent/low semiotic intent, etc.).

From this position three key tasks can then be pursued. First can be a reflection on the values chosen for each axis, with the option of changing these (and repeating the exercise) so able to

Figure 24.2   Students using a matrix to generate design responses

develop further useful options. Second, reflecting on the quadrant where design solutions are over- or under-represented and again reflecting on the reasons why. Third, consider how, for instance, a strong site-specific project that builds ecological value could be shifted/designed/developed to also build strong social value and/or can programmatically work across a number of sites. The strength of this is to both challenge design thinking – 'imagine this also as this' – and to also allow more collaborative actions: none of steps 1 to 3 require the person developing the concept to undertake any or all of the phases.

Probes

The design process sometimes feels like an unstoppable juggernaut, heading for its destination. There is a sense of urgency and assumed inevitability about maintaining forward momentum, and students are reluctant to explore 'sideways', wanting instead to quickly find a solution. Design probes can counter this urge and allow students to use their agility to undertake rapid small investigations, often prompted by the possibilities yielded from the Quattro Stagioni. Multiple probes again emphasise that designing is a task of expanding the imaginative possibilities in the world, rather than eliminating possibility in the drive to solve a problem. Probes, often developed as a series of quick hand drawings, can then be, in turn, analysed and explored, leading to new directions for designing. Such probes do not have to be undertaken individually. We find these especially effective when a group of three or four students is given a set task in a pre-given timeframe. The collective output of a number of these groups can then be examined using the Quattro Stagioni process.

Grounding

As a discipline, landscape architecture is fundamentally grounded in place, and the DesignLab approach looks to both the universal and the local. Questioning, as discussed above, might lead to a problem being unpacked into a broad universal context, becoming an investigation of a general concern, for example storm water, agriculture, or revegetation. Important for the DesignLab approach is the grounding of these broad designerly probes into a local context. And the opposite can also be true, where questioning may focus down into a very fine grain, and can benefit from being related back out to a broader context, as part of a 'so what?' interrogation.

Grounding can be seen as analogous to a testing stage, which again resonates with the laboratory model, and brings the abstract to the concrete, using a depth of local knowledge across a range of disciplines to explore the potential of the possibilities that evolve through designing.

Communicating

As with any designing, representation and communication are crucial as part of the DesignLab approach. Collaboration requires rapid and effective communication between those involved. For design students, hand drawing is actively encouraged, especially during the generative phases. Hand drawing can be fast, it can vary significantly in size (a thumbnail versus a wall), and can be easily co-created (Figure 24.3). While there are many digital platforms for co-creation and collaboration, during the fluid stages of design the digital environment can have the effect of designs appearing to be finished, because there is a sense of resolution to the images. This can limit the explorative process, with a tendency to keep things neat rather than messy and evolving.

Working with multiple disciplines and stakeholders amplifies the need for clear communication. This is both through the project development process and as outputs are framed so underlying values are made available. This too follows a creative process as determining what and how to communicate is designed and prototyped. Communicating with science and humanities collaborators and with lay audiences places extra demands and enriches the thinking about the outcomes of the designing process. The communication phase remains interactive and co-creative, for example encouraging collaborators to add to designs through the use of post-it notes, or as a means of commenting and critiquing. In the latter stages of the process, digital tools can be highly effective, and include the use of infographics, datascapes and so on. Digital tools can also offer efficient ways of expressing multiple designs for the same situation – photomontage techniques, for example, invite different readings and an opening-out rather than closing-down of possibilities.

Overview

The DesignLab approach re-thinks and challenges some of the conventional studio teaching strategies. It also offers a challenge to some of the broader institutional constraints that can prove limiting to the adoption of approaches such as collaboration. The 'laboratory' terminology signals both a place for experimentation and a mode of working. Just as a science laboratory has a configuration that enables experimentation, so too does the DesignLab, with a large and flexible space that can accommodate groups working together, 1:1-scale exercises where bodies become markers (as in the Quattro Stagioni), and where there is ample surface area for work to be displayed and interacted with.

While the DesignLab approach is offered here as a teaching strategy, it also speaks to broader professional and disciplinary contexts. It is an approach that resonates with professional

Figure 24.3 Sketches by Jorden Derecourt showing a range of rapid hand drawing techniques including sketches, diagrams, cartograms and maps as part of a taught design research studio project

settings, encouraging questioning of briefs, for example – rather than accepting a brief that is given by a client. Professionals may be cautious of appearing to be less efficient in the adoption of a questioning approach, but, as demonstrated with the Jefferson Memorial example outlined above, it can lead to much more effective outcomes. From a disciplinary perspective, the DesignLab is the core of design-directed research. Rather than appropriating methods from other disciplines – interviewing, surveying, monitoring – design-directed research recognises that designing itself is research. In this designing is the means by which new knowledge and new possibilities are generated. As Lily Chi asked:

> In what ways can design work's very specificity and finitude offer a medium of investigation for questions of broad concern? How do the creative and discursive interact? How does individual imagination figure in the deliberation of sociocultural matters? What role does the created artefact play in the conjectural process? How, in short, can design as design be practised – and read – as a pursuit of knowledge, understanding?
>
> *(Chi 2001: 250)*

A recent themed issue of *Landscape Review* (2018), guest-edited by Mick Abbott and Paul Roncken, focusses on design-directed research, and includes a range of international examples, including studio and 'real world' projects.

References

Abbott, M. and Roncken, P. (2018), 'Editorial: A time for designing', *Landscape Review*, 18(1): 1–3.

Blanchon, B. (2016), 'Criticism: the potential of the scholarly reading of constructed landscapes. Or the difficult art of interpretation', *Journal of Landscape Architecture*, 11(2): 66–71.

Buchanan, R. (2001), 'Wicked problems in design thinking', *Design Issues*, 8(2): 5–21.

Chi, L. (2001), 'Introduction: design as research', *Journal of Architectural Education*, 54(4): 250.

Ditkoff, J. (2013), 'Why you need to ask why', *The Huffington Post*, 14 February, www.huffingtonpost.com/mitch-ditkoff/why-you-need-to-ask-why_b_2681958.html, accessed 13 March 2013.

IDEO (2003), *IDEO Method Cards: 51 Ways to Inspire Design* (Palo Alto, CA: IDEO).

Larivière, V., Gingras, Y. and Archambault, É. (2006), 'Canadian collaboration networks: a comparative analysis of the natural sciences, social sciences and the humanities', *Scientometrics*, 68(3): 519–533.

Oles, T. (2014), *Go with Me: 50 Steps to Landscape Thinking* (Amsterdam: Architectura & Natura Publishers).

Owen, C. (2001), 'Structured planning in design: information-age tools for product development', *Design Issues*, 17(1): 27–43.

25

Studio-based landscape design teaching

Davorin Gazvoda

Introduction

There are sufficient relevant texts that attempt scientifically to define the design process, to assist in the discussion on how to teach design studios: Zeisel (1981), Steinitz (1990, 2012), Rowe (1987, 2002), Motloch (2001), Ogrin (2011), Freire (2012), van den Brink et al. (2017), to name just a few of the contributions that are most often cited.[1]

However, very often teachers of landscape design studios do not delve into the aforementioned literature but simply practice their own teaching methods and try to find the best possible method. There are probably as many approaches to teaching as there are registered teachers of design in the world. They all use very different methods on a daily basis in their work, are highly operative, but they do not engage in theoretical debate about what the design process is and how to teach landscape design.

This chapter tries to be somewhere between a more general or theoretical discussion and an operative work. It tries to shed light on the characteristics of a workshop–studio approach to teaching landscape design, particularly in a creative phase of students' work, and analyses what kind of teacher profile is most suitable in a landscape design studio.

Discussions concerning whether a studio is one of the best forms of direct educational work with students have already been going on for some time. Project-based learning (PBL) is one of the most generally established learning methods in modern education. It is characterised by a more relaxed and individual approach on two levels: the student's personal interest in solving the problem and the teacher's individual work on the workshop level instead of the '*ex-catedra*' approach.[2] Through the personal approach, the student gets in-depth knowledge by investigating and understanding problems, which generally derive from the real world. This general definition of PBL is similar to the workshop concept of work in design or architectural seminars (design studios), which is in fact one of the most used forms of study in design professions and is based on the old relation between master and apprentice. Therefore, the great majority of European schools of landscape architecture had no problem with introducing PBL, which was the desired method of study under the Bologna modernization of studies. It can actually be said that the Bologna modernization confirmed studio teaching in landscape architecture study programmes. There can thus no longer be any dilemma about whether design studios are necessary or not. The ECLAS tuning document also states this:

> Studio learning is at the centre of landscape architecture education: 40 to 60% of student's workload is reserved for studio based learning. Students work either individually or in small groups to develop design and planning approaches, to train in communication and to gain management skills, to apply a number of different techniques and technologies, etc.
>
> *(Bruns et al., 2010)*

The aim of this chapter is not to compare European schools of landscape architecture and different teaching practices of landscape design studios. Examples (also graphic) of good teaching practice obtained through the author's almost thirty years of pedagogic practice (at domestic schools and abroad) will be used in the contribution, highlighting the best products of students, which illustrate the presented pedagogic approach to the maximum extent and that was previously discussed in a paper on landscape architecture education (Gazvoda, 2002). In the presented cases, the emphasis is on the creative phase of the design process, in which students must generate new landscape designs and offer an interesting final design. It is characteristic of all examples that students quickly learn functional and technical requirements (students of the 1st year of BSc LA from Ljubljana), they can perform excellent analyses, but they have problems with creativity (workshop in Ljubljana and studio in Shenzhen, both in an MSc programme). The last two examples also have in common that the teacher and students are from different countries (international workshop) and are more relaxed, and the creative phase is effective for continuation of the project.

In parallel with the description of concrete examples of studios, attention is also devoted in the chapter to the role of the teacher in the learning process of landscape design, all of which could help fellow educators and landscape architects without special pedagogical and andragogic skills for working with students.

Characteristics of landscape design studio

A profile of a landscape architect we would like to cultivate can be described as:

> The responsive, creative designer must be an ambidextrous thinker. Such a person, capable of dealing effectively with both intuition and logic, is centred and able to think holistically and integratively. To help the student become more centred, creative, capable of perceiving relationships, and insightful is one of the primary tasks of design education.
>
> *(Motloch, 2001: 38)*

Although Motloch's definition of a designer could apply to all design professionals, it is especially important for landscape architects. Landscape architecture is in fact a highly interdisciplinary profession in which, depending on the definition of the problem, the field of work constantly shifts from a more research analytical approach through to creative solutions for creating new landscapes. Steinitz tried to define the design process more explicitly thirty years ago (1990), and he continued articulation of the planning process and trying to analyse it in more detail in his work *A Framework for Geodesign* (2012). This exposed the dimension of spatial issues and approaches to solving them:

> Large projects have a design emphasis on tactics and organization, or how different elements relate to each other. Smaller size projects emphasize details and expression, what something looks and feels like. These emphases in design are very different and relate both to size and scale.
>
> *(Steinitz, 2012: 22)*

Landscape design studios must also be placed within this span, which typically start on a large scale (smaller problems in a small, more manageable space) and end on a small scale (in a large space, with a regional planning approach to the solution).

Explaining things through a studio is already important in the early years of a BSc degree, when students are still inexperienced and must master a great deal of basic knowledge, organized into subjects, which are usually presented through lectures and exercises. A project-oriented approach and introductory design studios are normally linked to a smaller space, but the problems are complex enough to allow the processing of individual thematic groups of the landscape. Tasks are usually specific and derive from social needs that are manifested in space (e.g., project of a family house with a garden, with defined clients and their needs).

There is great stress on presentation and mastery of the design process; students must master the presented/recommended method and slowly begin to develop their own personal approach to design. The final result is less important; mastering procedures and creativity is more so.

These studies are characterized by a teacher who must stick more closely to the curriculum (defined pedagogical goals) because students must learn the basics, so the methods used are often more likely to be lectures and short exercises. It is important to use examples (including specific comparative analyses), an analytical approach is highlighted, which is in any case always needed everywhere and is one of the basic main features of the landscape architectural approach – how to solve spatial problems. Students need to understand what is wrong in their proposals and duly supplement, correct and improve them.

Small groups of students are usually involved in landscape design studios at MSc level or in higher years of the study of landscape architecture. In addition to a smaller number of students, a teacher's work is made easier by the fact that students are older, more experienced, some have even had practical experience, and they are highly motivated. In addition to the intrinsic motivation,[3] an additional motivation is the fact that in some state schools, and especially private schools, students pay high tuition fees, which increases their motivation to study.

Project work is most important and students work independently. Because they are used to a studio workshop method of work from previous years, they can immediately focus on identifying and tackling the problem. The problems may be the same as in the lower years but the level of their treatment is more demanding. Many problems are very specifically oriented, which in the design sense are solved from a very narrow point of view, but in a good sense of the term. Ideally, the problems come from practice, the clients are known, who express their needs and require-ments, and the specific location is also known, which the participants can see on the ground.

These studios often host colleagues from practice, not only landscape architects but also other experts, whose knowledge can supplement the view of solving the problem, depending on its complexity. At private schools, especially in the US, students are drawn to well-known landscape designers from practice, who often lead master studies. Students are familiar with the work of their bureaux, appreciate them, or even admire and want to be part of their success story, when visiting teachers, for example, are award-winning architects in that year. Their approach to landscape design is characterized as 'authorial'. The final student products may be very correct in the design sense, even excellent (in the design compositional sense), but design-ers from bureaux may affect students with their design philosophy too pronouncedly. In doing so, a number of aspects and the interdisciplinarity of landscape architecture, especially its ability to understand the wider (read: real) social problems, is lost. This can be and often is a problem with these 'big names' of studios.

Students can thus also entirely uncritically follow the teaching methods of professionals from practice who, despite their operational and technical skills, often do not have a sufficiently critical thought process, firstly towards their own projects but also towards those of others and they often do not see the design solutions of students in the wider theoretical framework. If one adds to this a lack of empathy, sensitive intelligence and communication and teaching skills, the results can be questionable.

Teachers of advanced landscape design studios must be flexible and must react to unforeseen situations. With advanced studios, cooperation is usually with specific clients (normally municipalities, local communities and others), so the theme and the selection of specific locations is determined in real time. The teacher must quickly and effectively adapt to the problem, clearly define it, articulate and establish a suitable approach. Students must then redefine the problem themselves, adjust their method and set about addressing specific work in cooperation with the client. In such a situation, the teacher must constantly adapt and prepare lectures that support the studio work. Sometimes a teacher must master a theme with which she or he was not previously familiar, to the extent that she or he can competently teach it and direct the students in their work. These studios are thus demanding for teachers but attractive for students. The repetition of the same exercises throughout the year is a lazy way of teaching practice, which evidently leaves a bad aftertaste for both students and teachers (students feel that the teacher has fallen into a routine and they are less motivated). 'Most teachers know that when students feel passion for a topic, they will seek out the tough problems, rather than the easy ones, and work harder to solve them. And best of all, they will have fun doing it' (Thomas and Brown, 2011: 80). With new problems and new locations, the teacher must try harder, at some point becoming a colleague of the student (associate) to jointly solve the problem, not for the grade but for the client. It is therefore important to have a specific client or, for example, to enter competitions with students. The motivation thus comes from the client or invitation to tender (with the promised reward for the best), and the teacher is jointly responsible if the client is not satisfied with the final product, or if there is no success in the competition.

Examples of three selected landscape design studios

It is possible to infer from the above described the already-mentioned range between large and small scale, which establishes the approach to solving spatial problems. It is the integration of the two poles of landscape architecture: an analytical articulated planning approach based on a transparent vision of physical reality (the world in layers) using a systematic approach. A studio begins with a definition of the procedure, methods, necessary information. It is followed by a transition to the design in which students simply continue their work and concretely design the defined use at selected locations. The scales are from 1: 25,000 (depending on the size of the area) to a specific proposal at a scale of 1: 1,000 or 1: 500 (for the chosen location). The advantage of this approach is that students master the space, understand the problem, space, and the client's wishes if they are specific and known.

How important the scale is and how it can affect the understanding and planning of space, not only in planning but also in design measures, was also shown by a student landscape architecture workshop 'Out of Scale', which was held at the Ljubljana school by colleague Udo Weilacher from TU Munich (Weilacher et al., 2013). Students were required to determine the personality of a 4-cm-tall human figure, and describe the activities and interests that the fictional person was realising in the space. The figure was then set in an existing space, photographed, and the spatial patterns in which the figure found itself were interpreted at three different scales. This was followed by a description of the possible problems that could occur if an attempt was made to provide solutions for the chosen user space, and both the problems and the finally envisaged spatial changes were illustrated on a final map to a scale of 1: 5,000 (Figure 25.1).

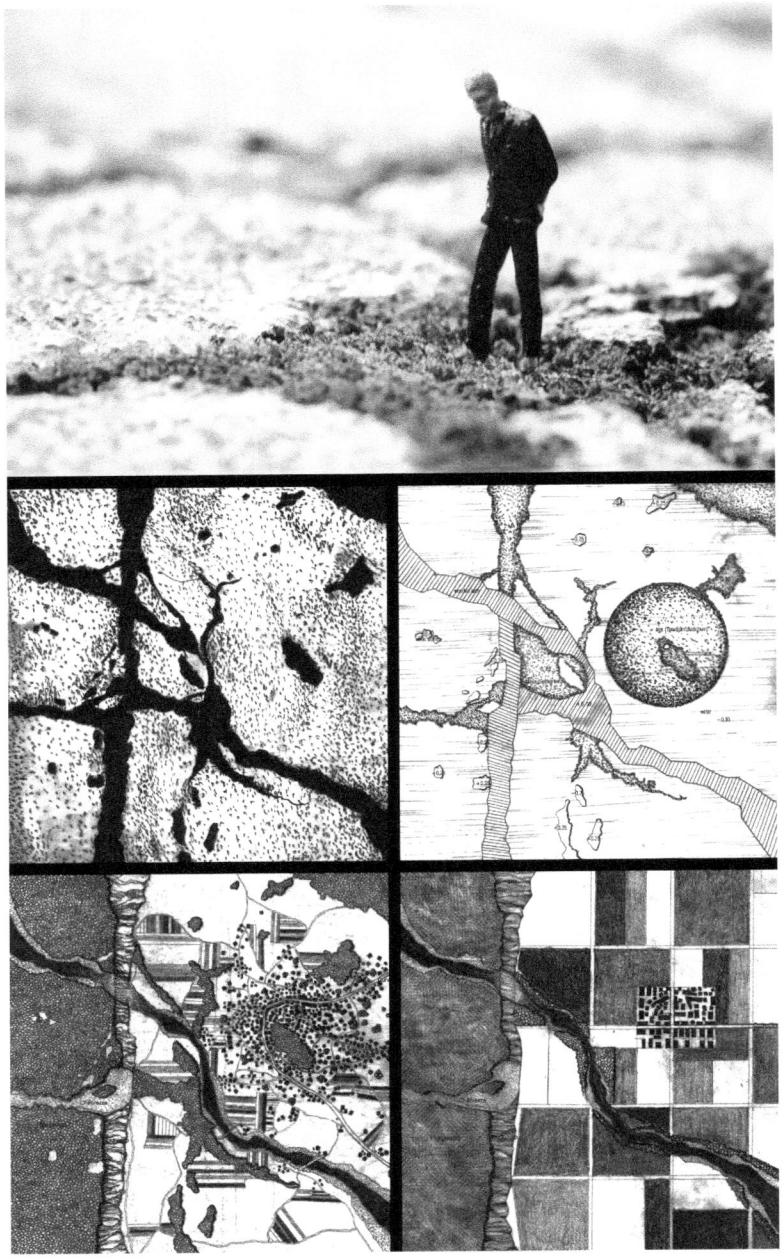

Figure 25.1 Workshop 'Out of Scale' project: The Great Discovery (Weilacher et al., 2013). In the presented project, the fictional person is businessman Ben, who sets out on a walk out of town to discover a new landscape that opens before him. The photograph of the man was taken on cracked ground and illustrated to a scale of 1:2. This was followed by a proposal of a garden to a scale of 1:100 and the space is already set in the urban and cultural landscape to a scale of 1:5,000, first as a presentation of the 'existing' landscape and then the envisaged spatial changes, which on the far right of the illustration shows the de-urbanized landscape, in which farmland dominates because of the need for food

This figure is shown in colour in the plate section of this book.

Figure 25.2 An example from Introductory Design Studio – improving the creativity of students with the insertion of an abstract artistic exercise: colour images composed of garden elements from the first design of the house garden (Gazvoda, 2008–2014)

This figure is shown in colour in the plate section of this book.

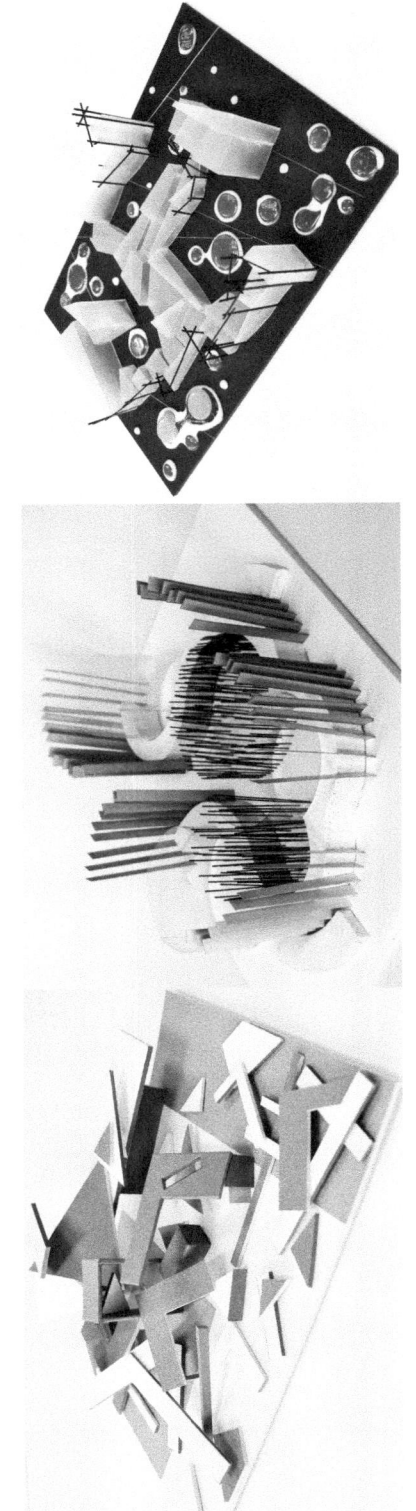

Figure 25.3 Abstract 2D shapes grow into 3D elements, from which an abstract space is composed. Students make a model (Gazvoda, 2008–2014)

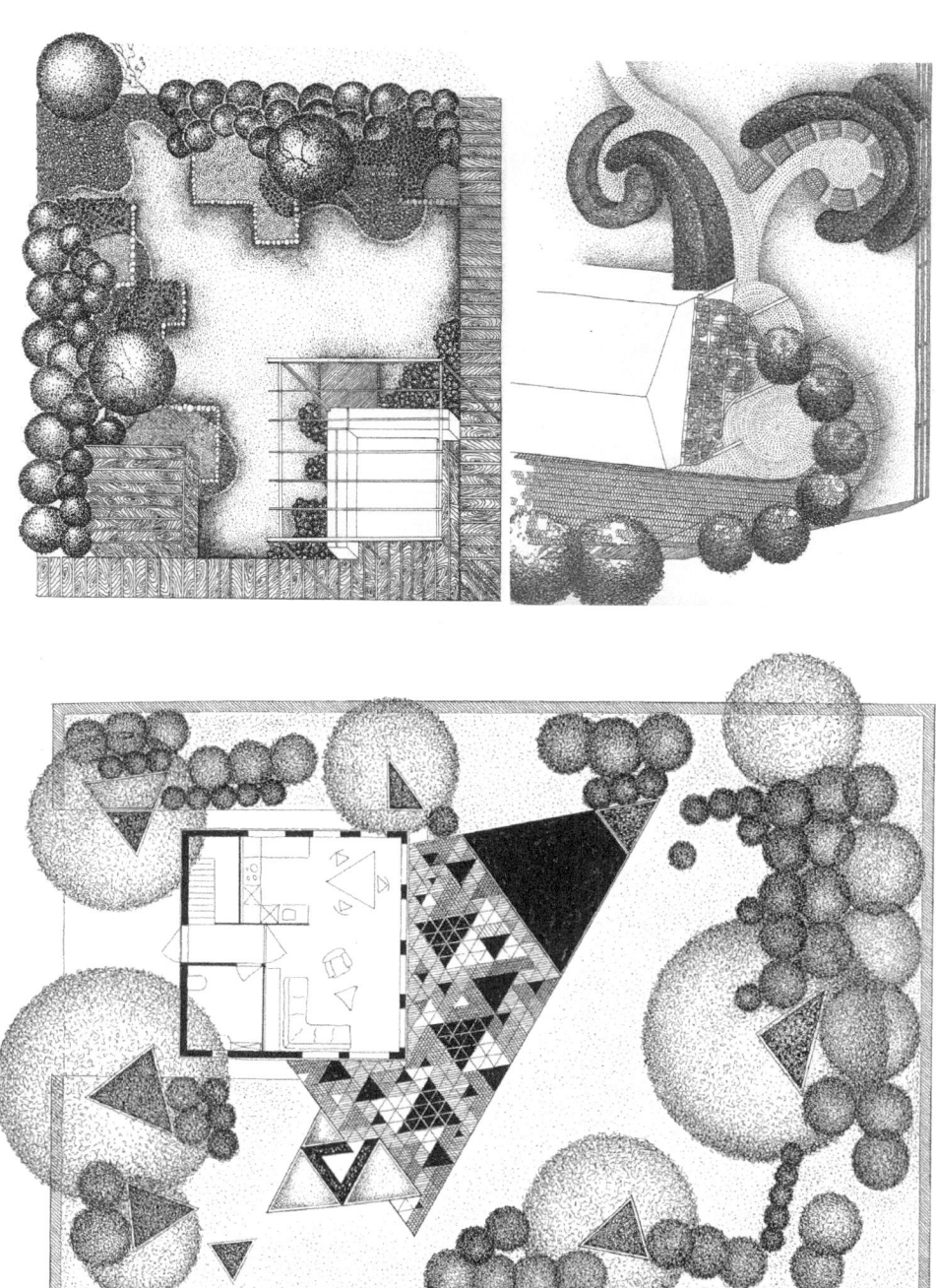

Figure 25.4   The final result is a more ambitiously designed garden, with a larger number of spatial elements, with more interesting concepts, varied forms and more complex compositions (Gazvoda, 2008–2014)

The change in scale shown in the workshop case forces students creatively to provide spatial changes in the context of substantive changes. A similar exercise in creativity is also to be welcomed at all levels of study, either as a separate exercise or as a step in an otherwise long and more complex design process. To illustrate, the following two examples are also interesting: one from Introductory Design Studio in Ljubljana and the other from a workshop at the University of Beijing in the framework of the studio Landscape Design Topics.

When working in the Introductory Design Studio in the first year of studies, BSc students fairly quickly understand and learn the functional and, to some extent, the technical requirements of a garden of a family house. They have more problems in determining the forms of garden elements, the compositional rules of use of these elements in space and conceiving garden elements (how to 'spatialize' programme requirements). After the first variant of the solution of a freestanding single-family house with a garden was handed in, the students focussed on purely artistic practice, in which they had to create from their already used forms a completely new artistically abstract solution, to add new forms and their combinations, in short, to create a colour abstract image that no longer has any connection with the garden (Figure 25.2). Then, from the elements of this painting, which is perceived as an abstract layout of the space, they determined heights, drew them in outline, created a 3D presentation of this space (axonometric or perspective view) and, at the same time, a model of the same abstract space (Figure 25.3). Only then did they start to look for possibilities of materialization of selected abstract spatial elements in their specific layout as garden elements. From the garden elements, they then composed a new garden beside the house, which was usually a more interesting and complex design than the first variant (Figure 25.4).

This kind of exercise is shown to be useful in another two cases: when working with students in programmes that are not true landscape architecture programmes, but at selected schools where such studies are merely introductory (so students do not have adequate landscape architectural background knowledge), and with students who come from a completely different cultural and socio-political environment to the teacher. In the first case, the students in question are already to some extent moulded as foresters, agronomists, engineers of horticulture, etc., and in the second case, landscape architecture students may be accustomed to different work and the problem is both language and cultural differences, and the educational approach is therefore more demanding, since pedagogical methods must be clearer and understandable to all. Although with different preliminary knowledge from a BSc degree, these students are already fully formed and skilled in a variety of aspects of spatial problems, they master spatial analysis, they can be excellent specialists, for example, with plants and the like, but they have a problem in creativity and in synthesizing the knowledge acquired into a final design solution, which I noticed, for example, with Chinese students throughout my ten-year involvement in teaching landscape design studio at Beijing University. The Chinese students were very industrious, studious, and research-oriented, but with design they quickly resorted to known and previously used design examples. Resetting here involved interpretation of the location through freehand drawing, its abstraction, with added forms from the space, the development of a characteristic shape grammar (taken from the groundplan), which was followed by synthesis of the predetermined design and spatial elements into a new composition (in this case the creation of a new campus). In the final phase, of course, students had to remodel these elements and conceive them with use, functionality, evocative power, symbolism – in short, to make a final complex project (Figures 25.5 and 25.6).

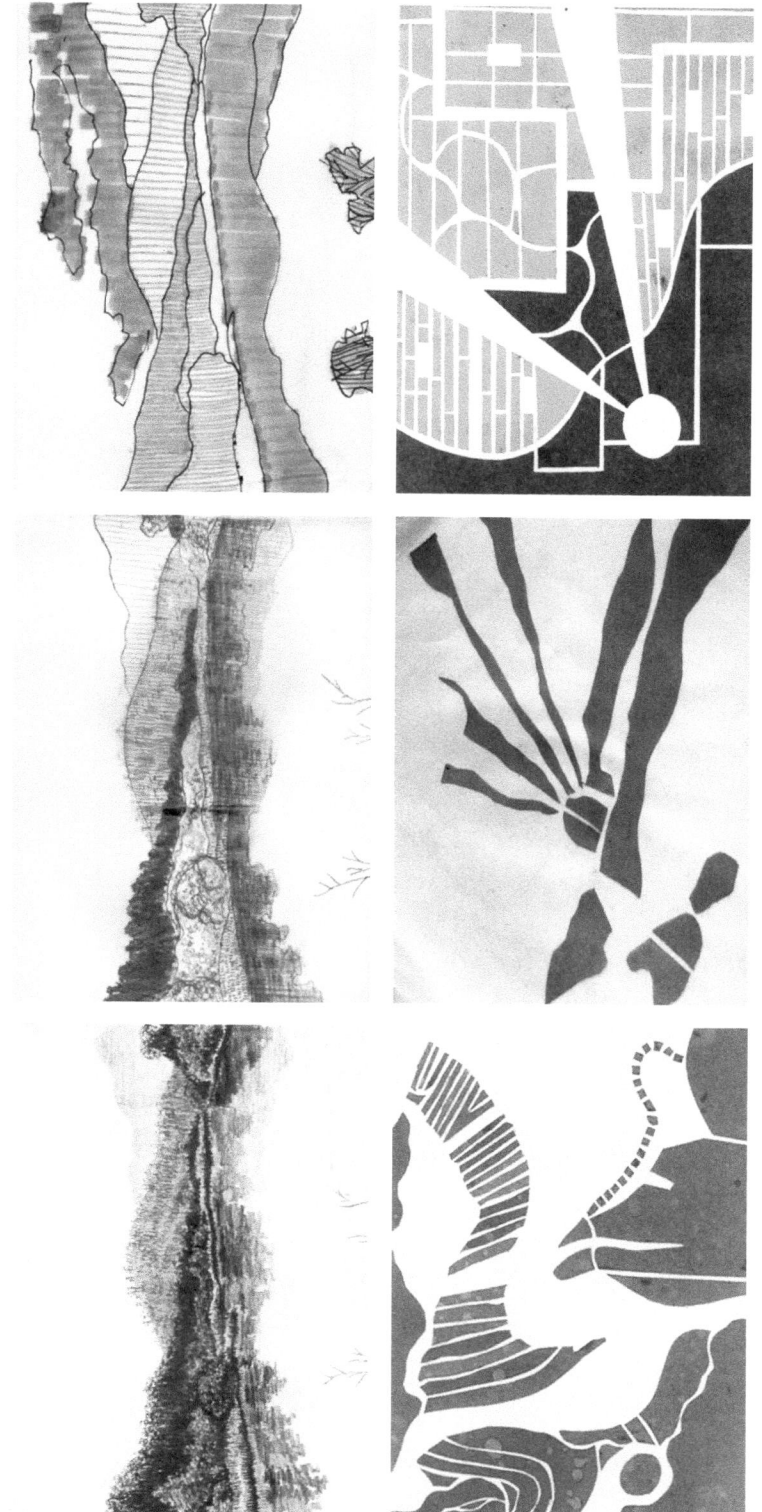

Figure 25.5 Landscape Design Topics Studio 2006. The process of generalization and abstraction of space begins with sketches of the location. The next step is the abstraction of existing spatial structures into a new composition, which serves as a basis for the ground plan of the new student campus (Gazvoda and Drašler, 2006)

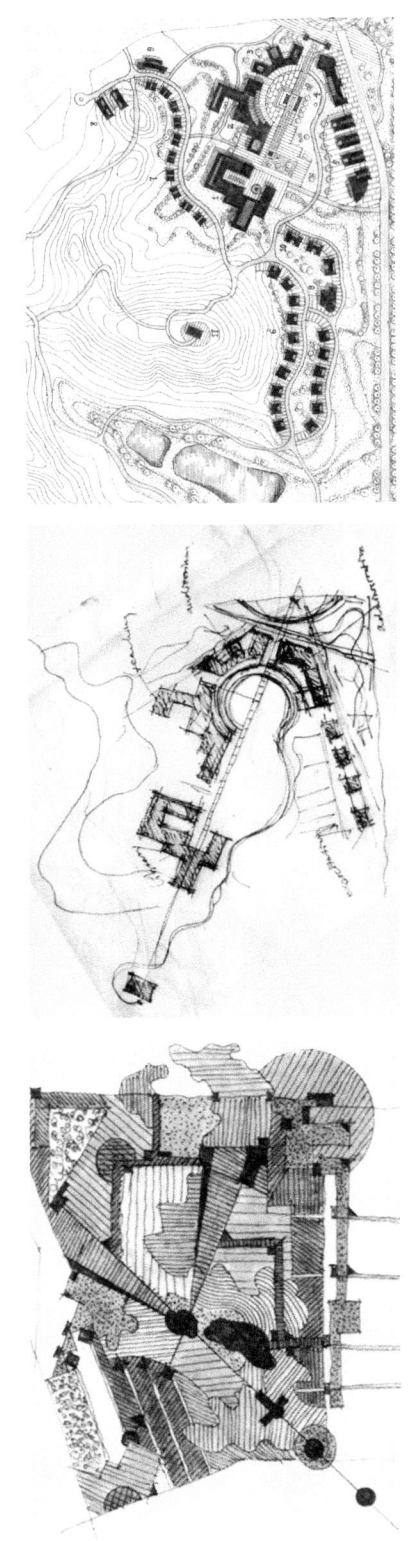

Figure 25.6 After three subsequent steps, the space is a complex composition, which must be arranged into a new student campus with all the required components: faculty, dormitory, dean's office, student centre, etc. (Gazvoda and Drašler, 2006)

Digitization of the learning process

After experience of working with students at various levels of study and in different countries, it is worth mentioning a few features about which teachers of landscape design should be vigilant. The first is the already mentioned copying of well-known already-seen solutions. It is therefore useful to provide lectures with demonstrated cases only in the middle of studio work, after the students have already sought their first independent solutions. This avoids students remembering in advance spatial solutions already seen (models) and using them, copying them into their solutions, even though subconsciously. This approach is difficult today because coming generations of students are accustomed to clear instructions and they want a framework, objectives to be promptly set about the given methods, and examples that can be found immediately with a few clicks. They demand study materials and they want lectures.

The second difficulty is that, despite the fact that at lower levels of schooling they have always had some sort of PBL and should be independent in solving problems, students constantly expect clear instructions. Perhaps this is a problem of a particular school or generation of students and cannot be generalized, but it is increasingly evident that this is probably a problem of motivation of the millennium generation in general, which feels safer with clear instructions and rules, and is in fact less independent than you would expect.

The third problem is the use of digital media in the learning process. Much of our work is dependent on the use of digital media, as required by the size of the space and the amount of spatial data. Digitization in the teaching process and the use of new emerging media are also characteristic of landscape architecture. Replacing the 'learning-based approach' with 'learning environments' enabled by new technologies appears to be an effective approach, because they allow unlimited access to an unlimited amount of information about anything (Thomas and Brown, 2011). Is this so-called 'new culture of learning', with an emphasis on digitization, really such a pronounced advantage, or may too fast and too much digitization be a weakness? The German psychiatrist and psychologist Manfred Spitzer talks about short-term memory, training in spatial orientation and other aspects of digitization of our education, which clearly also applies to students of landscape architecture, in his book *Digital Dementia*.[4] He draws attention to a loss of spatial orientation arising from reliance on digital media, GPS devices and guidance through space, so that we do not remember and thereby lose our sense of direction. In addition, he also cites the fact that we remember less well or do not remember at all information if we know that it is stored somewhere and can always be found (Spitzer, 2016: 94–95). In recent years I have noticed that students of landscape architecture already have poorer spatial representation when they come to the faculty. While studying, when they are already using spatial data, this may improve but, in comparison with older generations, they are less resourceful in space. They no longer grasp space as a whole and cannot remember it. The main problem is clearly in the fact that they have available numerous thematic maps, a variety of spatial data, but they do not remember them and do not have a sufficiently good grasp of the space in their heads. They thus often neglect certain spatial characteristics or even interpret the space incorrectly. The role of the teacher here is focussed on constantly drawing attention to what they know, or should know, about the space and that they constantly take this into account. Despite the problems outlined in the application of digital media, such application cannot of course be restricted as it has become an indispensable tool, and therefore the teacher must insist that, even in the early stages of a project, the students commit the space to memory and know how to evaluate it without the aid of computers, to present the first concepts by hand drawings. Freehand drawings, sketches, the use of design and other diagrams, schemes and so on still remain the medium of communication in daily interaction between teachers and students. This is not an orthodox view of the approach to design, but simply the fastest transmission of thought through pencil on paper, especially in the early stages of a project. This is, of course, then followed by computer models supporting the following planning phase through to the final presentation of solutions, whereby students today often master digital media better than the professor (especially the use of new software packages with a focus on virtual reality, 3D simulations, animations and the like). With the growing digitization of the learning process, the role of the teacher in the studio is clear: to ensure that students are able to connect all relevant information, that they learn how to evaluate the mass of data available to them and choose suitably. Students need to understand that the new media are an aid to the designer but are not necessarily an advantage for a quality solution, and that too-rapid use of digital presentations may be too superficial, to the detriment of substantive solutions to spatial problems.

Conclusion

Small groups in studios, individual work with students, solving specific spatial planning problems, cooperation with clients (local communities, municipalities, civil initiatives, and groups of various stakeholders in spatial planning) and presentation of results to interested publics are positive characteristics of modern design studios. An attempt is thus made to transfer practice into the study process, which is organised in this case similarly as in a modern design office, in which effective dialogue between the project manager or principal designer and the designer is crucial.

The ideal teacher of a design studio is therefore a knowledgeable and theoretically well-versed landscape designer who, in addition to design itself, also masters the teaching of theoretical aspects, is a skilled art critic and who knows how to recognise the potential of an individual student. Concretely, this means that a teacher should be first formed at the faculty through the established process of advancement and obtaining references, though she or he must also have completely practical design office experience and work, which can also follow later in her or his career. Why such weight to the proposal that a teacher should come primarily from the academic sphere and not from the design office? The pedagogic approach and ability to communicate with students are crucial. Sometimes it is simply necessary to understand that the final product or design solution itself is not as important as mastering the design process, with all its particularities. A good teacher must recognise what the students have learned and what kind of development they have achieved through the workshop approach. In this context, landscape architecture is a specific discipline that, because of the explicitly interdisciplinary character of the problems that it resolves, enables entirely specific solutions in the teaching process, which are often not carried out in the world of practice because of organisational or financial difficulties, but they provide important starting points for the development of the profession. Through work in a design office, students can learn many technical skills, presentational techniques, communication with clients and, not least, market laws or the demands of the market. Therefore, in the educational process, with good teachers, it is preferable to devote more time to understanding social problems, which are solved in space with respect for natural processes, functional needs and the demands of users of space and spatial aesthetics.

Notes

1 Anyone interested in a deeper insight into new methods of teaching and learning styles can also choose among various texts on more general methods of learning and teaching, e.g., Wiggins and McTighe (2005) or Thomas and Brown (2011).
2 'Project Based Learning is a teaching method in which students gain knowledge and skills by working for an extended period of time to investigate and respond to an authentic, engaging and complex question, problem, or challenge' (www.bie.org/about/what_pbl).

3 Intrinsic motivation: inclusion in a particular activity for pleasure and satisfaction at its implementation – without external or internal pressures and without expectation of reward.

4 In the paper the English translation of Spitzer's book is used, which was originally published in German in 2012 entitled *Digitale Demenz. Wie wir uns und unsere Kinder um den Verstand bringen* (Munich: Droemer Verlag).

References

Bruns, D., Ortacesme, V., Stiles, R., de Vries, J., Holden, R. and Jorgensen, K. (2010) *ECLAS Guidance on Landscape Architecture Education*. European Council of Landscape Architecture Schools, internal report [online] www.unideusto.org/tuningeu/images/stories/Summary_of_outcomes_TN/ECLAS_Guidance_on_Landscape_Architecture_Education.pdf.

Freire, M. C. M. (2012) 'Towards a different approach in teaching landscape design. A cross educational, cultural and disciplinary strategy', in: *Proceedings of 5th WSEAS International Conference on Landscape Architecture (LA'12)*, University of Algarve, Faro, Portugal 2 May, pp. 66–71.

Gazvoda, D. (2002) 'Characteristics of modern landscape architecture and its education', *Landscape and Urban Planning* [print ed.], vol. 60, no. 2, 117–133.

Gazvoda, D. (2008–2014) personal archive of Introductory Design Studio, various student works, Ljubljana: University of Ljubljana, Biotechnical Faculty, Department of Landscape Architecture.

Gazvoda, D. and Drašler, A. (2006) *Landscape Design Topics: Campus Design in Shenzhen*, Beijing, Shenzhen: Beijing University, Graduate School of Landscape Architecture, workshop and studio report.

Gazvoda, D., Mlakar, A. and Marušič, J. (2004) 'Landscape planning and design of the countryside: approach and teaching practise', *Landscape 21 (Ljubljana)*, vol. 1, no. 1, 44–51.

Motloch, J. L. (2001) *Introduction to Landscape Design*, 2nd edition, New York: Wiley.

Ogrin, D. (2011) *Krajinska arhitektura*, Ljubljana: University of Ljubljana, Biotechnical Faculty, Department of Landscape Architecture.

Rowe, P. (1987) *Design Thinking*, Cambridge, MA: The MIT Press.

Rowe, P. (2002) 'Professional design education and practice', in: A. Salama, W. O'Reilly and K. Noschis (Eds) *Architectural Education Today. Cross-Cultural Perspectives*, Lausanne: Comportments and authors.

Spitzer, M. (2016) *Digitalna demenca. Kako spravljamo sebe in svoje otroke ob pamet*, Celovec: Mohorjeva.

Steinitz, C. (1990) 'A framework for theory applicable to the education of landscape architects (and other environmental design professionals)', *Landscape Journal*, vol. 9, no. 2, 136–143.

Steinitz, C. (2012) *A Framework for Geodesign: Changing Geography by Design*, Redlands: ESRI Press.

Thomas, D. and Brown, J. S. (2011) *A New Culture of Learning: Cultivating the Imagination for a World of Constant Change*, CreateSpace Independent Publishing Platform.

Van den Brink, A., Bruns, D., Tobi, H. and Bell, S. (2017) *Research in Landscape Architecture*, Abingdon, UK: Routledge.

Weilacher, U., Bajc, K., Kučan, A. (ed.), Abram, Ž. et al. (2013) *Out of Scale: Landscape Design Workshop Ljubljana*, Ljubljana: University of Ljubljana, Biotechnical Faculty, Department of Landscape Architecture, report. What is Project Based Learning (PBL)? [website], www.bie.org/about/what_pbl.

Wiggins, G. and McTighe, J. (2005) *Understanding by Design* (2nd expanded edition), Alexandria: Association for Supervision and Curriculum Development.

Zeisel, J. (1981) *Inquiry by Design: Tools for Environment-Behavior Research*, Cambridge: Cambridge University Press.

26

Reaching out in teaching landscape

Engagement and service from the studio

Peter M. Butler

Introduction

Service-learning is an educational approach that combines teaching with community service in order to provide a learning experience while meeting societal needs. Developing centers for service-learning within universities and academic units supports community engagement. The design fields seek to satisfy many of the desired goals of service-learning when developing models for engagement. Among these are building capacity for student learning and leadership; providing experiences that highlight and strengthen inclusive communities; creating authentic experiences for community members and students in collaborative projects; and promoting principles of democracy, compassion and cultural diversity through civic engagement (adapted from West Virginia University 2017). Service-learning activities build soft skills in students: listening, interpreting, expression, and communicating in multiple modes. In underserved communities, service-learning builds partnerships with the academy and capacity in moving society towards positive alternative futures. The theory and practice of participatory design is tied to multiple traditions and methods as developed by anthropologists, artists, activists, community and regional planners, conservationists, and action researchers among others. Further cross-pollination through trans-disciplinary projects will provide for a more rich, deep, and meaningful experience for community members and students. Professionals and academics creating flexible multi-disciplinary teams are better prepared to address societal needs and solve the wicked problems of today.

Landscape architecture is a place-based (Grabasch 2015) and project-based practice (Gruenewald & Smith 2014). Regardless of scale and context (LaGro 2011), designs engage people within *space* and seek to create *place*. Mary Miss alluded to the concept that landscape architects knit, weave, or sew sites to their context (Miss 2014) – the physical geomorphological and hydrological elements, cultural and historic traditions and norms, and the living context of the people. In the pursuit of engaging the social aspect of context, this *stitching* is particularly important in establishing participative processes and creating approaches to design in working with disenfranchised and underserved populations (Thering 2007; Sanoff 2010; Hester 2006) through service-learning (Angotti, Doble & Horrigan 2012; Horrigan et al. 2014). Participative practices are central to a 'conscious' practice within an engaged model of the landscape architectural profession that promote sustainability, social justice and inclusive design.

The following narrative describes specific ways to conduct the practice of service-learning in landscape architecture studios. It defines broad learning outcomes for courses as activities and desired impacts for communities and students within the classroom. Modes of engagement for classes that seek interaction with community groups are described in detail, as well as the processes and products of those engagements. The modes and processes described in the narrative are not comprehensive as many practitioners and faculty are experimenting with novel approaches. These processes must be considered malleable to adapt to a variety of situations, community groups and sites. The author's reflections are included as a means to identify potential limitations and opportunities that may arise through the process of integrating service-learning activities into the studio environment.

Defining learning outcomes for service-learning courses

Training future landscape architects in participative modes of design through democratic processes (Horrigan & Bose 2017) is an intrinsic element of pedagogy within the field. Necessary skills in practice may be identified within service-learning courses as expected learning outcomes. Measurable outcomes (Hester 2012) from documented projects demonstrate the value of participative approaches. Comprehending outcomes from a broad set of perspectives – the community, collaborators, academy and students' learning outcomes – builds scholarship for faculty that work in the realm of engaged scholarship. Impacts on students in service-learning courses include the raising of awareness of socio-economic, environmental, political and other historic and cultural contexts. Exposing and leading students through a variety of participative practices then forms an expanding toolbox of strategies in working with people. New generations of landscape architects may then be positioned to lead in the practice of engaged community-based design. Evaluation of and reflection on pedagogical models of participation has become a significant component of design research. In following previous models of evaluation, the field moves 'up the ladder' established by Arnstein (1969) from *manipulation, therapy and consultation* towards *citizen empowerment* (Arnstein 1969; Halprin 1969) and *transactive* design (Hester 2006).

The following learning outcomes may be considered broadly as a basis for developing a service-learning course: define and create "community" at multiple scales; understand and develop methods of gathering community input; interpret and analyze community input; apply community input to the planning and design process; and design and plan with and for the public.

Depending on the context for the service-learning course, learning outcomes can be framed locally and globally to demonstrate the potency of design in impacting communities. These outcomes have been identified through practice. They are generalized in this chapter and should be refined depending on specific desired outcomes in courses. Primary to learning outcomes though is the need to *define and create "community" at multiple scales*. As projects will include collaboration, a positive beginning point in the course is working with students in the class to become more collaborative. Achieving and building a sense of community in the classroom can be undertaken in a variety of ways – through quick teamwork design projects; role-playing in defining participant contributions to an overarching effort or goal; or team-building through an external activity like an adventure course that requires cooperation in completing a task or physically solving and reaching a goal.

Another core learning outcome in developing a service-learning approach is to *understand and develop methods of gathering community input*. A single studio course in community design will not usually allow for the practice of all participatory methods, though exposing students to the multitude of approaches is important. A strategy for teaching the breadth of methods to students is to provide them with the overall project requirements – a small town community park, for example. And to share with them the variety of potential approaches that would allow for accessing community input – survey, workshop, charrette, focus groups, etc. Students then work through the process of identifying and selecting the approach that they would like to apply to the project. In a large studio class of twenty students there could be six or seven small teams that would each develop an approach

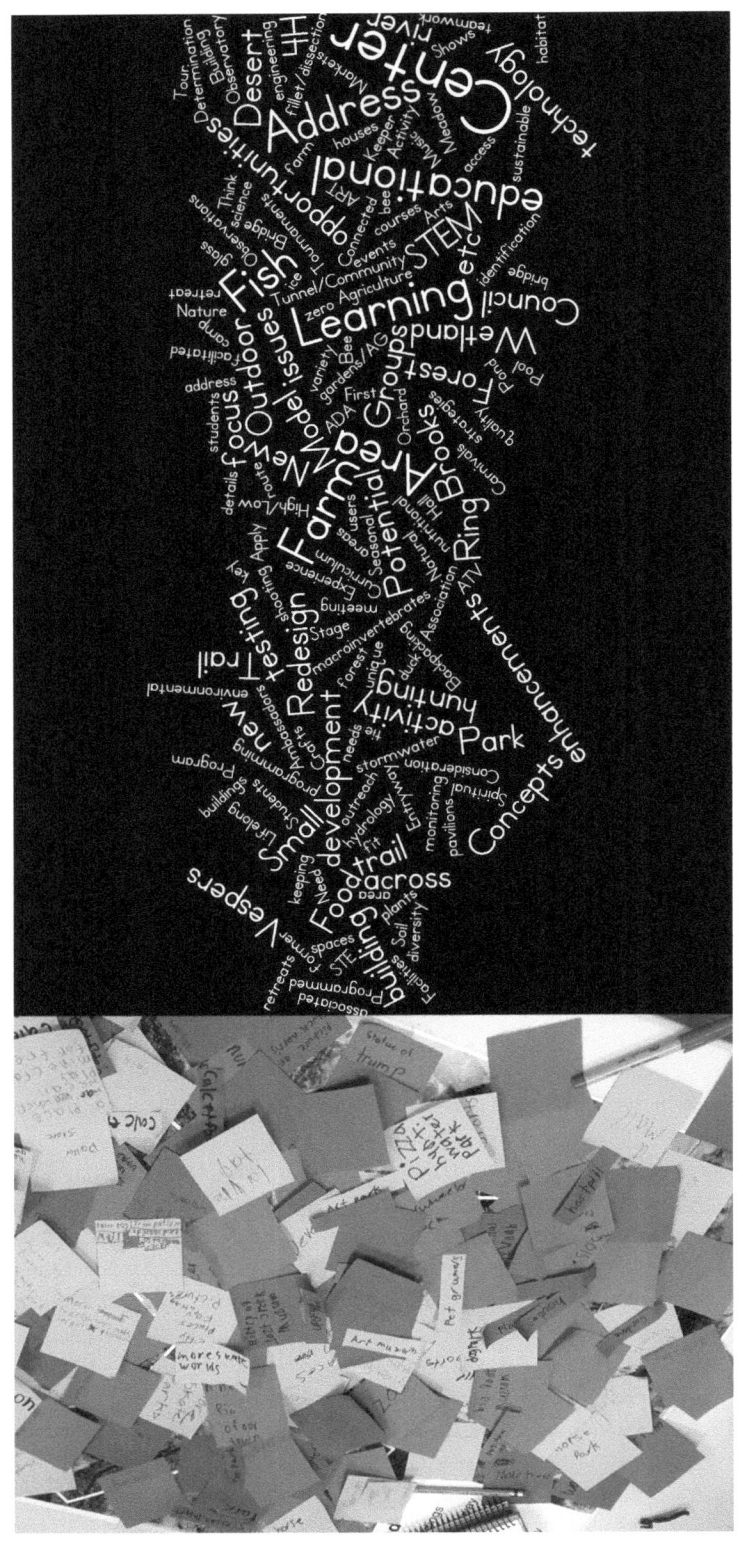

Figure 26.1 The collection of information at a meeting with community members includes a massive amount of text. The organization, graphic expression and distillation of collected data is a key activity for student learning (image credit: author and author's students)

to participation. With a diversity of approaches a high level of peer teaching and learning occurs. Teams have the ability to share with one another the development of their approaches and, through critique and input from peers and instructors, to refine their methods.

The execution of the variety of approaches then leads to *interpreting and analyzing community input*. The compilation of survey results in spreadsheets and ranking of preferences, for example, requires time and thoughtful organizational skills. Representing collected information is another design project in and of itself. Developing creative and clear graphic modes of communicating complex information to diverse groups of stakeholders is an important skill for landscape architects (Figure 26.1). Immersing students in the activity leads to skill development. Interpretation of community input is a challenge for the design student trained to develop his or her own concepts individually. Community preferences may be counter to student perceptions. Another level of interpretation of community input includes the translation from language to form, space, and experience. Character-defining language collected from a community process may include such abstract terms as welcoming, beautiful, pleasant, etc. As students may interpret these terms differently than community members, it is important to maintain clear communication with community partners and provide multiple iterations for review and comment.

Applying community input to the planning and design process is another learning outcome. As information from community participation may consist of text and rough sketches, the interpretation and application of input creates new opportunities and challenges. Values, concerns, goals and objectives, become spatial and integral to the design context. Student comprehension of the architecture, landscape, development patterns, materials, and textures, as well as the social concerns garnered from participative process, defines a context-sensitive design strategy. The inventory and analysis phase of a project completed by individual students or teams will uncover these endemic characteristics. While the language and sketches created through the participatory process provide guidance in creating design alternatives, knowledge of and sensitivity to the existing features of place ties plans to local tradition.

An overarching learning outcome for service-learning design studios is to *design and plan with and for the public*. As mentioned in the introduction, projects seek to reach high levels of Arnstein's "Ladder of citizen participation" (1969) *citizen control, delegated power, and partnership*. Levels to be avoided include *placation, consultation, informing, therapy and manipulation*. The semester schedule for student service-learning projects creates a challenge in developing community capacity that leads to citizen control. Partnerships are often the result of a service-learning course with faculty members continuing the relationship with stakeholders. Applying a diversity of methods within the structure of the course provides for the highest levels of comprehensive engagement.

Modes of engagement in service-learning design studios

The semester schedule creates a need for order and efficiency in delivering service-learning courses. The choice of modes of engagement depends on the 'who, what, where and when' that the engagement involves. The purpose of the project is central to developing service-learning projects, as well as concerns of student numbers, course sequencing and outcomes, skill level and maturity of students, and capacity of faculty. A variety of approaches have been developed in the last half century including: survey/focus group; charrette; workshop; design/build; in situ design; 'team' approaches; and participatory art installation (NCI 2017; Sanoff 2000; Mehrhoff 1999). While this set of distinct processes is not comprehensive, the processes reflect contemporary convention in approaching service-learning projects through participatory design. Each mode benefits student learning by providing training in the variety of methods. Each also can be defined by specific generative artefacts that are produced with design projects: text, sketch, mapping, models, drawings, etc.

Overarching unifying processes within each activity rely on Tuckman's (1965) group process and includes the *preforming* (preparation and planning), *forming* (group and partnership building), *storming* (activities and engagement), *norming* (decision-making and consensus building), *performing* (articulating and presenting the vision), and *interpreting* (applying to real-world context for implementation) (Tuckman 1965). These processes may be repeated multiple times through a variety of participatory activities and the overall iterative design process.

In preforming – preparing and planning for a project – a *survey/focus group* may be used as a tool to access broad participation from a community, neighborhood, or stakeholder group. Administering the survey before the beginning of the course will provide a baseline of information from which students may begin. The development of the survey tool for multiple uses and contexts is important, especially when faculty wish to develop research tied to service-learning. The repetition of the same survey produces data that may then be interpreted and published in order to create a knowledge base for future service-learning courses. A typical survey will be tied to previously identified goals and objectives as generated through initial project planning.

1   What do you perceive as opportunities for programming/activities for the space?
2   What would be the desired character/aesthetics of the space?
3   What do you perceive as specific opportunities/concerns in the development of the space?
4   What do you see as potential innovations that could occur in the space?
5   Describe what the site should 'feel' like?
6   Can you tell us a story about the site and how it fits into the history of the community?

These same, or similar, questions may be used in a *focus group* format. Focus groups may be a preferred method of collecting information if the community partner consists of a small organization that meets regularly and would welcome faculty and students into their meetings and events. Because the questions are open-ended, discussions in a focus group setting can expand responses from participants (in comparison with the survey) and generate discussion, argument and consensus. Outcomes of focus group meetings may create a strong sense of purpose and a concentrated set of goals and objectives for the organization, and, thus, for the student projects.

The survey and the graphic expression of survey results may then lead to the *charrette* – an interactive visual design exercise giving voice to engaged stakeholders and providing a forum for contribution and collaboration – forming and storming. Charrettes may last one day, two days or an entire week with a variety of participants joining at different points of the process. Preparing large format base maps, site photographs and a digital model of the site before the charrette event is important. Consolidating and presenting existing conditions information including inventory and analysis is also necessary, along with the survey response summaries so that all participants have a basic knowledge of site conditions. A core outcome of the service-learning course may be to train students in the necessary skills and knowledge to prepare for and lead charrettes – honing skills in quick sketching, mapping, verbal communication, listening, and iterative design process. As experiential learning, charrettes include the quick generation of ideas in an interactive setting, which includes community stakeholders, students, faculty, practitioners, alumni, etc. (Figure 26.2). The products of the charrette process are plans, sections, perspectives, writing, models, etc. (Figure 26.3). In a large studio with approximately twenty students, a team approach can generate multiple solutions to the same design problem. Following the charrette, students may have the opportunity (one or two weeks) to refine the products of their team designs and to generate a project report (Figures 26.4 and 26.5). As an artefact documenting the event, the project report summarizes and reflects on the experience of the activity and provides a set of new drawings and writing that captures the project more completely for the community partner.

With all of the team projects delivered to the community partner, and allowing them time to review the products, the next step is a *workshop*. The workshop is a time to digest the team products and to identify priorities for further design development – norming and performing.

Figure 26.2 A design charrette including landscape architecture students and alumni practitioners with community stakeholder input (image credit: author)

Community partners may select different elements from different team projects, or they may see a particular circulation pattern in one design and want to combine that design with a threshold from another. The main goal of the workshop is, then, to create a single plan with students that can become the vision of the project. Ranking of different designed elements and prioritizing next steps occurs at the workshop phase and developing a plan for project funding, implementation and phasing. If the collaboration has been successful, in reference to Arnstein's ladder, then *citizen control* or *delegated power* will occur at this point. Alternatively, a *partnership* is established for continued technical assistance from faculty members or a practitioner.

Adding benefits to service-learning experiences

Design/build studios may work for an entire semester with a community client through the same or similar participatory design processes as outlined above, with the goal of landscape installation. These courses develop skills in working through construction detailing and materials experimentation with applied experience in construction and may be tied to a construction technology curriculum. Especially in areas that are socially and economically depressed, installation of a landscape sets a seed for future revitalization. Activist landscape architecture faculty (Kyber, Lawson, Winterbottom, etc.) assist with the vitalization of neighborhoods through design/build projects (Figure 26.6). The projects often depend on local and student labor, as well local found and collected materials.

Distance and travel is always an issue in performing service-learning. Faculty (Badenhope, Kyber, Hester, Yuill, etc.) have addressed this issue by creating courses that are sited within communities often adjacent to particular project sites. *In situ design* establishes a temporary presence in the community creating a hub for design activities to occur. Student experiences are immersive as

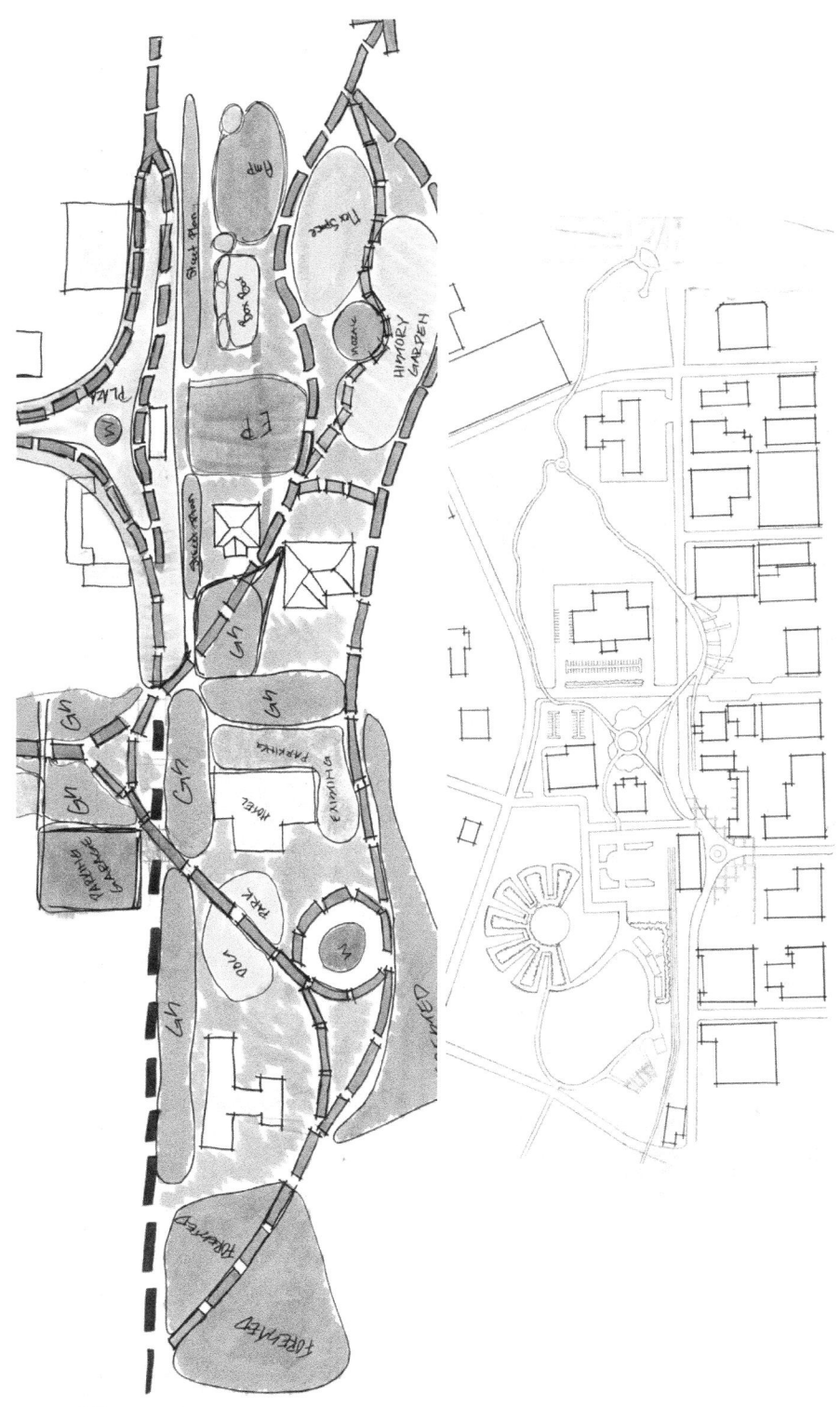

Figure 26.3 Resulting sketches and diagrams from a charrette process that develops student skills in communication, collaboration, quick iteration and sketching (image credit: author's students)

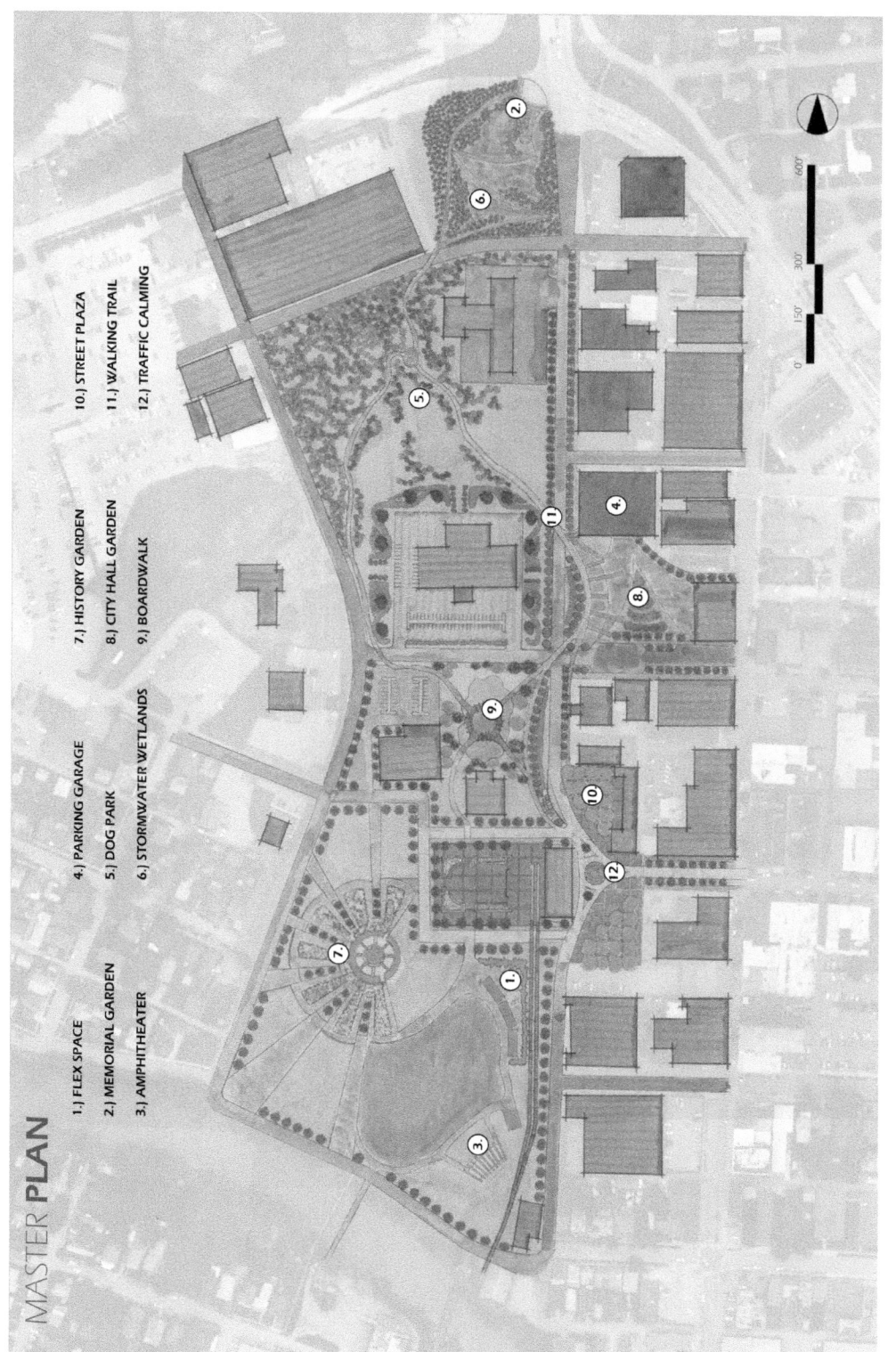

MASTER **PLAN**

1.) FLEX SPACE

2.) MEMORIAL GARDEN

3.) AMPHITHEATER

4.) PARKING GARAGE

5.) DOG PARK

6.) STORMWATER WETLANDS

7.) HISTORY GARDEN

8.) CITY HALL GARDEN

9.) BOARDWALK

10.) STREET PLAZA

11.) WALKING TRAIL

12.) TRAFFIC CALMING

Figure 26.4 Digitizing and refining the products of the charrette for presentation to the community partner allows for reflection and a more professional final product (image credit: author's students)

HISTORY **GARDEN**

The idea behind the history garden was to use the original shape of the round house to construct a raised planter and meandering path. The path connects with the street and all other walkways to keep circulation flowing. The center is a sunken area with another raised planter and large specimen tree for shade. At night, the sunken area is illuminated by LED lights to create a fun and exciting area for night life.

Figure 26.5 During the intense and limited period of the charrette it is difficult to generate digital illustrations of site designs. The post-charrette refinement of products allows for more detailed illustrations of proposed conditions (image credit: author's students)

they become a part of the community. The spaces serve as laboratory and gallery, office space in a professional environment and a gateway to participation for local residents. Challenges with this method of service-learning engagement include costs of facilities as well as scheduling students to be present within the 'storefront' during working hours of the day and beyond. The benefits of the depth of engagement allow for more expansive experiential learning and a deeper partnership.

In professional practice today, landscape architects rarely work alone, but rather are integrated with cohorts of allied professionals. Within each of the participatory methods inviting experts to contribute to the design process adds benefits to student learning as well as community outcomes. Allied professions that have traditionally been included in 'team' approaches (Mehrhoff 1999) include architects, engineers, public historians, public administrators, planners, lawyers, sociologists, resource managers, and ecologists, among others. The composition of the team depends on the context of the project. Enriching the process while not overwhelming the functioning of the effort is a key balance to achieve. Collaboration with multiple disciplines (de la Pena 2013) beyond conceptual design is also important in developing capacity for long-term continuation of projects towards success.

The *artefact* (Lawson et al. 2011) approach, or the production and installation of a work of art and design, has become not only a form of engaged production but also an important result of service-learning courses. Public artists can play a central role in the design processes. As a component of a larger effort, public artists bring innovative design strategies to engagement. As interpreters of culture and fabricators, public artists may be involved at any point of the process, especially if the community partner desires an installation to spark change.

Experimentation and new frontiers

Traditional approaches, as tested within the classroom environment, have demonstrated proven results for stimulating community development progress as well as for enriching students' experiential learning. Innovative approaches, primarily based on new technologies, seek to bridge the real and perceived divide between expectations of the community and the desired outcomes for students, while exposing both groups to new tools. Digital methods (Figure 26.7) provide access to participation that traditional modes may not allow. Innovation generally occurs within the context of the established, though practical concerns demand that strategies remain flexible in approach depending on the context. *Gaming* (Hou 2010, 2011, 2013; Butler et al. 2013) has increasingly become an important component of engagement with youth. The use of *virtual reality* environments within a digital three-dimensional environment for the testing and critique of community participants is becoming a more accessible practice. *Online survey tools* reach a broader audience and include mapping exercises that are easily accessed by community members. With digital tools, data collection and expression is more streamlined and polished as a product of participation. With these technological innovations, the core goals of service-learning in design studio through engaged scholarship remain the same – to provide an authentic and impactful learning experience for students while providing a meaningful and productive experience for community partners.

Conclusion

Landscape architecture studios can have a positive impact on communities through service-learning activities, especially as a component of university centres. The opportunities for collaboration across disciplines within the academic environment are boundless. Trans-disciplinary teams of faculty, professionals and students together with community members have the ability to create positive change within a multitude of contexts. The skills developed by students in engaging with these processes

Figure 26.6 An example of an activist design/build class project for a community garden constructed in the socially and economically compromised community of Osage, WV. Ashley Kyber, primary designer and coordinator. Mural created through the work of Eve Faulkes, Professor of Art and Design, West Virginia University, and her students (image credit: author)

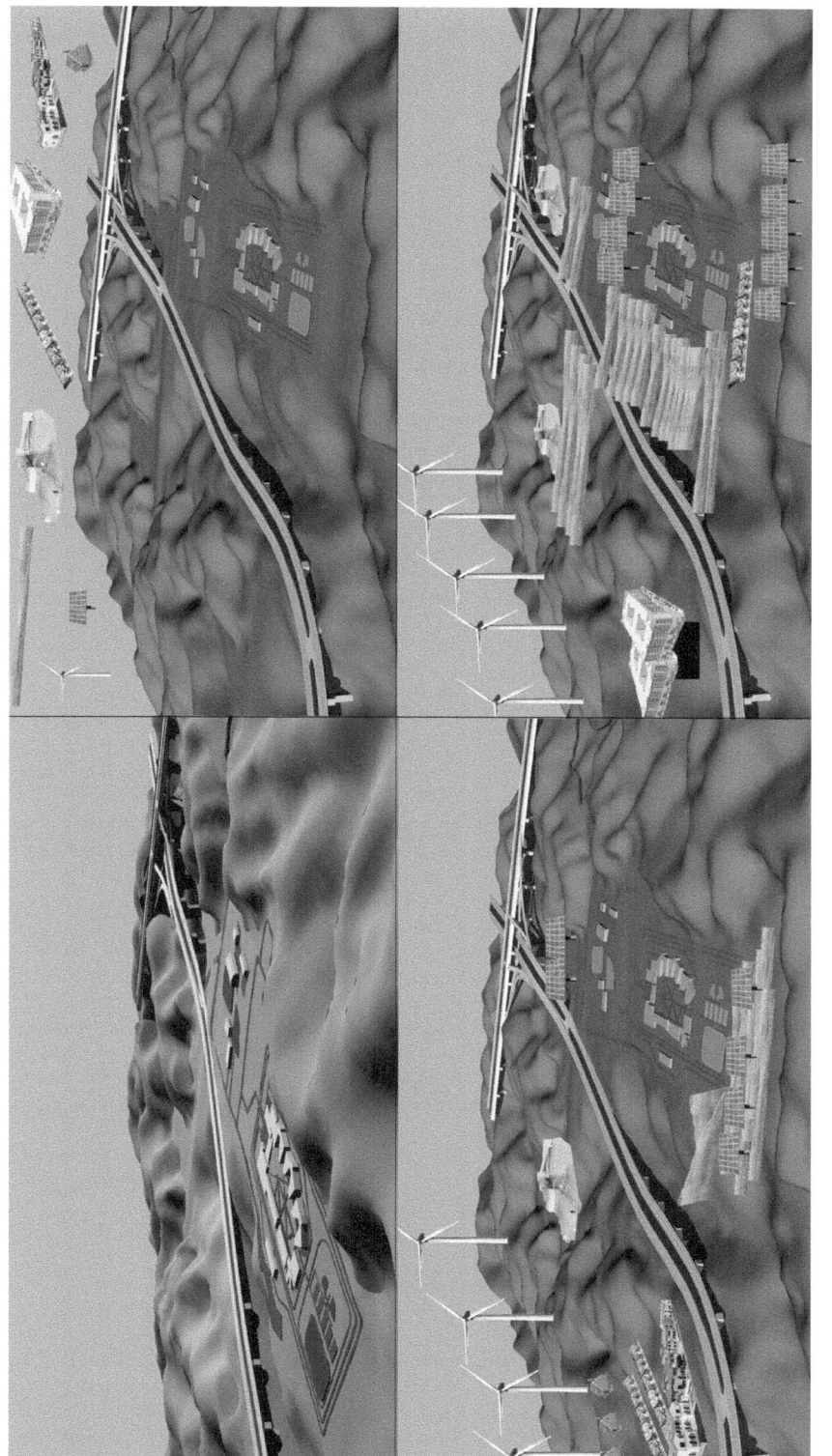

Figure 26.7 An exercise in participatory digital gaming in working with youth on the design of an alternative energy park (image credit: author)

This figure is shown in colour in the plate section of this book.

are essential to promote the goals of a conscious landscape architecture profession, to promote social and environmental justice, and to create partnerships between the academy and communities that attack the wicked problems of contemporary society.

References

Angotti, T., Doble, C. & Horrigan, P. (Eds) (2012), *Service-learning in design and planning: educating at the boundaries* (Oakland, CA: New Village Press).

Arnstein, S. R. (1969), 'A ladder of citizen participation', *Journal of the American Institute of Planners* 35.4: 216–224.

Butler, P., Campbell, A., Chu, J. & Riley, A. (2013) 'Making visible alternative futures on mine-scarred lands in Appalachia', *The Visibility of Research* (Charlotte, NC: Architectural Research Centers Consortium (ARCC)), 88–95.

de la Pena, D. S. (2013), 'Experiments in participatory urbanism: reform and autogestión as emerging forms of urban activism in Barcelona', PhD dissertation, University of California, Berkeley, available online at: http://digitalassets.lib.berkeley.edu/etd/ucb/text/delaPena_berkeley_0028E_13352.pdf, accessed 10 December 2017.

Grabasch, G. (2015), 'Landscape architecture – a profession of "place" – how do we profile a capricious profession', Brussels: The International Federation of Landscape Architects, available online at: http://iflaonline.org/2015/10/landscape-architecture-a-profession-of-place, accessed 10 December 2017.

Gruenewald, D. A. & Smith, G. A. (Eds) (2014), *Place-based education in the global age: local diversity.* (Abingdon, UK: Routledge).

Halprin, L. (1969), *The RSVP cycles* (New York: Brazille).

Hester, R. T. (2006), *Design for ecological democracy* (Cambridge, MA: MIT Press).

Hester, R. T. (2012), 'Scoring collective creativity and legitimizing participatory design', *Landscape Journal: Design, Planning, and Management of the Land* 31.1: 135–143.

Horrigan, P. H. & Bose, M. (2017), 'From social trustee towards democratic professionalism in landscape architecture', Paper presented at Defining Landscape Democracy Conference. Oscarsborg, Norway: Norwegian University of Life Sciences, Center for Landscape Democracy.

Horrigan, P. H., Bose, M., Doble, C. & Shipp, S. (2014), *Community matters: service-learning in engaged design and planning* (Abingdon, UK: Routledge Earthscan).

Hou, J. (Ed.) (2010), *Insurgent public space: guerrilla urbanism and the remaking of contemporary cities* (Abingdon, UK: Routledge).

Hou, J. (2011), 'Differences matter: learning to design in partnership with others', in T. Angotti, C. Doble and P. Horrigan (Eds) *Service-learning in design and planning: educating at the boundaries* (Berkeley, CA: New Village Press), 55–69.

Hou, J. (2013), 'Transcultural participation: designing with immigrant communities in Seattle's International District' in J. Hou (Ed.), *Transcultural cities: border crossing and placemaking* (Abingdon, UK: Routledge), 222–236.

LaGro Jr, J. A. (2011), *Site analysis: a contextual approach to sustainable land planning and site design* (Chichester, UK: Wiley).

Lawson, L., Spanierman, L., Poteat, V. P. & Beer, A. M. (2011), 'Educating for multicultural learning: revelations from the East ST. Louis Design Studio', in T. Angotti, C. Doble and P. Horrigan (Eds) *Service-learning in design and planning: educating at the boundaries* (Berkeley, CA: New Village Press), 70–85.

Mehrhoff, W. A. (1999), *Community design: a team approach to dynamic community systems* (Vol. 4). (London: SAGE Publications).

Miss, M. (2014), 'Greenwood Pond: double site', *Art and the landscape: the Cultural Landscape Foundation's 2014 Landslide*, available online at: http://tclf.org/sites/default/files/microsites/art-landscape/greenwood-pond.html, accessed 10 December 2017.

National Charrette Institute (NCI) (2017), *NCI Charrette System*(tm), available online at: http://charrette institute.org/theory, accessed 10 December 2017.

Sanoff, H. (2000), *Community participation methods in design and planning* (New York: Wiley).

Sanoff, H. (2010) *Democratic design: participation case studies in urban and small town environments* (Saarbrücken: VDM Publishing).

Thering, S. (2007), 'A practical theory based approach to action research in survivor communities', *Journal of Extension* 45.2, available online at: www.joe.org/joe/2007april/a3.php, accessed 10 December 2017.

Tuckman, B. W. (1965) 'Developmental sequence in small groups', *Psychological Bulletin* 63.6: 384–399.

West Virginia University (2017) West Virginia University Center for Service and Learning, available online at: https://service.wvu.edu/about, accessed 10 December 2017.

27

Cultivating the city
Instilling urban design in landscape architectural education

Karl Kullmann

Introduction: manifold urbanism

From the earliest stages of their education, students of landscape architecture learn that the agency of a site is inseparable from the dynamics of its context. Where this context once adhered to modern distinctions between culture and nature, today it is more likely to be defined by a condition that is simultaneously urban and landscape (Brenner 2014; Dettmar and Weilacher 2003). Within this paradigm, even a rural setting or wilderness becomes defined in relation to global flows of energy, information, capital and human migration. In a mechanistic sense, cities are simply the engines that drive these flows, siphoning in energy and expunging waste (DeLanda 1997). In the other very corporeal sense, cities remain places where people carry out their daily lives and create individual and collective meaning through time (Sennett 1993; De Certeau 1984; Jacobs 1961).

Urban design negotiates the space between these often-contradictory aspects of the urban condition. Most deliberately positioned as a cure for the dehumanizing ills of the modern metropolis, *traditional* urban design offers a template for compact walkability and built form drawn from pre-industrial cities. As part of this template, the significance of the public realm—and the street in particular—is revived from its marginalization within the modern city (see Jacobs 1993). Nevertheless, traditional urbanism adheres overall to a town-and-country model that positions landscape outside of the city (Duany 2002). In contrast, *landscape* urbanism rejects nostalgic notions of landscape as a mere witness to the contemporary metropolis. The figure/ground plans of traditional urbanism are inversed as the city is reimagined as complex ecological systems and infrastructures (Pollak 2006; Corner 2006; Bullivant 2006; Waldheim 2002).

Although these three approaches to urbanism—modern, traditional and landscape—define the past half-century of urban design discourse, all exhibit strengths and weaknesses (Heins 2015). For example, while traditional urbanism is demonstrably useful at the local neighborhood scale, it is less credible when the scope is broadened. Conversely, while landscape urbanism is effective at the regional structural scale, it has less to say about the pragmatics of dwelling (Duany and Talen 2013). Moreover, while modern urbanism assimilates technological innovation into the city, it makes less allowance for the impulses of individual place making (Kullmann 2015b).

Despite abundant rhetoric to the contrary, these overlapping and intertwining urban design doctrines demonstrate that no single approach offers a complete self-contained account of city making. The implications for urban design education in landscape architecture are manifold. First, effective learning necessitates pedagogy that acknowledges that aligning with a single urban design doctrine is likely to be limiting—and possibly precarious—over the longer term. Second, a truly landscape approach to urban design is akin to the landscape itself; it involves redundancy, overlap and untidiness. And third, since landscape frameworks are hard-wired to accommodate such disparate elements, landscape architecture is an apposite basis from which to learn and practice this manifold urbanism.

Motifs for landscape/urban pedagogy

This chapter explores the diverse terrain of urban design as it pertains to landscape architectural pedagogy. Towards the goal of cultivating a general sense of the city in landscape architectural education, a series of diverse themes relevant to both traditional landscape architecture and contemporary urbanism are overviewed as overlapping motifs. The motifs are sampled from a larger pool of themes that comprise a survey course introducing landscape/ urban theory to graduate students of landscape architecture. The didactic purpose of discussing these topics is to provide students with a diverse range of lenses through which to perceive the varied circumstances encountered when designing urban sites. The order of the following themes adhere to the chronology of their introduction within the theory course.

The contested landscape of alternate urbanisms

I begin the course by discussing the historical territories and trajectories of the various disciplines invested in urban design. The origins of landscape architecture schools are as diverse as the field itself (Baird and Szczygiel 2007). Those schools prioritizing creative expression tended to emerge from art, architecture or design departments, while those more grounded in the sciences tend to trace their origins to agricultural, horticultural, forestry, or geography departments. With the exception of landscape architecture programs closely associated with geography or architecture, these origins are unlikely to comprise significant exposure to urban design topics. Those that do are most likely to have been repositioned within a design college that brings disciplines invested in spatial design together under a single umbrella.

A typical arrangement comprising the three disciplines of landscape architecture, architecture and urban planning is augmented by urban design, which, although not a department in itself, operates at the intersection of the other three disciplines. In theory, this arrangement embodies the archetypal disciplinary model, wherein the confluence of landscape architecture, architecture and urban planning crystallizes into urban design (Schurch 1999). Nevertheless, although educational institutions are routinely structured along the lines of this Venn diagram, disciplinary culture rarely aligns so neatly (Figure 27.1a). Each discipline projects its own distinct version of urbanism that is based firmly in that discipline's value system (see Kullmann 2016a; Swaffield 2002).

These alternate urbanisms are often vociferously oppositional, as the disagreements between traditional urban design and landscape urbanism exemplifies (see Duany/Waldheim 2011). To cloud matters further, architectural and landscape architectural versions of landscape urbanism differ in subtle but nevertheless significant ways (Allen 2009; Waldheim 2002; Walker 2014, 1998). In practice, the interactions amongst disciplines invested in urban design are far more competitive and dynamic than the static symmetrical Venn diagram suggests (Figure 27.1).

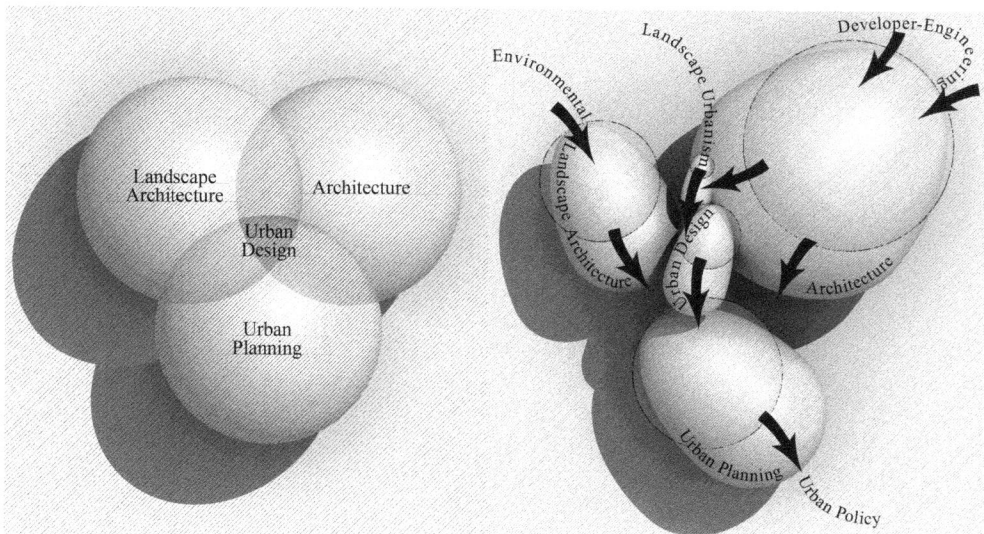

Figure 27.1 The city of disciplines. Disciplinary convergence: diagram illustrating the relationship between landscape architecture, architecture, urban planning, landscape urbanism and urban design, (1a: left) as a static Venn diagram, and (1b: right) as a dynamic model of competitive interactions. (Drawing © 2016 Karl Kullmann)

When teaching urbanism within landscape architecture, establishing priorities within this milieu can be disorienting. In the absence of a credible methodology with which to navigate the shifting disciplinary terrain of urban design, familiarity with the basis of each approach is a solid place to start. For example, learning the conventions of traditional urban design through the fundamentals of building typologies, street design and transportation remains useful in many contexts. Similarly, familiarity with the instrumentality of urban infrastructure, policy and code is often revelatory for site-focussed students (Figures 27.2 and 27.3).

The deconstructed landscape of urban decline

Somewhat counterintuitively, the course pedagogy shifts to a context that deconstructs the fundamentals of urban design. In urban districts and settlements experiencing shrinking populations, decline typically occurs in a piecemeal and dispersed manner. Over time, as services are decommissioned and surplus structures demolished, the urban landscape becomes increasingly perforated (Figure 27.4). This process is so contrary to the paradigms of progress and growth that underpin urban design and planning are constructed that most efforts explicitly or implicitly seek to reverse decline (Lynch 1981).

In the absence of a field that specializes in addressing decline, landscape architecture often defaults into the role of *reverse* urban design. The application of landscape architecture is partly pragmatic in the sense that the process of urban deconstruction usually creates new open landscapes that require some form of attention (Dettmar 2005). More intrinsically, landscape architecture's established emphasis on articulating and steering processes of emergence and decay more closely aligns to the challenges associated with decline than disciplines that are more calibrated towards growth, control, and progress (see Kullmann 2014b; 2013).

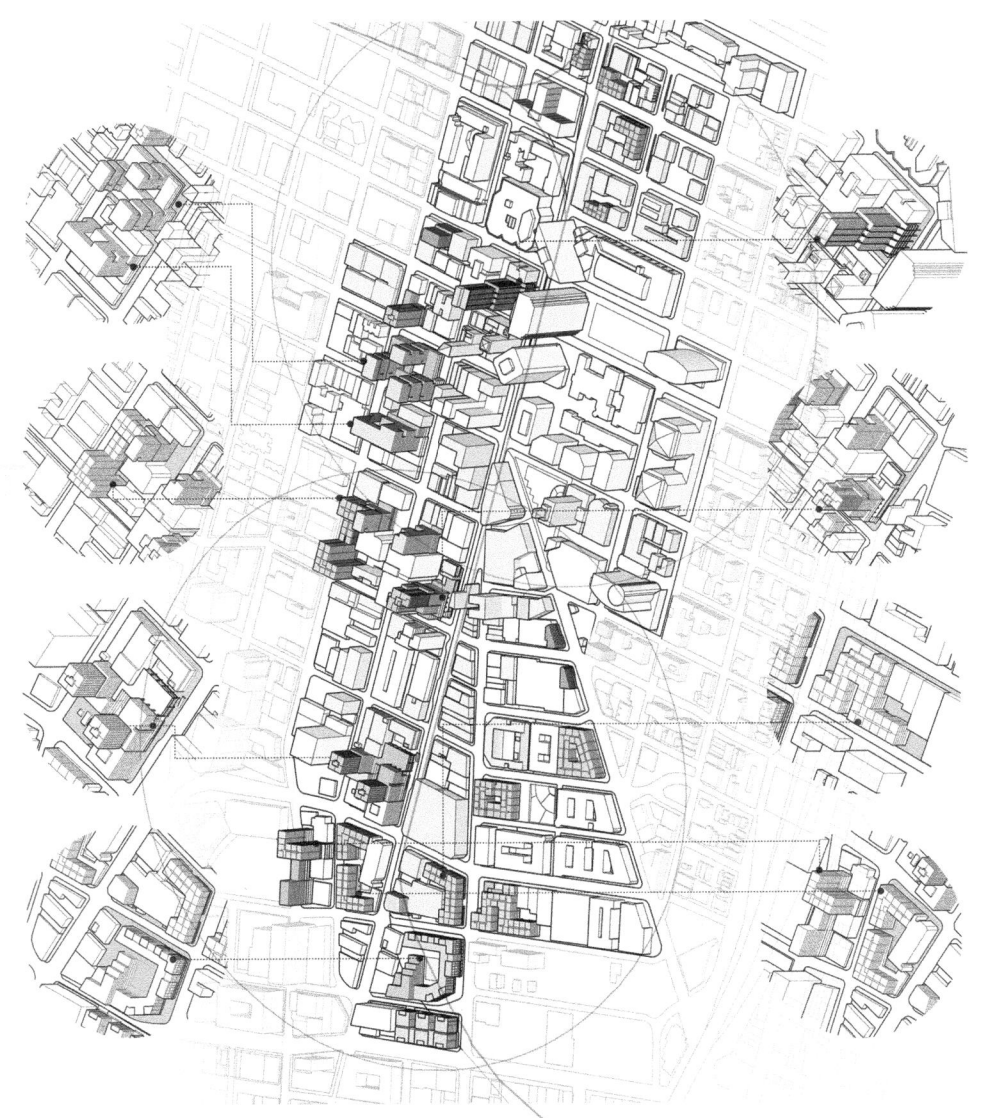

Figure 27.2 Student urban design project focussing on building typology in downtown Oakland, California. (Drawing © 2015 Yueyue Wang, reproduced with permission)

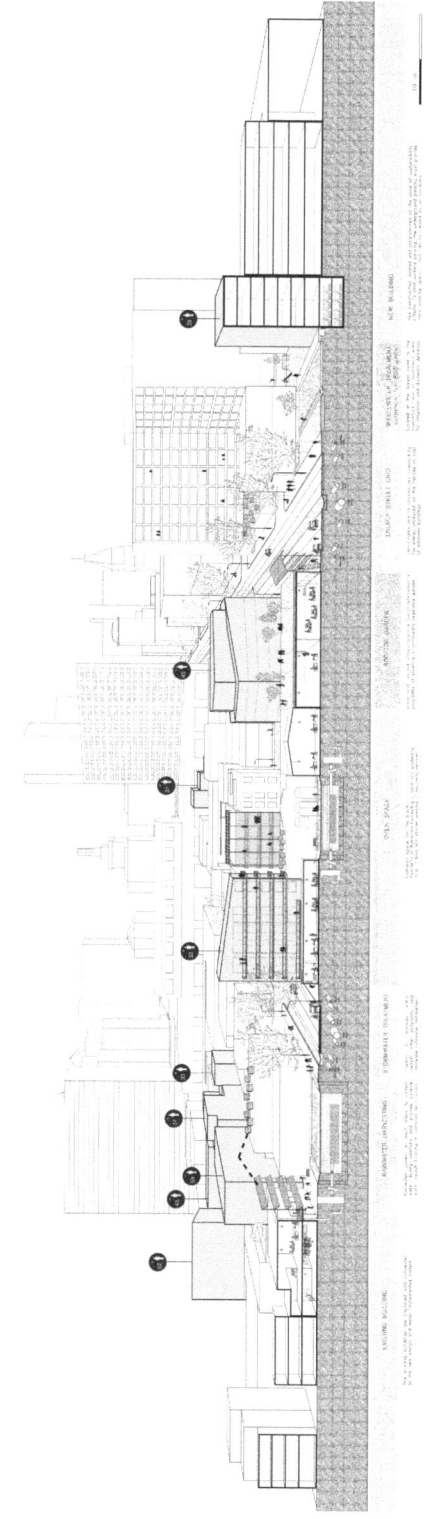

Figure 27.3 Student urban design project focussing on self-sufficient development in downtown Oakland, California. (Drawing © 2015 Marine Oudard, reproduced with permission)

Figure 27.4   The retreating city. Remains of the Detroit inner neighborhood of Islandview, Michigan. (Image © 2016 Karl Kullmann)

Extending this reasoning, exploring a landscape-based approach to urban decline is an effective point of departure in landscape architectural urban design education. Injecting urban design into landscape architectural education through the reverse process of urban decline vividly reveals how the topic is fundamentally grounded in landscape and pertains to far more than buildings and streets. Moreover, the topic demonstrates landscape interconnectedness at multiple scales, whereby decline and growth are revealed as part of the same dynamic process of population flows (Oswalt 2006).

The linear landscape of urban infrastructures

Shifting focus from human to hydraulic flows, the invention of pipes (for potable water, sewage and storm water) and transportation systems historically enabled urban densification and expansion. The civic value originally invested in this infrastructure evaporated through the twentieth century as the utilities and networks that service the city became increasingly efficient, mundane and invisible (Kullmann 2012b; Morrish and Brown 1995). Today, infrastructure has a more bimodal relationship with urbanism, with utilities and networks as likely to rupture the urban fabric as to support it (Strang 1996). The linear spaces that result may positively enable a transition, threshold, or activity corridor, or conversely may negatively enforce an impermeable boundary or linear void (Smith 1999).

Within these extremes, residual linear spaces exhibit a variety of evident and latent characteristics that are contingent on infrastructural origins, urban context, landscape condition, policy, regulation and market forces. Typologically, linear landscapes can be: a *filter* that selectively edits through-flows; a *program sink* that accommodates precisely defined uses; a *conduit* that channels rapid non-vehicular movement; a *suture* that stitches an urban rupture; a *stage* that links a necklace of events; a *pedestal* for observing external spectacles; and/or a *thicket* that impedes passage in any direction (see Kullmann 2011) (Figure 27.5).

Harnessing the typology of residual linear spaces into a landscape infrastructure provides students with an alternative framework for seeing, mapping, and engaging the city. To be certain, the term *infrastructure* has been stretched in urban design discourse and education to encompass almost anything and everything that is multiplied across the city. Given that the *infra* in infrastructure means below, the term pertains more concisely to the ground or underlying structure of the city. Through the medium of the landscape, an infrastructure of the ground becomes of integral value to urbanism. Far more than a passive green counterbalance to the 'real' city of buildings and pipes, the landscape's service to the city becomes as quantifiable as floor space or optical fibre.

The rough landscape of topographic urbanism

Returning from the underground to the surface reveals the strong predisposition of conventional urban design towards flatlands. Historically, this bias can be traced to the capacity for cities with control of their hinterlands to prosper on navigable river floodplains (Rykwert 1976). Conversely, hill towns were an option of last resort that sacrificed access to resources for defensibility. Given the influence of prominent European and American cities on present-day traditional urban principles, it follows that urban design typically presumes level ground, with prominent landform typically relegated to scenic backdrops or landmarks within the urban zone (see Duany 2002). Perhaps surprisingly, landscape urbanism also inadvertently preferences the flatlands through an emphasis on post-industrial sites, which are generally located on reclaimed waterfronts.

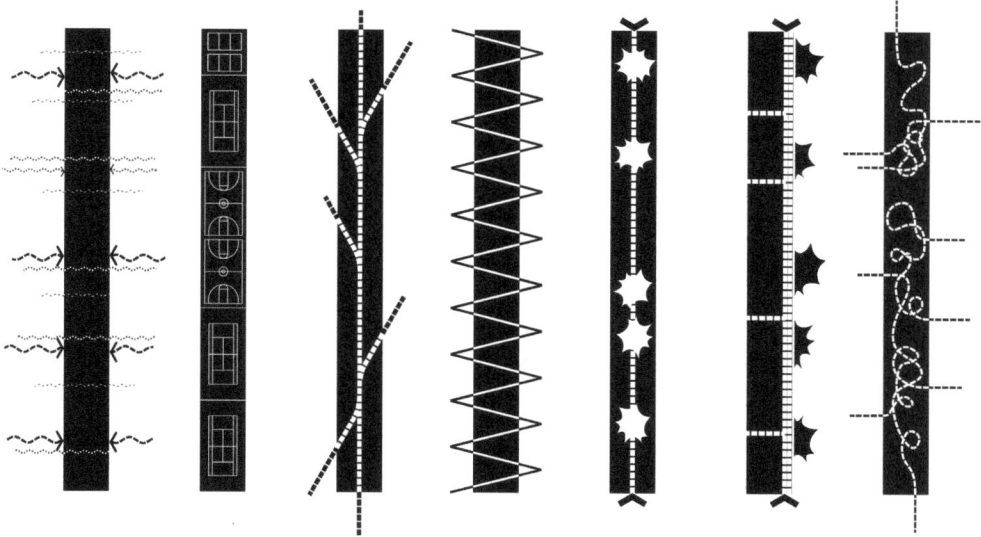

Figure 27.5 The city of conduits. Linear landscape typology (left to right): (a) filter; (b) program sink; (c) conduit; (d) suture; (e) stage; (f) pedestal; (g) thicket (Drawing © 2011 Karl Kullmann)

In the contemporary landscape/urban condition, this oppositional relationship between topography and urbanism dissolves. As urban populations increasingly coastalise on a planetary scale, highly variable coastal terrain becomes a significant setting for twenty-first century urbanism (see Engelman 1997). The coastal metropolis is often characterized by distorted urban morphologies and untamed geomorphic expression in the form of geological outcroppings or complex sand dune systems. Indeed, some of the world's most memorable coastal cities derive their distinctiveness from the pragmatic negotiation between inhabitation and terrain (Kullmann 2017f; Lipsky 1999) (Figure 27.6).

In addition to historical influences, the technical complexity of designing with landform contributes to the continuing urban design bias towards the flatlands. The three-dimensional nature of topography is not fully comprehended in the plan, section and axonometric drawings that typify urban design delineation. Given that topographic training is central to landscape architectural education (through site technology and design studios), students are pre-primed for generating a topographic approach to urban design. This topographic approach deploys three-dimensional modelling to tie the morphology of the ground to the morphology of the urban fabric (see Kullmann 2017d; 2014a). The result is urban form that enhances—rather than marginalizes or flattens—the topographic character of its setting.

The inflective landscape of urban intervention

Complex, disorienting and extending well over the horizon, the contemporary city confounds even the most comprehensive maps and models. This city is shaped more by capital and code than by the designer's masterplan (Dagenhart and Sawicki 1992). Without material control of

Figure 27.6 The topographic city. Contour signature for north-eastern quarter of San Francisco. (© 2016 City of San Francisco OpenData, reproduced in accordance with non-commerical use policy)

the metropolitan fabric, urban design customarily exerts agency at the project, or site, scale. This reality regularly leads to siloed urban developments that contradict urban design's meta-objectives of connectivity and integration. Landscape architecture also negotiates this disjunction between scale of agency and scale of intent. Whereas the scope of the landscape imagination is unlimited, the scale of physical intervention in the landscape is encapsulated in the archetypal enclosed garden (see Aben and de Wit 1999).

The garden and the city are customarily situated antithetically. Through the use of a physical frame, the garden historically functions as a locus of respite and retreat from the noisy disorientation of the surrounding urban milieu (see Harries 1989). Nevertheless, as the city becomes more expansive and dislocating—and the garden becomes more private and withdrawn—this relationship becomes strained (see Marcuse 1997; Kullmann 2017b; 2016b; 2012a). When the rigid frame is reassessed, the urban public garden assumes more catalytic capabilities in urban design. In place of the walled enclosure, a topographically formed semi-permeable threshold recalibrates the urban/landscape relationship (Figure 27.7). Rather than being removed from the tapestry of the city, the garden takes the form of a continuous fold or inflection in the urban fabric (see Cache 1995).

By leveraging the familiar medium of the unassuming garden (the design of which is covered in most landscape architectural programs) students discover a potent mechanism through which to act in the city. Rediscovering the corporeality and agency of the garden enables students to reimagine this archetype as an urban catalyst as opposed to a self-contained enclosure. The creative work of Shusaku Arakawa and Madeline Gins is relevant to this enterprise, whereby the 'open containment' of the garden becomes a kind of mind-body training ground for negotiating the accelerating, disorienting and immersive qualities of the contemporary urban condition (Arakawa and Gins 1994).

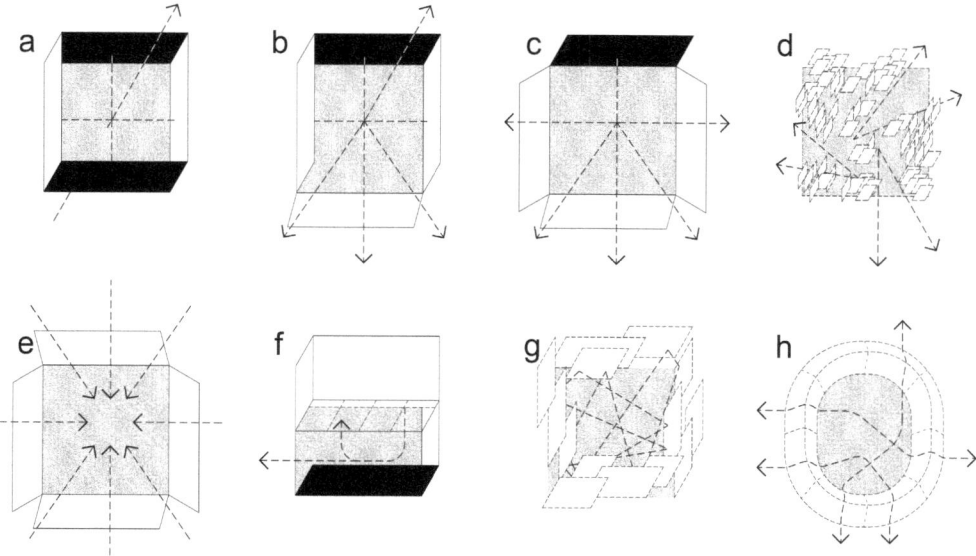

Figure 27.7   The city of inflections. Historical typology of garden enclosure and orientation: (a) medieval cloister garden; (b) renaissance garden; (c) baroque garden; (d) picturesque garden; (e) hunting park; (f) modern garden; (g) postmodern garden; (h) inflected garden. (Drawing © 2016 Karl Kullmann)

The de/programmed landscape of uselessness

Just as the inflected garden embraces some of the activity of the city, the open-ended characteristics of landscape also potentially cross-pollinate back into the city. Whereas the mandate of urbanism and architecture is to be demonstrably useful and programmable, landscape has a more historically complex relationship with utility.

Despite origins in the working landscape, strong associations with scenography led to landscape being positioned as 'other' outside of the legitimate business of the city (Casey 2002; Cosgrove 2004). Certainly, the usefulness attributed to landscape has historically waxed and waned, with greater emphasis under modernism and less emphasis in the Beaux Arts and postmodernism. More recently, performance and program emerged as mechanisms for recovering landscape usefulness and agency. Urban public spaces of all types became settings for programmatic activation and ecological function.

Elevating landscape functionality and agency retrieves landscape architecture from its diminished role as a decorative veil to the industrialized world (Corner 1999). However, with public parks and other spaces now obliged to appear as continuously useful as the neoliberal cities in which they are set, the preoccupation with program and usefulness risks smothering more ephemeral landscape qualities. In the sense that it reduces the potentiality of a site, the highly tuned and programmed space is likely to be less robust and adaptive once the physical or cultural conditions change (see Overmeyer 2007). The increasing reliance on specialized—but rapidly superseded—infrastructure and props to service the escalating pressure placed on urban eventscapes epitomizes this situation (Figure 27.8).

In this context, 'depressurized' urban design approaches may incorporate or emulate the landscape 'openness' that is often a feature of large parks and, increasingly, urban wastelands

Figure 27.8 The de/programmed city. Example of a discarded urban prop in Berlin's Mitte disctrict. (Photograph © 2001 Karl Kullmann)

(see Lynch 1972). As a landscape approach, openness is familiar to students of landscape architecture who routinely study precedents and design projects that fit within this framework. Gasworks Park in Seattle, Washington, and what remains of No-Mans-Land in Berlin memorably express this openness through topographic strategies. Existing topography influences programmatic choices on its surface, and, while programming and props tend to come and go, topography often remains significantly intact over time (see Kullmann 2015a). By applying observational study methods to open landscapes, students witness the behaviors and patterns of urban actors in diverse settings across the city (see Whyte 1980).

The elastic landscape of urban imaging

How the city is observed directly influences how it is imaged, imagined and designed. For this reason, representation is as integral to the spatial cognition and mental imaging of urban environments as direct, grounded experience. The history of urban representation charts progressively higher vantage points, as the cartographer's eye in the sky passed from hilltops and cathedrals, to camera-equipped balloons, kites, pigeons, and planes (see Cosgrove 2008). Situated at the apex of this skyward journey, the satellite reveals cultural and natural patterns and associations on the ground. Nevertheless, the satellite's abstracted Cartesian perspective exudes a seductive clarity that selectively skims over the scruffiness of everyday life. Even with familiarization and steadily improving image resolutions, abstract planimetric forms routinely fail to resonate with an individual's perception of their place in the world. The recurring popularity of more immersive angles such as the pre-Cartesian bird's eye view is a reaction to this lingering apprehension (Söderström 1996).

High fidelity drone-based imaging technologies are poised to harness the imaginary qualities of the bird's eye view and realign the satellite's distanced perspective. Whereas satellite mapping illuminates large-scale landscape systems and associations, low altitude drone mapping illuminates small-scale landscape details and nuances (Rekittke et al. 2013) (Figure 27.9). As it continues to proliferate, drone imaging and mapping is likely influence how individuals view, image and cognitively map their urban environments. Placed in the context of urban design, this new perspective is potentially transformative. Just as widespread access to the satellite's expansive view stimulated an ecological approach to urbanism, the drone's close-in view is potentially instrumental in refocusing urbanism towards the 'behavioral' scale at which people interact with cities (see Birtchnell and Gibson 2015; Kullmann 2017e; 2017c; 2017a; 2014c).

The applications and implications of this new technology are highly relevant to landscape architectural and urban design education. Offering image fidelities that are comparable to the world as perceived at eye level, drones provide students with a bridge between the aerial city of plans and the grounded city of everyday life (see de Certeau 1984). Moreover, whereas satellite imagery and mapping is received from government and corporate sources, drone surveys are created in person on site (Girot and Melsom 2014). This direct involvement in the urban imaging process is revelatory for students who often struggle to reconcile the isolation of the studio and digital environments with the real world that they study and design.

Conclusion: everything is landscape

As both figure and ground (see Meyer 1997), landscape is the setting for urbanism, but also cultivates and structures urbanism. As an apposite framework for filtering these landscape approaches, landscape architecture does not codify a singular doctrine of urbanism per se, but rather accommodates a range of approaches. The urban agency of landscape architecture is

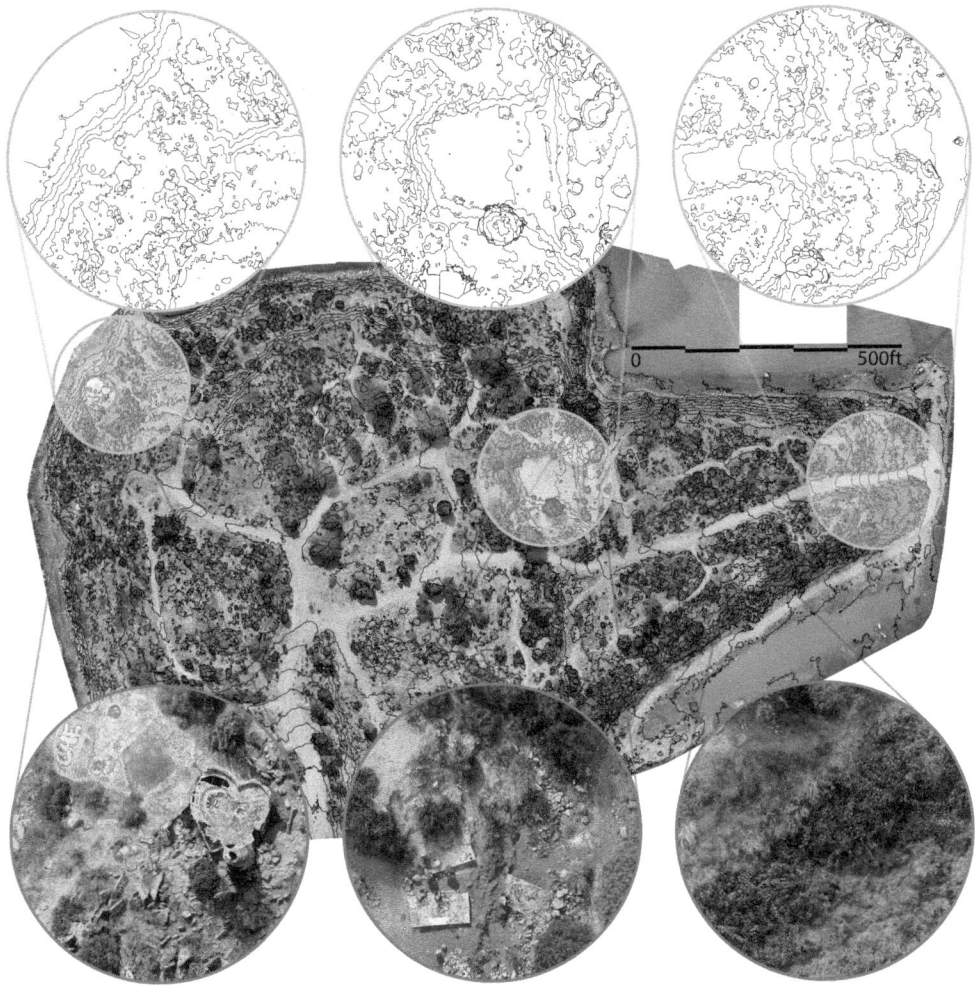

Figure 27.9   The imag(in)ed city. High fidelity drone mapping of the Albany Bulb landfill site on San Francisco Bay, California. (Image © 2016 Karl Kullmann)

grounded in the field's experience with the simultaneously pervasive and evasive nature of landscape. Pedagogically, this equates to a spectrum of overlapping approaches to urbanism that is not fully encapsulated in a single design studio or theory course, but ideally infiltrates throughout landscape architectural education.

References

Aben, R. and de Wit, S. (1999), *The Enclosed Garden* (Rotterdam: 010 Publishers).

Allen, S. (2009), 'Beyond Landscape Urbanism', *Lotus International* 139: 112–113.

Arakawa, S. and Gins, M. (1994), *Architecture: Sites of Reversible Destiny* (London: Academy Editions).

Baird, T.C. and Szczygiel, B. (2007), 'Sociology of Professions: The Evolution of Landscape Architecture in the United States', *Landscape Review* 12/1: 3–25.

Birtchnell, T. and Gibson, C. (2015), 'Less Talk More Drone: Social Research With UAVs', *Journal of Geography in Higher Education* 39/1: 182–189.

Brenner, N. (ed.) (2014), *Implosions / Explosions: Towards a Study of Planetary Urbanization* (Berlin: Jovis Verlag).

Bullivant, L. (2006), 'The Thickening Ground: The Landscape Urbanism Graduate Programme', *Architecture + Urbanism* 426/3: 122–127.

Cache, B. (1995), *Earth Moves* (Cambridge, MA: MIT Press).

Casey, E.S. (2002), *Representing Place* (Minneapolis, MN: University of Minnesota Press).

Corner J. (1999), 'Recovering Landscape as a Critical Cultural Practice', in J. Corner (ed.), *Recovering Landscape* (New York, NY: Princeton Architectural Press), 1–26.

Corner, J. (2006), 'Terra Fluxus', in Charles Waldheim (ed.) *The Landscape Urbanism Reader* (New York, NY: Princeton Architectural Press), 21–34.

Cosgrove, D. (2004), 'Landscape and Landschaft', lecture delivered at the Spatial Turn in History Symposium, German Historical Institute, February 19, 2004.

Cosgrove, D. (2008), *Geography and Vision* (New York: I.B. Taurus).

Dagenhart, R. and Sawicki, D. (1992), 'Architecture and Planning: The Divergence of Two Fields', *Journal of Planning Education and Research* 12: 1–16.

De Certeau, M. (1984), *The Practice of Everyday Life* (Berkeley, CA: University of California Press).

DeLanda, M. (1997), *A Thousand Years of Nonlinear History* (New York, NY: Zone Books).

Dettmar, J. (2005), 'Forests for Shrinking Cities?', in I. Kowarik et al. (eds) *Wild Urban Woodlands* (Berlin: Springer-Verlag), 263–276.

Dettmar, J. and Weilacher, U. (2003), 'Baukultur: Landschaft als Prozess / Landscape as process', *Topos* 44: 76–81.

Duany, A. (2002), 'The Transect', *Journal of Urban Design* 7/3: 251–260.

Duany, A. and Waldheim, C. (2011), 'Closing Plenary: Charles Waldheim and Andres Duany Discuss Landscape Urbanism', Congress of the New Urbanism, www.cnu.org/closecnu19.

Duany, A. and Talen, E. (eds) (2013), *Landscape Urbanism and its Discontents* (Gabriola Island, Canada: New Society Publishers).

Engelman, R. (1997), 'Earthly Dominion: Population Growth, Biodiversity, and Health', in F. Grifo and J. Rosenthal (eds) *Biodiversity and Human Health* (Washington, DC: Island Press), 39–59.

Girot, C. and Melsom, J. (2014), 'The Return of the Aviators', *Topos* 86: 102–107.

Harries, K. (1989), *The Broken Frame: Three Lectures* (Washington DC: The Catholic University of America Press).

Heins, M. (2015), 'Finding Common Ground Between New Urbanism and Landscape Urbanism', *Journal of Urban Design* 20/3: 293–302.

Jacobs, A.B. (1993), *Great Streets* (Cambridge MA: MIT Press).

Jacobs, J. (1961) *The Death and Life of Great American Cities* (New York, NY: The Modern Library).

Kullmann, K. (2011), 'Thin Parks / Thick Edges: Towards a Linear Park Typology for (Post)infrastructural Sites', *Journal of Landscape Architecture* 6/2: 70–81.

Kullmann, K. (2012a), 'De/framed Visions: Reading Two Collections of Gardens at the Xi'an International Horticultural Exposition', *Studies in the History of Gardens and Designed Landscapes* 32/3: 182–200.

Kullmann, K. (2012b), 'Green-Networks: Integrating Alternative Circulation Systems into Postindustrial Cities', *Journal of Urban Design* 18/1: 36–58.

Kullmann, K. (2013), 'Design for Decline: Landscape Architecture Strategies for the Western Australian Wheatbelt', *Landscape Journal* 32/2: 243–260.

Kullmann, K. (2014a), 'Towards Topographically Sensitive Urbanism: Re-Envisioning Earthwork Terracing in Suburban Development', *Journal of Urbanism* 8/4: 331–351.

Kullmann, K. (2014b), 'Red Loops, Green Links: Park Rabet and Urban Decline in East Leipzig', *Studies in the History of Gardens and Designed Landscapes* 34/4: 259–274.

Kullmann, K. (2014c), 'Hyper-Realism and Loose-Reality: the Limitations of Digital Realism and Alternative Principles in Landscape Design Visualization', *Journal of Landscape Architecture* 9/3: 20–31.

Kullmann, K. (2015a), 'The Usefulness of Uselessness: Towards a Landscape Framework for Un-activated Urban Public Space', *Architectural Theory Review* 19/2: 154–173.

Kullmann, K. (2015b), 'Grounding Landscape Urbanism and New Urbanism', *Journal of Urban Design* 20/3: 311–313.

Kullmann, K. (2016a), 'Disciplinary Convergence: Landscape Architecture and the Spatial Design Disciplines', *Journal of Landscape Architecture* 11/1: 30–41.

Kullmann, K. (2016b), 'Concave Worlds, Artificial Horizons: Reframing the Urban Public Garden', *Studies in the History of Gardens and Designed Landscapes* 37/1: 15–32.

Kullmann, K. (2017a), 'The Satellite's Progeny: Digital Chorography in the Age of Drone Vision', *Forty-Five: Journal of Outside Research*, http://forty-five.com/papers/157.

Kullmann, K. (2017b), 'The Garden of Entangled Paths: Landscape Phenomena at the Albany Bulb Wasteland', *Landscape Review* 17/1: 58–77.

Kullmann, K. (2017c), 'The Drone's Eye: Applications and Implications for Landscape Architecture', *Landscape Research*: 43/7: 906–921.

Kullmann, K. (2017d), 'Re-Envisioning Suburban Terracing: Development Scenarios for a Sandy Coastal Site', *Landscape Journal* 36/1: 15–36.

Kullmann, K. (2017e), 'The Mirage of the Metropolis: City Imaging in the Age of Digital Chorography', *Journal of Urban Design*: 23/1: 123–141.

Kullmann, K. (2017f), 'Hong Kong, Grounded', *Places Journal*, https://doi.org/10.22269/170502.

Lipsky, F. (1999), *San Francisco: The Grid Meets the Hills* (Paris: Editions Parentheses).

Lynch, K. (1972), 'The Openness of Open Space', in G. Kepes (ed.) *The Arts of Environment* (New York, NY: Braziller), 108–124.

Lynch, K. (1981), *Wasting Away* (San Francisco, CA: Sierra Club Books).

Marcuse, P. (1997), 'Walls of Fear and Walls of Support', in Ellin, N. (ed.), *Architecture of Fear* (New York, NY: Princeton Architectural Press), p. 103.

Meyer, E.K. (1997), 'The Expanded Field of Landscape Architecture', in G.F. Thompson and F.R. Steiner (eds) *Ecological Design and Planning* (New York, NY: Wiley), 45–79.

Morrish, W. and Brown, C. (1995), 'Putting Place Back into Infrastructure', *Landscape Architecture*, 85/6: 50–53.

Oswalt, P. (2006), 'Introduction', in P. Oswalt (ed.), *Shrinking Cities Volume 1: International Research* (Ostfildern: Hatje Cantz), 66–73.

Overmeyer, K. (ed.) (2007), *Urban Pioneers: Temporary Use and Urban Development in Berlin* (Berlin: Jovis).

Pollak, L. (2006), 'Constructed Ground: Questions of Scale', in C. Waldheim (ed.) *The Landscape Urbanism Reader* (New York, NY: Princeton Architectural Press), 125–140.

Schurch, T.W. (1999), 'Reconsidering Urban Design', *Journal of Urban Design* 4/1: 5–28.

Sennett, R. (1993), *The Conscience of The Eye* (London: Faber & Faber).

Smith, K. (1999), 'Linear Landscapes: Corridors, Conduits, Strips, Edges, and Segues', *Harvard Design Magazine*, Winter-Spring: 77.

Söderström, O. (1996), 'Paper Cities: Visual Thinking in Urban Planning', *Ecumene* 3/3: 249–281.

Strang, G. (1996) 'Infrastructure as Landscape', *Places* 10/3: 8–15.

Swaffield, S. (2002), 'Social Change and the Profession of Landscape Architecture in the Twenty-First Century', *Landscape Journal* 21/1: 183–189.

Rekittke, J., Paar, P., Lin, E. and Ninsalam, Y. (2013), 'Digital Reconnaissance', *Journal of Landscape Architecture* 8/1: 74–81.

Rykwert, J. (1976), *The Idea of a Town* (London: Faber and Faber).

Waldheim, C. (2002), 'Landscape Urbanism: a Genealogy', *Praxis* 4: 10–17.

Walker, P. (1998), 'Commentary', *Landscape Architecture* 88/2: 74–79, 90–91, 93–94.

Walker, P. (2014, March 31), 'ASLA-NCC Legacy Lecture: A Conversation With Peter Walker', interviewed by Charles Birnbaum (Berkeley, CA: Wurster Hall).

Whyte, W.H. (1980), *The Social Life of Small Urban Spaces* (Washington DC: The Conservation Foundation).

28

Teaching landscape construction as part of a holistic design process

Ingrid Schegk

Introduction

The aim of this contribution is to both describe and discuss a holistic, contextual and process-oriented approach to teaching landscape design and construction. The construction part in particular – including planning, detailing and technical execution – can be seen as the crucial unit for transforming the landscape.

The construction is generally the last phase of the design process, and definitely plays an important role for the project's success after realisation. Nevertheless, it is not always easy to simulate this phase in a purely academic environment and outside of real practice. Although it is a usual academic custom to teach landscape construction in specific modules and courses, the strong interrelation between the conceptual phase of design and the execution planning means that the construction process is very relevant to ensure satisfying design results. Hence, landscape construction has to be seen as a part of a holistic design process, creating sustainable design solutions in various contexts, respecting complex spatial frame conditions as well as particular cultural and social identities.

This ambition raises a couple of methical and didactic questions: how can the creative process of landscape design and the outcomes of its different phases be described? How can the awareness of students for complexity and holistic thinking be trained? Which approved and innovative teaching methods are suitable? Which teaching elements, especially regarding landscape transformation and construction, can be implemented?

The following chapter tries to give some answers to these questions. It begins with the general approach of interpreting landscape transformation as a circular process. The second section will discuss different methods of teaching landscape construction, most specifically the relationship and the interaction between deductive/inductive teaching and learning approaches. Finally, the implementation of the methodology in landscape education is shown and elaborated upon through different types of typical academic contents and students' work results.

General approach: design as a circular process

The design methodology described as follows is an outcome of the author's research and reflections on teaching landscape design studios and construction classes at university, as well as on practical experiences as a freelance landscape architect.

This approach defines the design process as a cycle with four main design steps, represented by four quadrants without sharply defined borders between each other (Figure 28.1). *Context, concept, composition* and *construction* can be seen as key components of this cycle, regardless of the type, scope and scale of the planning task. They are the intermediate results of a process consisting of conception, *conceiving* (narrating) based on a specific context, *composing* (inventing) and *constructing* (shaping). The process ends with the physical integration of the construction into the context as well as its interpretation by the users. This step of *establishing* the design decides its functional, social and ecological success.

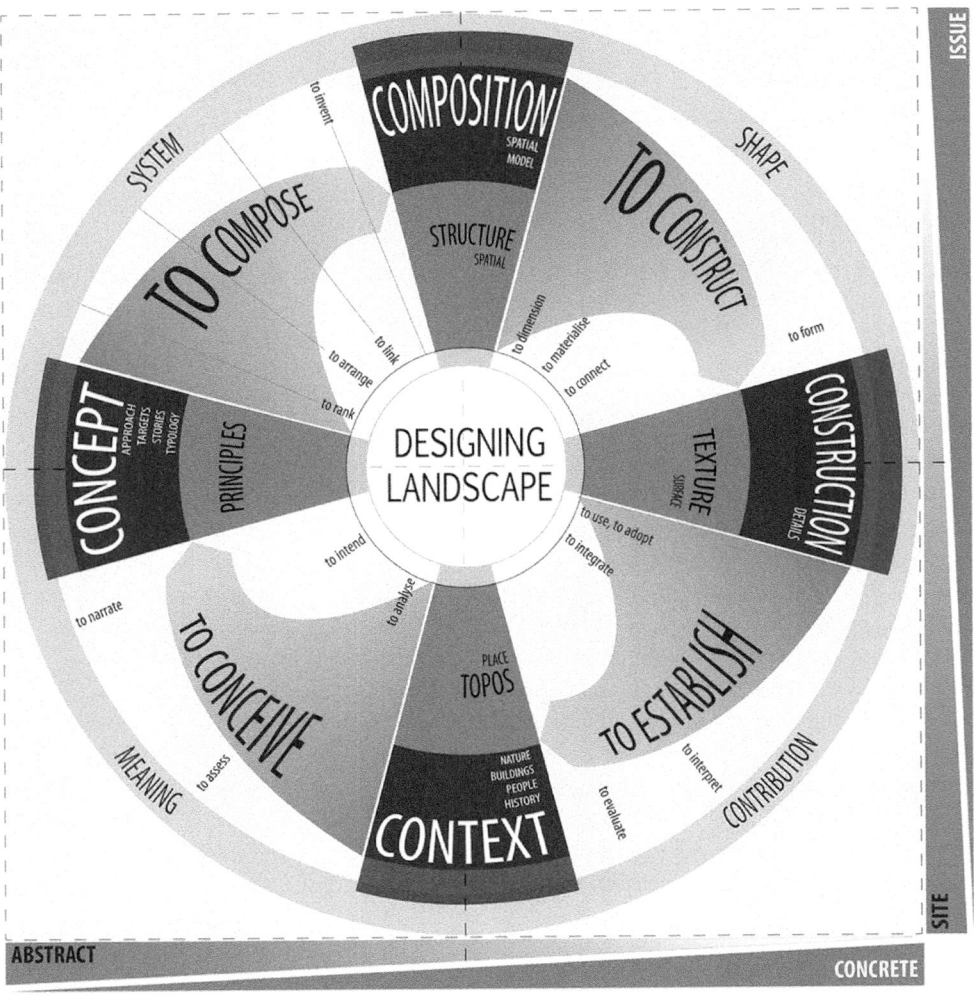

Figure 28.1   Model of the design process as a cycle with the four main quadrants context, concept, composition and construction; the frame is defined by the four 'corners' abstract and concrete, site and issue

From context to composition

In landscape architecture, the physical and intangible *context* defined by a particular site or place (*topos*) represents a key component of design, regardless of the extent to which it is ultimately considered in the final design. Context has many dimensions, from local to global, and many layers, such as political, historical, cultural, social, economic and geographic (Carmona 2010: 47–71).

The design-methodological path from context to *concept* includes important planning steps. These involve analysing and evaluating, from the planning of objectives and the statement of a site's potential, to visions for future developments. The concept is the first and most abstract intermediate step in the design process, shown as comprehensive ideas or principles. It can be illustrated in different ways depending on the type and scale of the planning task (e.g., diagrams, typologies, scenic images, storyboards, etc.).

The term *composition* can be defined as the interaction of system with design (Engel 2002: 65ff.) and/or content (originating from the concept) and form (leading to the construction). As a substantiation of the concept, the composition must integrate 'programmatic, physical, technical or constructional qualities' into a spatial structure (Steenbergen 2008: 17). In this sense, the *composition* is an issue-oriented spatial arrangement based on the concept. The design step from the concept to the composition includes the arrangement of points, linear, two-dimensional or spatial elements (Loidl and Bernhard 2003: 14ff.) and the linking of these elements. This is similar to a system configuration. The composition is illustrated by images, plans and models. Accordingly, an important methodological instrument for this compositional design phase is sketching and drawing. Both play 'a crucial role in this iterative process of reflection and intuitive action' (Steenbergen 2008: 23ff.).

Construction as a contextual contribution

Construction is the concretisation of the composition. It prepares for the implementation of design into reality. Classical Vitruvian quality criteria dictate that the construction must bring together function or utility (*utilitas*), pleasing form or beauty (*venustas*) as well as technical durability and reliability (*firmitas*). Nowadays perhaps this criterion can also be understood in the wider sense of sustainability. Construction primarily refers to shaping, or materialising, dimensioning and connecting modules and parts into a useful, stable form. Strictly speaking, only the surface of the construction (which can also be addressed as 'texture') interacts visibly with the environment and is available for use by people. In spite of this, the constructive elements that make this surface possible (such as invisible foundations) are part of the design. The construction process also requires the development and visualisation of these secondary components. Construction signifies 'decomposition and detail design', which is the 'rational breakdown' of the composition into its 'determining functions and constituent parts'. As 'regressive composition', it defines the object, the task, the scale and the form of presentation and visualisation by drawings and models of the individual parts. It ensures that the breakdown and dissection attest to the composition as a whole (Engel 2002: 95, 100).

The construction with its space-defining exterior is the monitor to an extent, the screen for the underlying composition and the concept. At the same time, it is also part of the physical context. Accordingly, the construction can be understood as the most concrete component of the design process, balanced and balancing between the issue and the site.

The design methodology approach could end with the implementation of the construction in the established context at this point. However, an integrated planning and design approach requires another, fourth, phase that connects the construction with the context as the starting point and closes the design process cycle. The construction (as the exterior surface of a newly created topology and topography) is analogously positioned as a contribution to the context. This means it is used and integrated into the physical and non-physical context and must prove itself. It is evaluated and interpreted by users and planners with regard to functionality and capacity to please. Its importance is attributed and given a positive or negative image. It becomes part of the narrative. In this spirit, the construction, as an expression of composition and concept, shapes the identity of the place and becomes part of the context and/or redefines it (Schegk 2014: 17).

The outlined design cycle can also be approached the other way around. In a number of respects, the context forms the basis for the construction as the most concrete component of the design. It is responsible for the implementation and development of the composition as a substantiation of the concept, which provides new content for the context.

This circular description of the design process, which implements continual development and change of designed landscape by use, maintenance, ageing, etc., is, in some aspects, comparable with the linear model of the design process published by the German urban planner Sophie Wolfrum. In this model the design process is represented by a chain leading from the starting box 'problem and concept' through 'designing' to 'design' (comparable with the *composition* in the circular model) and forward through 'building' to the final box 'work' (the realised *construction*), which is an integrated part of the 'situation' (the context) where the *work* meets the users, here called 'actors' and 'recipients'. The author compares her model with the 'concept of performativity in theatre studies', which leads from the 'idea' through 'writing' to the 'piece' (which means the play) and further through 'staging' to the 'stage' meeting the 'audience' during the 'performance' (Wolfrum 2014: 146). Even though this linear approach fits better with urban design than with landscape design, it is able to suggest opposite images. The comparison of the designed context to a permanent performance on a stage is a very helpful one with which to consider all the different functional, social, cultural, aesthetic and technical demands on the construction.

Application and appraisal

Teaching landscape design by means of this circular design process model has been applied and tested by the author together with different colleagues during various design classes and project studios, in Bachelor programmes as well as in Master's programmes (Schegk 2014). The experience with this approach shows some advantageous characteristics.

- The circular process works like a wheel, moving forwards and backwards as necessary. It represents the design process within students' projects, and demonstrates the necessity to sometimes go backwards again, from a planning idea back to a more analytical consideration, or from a concrete design step back to a more abstract one.
- It is very helpful when students are always conscious of their current position on the design wheel and know in which quadrant they are working (abstract and site-oriented, abstract and issue-oriented, concrete and issue-oriented, or concrete and site-oriented). This helps in choosing suitable design and visualisation methods. Accordingly, it can evoke an interesting learning effect to trace the course of a design during the process and maintain a kind of design logbook.
- In this approach, landscape construction is not a separated additional task, which follows the final design result, but rather an integrated part of every unique design process depending on contextual, conceptual and compositional considerations and design decisions. It is adequate to get elementary knowledge in the specific field of landscape construction as well as to train constructive skills and competences in preparative or accompanying study units.

Teaching methodology: deductive versus inductive methods

Against the theoretical background explained above, teaching landscape construction as the most concrete part of the landscape design process offers differing potential for various types of didactic techniques – both deductive 'top down' methods, and inductive 'bottom up' methods.

Deductive teaching methods

> Deductive teaching (also called direct instruction) [. . .] is based on the idea that a highly structured presentation of content creates optimal learning for students. The instructor using a deductive approach typically presents a general concept by first defining it and then providing examples or illustrations that demonstrate the idea.
>
> *(Grumbine, Hecker and Littlefield 2005)*

According to this definition, deductive teaching methods such as lectures can help students gain 'explicit' (Jonas 2014: 85) theoretical knowledge, for example about constructive principles, material properties, as well as important technical rules and standards. The success of this 'top down' method is crucially related to a conclusive structure of the lecture. Different relevant literature sources present different concepts of structuring the field of landscape construction, such as element-based or process-based approaches.

Element-based structures focus on the different 'families' of constructive elements in landscape architecture, each including knowledge about appropriate materials. The family of solid construction elements is represented by heavy stone volumes in interaction with soil and water, such as stone or concrete walls and stairs; the family of frame construction elements are made of linear timber and/or steel components such as pergolas and pavilions, decks, platforms or fences. The family of typical landscape construction elements are built as an arrangement in different laminar functional layers like pavements, water features or green roofs including geo-synthetic barriers and geo-textiles (Schegk and Brandl 2012). Some other element-oriented approaches present materials (stone, concrete, brick, metals, timber, polymers, soil and plants) and elements such as earthworks, retaining structures, steps, ramps, walls, pergolas, small bridges, walkways, water features and green roofs separately (Holden and Liversedge 2011; Zimmermann 2011). Alternatively, landscape construction can be taught in a process-oriented manner as a workflow, e.g., in the case of stone constructions, beginning with the stone genesis to the exploitation and treatment of natural stone products through to construction and maintenance works (Schegk 2016). In both cases, it is important to visualise and exemplify the theory by case studies, sketches, details and/or field trips to landscape companies, quarries, construction sites and realised projects (Figures 28.2, 28.3 and 28.4).

In landscape architecture, deductive teaching and learning represents only one possible method with which to gain professional competence, which is much more than pure knowledge. Inductive methods appear to be even more important than deductive approaches.

Inductive teaching methods

> Inductive teaching (also called discovery teaching or inquiry teaching) is based on the claim that knowledge is built primarily from a learner's experiences and interactions with phenomena. [. . .] The teacher's role is to create the opportunities and the context in which students can successfully make the appropriate generalizations.
>
> *(Grumbine, Hecker and Littlefield 2005)*

Figures 28.2–Figure 28.4   Valuable additions to deductive teaching: field trips to quarries and stone work companies and construction sites, experiencing material (here crushed-glass-gravel)

Inductive teaching in landscape architecture works with a wide range of various methods. These include case studies and construction surveys, observations and enquiries, project studios, working with analogue or digital models and practical experiments or workshops. Inductive teaching affects problem-based learning and creates 'implicit knowledge' (Jonas 2014: 85). The part of the teacher is more like that of an adviser or tutor who helps to support the process, rather than promoting a certain result. In the field of landscape construction there is often no right or wrong, only better or worse solutions to complex problems. Inductive teaching is indispensible for imparting this awareness.

A good start for implementing inductive methods into a first year landscape construction programme could be to use case studies and surveys of realised constructive elements or details, such as benches, walls, steps and stairs, pavements, small courtyards, pavilions or pergolas. This includes studying and recording their proportions, material, and exposition as well as planned and spontaneous functions. Students should use only simple measuring devices and sketch books and complete their studies with a comparison of similar objects and/or critical evaluations (Figures 28.5 and 28.6).

Reflexive studying

When considering the advantages and disadvantages of deductive and inductive learning styles, it seems that there is no clear preference.

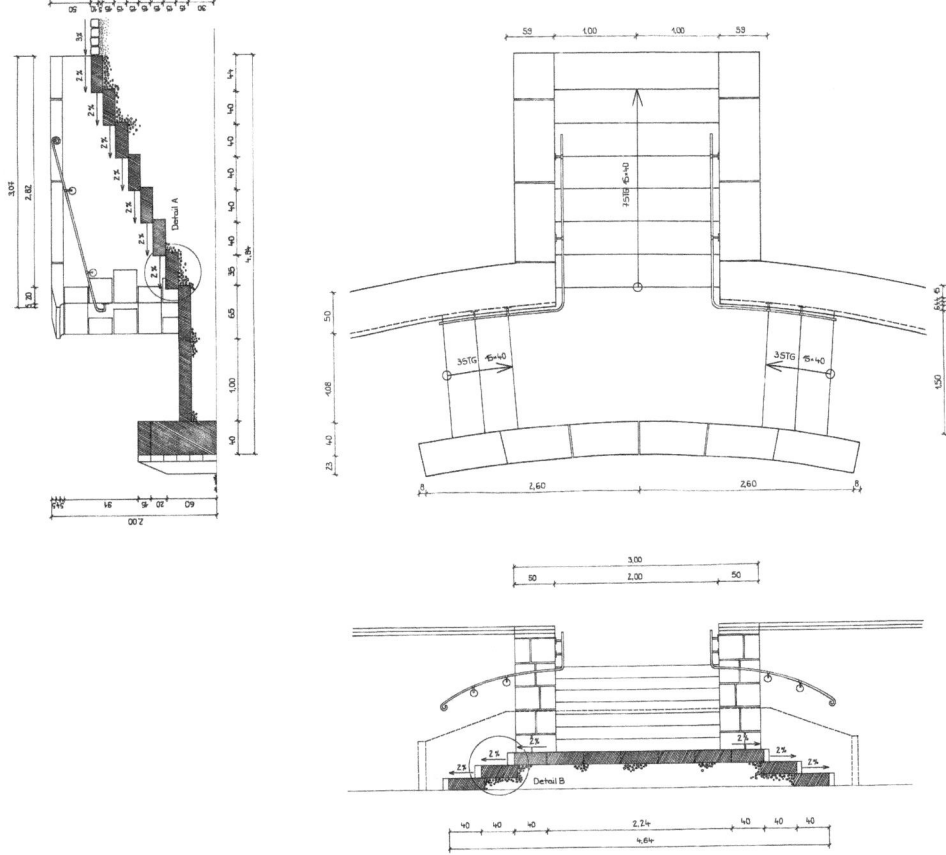

Figure 28.5 Inductive method: scalalogy – staircase research or, in brief, 'knowledge of the staircase', offers interesting opportunities for case studies in landscape construction. First exercise: construction survey of an outside staircase (drawing: Ariane Kreß)

> Some students learn best through an inductive approach; some learn best through a deductive approach. Inductive learners like making observations and poring over data looking for patterns so they can infer larger principles. Deductive learners like to have the general principles identified and prefer to deduce the consequences and examples from them. These are often the same learners who prefer more structure in general.
>
> *(Grumbine, Hecker and Littlefield 2005)*

Experiences in teaching landscape design and construction show that it seems to be most successful to evoke the concurrence and interdependency of both methods.

In accordance with the 'reflexive design' approach, described in the book of the same name (Buchert 2014), as well as the definition of 'reflexive' research strategies as an interaction of deductive and inductive strategies (Deming and Swaffield 2011: 9), the term of 'reflexive' studying and learning methods can be introduced. 'Reflexive' methods allow the interplay of deductive knowledge-based teaching/learning and inductive experience-based or inquiry teaching/learning. Deming and Swaffield use the term 'abduction' to 'describe a way of creating knowledge that

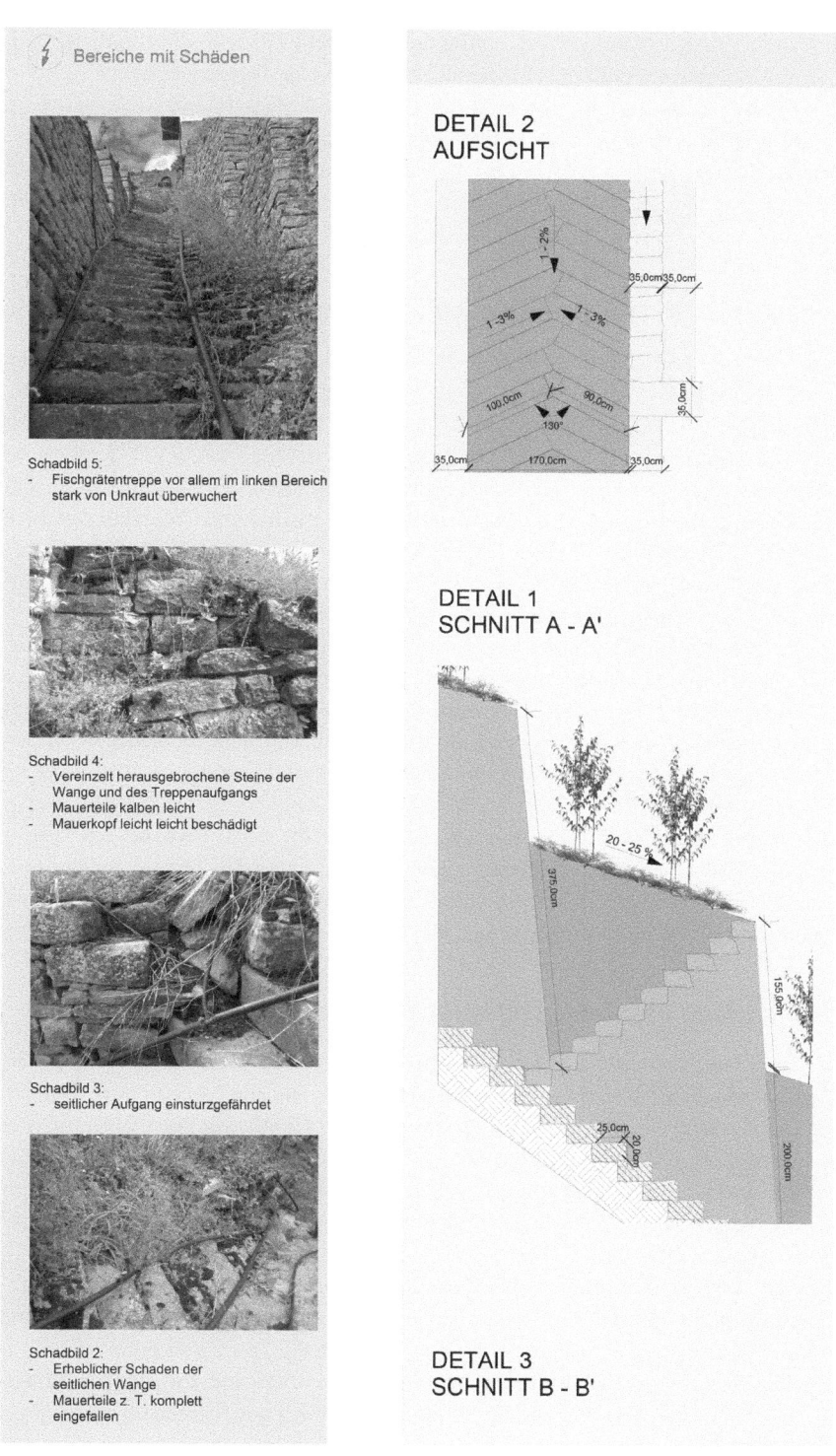

Figure 28.6a and b   Scalalogy in the cultural landscape: complex case study of an unique herringbone staircase in a traditional German vineyard; the construction combines the functions of steps and gutter (drawing: Corinna Wolfmüller)

is neither inductive nor deductive' (Deming and Swaffield 2011: 8). A combination of both, inductive and deductive, could thus be defined as reflexive or *abductive* studying.

Whatever the name used for the method, a successful teaching approach has to combine deductive and inductive elements and make them interact. Every landscape project is unique and requires a unique design solution involving context, concept, composition and the construction in particular. Learning and studying how to create such a design solution needs an intensive process of applying learned knowledge, reflection and research.

Implementation in landscape education: examples and student work

Trying to implement reflexive teaching methods and the design approach described before in a landscape construction class brings up some further questions: what is an appropriate degree of complexity, of abstraction, of concretion? How and how far can we simplify constructive problems? How important are technical standards and standardised solutions? How can we motivate students to create context-oriented constructions without copying the existing? How can we motivate students to invent innovative, sustainable constructions complying with technical requirements?

At this point two partial aspects of construction are to be introduced: *re-construction* and *de-construction*. *Re-construction*, which means here exploring approved typologies in different contexts, enables students to understand and evaluate existing structural elements related to landscape. *De-construction*, which can be defined as exploring every single part of the whole unit, can stimulate students' competence to recognise and develop design and constructive principles. Both are representing 'reflexive' learning methods. Both can help to manage complexity, to understand technical functions and requirements of structural elements as well as to develop a personal methodology with which to invent new ones.

The following examples of students' works and projects try to illustrate the 'reflexive' teaching and learning approach in landscape construction – including aspects of re-construction and de-construction – as a crucial part of the whole design process. Different themes – terrain modelling, 3D-detailing and practicing – are presented, all of which train different skills and competencies.

Terrain modelling as a basic skill in landscape construction

Terrain modelling, also called 'grading', is one of the crucial aspects of landscape design. It can be seen as the direct interaction between the context and the construction. Together with planting and dealing with vegetation, grading is one of the most specific competencies of landscape architects. Students must be able to design using contour lines and to be conscious of the effect interventions have on any sloped terrain (Petschek 2014). Grading exercises combine deductive and inductive teaching methods and can include re-constructive and de-constructive aspects.

Moreover, the design of terrain models exemplifies perfectly the holistic and circular process work: based on an obvious physical *context* (slope, shape, exposition, etc.) as well as an intangible *context* (e.g., vistas, history, etc.) students develop a *concept*, e.g., for better perceptibility, usability or barrier-free connectivity. Then they create a *composition* of ups and downs, plain terraces, paths and steps. Contour lines allow visualising this composition in a quite abstract but very precise way (Figure 28.7). Building a simple elevation model with cardboard layers means to de-construct the terrain and helps to investigate and understand the design interventions (Figures 28.8 and 28.9). In the final design step, the spatial composition has to be translated in concrete *construction* elements such as retaining walls, pavements, ramps, staircases, planted embankments, etc.

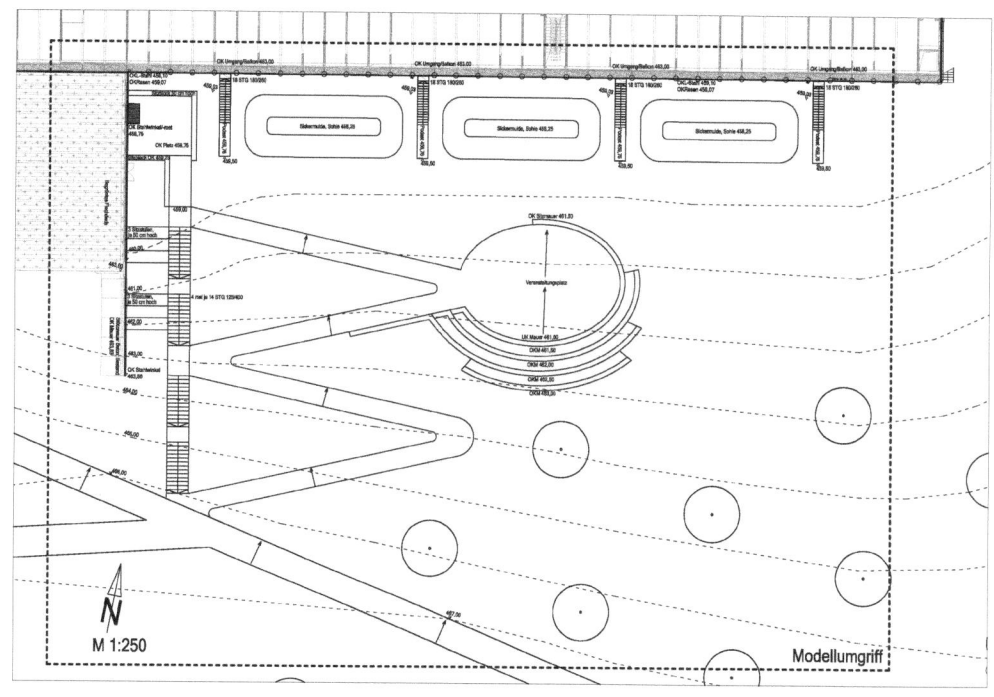

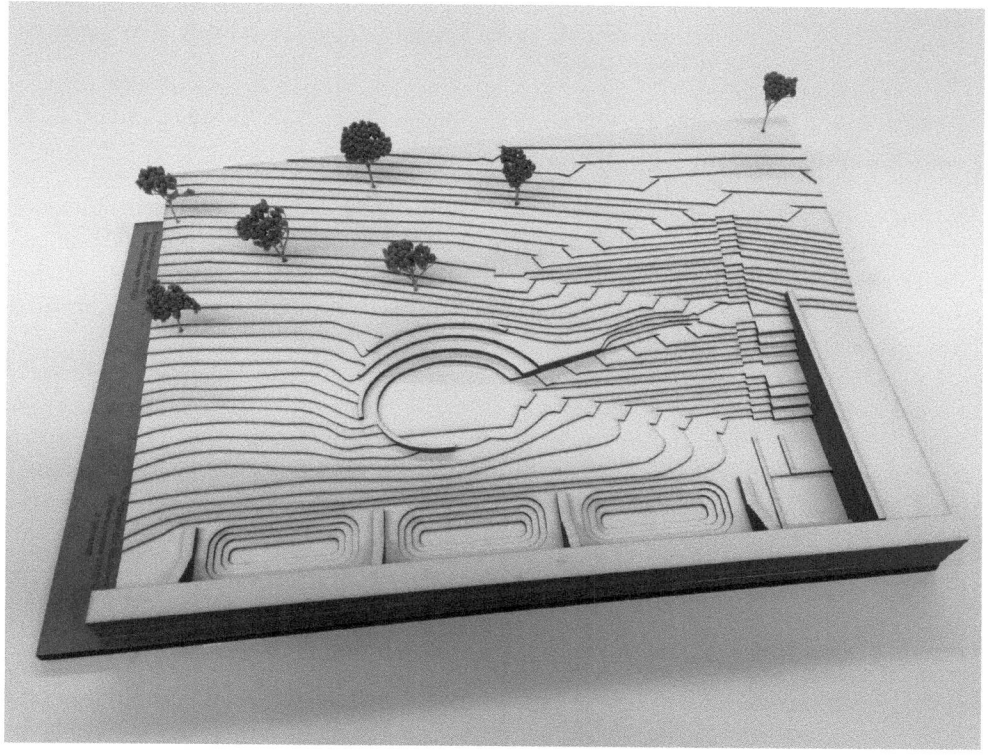

Figures 28.7–Figure 28.9 (continued)

These figures are shown in colour in the plate section of this book.

Ingrid Schegk

(continued)

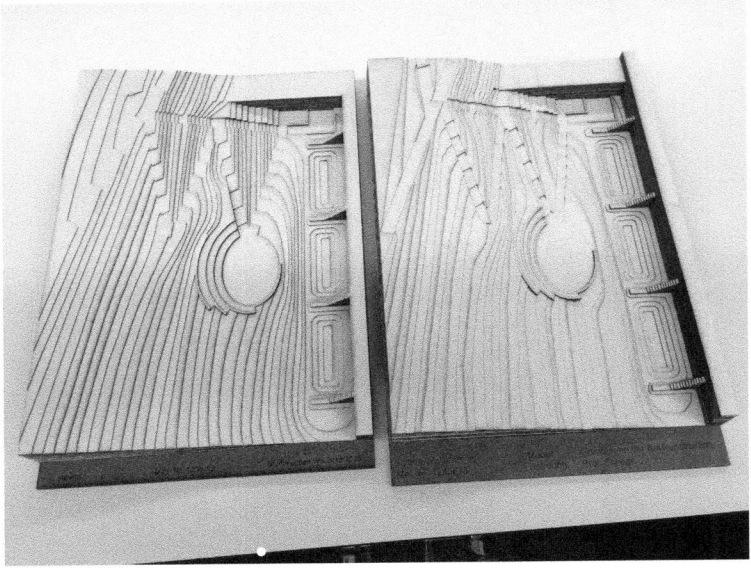

Figures 28.7–Figure 28.9 'Reflexive' terrain modelling exercise: students have to match the contour lines with the planned interventions (red lines) and build a model of the result; they have to follow certain construction rules such as steady-going slope of the paths, etc. (drawing: Ingrid Schegk; models: Saskia Schrader, Sabine Stockbauer, Kim Sander, Hanna Waschek)

Exploring and detailing with 3D-models

Detailing is the concretisation of the composition. It involves its single parts or elements within a larger scale. As part of the construction process, detailing makes the design buildable and prepares its execution. Detailing demands not only technical skills. In fact, it is a distinct creative phase of the design process including contextual, conceptual, compositional and constructive considerations in a detailed scale.

Three-dimensional models offer excellent opportunities to train this. Using digital and analogue models, students are able to visualise existing situations, or can reconstruct historic ones (Figure 28.10). They can also help to de-construct in order to explore the spatial impacts of interventions, to analyse the functionality of statically loaded construction, to study the interaction of constructive elements and to understand different options for material use (Figures 28.11–28.15). In this spirit, detailing is a good example for 'reflexive' learning combining deductive and inductive methods.

Learning by doing – practical workshops

Practical workshops and educational or experimental construction sites are always very welcome additions to lectures and studios. They allow bottom-up perception and reflection of landscape construction and create tangible results. The interrelation between context, concept, composition and construction as part of the new context is obvious.

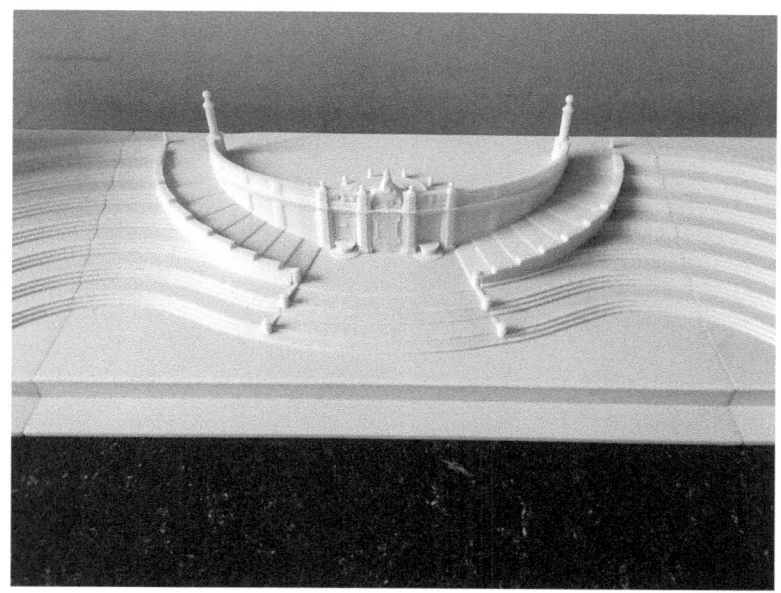

Figure 28.10 Reflection through re-construction: 3D-printed digital model of the former baroque harbour 'Porto di Ripetta' in Rome (research and digital model: Florian Martin, Alexander Schmidl; 3D-print: Matthias Thoma)

Figures 28.11–Figure 28.14 (continued)

(continued)

Figures 28.11–Figure 28.14 Reflection through (de-)construction: sustainable design and construction explored by analogue models of swimming pavilions (some using bamboo) and a forest watch tower (using timber) (models pavilions: Regina Bauer, Katharina König, Alexander Kugler; model watch tower: Andy Steber)

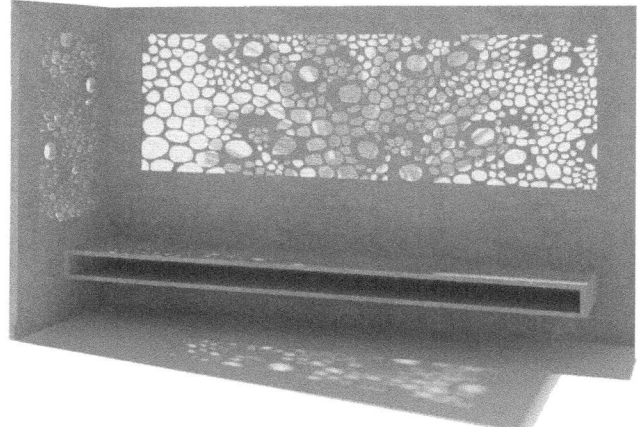

Figure 28.15 Reflection through the interplay of design and construction: digital 3D-model of precast screen elements between private terraces/small gardens in a dense urban housing environment; the material is bamboo-concrete, the semi-transparent window shows the microscopic structure of the bamboo cane; the project elaborated in an elective course 'Design and Construction' (Bachelor programme in Landscape Architecture) was a contribution to the Concrete Design Competition under the title 'Elegance' (design and model: Annemarie Haselhuhn, Lisa Hauner, Andrea Wachinger)

Suitable 'reflexive' workshop themes are traditionally sustainable manual techniques, such as dry stone walling, cobblestone paving (Figures 28.16–28.18) and rammed-earth construction. They can also involve experimental topics like using bamboo for building or working with recycled materials. The realisation of one's own ideas can prove a very valuable experience. Students learn to develop their construction ideas beginning with first conceptual sketches, moving on to detailed drawings and models and then finishing with the building process itself. Collaborating in a workshop fosters a strong group sentiment amongst the students, and trains their abilities in project management and in critical evaluation of different solutions. Finally, there is always a unique elation when building something with one's own hands. During a dry stone walling workshop, a student once expressed: 'What a feeling of success, when a stone fits after the first try!'

Conclusion

Teaching landscape construction as the crucial unit for transforming the landscape has to combine design methodology and teaching methodology. Both topics are complex and multi-dimensional; simple rules or recipes for 'how to design?' or 'how to teach?' do not exist. Rather, it is the ever-new experiences based on some published theories and methods that help us to develop successful approaches in both areas.

It seems to be relevant that students recognise landscape construction as a key component of a holistic and process-oriented design approach. In this approach, the physical and intangible context plays a crucial role for the (abstract) design concept, the spatial composition and the construction itself, which represents the most concrete design component.

Against this background, teaching requires a good balance between deductive knowledge-based and inductive experience-based teaching methods, which can be summarised as 'reflexive' methods, based on students' reflection and intuition.

Figures 28.16–Figure 28.18   Reflection through de-construction: dry stone walling workshop in Bamberg/Franconia and Vorarlberg/Austria (all photos: Ingrid Schegk)

The practical implementation of reflexive methods includes exercises of re-construction and de-construction. Both terms describe abstractly the exploration of constructive typologies as a whole and as an interrelation of its different single constructive elements. Particularly proven examples of such exercises are:

- terrain modelling respectively grading studies combining the abstract tool of contour lines with very concrete landscape interventions and constructions;
- detailing exercises using analogue and digital 3D-models that allow a holistic design approach in a detailed scale and a deeper understanding of constructive effects;
- practical construction workshops that ensure the students' direct contact with the site and the material, as well as promoting social skills.

References

Buchert, M. (ed.) (2014), *Reflexive Design. Design and Research in Architecture* (Berlin: jovis).

Carmona, M. (2010), *Public Places – Urban Spaces. The Dimensions of Urban Design* (Amsterdam: Elsevier).

Deming, M. E. and Swaffield, S. (2011), *Landscape Architecture Research: Inquiry, Strategy, Design* (Hoboken: Wiley).

Engel, H. (2002), *Methodik der Architekturplanung* [Methodology of architectural planning] (Berlin: Bauwerk).

Grumbine, R., Hecker, L. and Littlefield, A. (2005), 'Using Varied Instructional Techniques: Inductive and Deductive Teaching Approaches', in National Institute – Landmark College (ed.), *Biology Success! Teaching Diverse Learners*, www.pagegifted.com/uploads/1/1/6/0/11600328/inductivedeductive.pdf, accessed 16 April 2017.

Holden, R. and Liversedge, J. (2011), *Construction for Landscape Architecture* (London: Laurence King Publishing Ltd).

Jonas, W. (2014), 'Research for Uncertainty. Reflections on Research by Design', in Buchert, M. (ed.), *Reflexive Design. Design and Research in Architecture* (Berlin: jovis).

Loidl, H. and Bernard, S. (2014), *Opening Spaces: Design as Landscape Architecture* (Basel: Birkhäuser, Part of De Gruyter).

Petschek, P. (2014), *Grading* (Basel: Birkhäuser, Part of De Gruyter).

Schegk, I. (2014), 'Designing Identity – Methodical Approach', in Schegk, I. and Gruber S. (eds), *Nuovi paesaggi per/New Landscapes for Ostia* (Freising-Weihenstephan: Faculty of Landscape Architecture), 16–17.

Schegk, I. (2016), *Natursteinarbeiten im Garten- und Landschaftsbau* [Stone works in landscape construction] (Stuttgart: Ulmer).

Schegk, I. and Brandl, W. (2012), *Baukonstruktionslehre für Landschaftsarchitekten* [Structural design theory for landscape architects] (Stuttgart: Ulmer).

Steenbergen, C. M. (2008), *Composing Landscapes. Analysis, Typology and Experiments for Design* (Basel: Birkhäuser).

Wolfrum, S. (2014), 'Performative Urbanism – Designing the City', in Buchert, M. (ed.), *Reflexive Design. Design and Research in Architecture* (Berlin: jovis).

Zimmermann, A. (2011), *Constructing Landscape. Materials, Techniques, Structural Components* (Basel: Birkhäuser).

29

On-site learning

Simon Colwill

Introduction

This chapter discusses landscape construction teaching methods that focus on learning through on-site learning activities. These student assignments use built landscape works as the source of enquiry and learning.

The current generation of students has grown up with an almost endless availability of digital information. In an ever-more complex world, taking students out of the classroom, away from their desktops and laptops and into the field, has become more important than ever. Educators therefore need to develop new teaching methods that engage students in the learning process, increase their attention and motivation, and promote active listening, refection, problem solving and creative thinking.

Built landscape is a dynamic system influenced by factors such as material selection, weathering, use and abuse, succession and maintenance. In order to understand this complexity, construction teaching in the classroom needs to be accompanied by on-site learning activities and assignments that link theory with practice by engaging the students in active learning.

Case studies of courses at the Technische Universität Berlin (TU Berlin) and Harvard Graduate School of Design (Harvard GSD) will illustrate the significance of integrated field-learning activities. Both schools use the site as an essential source of knowledge in their methods of teaching and combine classroom teaching with a broad range of on-site learning activities.

On-site learning

> Not having heard something is not as good as having heard it; having heard it is not as good as having seen it; having seen it is not as good as knowing it; knowing it is not as good as putting it into practice.
>
> *(Xunzi [Teachings of the Ru], trans. J. Knoblock 1988: Book 8, Chapter 11, p. 81)*

A research project at the TU Berlin entitled 'Landscape architecture and the time factor: construction research on the contextual change of built landscape elements and the development of optimisation strategies' is currently developing a low-threshold and non-destructive cyclic monitoring method for identifying frequently occurring points of weakness and patterns of change to built landscape works through field research. The method being developed allows practitioners to monitor the development of built works after completion and provide clients with recommendations for optimisation. This cyclic monitoring method enables 'lifelong learning' from built works throughout one's academic and professional career. The research project is running hand in hand with teaching, allowing for continuous curriculum improvement and for students to focus on the core themes of the investigation through seminars, workshops and thesis topics. The initial findings highlight frequently occurring points of weakness in landscape detail design caused by *contextual factors*, *component quality* and *operating conditions* throughout the project cycle (Figure 29.1).

The repetitive nature of these weaknesses underlines a distinct lack of knowledge within the profession of the processes influencing change through time. These results point towards education as one of the key priorities for improving the understanding of weathering, temporality, durability and time-based change within the profession, and, therefore, for optimising the durability and sustainability of contemporary landscape architecture projects (Colwill 2016: pp. 399–400).

On-site assignments that engage students in analysing the built environment and critically reflecting on what they are experiencing significantly enhance construction teaching methods. This provides the students with multifaceted information that is often difficult to convey in the classroom. They combine otherwise separately taught course content such as planning, design, context, scale, proportion, material characteristics, haptic and optical qualities, together with the influences of weathering, use, maintenance and durability over time. This enables integrative learning in all fields of landscape architecture, urbanism, sociology of space, climatology, construction, maintenance and management.

These field activities are vastly enriched when accompanied by the project designer, construction or maintenance firm and/or client, together with the design and construction drawings. The first-hand experiences of project stakeholders enable, for example, discussion on contradictions

Category	Cause criteria
Context	**Site and contextual factors** Change/decay due to degree of exposure, aspect, access and circulation etc.
Component Quality	**Design and detailing factors** Change/decay due to due to quality of design, detailing and durability features
	Material specific factors Change/decay due to material suitability and/or quality.
	Implementation factors / workmanship Change/decay due to quality of implementation, workmanship (conformance with construction standards and guidelines)
Operating conditions	**Environmental processes / weathering** Change/decay due to environmental processes
	User actions / usage Change/decay due to intensity of use and/or misuse (physical stress caused by humans, animals, plants, vehicles etc.)
	Maintenance and repair Quality and frequency of maintenance and repair
	Force majeure Level of impact of incidents such as flooding, fire, riots etc.

Figure 29.1   Identifying the causes of change to built landscapes over time (based on Kirkwood 1999: pp. 166–177 and Colwill 2016: p. 398)

between design intention and construction, key problems and solutions during the planning and construction phase, together with issues of performance over time. Guest lectures from designers and industry experts bring professional practices and new perspectives from the 'real world' into the classroom. The key aim hereby is to establish a dialogue between academics and practitioners, linking theory to practice, taking students to the field and bringing professionals to the classroom for mutual benefit.

The site itself is an invaluable source of knowledge at each stage of project development:

> *Prior to construction* the existing topographic features of the site can be investigated, critical issues such as existing structures and vegetation evaluated, the character and genius loci (the distinctive atmosphere) of the site experienced, and the impact of development deliberated.
>
> During the *construction phase* students learn from the scale and complexity of the construction site and gain a feel for craftsmanship, construction techniques, foundations, detail design and materials, much of which are no longer visible after completion.
>
> In the *post completion phase* students experience built landscape as a dynamic evolving system interacting with the natural environment and patterns of use. This also allows reflections on the design, the vocabulary of landscape detail, the durability of materials, and the processes of change through time. 'Reflection is an important human activity in which people recapture their experience, think about it, mull over & evaluate it. It is this working with experience that is important in learning' (Boud et al. 1985: p. 43).

There are, however, two major hurdles regarding landscape technology field trips. First, learning in one context does not easily transfer to another; therefore it is essential that students experience a broad range of projects and detailed design approaches. Second, taking students out of the classroom is becoming increasingly difficult within academic institutions, especially with regard to building sites, due to increasing amounts of safety management issues and the administration necessary.

Methods

The teaching methods developing from this research aim to improve learning by involving students in on-site surveys, analysis and evaluations of 'real' projects and construction details after completion. This enables students to experience built landscape as a dynamic evolving system interacting with the natural environment, patterns of use and maintenance regimes within an academic context. These teaching methods follow the 'Experiential Learning Cycle' model of learning through experience and discovery developed by the educational theorist David A. Kolb (1984). The model employs a learning cycle that generally begins with *concrete experience* (doing, having a specific experience, e.g., on field trips or on-site assignments), moving to *reflective observation* (review, reflect and discuss the information gathered from different perspectives before making a judgement), then to *abstract conceptualisation* (draw conclusions, learn and develop a clear understanding of the theory) and finally to *active experimentation* (applying what you have learned to new situations) (Figure 29.2).

The most effective learning takes place when learning involves all four stages of the cycle. Kolb describes experiential learning as 'the process whereby knowledge is created through the transformation of experience. Knowledge results from the combination of grasping and transforming experience' (1984: p. 41).

It is generally accepted that people learn in different ways; whereas some students achieve through classroom activities, others can grasp complex theory and concepts through interaction

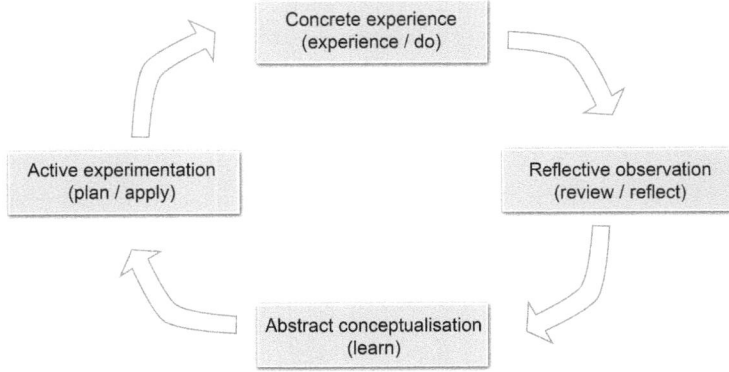

Figure 29.2 The Experiential Learning Cycle (based on Kolb 1984)

with real-life situations. There are many models and theories on learning preferences; the VARK model developed by the educational developer Neil Fleming presents four different learning strengths and preferences (Fleming 2012: p. 1).

> *Visual learners* learn from what they observe. They prefer learning from images, drawings, diagrams, charts, graphs, mind maps, etc.
>
> *Aural learners* (or auditory learners) learn from what they hear. They prefer learning through lectures, discussions, podcasts, oral presentations, etc.
>
> *Read/write learners* learn from read or written words and by taking notes. They prefer learning through books, texts, essays, etc.
>
> *Kinesthetic learners* learn from what they touch, feel and do. They prefer learning through multi-sensory experiences such as field trips, real-life examples, hands-on projects, etc.
>
> *(based on Fleming 2012: p. 1)*

Many learners show a strong preference for one of these learning styles, while others are multimodal and have any combination of two, three or four preferences. Multimodal learners are flexible about how they learn, however, to improve learning, various modes of learning are often necessary (Fleming 2012). Research from J. Sarabdeen (2013: p. 1) states that for multimodal learners: 'The practical implication is that the trainers should adopt various learning strategies to achieve the learning objective'. Fleming and Baume (2006: p. 5) suggest that 'Teaching often reflects the teacher's preferred teaching style rather than students' preferred learning styles'.

The results of a learning preference survey at California Polytechnic State University from 2010–2012 showed that the highest preference amongst 85 architectural students is visual (48%), followed by kinaesthetic (26%), aural (14%) and then read/write (12%). Furthermore, roughly 40% of all students 'would be defined as having multiple preferences that include both Kinaesthetic and Visual' (Nelson and Lawson, 2013). These results enable educators to use teaching methods that reflect the learning strengths and preferences of specific groups of learners in a course in order to increase learning outcomes. Integrated field learning assignments are mainly kinaesthetic and visual learning activities, and thus address a large proportion of architectural students' preferential learning styles.

Case studies

The following case studies from Harvard University and TU Berlin aim to show how diverse teaching methods involving site interactions ensure a thorough understanding of construction technologies and techniques. Both of these schools use the site as a source of knowledge through field trips and field research assignments.

Case study 1 – Harvard Graduate School of Design

Course title: 'Landscape Technology as Design: Material, Tectonics and Time'.

Programme: Master of Landscape Architecture.

The course is supervised by Professor Niall Kirkwood, a Professor of Landscape Architecture and Technology at Harvard Graduate School of Design (Harvard GSD) since 1992, and Alistair McIntosh, a lecturer with over 35 years of landscape practice and teaching experience.

Before teaching at Harvard GSD Niall Kirkwood worked in landscape architecture and architecture private design practices in the United Kingdom and the United States for 18 years and gained hands-on practical experience through supervising the field construction of built landscapes, infrastructures and buildings in Ayrshire Scotland, London, Barcelona, Columbus Ohio and New York. One of his many fields of research is landscape detail design, traditional and emerging construction technologies, and the on-going durability of built landscapes. His books, entitled *The Art of Landscape Detail: Fundamentals, Practices and Case Studies* (1999) and *Weathering and Durability in Landscape Architecture: Fundamentals, Practices, and Case Studies* (2004), provide pioneering information on the theories, approaches, and practices of landscape detail, together with the weathering, durability, and physical changes in the designed landscape over time. His teaching methods reflect this research by employing a diverse variety of methods and techniques in order to address the complexity of landscape detail design.

The objective of the course is to develop:

> a critical understanding of both tested and emerging practices of detail design and construction in landscape architecture, address the interdependence between site, design, technology, tradition and innovation in the making of landscape architecture and how this can inform function and expression in landscape design work at a range of project scales.
>
> *(Kirkwood and McIntosh 2017)*

The course is split into two main components, 'The Indoor Classroom' involving a series of lectures and workshops, followed in the second half of the semester by the 'The Outdoor Classroom' with field trips to a wide range of historic and contemporary built landscapes. The individual course assignment runs parallel to the classes throughout the semester.

The Indoor Classroom – a series of lectures, discussions and interactive workshops

The lectures focus on issues of landscape technology, materials and construction, detail vocabularies and tectonic syntax, weathering and durability, structural principles and soft engineering. Accompanying workshops aim to demonstrate how the above concepts are integrated into practices of design development. Class participants engage in an interactive analysis of case

Figure 29.3 The Outdoor Classroom. Active discussion of detail design and construction issues. Harvard GSD (photo: N. Kirkwood 2016)

studies using the diagnostic section. The concepts and methods introduced in the Indoor Classroom form the basis for students to analyse and comprehend what they physically experience during the field trips.

The Outdoor Classroom – a series of field trips

The field trips, or 'Outdoor Classrooms', address a wide range of approaches to landscape design and construction. The sites are selected to allow students the opportunity to observe and engage in a wide range of landscape programmes, detail languages, material applications, design form and expressions, from varied landscape architecture offices.

The course assignment consists of three main parts:

Part 1. A technological critique of a landscape architecture project from the last 25 years is carried out. Particular focus is placed on the application of detail design at a range of scales and tectonic applications.

Part 2. This involves the research, design and reverse engineering (derivation of detailed information on design, construction and operation from an existing object) of a detail design landscape prototype that must be described in a material and tectonic manner over time. The *detail design prototype* is of a complex nature consisting of a variety of interrelated natural and constructed boundaries, transitions, surfaces and objects derived from a diagnostic section. The *diagnostic section* is a *research and development* tool involving both technical design analysis and the development of optimisations. Built elements are broken down through *reverse engineering* into their constituent parts in order to comprehend how they

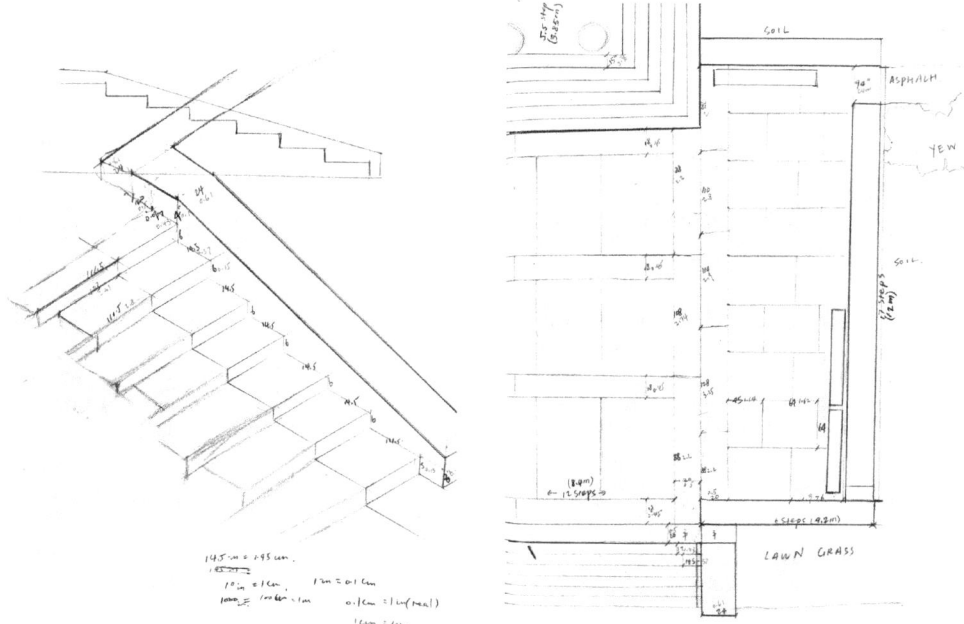

Figure 29.4 Initial site investigation sketches from Part 1 as a basis for the diagnostic section. X. Yuan, Harvard GSD (2016)

were constructed. Diagnostic evidence is also added to assess the current condition. This enables students to critically analyse built landscape works, derive constructional features and evaluate performance through time. The facts established in the same diagnostic section can then be used to inform the speculative development of new built works.

Part 3. The *detail design prototype* is now applied to a new geographic location taking into account the specific site topography, microclimate, soils, groundwater, availability of materials, labour and cultural context. The prototype needs to be modified to ensure the necessary performance over dedicated periods. Throughout the workshop and field exercise the 'Students learn and apply methods of observation that enable a critical understanding of existing built works and apply those insights to the productive development of their own landscape proposals from the conceptual to the detail scales' (Kirkwood & McIntosh 2017).

Case study 2 – Technische Universität Berlin

Course title: 'Landscape Construction and Materials'.

Programme: Master in Landscape Architecture.

The course aims to develop understanding of how initial conceptual ideas are transformed through design development processes into concrete landscape proposals whilst addressing the implications of the specific site, function, design, construction, materials and the dynamic nature of physical change over time. This involves the creative transformation of physical materials through techniques of landscape construction into a vocabulary of built landscape form.

Figure 29.5   Field trip to a concrete plant enabling in-depth learning of production techniques. TU Berlin (photo: C. Schellhorn, 2015)

Students need to develop a critical understanding of current, new and emerging methods of detail design and construction, a thorough knowledge of the qualities and properties of materials, together with a clear understanding of the factors influencing patination and deterioration. The course is split into a series of classroom-based learning activities involving lectures and seminars supported by on-site learning activities focussing on the detailed analysis of built landscapes.

Classroom learning – a series of lectures, seminars, discussions and workshops

The classroom learning activities involve a series of lectures and seminars held by university staff and visiting experts, focussing on developing knowledge on the interrelations between site design, detail design, building materials, construction detailing, structural engineering, maintenance and the processes of time-bound contextual change. These take place parallel to the progress of the field-based exercises. Guest lecturers are invited to present specific project case studies that further illustrate course content.

On-site learning – landscape forensics

Assignments within our construction seminar for master's degree students involve students in small groups going to 'real' landscape projects and analysing situations in detail before formulating a tailored response. This is set as a research question, the object of research being 'real' landscape projects. Students examine the current condition in relation to the surrounding context and reflect on interrelations between site design, detail design, building materials, technical implementation, maintenance and performance issues. Comparisons with images in publications at the time of completion, together with project descriptions or reviews, enable the students to identify time-bound changes to the built landscape, as well as discrepancies between design intentions and the built reality. Teacher support enables the students to 'read' and interpret the traces of wear and tear, weathering, maintenance and succession in order to determine, for example, patterns of use, misuse, maintenance and/or points of weakness. The factors influencing change through time are introduced in a classroom learning context prior to the on-site interactions, serving as a basis for 'reading' and interpreting the condition (from patination to deterioration) of the projects and detail elements under examination. An on-site lecture from a practitioner is also organised at a current construction site or recently completed landscape architecture project.

This on-site learning is based on what we call 'landscape forensics', which is a form of learning by examining the problems and failures arising on built landscape works through time. Furthermore, the location, spread and intensity of patination and decay allows specific vulnerabilities and weaknesses to be identified. Through analysing the *root causes of failures* methods for deterring future failure and enhancing durability can be derived. The method reflects on the entire design, construction and post completion phases of the project, together with the current state of maintenance. The cause criteria listed in Figure 29.1 form the basis for this analysis.

The course assignment consists of four parts:

Part 1. The students perform an on-site examination and critical analysis of a built landscape architecture project completed in the previous 20 years.

Part 2. This involves the technological critique of a landscape detail within the selected site through *reverse engineering* and interpretation. The built element including the surrounding context is analysed with regard to the appropriateness of the design, construction and materiality, together with the implications of functionality, location, weathering and durability. Points of weakness are identified that, due to their exposed position (corners, edges, etc.) or particularly high demands (intensively used surfaces, surfaces with ground contact, etc.), are exposed to greater levels of stress than other areas of the same element. The root causes of time-bound change are assessed according to the factors listed in Figure 29.1. In-use *condition assessment* takes place by analysing the differences between the current and original condition. This evaluation method is being further developed in the before-mentioned research project. Changes can be classified into those that are purely cosmetic and those that lead to a reduction in aesthetics, functionality, stability, and/or durability. Therefore, a qualitative assessment of the following factors is carried out:

Aesthetic condition – from the initial process of cosmetic patination to the latter phase of visual degradation.

Functionality – usability, function, process-related serviceability and safety.

Stability – the carrying capacity of the structure at the time of the survey.

Durability – the ability of the structure to withstand damaging impacts through expected service life, during scheduled use and maintenance.

The students produce a variety of texts, photo documentations, diagrams, sketches and detail drawings to present their results; an example is shown in Figure 29.6.

Part 3. An optimisation strategy is then developed within a classroom learning context for the selected landscape detail with regard to the specific requirements of location (weathering, use intensity, level of maintenance, etc.), use (form, material, etc.) and for deterring constructional and material vulnerability.

Part 4. The landscape detail is redesigned for a specific location using the knowledge acquired from the analysis in Parts 1 and 2 together with the optimisation strategy from Part 3. A complete set of design and construction drawings and a scale model are then produced for the optimised landscape detail following standards for architectural construction and working drawings as shown in Figure 29.7.

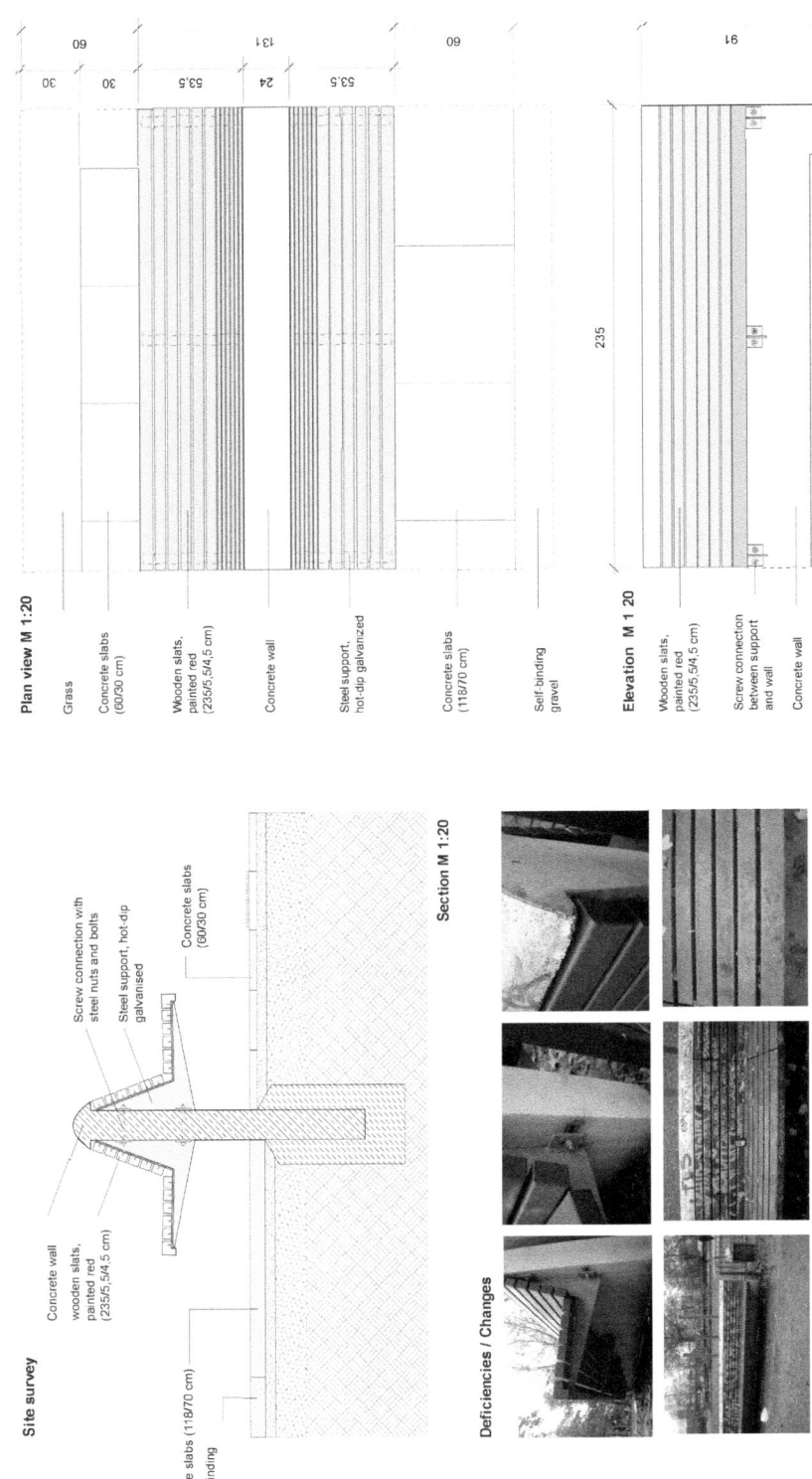

Site survey

Concrete wall
wooden slats,
painted red
(235/5,5/4,5 cm)

Screw connection with
steel nuts and bolts

Steel support, hot-dip
galvanised

Concrete slabs
(60/30 cm)

Section M 1:20

Granite slabs (118/70 cm)

Self binding
gravel

Deficiencies / Changes

Plan view M 1:20

Grass

Concrete slabs
(60/30 cm)

Wooden slats,
painted red
(235/5,5/4,5 cm)

Concrete wall

Steel support,
hot-dip galvanized

Concrete slabs
(118/70 cm)

Self-binding
gravel

Elevation M 1 20

Wooden slats,
painted red
(235/5,5/4,5 cm)

Screw connection
between support
and wall

Concrete wall

Figure 29.6 Excerpt from a submission for Part 2 – landscape detail critique and reverse engineering drawings. F. Karle, TU Berlin (2012)

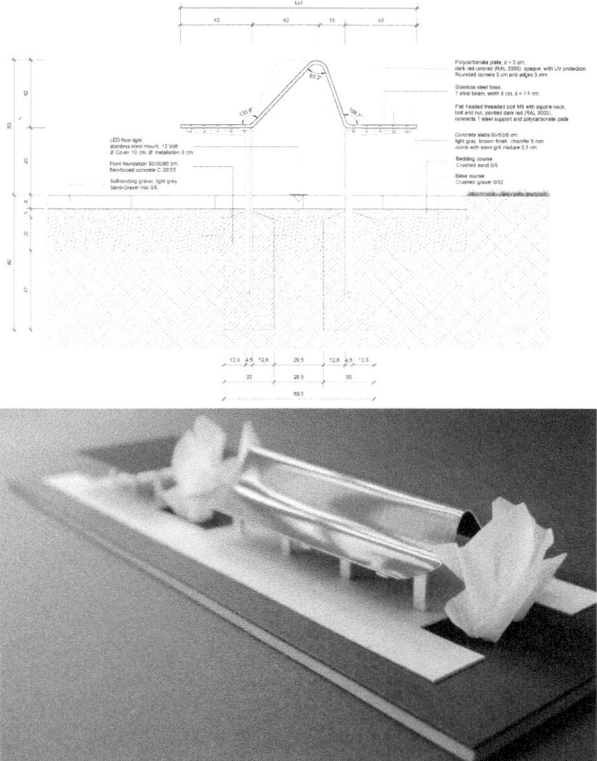

Figure 29.7   Excerpt from a submission for Part 4 – detail drawing and model of the optimised construction. F. Karle, TU Berlin (2012)

Discussion

The learning approaches presented here enable an integrative approach to teaching landscape construction by treating built landscape projects as research objects and engaging students in on-site research activities. Both courses use the site as an essential source of knowledge, a learning instrument informing the students on real-life situations in the dynamic realm of time and change. The learning objectives are to provide students with techniques for design exploration through critical observation, technical thinking, and for monitoring the performance of built landscapes through time.

Reverse engineering is a key teaching method of both courses and is based on a process of enquiry through observation and research. On-site observations of the current condition lead to the students posing questions regarding the design, construction, materials and the mechanisms of change. Individual research is then necessary to develop their knowledge, in order to analyse the site and its component parts in detail. The aim is to develop methods to critically analyse built landscape works, deduce the root cause of problems, evaluate performance through time, and develop optimisation strategies and solutions. Teachers who guide the students through the deductive process assist this process.

The course assignments not only aim to exercise and develop the tools, techniques and technologies of detail design practice in landscape architecture, but also to predict and adjust to factors that affect the durability of landscape architecture projects over time. During these on-site assignments, students confront all facets of a project simultaneously; they need to think, discuss and analyse built landscape before formulating a judgement and an optimal response. The processes of observation, *technical thinking*, reflection and causal research enable a more founded development of innovative solutions. The role of the teacher in this process is as an educational coach, guide, and mentor who, if necessary, recommends alternatives for ineffective practices and/or teaches possible alternatives. These teaching methods complement the more traditional techniques in lecture halls and seminar rooms.

These teaching methods attempt to equip students with the tools necessary for *lifelong learning* from monitoring the development of both their own built landscape architecture works and the works of others. The assignments demonstrate to students how knowledge from built landscapes can be extracted and interpreted to inform future projects. The case studies follow Kolb's (1984) cyclical model of 'experiential learning', from the on-site data collection (experience/do) to the analysis (review/discuss), the formulation of optimisation measures (learn) and the development of an optimised solution (plan/apply) (Figure 29.2). Through repetition of this research cycle, a spiral process of continual learning and optimisation (Figure 29.8) can be achieved. This process of *research and development* is similar to the monitoring methods currently being developed by the author within the previously mentioned research project.

The diversity of teaching methods also allow the courses to follow Neil Fleming's VARK model of learning, optimising learning outcomes through addressing the preferences of a wide range of learning types, which, in turn, often leads to increased group motivation.

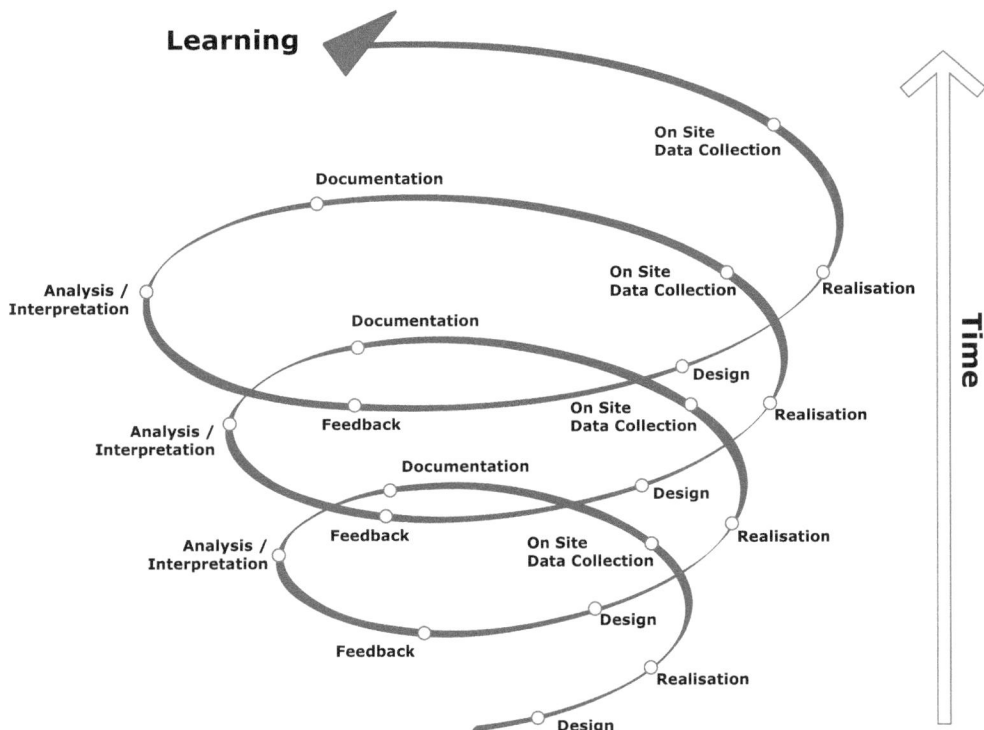

Figure 29.8   The spiral process of continual learning and optimisation (the author, 2013)

One of the bonus effects of these teaching activities is the passive learning that occurs. Observations of scale, form, materials and their surfaces, use, abuse, maintenance, and climatic interactions with the site allow 'real-world' insights into landscape architecture projects within an academic framework. The problem-solving 'reverse' assignments involving active learning and participation also enhance the learning experience by providing an activity in which the students can learn from each other. Boud, a professor of adult education, describes this 'peer learning' as the 'sharing of knowledge, ideas and experience between the participants' (2001: 3). The participants work collaboratively, give and receive feedback, and develop a wide range of skills. This engagement is reflected in the quality and diversity of the coursework.

Conclusion

The teaching and learning methods discussed in this chapter demonstrate a shift of emphasis in the pedagogical framework of teaching landscape technology towards landscape performance, change, temporality and monitoring. Field-based learning assignments form an essential component in understanding these complex relationships. The intensive on-site learning assignments involve observation, enquiry and critical reflection on what they are investigating, which triggers deeper active learning. Hickcox (2002) explains that field experiences are student-centred learning activities, enabling the application of ideas and concepts taught in a traditional classroom context to a specific environment that stimulates critical thinking and analysis. They provide students with the opportunity to contextualise their classroom learning in the 'real world' of the built environment, therefore linking theory and practice. Both case studies presented here aim to improve teaching practices, enhance student learning, increase student engagement, and better prepare students for the complex requirements of the profession. The teaching methods focus on the relationship between site, design, landscape technology, and the dynamic forces of weathering and usage over time. The depth and complexity of the student results demonstrate a multifaceted technological understanding of landscape architectural detail design.

Acknowledgements

I wish to thank Niall Kirkwood and Alistair McIntosh for their assistance in preparing this chapter and for giving me an insight into their teaching practices. The author gratefully acknowledges the financial support provided by the German Research Foundation (DFG).

References

Boud, D. (2001), 'What is peer learning and why is it important', in Boud, D., Cohen, R. and Sampson, J. Peer Learning in Higher Education. Learning From and With Each Other (Hoboken, NJ: Taylor and Francis).

Boud, D., Keogh, R. and Walker, D. (1985), Reflection. Turning Experience into Learning (Abingdon, UK: Routledge).

Colwill, S. (2016), 'Time, Design and Construction: Learning from Change to Built Landscapes Over Time', in Bridging the Gap, ECLAS Conference 2016, Rapperswil, Switzerland, Conference Proceedings.

Fleming, N. D. (2012), Teaching and Learning Styles. VARK Strategies. Revised (Christchurch: Neil D. Fleming).

Fleming, N. D. and Baume, D. (2006), 'Learning Styles Again: VARKing up the Right Tree!', Educational Developments, SEDA Ltd, Issue 7.4, November, 4–7.

Hickcox, L. K. (2002), 'Personalizing Teaching through Experiential Learning', College Teaching 50(4): 123–128.

Kirkwood, N. (1999), *The Art of Landscape Detail: Fundamental, Practices, and Case Studies* (Hoboken, NJ: Wiley).

Kirkwood, N. (2004), *Weathering and Durability in Landscape Architecture. Fundamentals, Practices, and Case Studies* (Hoboken, NJ: Wiley).

Kirkwood, N. and Mcintosh, A. (2017), 'Syllabus GSD 6242 Ecologies, Techniques and Technologies IV', Spring.

Knoblock, J. (1988), *Xunzi: A Translation and Study of the Complete Works, Vol. I: Books 7–16* (Stanford: Stanford University Press).

Kolb, D. (1984), *Experiential Learning: Experience as the Source of Learning and Development* (Englewood Cliffs, NJ: Prentice-Hall).

Nelson, J. and Lawson, J. W. (2013), 'Teaching Architecture, Engineering and Construction Disciplines: Using Various Pedagogical Styles to Unify the Learning Process', in *ASEE Annual Conference & Exposition Proceedings* (Washington, DC: American Society for Engineering Education), pp. 13907–13917.

Sarabdeen, J. (2013), 'Learning Styles and Training Methods', in *The Communications of the IBIMA (CIBIMA)*, Volume 2013, Article ID 311167, 1–9.

30

By land, by air, by sea

Jörg Rekittke and Yazid Ninsalam

Teaching implies research, and research implies teaching—in the context of our studio methodology. In the studio, students and teachers spend long hours in time-intensive studio sessions, critiques, and reviews. Hence, the best chance to achieve the required research progress, despite the categorical lack of time, is given by a simultaneous combination of teaching and research in a collaborative act by students and teachers. This is when research becomes teaching, and teaching becomes research. We take comfort in the categorical awareness that any research—independent of academic subject or professional affiliation—depends on the collection of particular sorts of evidence through the prism of particular methods, and every method has its strengths and weaknesses (Britten and Fisher, 1993). Likewise, the Merriam-Webster definition of research (2017)—1) a careful or diligent search, 2) a studious inquiry or examination (especially an investigation or experimentation aimed at the discovery and interpretation of facts, revisions of accepted theories or laws in the light of new facts, or practical applications of such new or revised theories or laws), and 3) the collecting of information about a particular subject—comprises design work and related analytical efforts. Our contribution is about the particular research form of field research or *fieldwork*. In a classic laboratory setting, the designed environment allows the perfect control of an experiment. Out in the field, in real life and in direct contact with the unadorned environment, the researching designer has to deal with an endless number of interconnected phenomena that are all in uncontrollable motion (Højlund Nielsen, 2007). Designers—a hard-to-define academic species—have to live with their nature of being limited scientists. Though time and again starting their work in the most scientific way, they reach a certain point where they can no longer resist a personally fixed idea—a point of no return. It is where an objective approach changes into subjective method, a form of dogmatic obsession, which can be frustrating for the rationalist, but is largely inevitable—and successful—in design practice. Designers deliberately lose their scientific integrity and consciously gamble away the secure position of rational impartiality. In so doing, they delightedly lapse into acting as scientifically incorrect stand-alone empiricists (Rekittke, 2015). In no way does this mean that we do not exhibit academic rigour. Mays and Pope (1995) point out that systematic and self-conscious research design, data collection, interpretation and communication are the basic strategy to ensure rigour in qualitative research. In the frame of our fieldwork, we create an independent account of method and data, and our production of a plausible and coherent explanation of the investigated phenomenon allows other researchers to analyse the same data in the same way, and to come to the same conclusions. Only via first-hand observation and adequate time spent in the field can the qualitative researcher become thoroughly familiar with the milieu under scrutiny—as well as the milieu of the researcher (Mays and Pope, 1995). A deep understanding of the research setting, even if not succeeded by any consequential design intervention, can result in convincing design-research outcomes. As Christophe Girot states in his essay 'Four Trace Concepts in Landscape Architecture', "the designer seldom belongs to the place in which he or she is asked to intervene" (Girot, 1999). *Outsider* designers can only acquire a true understanding of place directly from the field of action, which will enable them to act wisely and knowledgeably (Girot, 1999). In the essay 'Mandatory Fieldwork' (Rekittke, 2015) we adopt the attitude that in situ spatial and contextual data collection is the lynchpin of all landscape architectural design work. The essential qualitative research phases of fieldwork—Girot coined them *landing, grounding, finding* and *founding*—have to be taken by any landscape designer in the context of real-world projects, in academia and in the profession. We consider fieldwork not to be optional but mandatory for all our academic studio projects. In other words, we are convinced that *teaching landscape* works best in the landscape itself. Maybe we should formulate it the other way round—in the same manner. *Learning landscape* works best in the landscape itself. The obligatory integration of fieldwork and studio work, the contemporaneous combination of input as well as output by teachers and students, and the simultaneous execution of teaching and research, define the framework of our method.

Leaving the detachment and boundedness of the institutional academic studio room, and making the real urban landscape the paramount laboratory for design research, became a kind of trademark of our research team and the students who we teach and collaborate with. In our collective academic endeavours, we are compassing what we refer to as *landscape credibility*—by exposing ourselves and our students to various landscapes, physically and directly, on local as well as on the international scale. Mobility is key in our projects; therefore, we work as backpackers who carry all our lightweight equipment and portable tools into the field. Our fieldwork is carried out in mostly urban, often mega-urban environments in Southeast Asia, and the results have been drawn up under the name 'Grassroots GIS' (Rekittke and Paar, 2010). We continue to build on inexpensive technology, easy or free access to applied tools, geodata, georeferenced design data as well as open source, open standard and cost-free software and data storage possibilities. After we successfully gathered work experience even in underwater biotopes—crowded by snorkelers and divers—our activities meanwhile extend to all relevant layers of populated landscapes. We may claim that we became able to operate and teach by land, by air, and by sea.

By land

Our works have been published in conference proceedings, journals, and exhibition catalogues. However, there is still an unpublished piece of research that had been embedded in a research project titled 'Under-the-Urban-Canopy 3D', conducted by the authors. It was funded by the National University of Singapore, but remained unfinished because of moving abroad. Notwithstanding, we consider the related work and method an ideal example for our work by land. Equipped with rather expensive mobile scanner technology, this time we chose to work on a sample of the age-old primary forest patch of the Bukit Timah Nature Reserve in Singapore. A virgin forest leftover of a mature tropical rain forest—in the middle of the fully urbanized city state of Singapore. We focussed on the integration of data sets acquired from a terrestrial laser scanner, a state-of-the-art handheld close-range scanner (Figure 30.1), and a cinema-grade camera-equipped unmanned aerial vehicle system (UAV), to build a high-resolution 3D model

Figure 30.1   Terrain and vegetation scan with a state-of-the-art handheld 3D scanner (FARO
Freestyle 3D), Bukit Timah Nature Reserve, Singapore (Rekittke and Ninsalam, 2016)

This figure is shown in colour in the plate section of this book.

of a layered tropical rain forest sample. As our working hypothesis, we stressed the proverb of
not seeing the wood for the trees with the intent to reframe it into something like *very well seeing
the wood for the trees*. The plot in the Bukit Timah Nature Reserve had been chosen because it
represents a unique piece of land, and landscape, in the immediate vicinity of a big city, with
a maximum of vegetation mass and structural complexity. It is situated in the central tropics,
where the tree species variety can be a hundredfold higher than in a temperate forest in Europe
or North America (Ng, Corlett and Tan, 2011).

Functional limitations and challenges regarding the capture of spatial data sets from such
protected environment include accessibility, complexity, limited range of sensors and physical
disruptions. A terrestrial laser scanner typically acquires spatial information limited to the understory of a forest, inclusive of its lower canopy layers, whereas aerial photogrammetry is capable of
providing a survey of the canopy and emergent layers. We used a camera UAV, a terrestrial laser
scanner, and a handheld laser scanner to capture 3D pointclouds and thus facilitate the acquisition
of geometric and RGB-colour information of the tropical rain forest plot. We assessed the feasibility of using a multiple sensor approach to digitally model the structure of a tropical rain forest
plot, and aimed at testing the practical utility and deployment of the chosen sensors to overcome
the problem of visual occlusion. Our aim was to retrieve a complete 3D digital model sample of
a tropical rain forest plot (Figure 30.2). We gathered the main parts of the puzzle in the field, but
we still worked on a time-efficient method to assemble them into a seamless model. Nonetheless,
our experimentation allowed us to understand and assess the performance of each applied sensor
in terms of coverage, resolution, and visual fidelity in the reconstruction of a tropical rain forest
plot. First we used the terrestrial laser scanner to create a baseline scan of the site. The numerous
disruptions in line of sight, due to the heterogeneous nature of the forest, was complemented by
the use of the handheld scanner that was deployed for the detailed recording of the understorey.
Consequently, we took advantage of an image-based *structure-from-motion* (SfM) method by using
the photographs acquired by the UAV. The aerial manoeuvre was employed to acquire overlapping UAV image sets that document layers that are not within the range of the terrestrial laser
scanner and handheld scanner acquisition. The *FARO Terrestrial Laser Scanner* allows a range of 0.6
m up to 130 m, and provides a field of view of 360 degrees in horizontal direction and 300 degrees
in vertical direction. The resulting data set produces a full panoramic view with RGB-colour
data acquired from the embedded 70 megapixels optical device (Figure 30.3). The handheld 3D
laser scanner, *FARO Scanner Freestyle3D*, was deployed to capture landscape elements that are not

within the line of sight of the terrestrial laser scanner. This method is to complement detailed flora structures that may be found within the forest floor and shrub layer (Figure 30.4). With a scanning accuracy of 1 millimeter at 1-meter range, and an automatic flash mode, it enables users to scan objects at a close range in a variety of lighting conditions. A *DJI Inspire 1* UAV platform was deployed within the site to provide aerial footage and perspective of the site documented (Figure 30.5). It carries a full 360-degree rotation camera that captures 4K video and a 12 megapix-3d scan. It operates safely within a maximum flight time of 18 minutes. Besides the methodological experimentation with the forest-layer-specific tooling, our study built on a previous census of trees at the selected forest plot in 2006, where big trees of 30 cm or more in diameter were measured by hand throughout the entire forest of the Bukit Timah Nature Reserve. We compared the measurements derived from our scanning campaign with an analogue database that had been recorded by hand in 2006. The results demonstrate the successful deployment of a tool set that

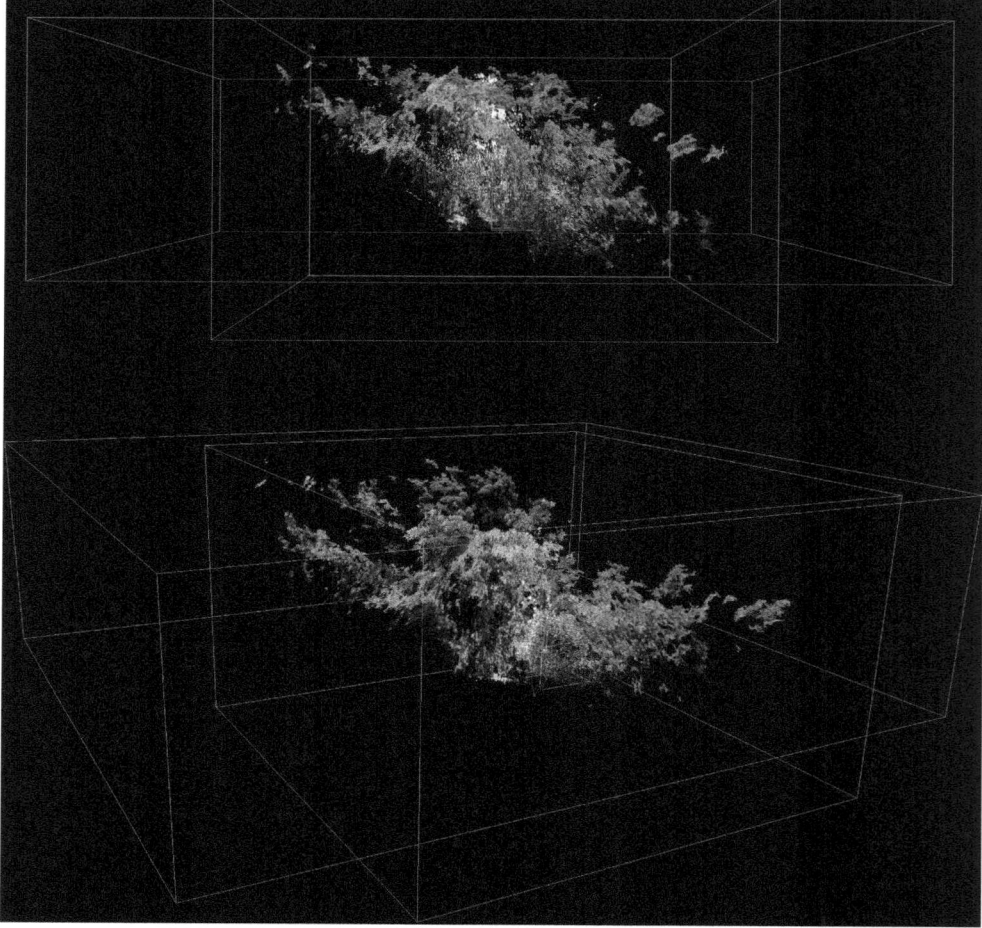

Figure 30.2 Complete 3D digital model of a tropical rain forest sample derived from scans by a terrestrial laser scanner, a close range handheld scanner, and a photogrammetrical artefact resulting from UAV camera flights (Rekittke and Ninsalam, 2016)

This figure is shown in colour in the plate section of this book.

is able to rapidly measure the complex environment and capture a substantial amount of spatial information that may be used to quantify the measurements of upper stem diameters, branch internodal distance and canopy dimensions (Watt and Donoghue, 2005) in a fraction of the time needed to manually survey the site. We did not develop and demonstrate this method in order to become able to redesign a pristine forest. But the aptitude to successfully analyse, measure and model the complex spatial structure of such a piece of forested land allows promising projections concerning new design precision and methodology for structurally much less complex built environments. *Quod erat demonstrandum.*

By air

One of our latest projects enabled us to tweak our landscape information-gathering methods by air, and, in addition, allowed us to expand our research sphere to operations by sea. So far, our experiences had been limited to applications by land, by air, and by inland watercourses. In the maritime ecosystem of the Bunaken Marine National Park, North-Sulawesi in Indonesia (Mehta, 1999), the landscape above the water's surface—parts of the Tanjung Kelapa coast, and the five islands of Bunaken, Manado Tua, Mantehage, Nain and Siladen—is deemed to be a minor matter. The main attraction in the related touristic business is the vast underwater world that is accessible

Figure 30.3   3D model component (see Figure 30.2) of ground level forest structure acquired from twelve single scans with a terrestrial laser scanner (Rekittke and Ninsalam, 2016)

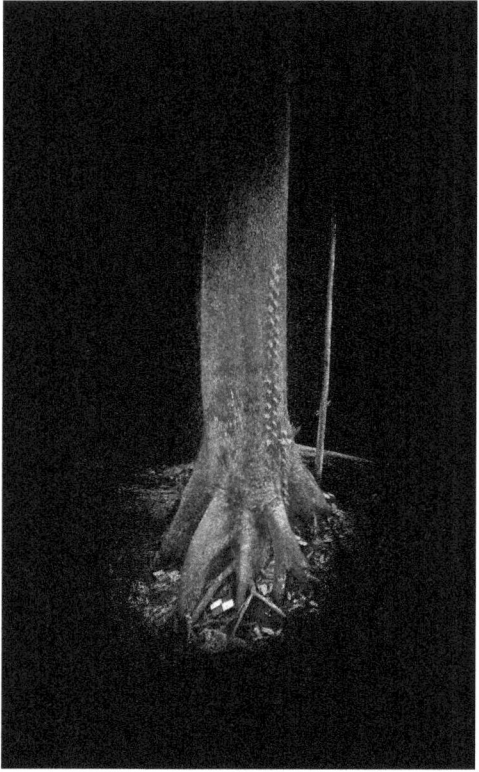

Figure 30.4   3D model component (see Figure 30.2) of detailed flora structure within the forest floor and shrub layer, acquired from one continuous close range scan with a handheld laser scanner (Rekittke and Ninsalam, 2016)

and visible only for snorkelers and divers. It is the part of the National Park that is most endangered by urban expansion and increasing visitor numbers. The related environmental stress, like massive pollution, soil compaction, trampling damage, and trophy hunting, causes biotope and diversity loss. We argued that it was a necessity to make the sensitive biotopes of the National Park visible in the form of demonstrative models, before developing any design or management proposals and recommendations. Our work by air delivered the indispensable overviews of the landscape system, but we did not hesitate to submerge ourselves in the warm waters of the Celebes Sea subsequently.

A look at the world map and the geographic location at the northern tip of North Sulawesi, Indonesia, suggest an initial conjecture of paradisiac conditions and remoteness. In reality, the world's end is overstrained. Divers and snorkelers are hunting and gathering globetrotters, and they start their journeys from any major airport on the five continents. Only the last mile constitutes a bottleneck as the spectacular core of the marine national park is only accessible by boat. The putative paradise comprises multiple environmental world records. In the reef system we find the highest number of mangrove species, the highest number of seagrass species, and the highest number of coral species—worldwide. The magnificent spectacle has an admission fee that applied to the past twenty-five years. Nevertheless, during a quarter of a century, hardly any project had been financed by the proceeds. Corruption successfully enriches a handful of thugs financially, and effectively degrades a whole region environmentally (Rekittke, 2017).

Figure 30.5   3D model component (see Figure 30.2) of the rainforest canopy acquired
from one continuous UAV camera flight. The photogrammetrical artefact, a
pointcloud, results from 178 single aerial images (Rekittke and Ninsalam, 2016)

The idea for an academic studio on the Bunaken Marine National Park, with landscape architecture students of the National University of Singapore, originated from the author's private family holidays in the National Park. The neighbour of their host on Siladen Island was the government official of Siladen, and, together with his extended family, he unleashed enormous fireworks on New Year's Eve. The next day, the aforementioned neighbour—responsible for the local enforcement of the National Park regulations—discharged his tonnes of fireworks trash directly into the vulnerable legally protected coral reef of the National Park. Such behaviour, in combination with increasing visitor numbers and related above- and under-water stress, affects the precious mangrove belts, seagrass meadows, reef flats, reef crests and the vertical reef walls. The negative effects of the local environmental unconsciousness and the conscious hedonistic exhaustion of the environment by foreigners have to be approached by a holistic analysis and an inclusive design strategy, also considering the set of problems faced underwater. Our wider studio aim was to develop necessary infrastructure components and essential forms of management on both land and water. We started by generating an aerial overview of a representative island-to-reef sample by a flight campaign with a camera-drone (Figure 30.6). The UAVs equipped with digital camera technology became quasi standards in the field of landscape data acquisition. We refer to these tools as *overview machines* (Rekittke and Ninsalam, 2016). The possibility of generating three-dimensional representations of reality in the form of dense pointclouds catalyses very useful digital tools and methods for landscape architects.

When working on large-sized landscape projects, the contemporary landscape designer is well advised to carry an overview machine in the field. Unrivalled overall views of high resolution, generating potentially eye-opening insights, emancipate the landscape architect from a long tradition of territorial myopia. We have had satellite imagery and Google Earth at our disposal for a long time, but we did not have military-grade (almost all high technology that can be utilised for planning and design has been initially developed for military purposes, any naiveté about this would be inappropriate), affordable, sharp-sighted eyes in the sky, portable in our backpack. Of course the civilian transformation of the aerial imagery into pointclouds leads to terrain models of significant precision. The model generation is quick, effective and rather simple (Figure 30.7). The ecosystem of the Bunaken Marine National Park comprises three main types of zones. The coastal secondary forest of the islands, where a building boom for illegal resorts happens. The intertidal mangrove belts around the islands, that become increasingly perforated by boat traffic to such resorts. And the sensitive intertidal seagrass meadows that extend from the mangrove belts to the steep coral walls of the reefs. Tidal change brings about the patterns of use of these zones, and defines their accessibility concerning specific fieldwork tools and methods. Our aerial overviews had been produced with two *DJI Inspire-1* UAV that were equipped with Zenmuse X3-FC350 cameras. The above-water part of the documented ecosystem transect (Figure 30.8) was spatially defined by the central island road on Bunaken to the east, and the underwater part by the coral wall of the island reef to the west. A total of 318 aerial images were captured at a frequency of one shot every five seconds, and 314 of these shots were calibrated with the help of the processing software *Pix4D*. The aerial overview of the site provided us with a comprehensive georeferenced base model, which we rounded off by the insertion of supplementary close-range terrestrial and underwater pointcloud models (Rekittke and Ninsalam, 2016).

Figure 30.6 Camera-carrying unmanned aerial vehicle (UAV, DJI Inspire 1) for aerial imaging (left). The image material is used for the generation of 3D models (right). The example shows a resulting model of a land zone and intertidal zone in the Bunaken Marine National Park, North-Sulawesi (Rekittke and Ninsalam, 2016)

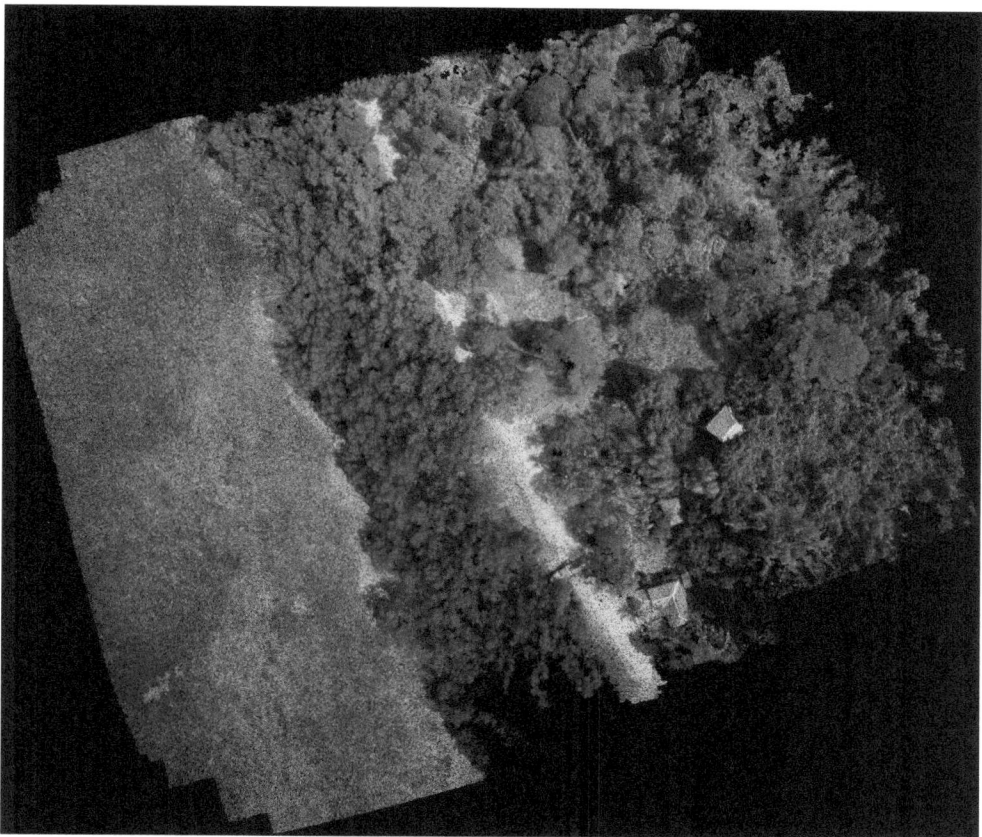

Figure 30.7 Complete 3D model of a site sample of Bunaken Island, in the Bunaken Marine National Park, illustrating the interconnected ecosystem components. From left to right: steep reef wall (black), inter-tidal reef flat, broad mangrove belt, beach strip, and secondary island forest (Rekittke and Ninsalam, 2016)

By sea

The marine component of the fieldwork in Bunaken was not a walk in the park. It's not a swimming pool, most students in our group had never snorkelled before, and not everyone could swim without floatation devices. That said, we successfully produced relevant model samples of the species-rich coral reef that is permanently underwater. As a next step we focussed on the modelling of the complex, exposed and vulnerable structures of the intertidal mangrove belt. Mangroves grow partly above and partly under the water's surface (Figure 30.9). The corresponding snorkelling campaigns between the mangrove roots—conducted by us, the researching academics—were eerie and taxing. Seen from an underwater perspective, mangrove environments seem to be not of this world. The roots and branches waved like fairy arms, the tidal change induced extreme currents, and the sighting of banded sea kraits, non-aggressive but one of the most venomous water snake species, admittedly caused palpitations. Anyway, mangroves epitomise the fluctuating conditions in an intertidal ecosystem, representing the alternating visibility and invisibility of marine landscapes (Figure 30.10). For the underwater photogrammetry

Figure 30.8 The pointcloud model represents the landscape inclusive of its vegetation and topography in a realistic and vivid way. The sample covers a hectare of land, can be dissected into cross-sections [upper part of figure], and freely moved, panned, and zoomed [lower part of figure] (Rekittke and Ninsalam, 2016)

campaign, we were equipped with an *Olympus TG-4* camera, off-the-shelf technology like all the tools we apply in the field. Our task was impacted by the changing clarity of the water, and the limitation of our equipment to a depth of 15 metres. The lightweight Olympus camera and its long battery life enabled us to conduct several one-hour-long underwater campaigns. Though being a low-cost solution, the camera delivered satisfactory imagery to digitally reconstruct a demonstrative model. The underwater and the above-water material were manually assembled in the 3D modeller *Rhino* (Figures 30.11 and 30.12). The detailed pointclouds of mangroves, acquired underwater, were embedded within the georeferenced landscape overview model acquired in the aerial campaign. Our models and transects typify the intertidal ecosystem of the National Park, illustrate the tidal change, represent all types of boats and their different

Figure 30.9   Photogrammetry campaign (top) with underwater camera (Olympus TG-4), in order to model the underwater aspect of a mangrove belt and seagrass meadow (bottom), Bunaken Marine National Park, North-Sulawesi (Rekittke and Ninsalam, 2016)

This figure is shown in colour in the plate section of this book.

draft, show tourists and their problematic behaviour, and include sea animal species, plants, corals, et cetera (Figure 30.13). They are detailed enough for all aspects of analytical thinking, subsequent design work, and related proposals for park management. What we could show by our study of Bunaken Island is that increasing visitor numbers cause cumulative environmental stress and damage. In this regard, one of the most problematic current practices is the individual boat transport to the many island resorts. On arrival and departure, each visitor tramples repeatedly through the sensitive intertidal zone, plus they also do so several times a day when going snorkelling and diving. In addition, all luggage and dive equipment are carried in this detrimental manner. Also underwater we find a lot of traffic, not unsimiliar to a *dive highway*, and every single dive tourist that we monitored during two weeks of fieldwork touched the reef many a time. But it was the officers of the Bunaken Marine National Park Office who took the cake, in our opinion. They displayed a vast collection of visitor photos showing grinning snorkelers in colourful wetsuits, holding on to the colourful reef. The photographing is offered by the park rangers; the photos can be purchased in the Park Office.

Our core project proposal is to diametrically oppose the current transportation practice, from a damaging sea-centred transport system to a sustainable island-centred transport system. Instead of leading visitors and carrying goods through the reef flats, seagrass meadows, and mangroves, a

Figure 30.10   Pointcloud models of the above-water part of mangroves (upper part of figure) and their underwater structure (lower part of figure) on Bunaken Island, Bunaken Marine National Park. The green points and frames represent all the single photos that had been taken above and under water. The photogrammetrical artefacts—pointclouds—result from these images (Rekittke and Ninsalam, 2016)

This figure is shown in colour in the plate section of this book.

Figure 30.11   Complete pointcloud model that combines all parts of an above-water mangrove vegetation sample with its underwater elements like stems, tillers, and roots (Rekittke and Ninsalam, 2016)

This figure is shown in colour in the plate section of this book.

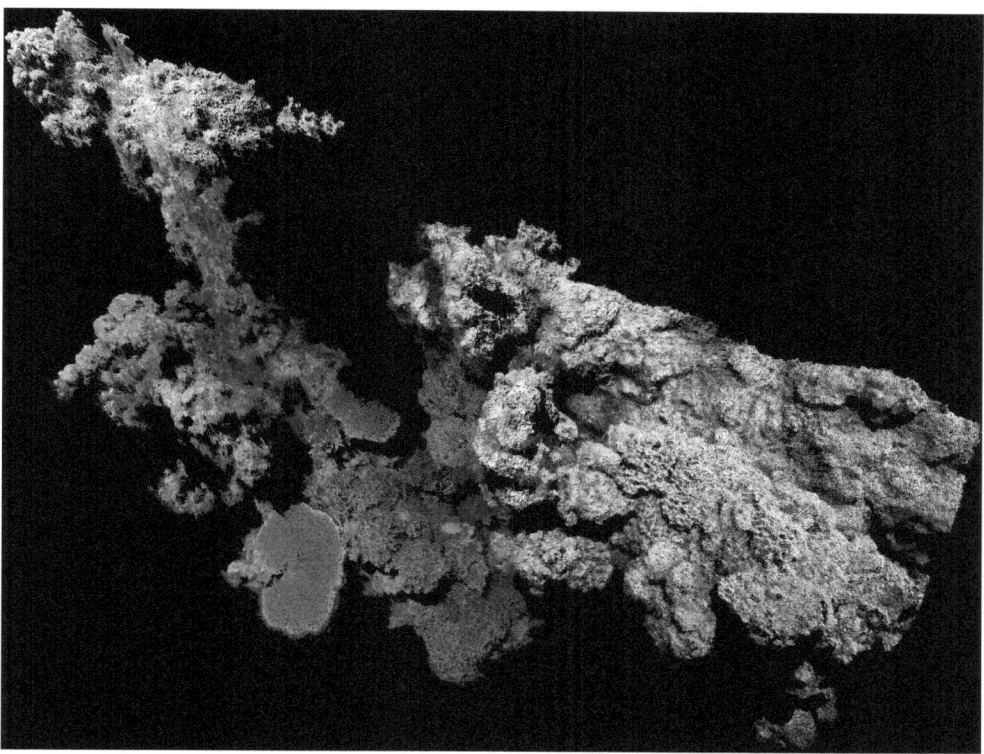

Figure 30.12   Pointcloud model sample of corals attached to a vertical reef wall in the waters of Bunaken Island, Bunaken Marine National Park (Rekittke and Ninsalam, 2016)

This figure is shown in colour in the plate section of this book.

few reef-compatible large jetties must serve as a contact point to the island and the underwater world. From these jetties, all dive and snorkel spots can be reached without putting any strain on the reef. The boat landings at the resort beaches will have to end, and the lost mangroves can be restored. Offshore, strategically placed mooring buoys reduce destructive reef anchoring of diving boats. Boat marker buoys indicate unnavigable and protected reef zones. Snorkel marker buoys, comparable with established swimming lane demarcation in a pool, stop snorkelers from damaging the reef with fins and hands. A new well-equipped park ranger team enforces the laws of the game. The current damage zone becomes a conservation zone, without compromising any of the touristic businesses in the National Park.

The logic of this concept is straightforward: If the corals and mangroves vanish—attractivity and fish stock decline. Without abundant corals and fish stock—dive tourists stay away. If the dive tourism collapses, many local resort and dive operators lose their income. Visitors to the Bunaken Marine National Park like to stay on the islands, they reach the region by air, and they come to snorkel and dive in the sea. For the landscape designer it makes great sense to operate at all levels, by land, by air, and by sea.

Conclusion

As researching landscape architects and urban designers, we regard our personal challenge to be intimately connected with that of our professional guild: to bring us closer to the settlement

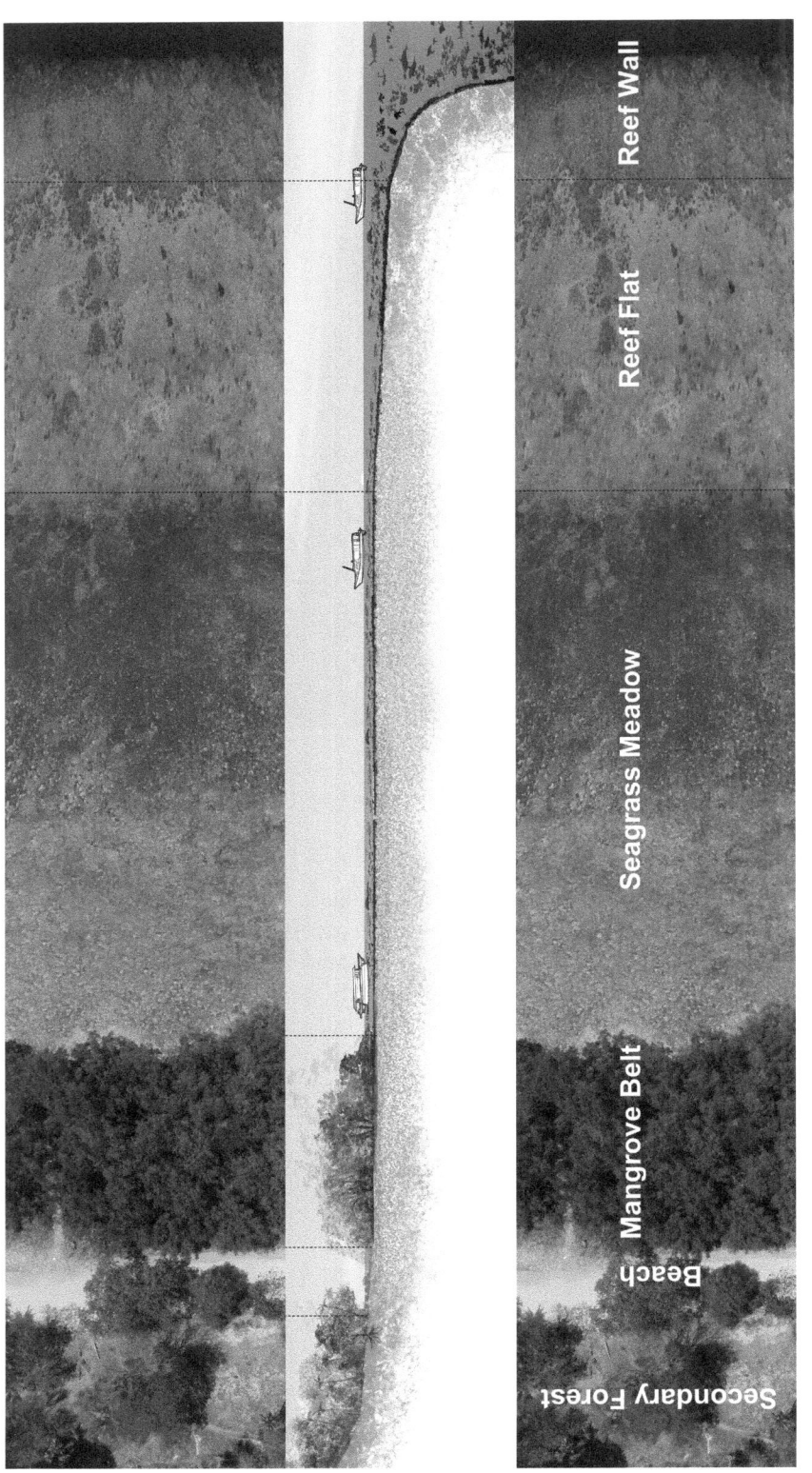

Figure 30.13 Our pointcloud models and the drived transects typify the intertidal ecosystem of the National Park, and are detailed enough for all aspects of analytical thinking, subsequent design work, and related proposals for park management (MLA Studio Rekittke, 2015)

This figure is shown in colour in the plate section of this book.

of the claim that we, the landscape people, are the ones who are prepared to deal with ultra-complex projects ranging from urban sites to regional ecosystems, from vibrant cities to resilient metropolitan ecologies, from degraded ecosystems to the challenge of climate change. Only by dint of a de facto integration of all those factors and layers that constitute what we denote as landscape, will we be able to achieve much more than just scratching the surface of the big problems that we compass to deal with. With an alliance of teachers and students, well equipped with methods and technology, we will be able to tackle genuinely big projects. The data packages that we deal with will be increasingly large; the teams that we field will be big, variegated and, finally, interdisciplinary. And building these kinds of teams, and tackling large complex problems together, will require an innovative act in the professional and academic reality from which we speak. Teaching and researching by land, by air, and by sea stands for the first small steps in a multidimensional approach that we consider necessary for the generation of landscape credibility and authentic landscape design work—characterized by boundlessness, timelessness, and infiniteness.

References

Britten, N. and Fisher, B. (1993), 'Qualitative Research and General Practice', *The British Journal of General Practice* 43/372: 270–271.

Girot, C. (1999), 'Four Trace Concepts in Landscape Architecture', in J. Corner (ed.), *Recovering Landscape: Essays in Contemporary Landscape Architecture* (New York: Princeton Architectural Press), 59–67.

Højlund Nielsen, M. P. (2007), 'Fieldworking in an Interdisciplinary Perspective', *NIAS Nytt, ProQuest Social Science Journals*, No. 01: 4.

Mays, N. and Pope, C. (1995), 'Rigour and Qualitative Research', *British Medical Journal* 311/6997: 109–112.

Mehta, A. (1999), *Bunaken National Park: Natural History Book* (Indonesia: National Research Management Program USAID) English Version.

Merriam-Webster Online Dictionary (2017, April), definition: Research, www.merriam-webster.com/dictionary/research.

MLA Studio Rekittke (2015), Studio Bunaken Marine National Park, National University of Singapore, Master of Landscape Architecture Programme. Studio leader: Jörg Rekittke. *Teaching staff: Yazid Ninsalam.* Students: Qiying Poh, Yibei Tao, Yunzi Sun, Anzhou Chen, Deepika N. Amonkar, Anushree Agarwal, Li Xuan Lim Cherlyn, Li Wen Tan Ashley, and Zhi Ning Cheok.

Ng, P. K. L., Corlett, R. and Tan, H. T. W. (eds) (2011), *Singapore Biodiversity: An Encyclopedia of the Natural Environment and Sustainable Development* (Singapore: Editions Didier Millet).

Rekittke, J. (2015), 'Mandatory Fieldwork', in *PAMPHLET, Issue 19, Field Instruments of Design* (ETH Zurich: gta publishers), 56–67 & plate V.

Rekittke, J. (2017), 'Bunaken – A Natural Paradise Under Strain', *Topos Reviews*, 10 April, www.toposmagazine.com/bunaken.

Rekittke, J. and Ninsalam, Y. (2016), 'Sliced Ecosystem: Modelling Transects of Vulnerable Marine Landscapes', *Journal of Digital Landscape Architecture*, Issue 01: 36–45.

Rekittke, J. and Paar, P. (2010), 'Grassroots GIS – Digital Outdoor Designing Where the Streets Have No Name', in E. Buhmann, M. Pietsch and E. Kretzler (eds) *Peer Reviewed Proceedings of Digital Landscape Architecture 2010 at Anhalt University of Applied Sciences* (Berlin, Offenbach: Herbert Wichmann Verlag), 69–78.

Watt, P. J. and Donoghue, D. N. M. (2005), 'Measuring Forest Structure with Terrestrial Laser Scanning', *International Journal of Remote Sensing* 26/7: 1437–1446.

Index

Page numbers for figures are given in *italics*, and for tables they are given in **bold**. Notes are given as: [page number] n [note number].

Index